Adaptation under Stressful Environments through Biological Adjustments and Interventions

Rajkumar Tulsawani • Divya Vohora
Editors

Adaptation under Stressful Environments through Biological Adjustments and Interventions

Springer

Editors
Rajkumar Tulsawani
Defence Institute of Physiology & Allied Sciences (DIPAS)
Defence Research and Development Organization (DRDO)
Delhi, Delhi, India

Divya Vohora
Department of Pharmacology
School of Pharmaceutical Education and Research (SPER), Jamia Hamdard
New Delhi, Delhi, India

ISBN 978-981-99-7654-6 ISBN 978-981-99-7652-2 (eBook)
https://doi.org/10.1007/978-981-99-7652-2

This Springer imprint is published by the registered company Springer Nature Singapore Pte Ltd.
The registered company address is: 152 Beach Road, #21-01/04 Gateway East, Singapore 189721, Singapore

Preface

Stress is an inevitable part of life that affects humans in various ways. Any stimulus (intrinsic or extrinsic) that triggers a biological response in the body may be termed "Stress." A stressful environment affects overall health impacting physical and psychological health and premature aging. Stress releases hormones that trigger physiological responses with the involvement of vital organs including the brain, heart, etc. leading to diseases. Depending upon the quantum and duration of exposure to stressors, our body produces an adaptive response.

Acute stress responses, in general, are adaptive in nature. However, excessive and uncontrolled acute stress often leads to illness. However, in case of extreme stress exposure or prolonged stress, cellular adaptive responses may not be adequate and require interventions to augment adaptive responses to prevent death. Adaptive responses to cope with stressful situations require the engagement of complex mechanisms. With appropriate stress management through pharmacological interventions, dietary modification or supplements, improvement in psychological response, etc., adaptive resilience can be achieved, and adaptation can be conditioned in a particularly stressful environment.

This volume incorporates a wide range of stressors and environmental stimulus, which directly or indirectly affects human health, such as hypoxic, high altitude, extreme stress, nutritional, oxidative, aging-related, and genotoxic stress. Each chapter in this book presents a particular stressor along with ways to attain adaptive responses and the best feasible approach for adaptation by presenting a distinct pharmacological therapy, including phytonutrients or herbal medicines, or other synthetic targets against a specific stressor. Further, a few chapters are also dedicated to non-pharmacological and dietary interventions for stress adaptation.

We hope that this volume will provide a comprehensive understanding of the impact of stress on human health and offer practical strategies to manage and adapt to stressful environments. The volume, with expertise from 51 scientists from 9 institutes, would hopefully be an indispensable tool not only for new researchers interested to work on stress and adaptive approaches but will also provide invaluable information to established researchers on various mechanisms and interventions.

Delhi, India — Rajkumar Tulsawani
New Delhi, India — Divya Vohora

Contents

Editors and Contributors

About the Editors

Rajkumar Tulsawani is a Scientist at the Defence Institute of Physiology and Allied Sciences, Delhi, India, since January 2006. Previously, he served at Defence Research and Development Establishment, Gwalior, India, from 2000 to 2006. He obtained his Ph.D. in Zoology from Jiwaji University, Gwalior, Madhya Pradesh, India, in 2007 and visited the University of Nebraska Medical Center, Omaha, USA, during 2009–2010 for postdoctoral research work.

Dr. Rajkumar Tulsawani has gained expertise in biomedical sciences and related research areas including pharmacology, toxicology, cellular biology, molecular biology, and analytical techniques. He has research experience in the identification of causes/mechanisms of cyanide and sulfur mustard poisoning (scheduled warfare agents) and the development of therapeutic agents to neutralize their toxic effects. Presently, Dr. Tulsawani's research program focuses on understanding the biological response to hypoxia including chemical and physiological hypoxia and high-altitude maladies, its clinical implication, and therapeutic targeting of hypoxia-responsive pathways. Additionally, he is pursuing research work on understanding the bio-effects and safety of high-energy lasers.

He is a recipient of many awards, notably the Gufic Prize, 2019; Director Award for S & T Coordination, 2019; Medal of Excellence from Air Force Research Laboratory (AFRL), 2018; DIPAS Laboratory Scientist of the Year, 2016; CSIR Young Scientist Award from Indian Pharmacology Society, 2005; Young Scientist Award from Society for Toxicology, 2005; and Best

Performance Award from Defence Research and Development Establishment, Gwalior, India, 2001. He is a recipient of the DST-BOYSCAST Fellowship for visiting the University of Nebraska Medical Center, Omaha, USA, in 2009–2010, and served as an external examiner at various universities. He has been associated with three patents (two granted; one at granting stage) and the development of five products: an antidote for cyanide poisoning, a natural performance enhancer, an adaptogen, herbal sanitizer and herbal sanitizing wipes/towels (for COVID-19 pandemic). Dr. Rajkumar Tulsawani has published over 50 research publications, and 2 book chapters, and has been a reviewer of many peer-reviewed journals.

Divya Vohora is a Professor of Pharmacology in the School of Pharmaceutical Education and Research (SPER) at Jamia Hamdard, New Delhi. She has over 24 years of teaching/research experience and has held positions as HoD, Pharmacology; Ass Dean, Student's Welfare, Director, DODL, and as In-charge, Pharmaceutical Medicine (joint collaborative doctoral program with Sun Pharma). She has over 200 publications in national and international journals of repute with over 4500 citations and an h-index of 38 as of Dec 2023. She is also the author/editor of 4 books published by Elsevier, Springer Nature, and CRC Press, USA. More than 80 research degrees (Master's/Ph.D./MD) in Pharmacology have been awarded under her supervision.

Prof. Vohora has received prestigious fellowships including Elected Fellow of the National Academy of Medical Sciences (FAMS), India; Fellow of the Indian Pharmacological Society (FIPS); Fellow of the British Pharmacological Society (FBPhS); and Fellow of the Royal Society of Biology, UK (FRSB). She has been listed among the Top 2% of World Scientists by Stanford University study for the whole career and for single-year impact in 2020–2021 and 2021–2022.

She is a recipient of several awards including the Dr. B. K. Anand Oration Award and Dr. Vinod Kumar Bhargava Award of the National Academy of Medical Sciences; Distinguished Woman Scientist Award of the Indian Society of Chemists Biologists; Dr. D. N. Prasad Memorial Oration Award of the Indian Council of Medical Research; Pharm Innova Award, Career Award for

Young Teachers by AICTE; Fast Track Award for Young Scientists by DST, Govt. of India; and Chandra Kanta Dandiya Prize in Pharmacology.

She is a member of the International Advisory Group of the British Pharmacological Society, UK, and is an expert member for research grants submitted to National Science Centre, Poland, and La Caixa Foundation, Spain.

She has worked as PI on several research grants funded by AICTE, ICMR, UGC, CSIR, and DST, Govt of India. Her major areas of research interest are neuropharmacology, particularly epilepsy, and neurodegenerative diseases including Alzheimer's disease and associated comorbidities. Her laboratory has identified novel targets for epileptogenesis and addressed several epilepsy-associated complications. New mouse models were developed for secondary osteoporosis in mice, and novel biomarkers for epilepsy were investigated with a focus on aromatase, histamine H3 receptor antagonism, and PI3K-mTOR pathway as some potential targets.

Contributors

Yasmin Ahmad Defence Institute of Physiology and Allied Sciences (DIPAS), Defence Research and Development Organization (DRDO), New Delhi, India

Faraha Ahmed Department of Pharmacology, School of Pharmaceutical Education and Research (SPER), Jamia Hamdard, New Delhi, India

Hanaan Kasim Ansari Department of Molecular Medicine, School of Interdisciplinary Sciences and Technology (SIST), Jamia Hamdard, New Delhi, India

Jyotsana Bakshi Defence Institute of Physiology and Allied Sciences (DIPAS), Defence Research and Development Organization (DRDO), Delhi, India

Rifat Ara Begum Department of Biochemistry and Molecular Biology, University of Dhaka, Dhaka, Bangladesh

Abhishek Bharadwaj Defence Institute of Physiology and Allied Sciences (DIPAS), Defence Research and Development Organization (DRDO), Delhi, India

Debojyoti Bhattacharyya Defence Institute of Physiology and Allied Sciences (DIPAS), Defence Research and Development Organization (DRDO), Delhi, India

Tirthankar Chatterjee Defence Institute of Physiology and Allied Sciences (DIPAS), Defence Research and Development Organization (DRDO), Delhi, India

Pooja Chaudhary Defence Institute of Physiology and Allied Sciences (DIPAS), Defence Research and Development Organization (DRDO), Delhi, India

Emran Kabir Chowdhury Department of Biochemistry and Molecular Biology, University of Dhaka, Dhaka, Bangladesh

Iti Garg Division of Pathophysiology and Disruptive Technology, Defence Institute of Physiology and Allied Sciences (DIPAS), Defence Research and Development Organization (DRDO), Delhi, India

Rajinder Kumar Gupta Defence Institute of Physiology and Allied Sciences (DIPAS), Defence Research and Development Organization (DRDO), Delhi, India

Ritika Jain Defence Institute of Physiology and Allied Sciences (DIPAS), Defence Research and Development Organization (DRDO), Delhi, India
Faculty of Life Sciences, Mandsaur University, Mandsaur, Madhya Pradesh, India

Tanuj Joshi Department of Pharmaceutical Sciences, Bhimtal Campus, Kumaun University, Nainital, Uttarakhand, India

Savneet Kaur Department of Molecular and Cellular Medicine, Institute of Liver and Biliary Sciences (ILBS), New Delhi, India

Razia Khanam Department of Biomedical Sciences, College of Medicine, Gulf Medical University (GMU), Ajman, United Arab Emirates

Anuja Krishnan Department of Molecular Medicine, School of Interdisciplinary Sciences and Technology (SIST), Jamia Hamdard, New Delhi, India

Sachin Kumar Defence Institute of Physiology and Allied Sciences (DIPAS), Defence Research and Development Organization (DRDO), Delhi, India

Anu Kumari Department of Pathophysiology and Disruptive Technologies, Defence Institute of Physiology and Allied Sciences (DIPAS), Defence Research and Development Organization (DRDO), Delhi, India

Manisha Kumari Department of Pathophysiology and Disruptive Technologies, Defence Institute of Physiology and Allied Sciences (DIPAS), Defence Research and Development Organization (DRDO), Delhi, India

Rekha Maithani Defence Institute of Physiology and Allied Sciences (DIPAS), Defence Research and Development Organization (DRDO), Delhi, India

Subhodip Majumdar Defence Institute of Physiology and Allied Sciences (DIPAS), Defence Research and Development Organization (DRDO), Delhi, India
Amity Institute of Biotechnology, Amity University, Noida, Uttar Pradesh, India

Ramesh C. Meena Department of Pathophysiology and Disruptive Technologies, Defence Institute of Physiology and Allied Sciences (DIPAS), Defence Research and Development Organization (DRDO), Delhi, India

Kamla Prasad Mishra Defence Institute of Physiology and Allied Sciences (DIPAS), Defence Research and Development Organization (DRDO), Delhi, India

Swaraj Mohanty Defence Institute of Physiology and Allied Sciences (DIPAS), Defence Research and Development Organization (DRDO), Delhi, India

Akash Kumar Mourya Department of Molecular and Cellular Medicine, Institute of Liver and Biliary Sciences (ILBS), New Delhi, India

Mukesh Nandave Department of Pharmacology, Delhi Pharmaceutical Sciences and Research University (DPSRU), New Delhi, India

Usha Panjwani Soldier Performance Vertical, Defence Institute of Physiology and Allied Sciences (DIPAS), Defence Research and Development Organization (DRDO), Delhi, India
Manav Rachna International Institute of Research and Studies, Faridabad, Haryana, India

Harsh Pawar Soldier Performance Vertical, Defence Institute of Physiology and Allied Sciences (DIPAS), Defence Research and Development Organization (DRDO), Delhi, India

Chhavi Rai Department of Pathophysiology and Disruptive Technologies, Defence Institute of Physiology and Allied Sciences (DIPAS), Defence Research and Development Organization (DRDO), Delhi, India

Richa Rathor Defence Institute of Physiology and Allied Sciences (DIPAS), Defence Research and Development Organization (DRDO), Delhi, India

Archana N. Sah Department of Pharmaceutical Sciences, Bhimtal Campus, Kumaun University, Nainital, Uttarakhand, India

Dolly Sharma Department of Pathophysiology and Disruptive Technologies, Defence Institute of Physiology and Allied Sciences (DIPAS), Defence Research and Development Organization (DRDO), Delhi, India

Poornima Sharma Defence Institute of Physiology and Allied Sciences (DIPAS), Defence Research and Development Organization (DRDO), Delhi, India

Ratnika Sharma Defence Institute of Physiology and Allied Sciences (DIPAS), Defence Research and Development Organization (DRDO), Delhi, India

Shubhangi Sharma Defence Institute of Physiology and Allied Sciences (DIPAS), Defence Research and Development Organization (DRDO), Delhi, India
Amity Institute of Biotechnology, Amity University, Noida, Uttar Pradesh, India

Hossain Uddin Shekhar Department of Biochemistry and Molecular Biology, University of Dhaka, Dhaka, Bangladesh

Mrinalini Singh Defence Institute of Physiology and Allied Sciences (DIPAS), Defence Research and Development Organisation (DRDO), Delhi, India

Priyansha Singh Department of Molecular and Cellular Medicine, Institute of Liver and Biliary Sciences (ILBS), New Delhi, India

Sarika Singh School of Allied Health Sciences and Management, Delhi Pharmaceutical Sciences and Research University (DPSRU), New Delhi, India

Swati Srivastava Division of Pathophysiology and Disruptive Technology, Defence Institute of Physiology and Allied Sciences (DIPAS), Defence Research and Development Organization (DRDO), Delhi, India

Smriti Sudha Defence Institute of Physiology and Allied Sciences (DIPAS), Defence Research and Development Organisation (DRDO), Delhi, India

Geetha Suryakumar Defence Institute of Physiology and Allied Sciences (DIPAS), Defence Research and Development Organization (DRDO), Delhi, India

Bhanuteja Thondala Soldier Performance Vertical, Defence Institute of Physiology and Allied Sciences (DIPAS), Defence Research and Development Organization (DRDO), Delhi, India

Dinesh Mani Tripathi Department of Molecular and Cellular Medicine, Institute of Liver and Biliary Sciences (ILBS), New Delhi, India

Rajkumar Tulsawani Defence Institute of Physiology and Allied Sciences (DIPAS), Defence Research and Development Organization (DRDO), Delhi, India

Rajeev Varshney Department of Pathophysiology and Disruptive Technologies, Defence Institute of Physiology and Allied Sciences (DIPAS), Defence Research and Development Organization (DRDO), Delhi, India

Shobi Veleri Drug Safety Division, ICMR-National Institute of Nutrition, Department of Health Research, Ministry of Health & Family Welfare, Government of India, Hyderabad, Telangana, India

Divya Vohora Department of Pharmacology, School of Pharmaceutical Education and Research (SPER), Jamia Hamdard, New Delhi, India

Ronika Yadav Defence Institute of Physiology and Allied Sciences (DIPAS), Defence Research and Development Organization (DRDO), Delhi, India

Vaishali Yadav Neurobehavioral Pharmacology Laboratory, Department of Pharmacology, School of Pharmaceutical Education and Research (SPER), Jamia Hamdard, New Delhi, India

Part I

Hypoxic, High-Altitude, and Extreme Stress: Health Implications and Interventions

Stress and Adaptogens

1

Rajkumar Tulsawani and Divya Vohora

Abstract

Stress is appropriately defined as a "nonspecific response of a body to any demand". Stress may be classified based on various factors including the type of stressor, duration of the stress response, and also on the type of influence on an individual. Upon exposure to stress and based on the type of stress, the body initiates an adaptive response that is governed by various pathological and genomic pathways. Furthermore, if exposure to stressors is prolonged for years in case of extreme stress, it may lead to a pathological state leading to chronic diseases such as cognitive deficit, accelerated aging, hypertension, and other cardiovascular diseases. Such incidences may require interventions such as nutraceuticals and medicines. Adaptogens are plant extracts that help the body adapt or adjust to chronic exposure to stress. They protect against stress-induced pathophysiological symptoms, accelerate mental functioning, and regulate homeostasis. In this chapter, we provide a basic overview of stress and how adaptogens play an excruciating role in its management. The role of selected adaptogens such as *Rhodiola imbricata, Hippophae rhamnoides, Ganoderma lucidum,* and *Cordyceps sinensis* eliciting various pharmacological actions by acting on different mechanistic pathways is described.

R. Tulsawani (✉)
Defence Institute of Physiology and Allied Sciences (DIPAS), Defence Research & Development Organization (DRDO), Delhi, India
e-mail: rktulsawani@yahoo.com

D. Vohora (✉)
Department of Pharmacology, School of Pharmaceutical Education and Research (SPER), Jamia Hamdard, New Delhi, India
e-mail: dvohra@jamiahamdard.ac.in

R. Tulsawani, D. Vohora (eds.), *Adaptation under Stressful Environments through Biological Adjustments and Interventions*,
https://doi.org/10.1007/978-981-99-7652-2_1

Keywords

Stress · Classification of stress · Stress management · Adaptogens

1.1 Introduction

Hans Selye defined stress as nonspecific exertion on the body disturbing body functions and adaptive response involving certain hormones, stress hormones, and defined it as general adaptation syndrome (Selye 1956). The general adaptation syndrome involves alarm reaction sensing stimulus and disturbing body homeostasis, resistance, adaptive response to the stressor and exhaustion, and adaptive failures upon prolonged exposure (Tost et al. 2015). Stress is often nonspecific and complex process and adaptive response is individual-specific. The active process of adaptation against stressors is known as 'allostasis' which involves the preservation of homeostasis via the nonlinear intervention of the autonomic system, metabolic system along with immune system (McEwen and Wingfield 2003). The cumulative effects of multiple stimuli and the dysregulation of the nonlinear mediation of allostasis lead to 'allostatic load' which is responsible for the 'wear and tear' of the brain and body causing damage to organs, tissues, and cellular functioning such as mitochondrial dysfunction and telomere shortening that is associated with accelerated aging.

A stimulus, a stressor, in a biological system initiates cellular events that result in cellular stress. Stressors can be of various types including chemicals, environmental conditions, or biological agents that may activate the hypothalamic-pituitary-adrenal (HPA) axis stress response (McEwen 1998). Cellular stress initiates response including stress hormone secretion, proliferation, and immune response to maintain the well-being state of body and mind leading to adapt the individual to the stressor (Cortez 1991; Cortez et al. 2008; Cortez and Silva 2020). Exposure to stress causes increased levels of hormonal mediators including cortisol and catecholamines such as epinephrine and norepinephrine that are regulated through the hypothalamic pituitary adrenal (HPA) axis along with the sympathetic nervous system (Ajala 2017) (Fig. 1.1).

Stressors impact mood, state of well-being, psychological distress, and health. In healthy individuals, acute stress may be adaptive; however, long-term stress in unhealthy individuals may cause illness (Schneiderman et al. 2005).

The stressors are multi-mechanistic in altering gene expression. They are responsible in the activation of epigenetic mechanisms and gene transcription through glucocorticoids, along with modifications of histones as well as methylation and hydromethylation of CpG residues in DNA, thereby repressing and activating genetic factors such as retrotransposons (McEwen et al. 2015). The primary target of stressful environment is the brain, and stimulation of glucocorticoids and excitatory amino acid neurotransmitters through stressors leads to alteration of the neuronal architecture by affecting the dendritic growth, synapse density, as well as inhibition of neurogenesis (Lupien et al. 2018). The three major regions of brain that

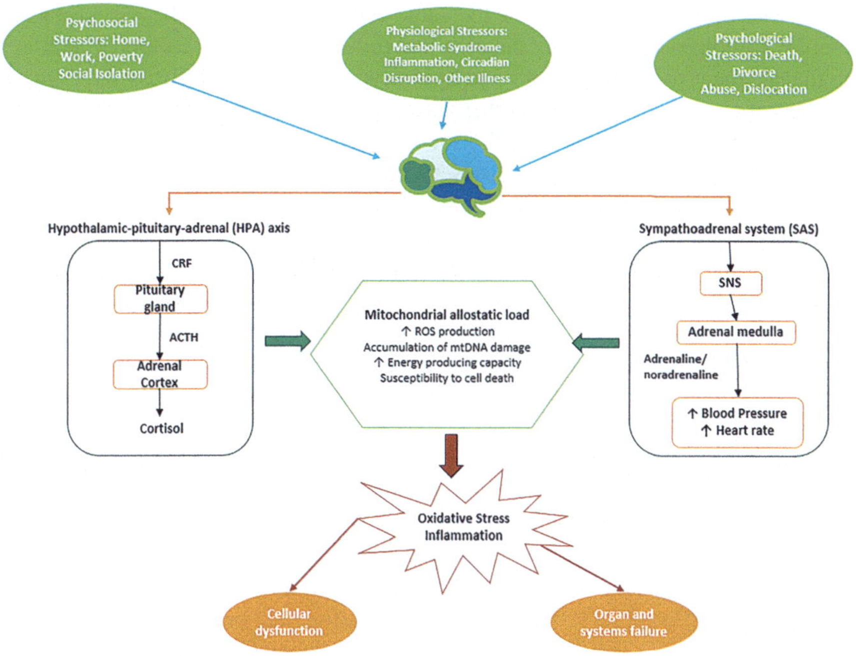

Fig. 1.1 Mechanism of chronic stress response system

play an important role in the regulation of the HPA stress response include hippocampus, amygdala, and prefrontal cortex (McEwen et al. 2015). The presence of glucocorticoid and mineralocorticoid receptors in the hippocampus evidently illustrates the effect of stress through glucocorticoids by shrinking the dendritic growth and spinal loss along with inhibition of neurogenesis in the dentate gyrus region (McEwen et al. 2016). In the amygdala, the effect is based on the type of stress such as increased spine density on basolateral neurons is observed during acute traumatic stress and chronic stress leads to dendritic extension and loss of spines. The effects of stress on the prefrontal cortex provide an insight to the memory loss associated with age as chronic stress leads to dendritic shrinkage and debranching of neurons (Arnsten 2009).

It is well-documented that stress stimulus is present in physiological and pathological processes (Prado et al. 2006) and is often mistakenly defined as illness. In actuality, stress is well defined as a stimulus inducing biological alarm and prolonged exposures may lead to a pathological state, i.e., illness or diseases. Moreover, anxiety and stress are defined as the same, but stress is a response to stimulus and may lead to anxiety (Silva et al. 2017; Rom and Reznick 2016; Sevillano-García et al. 2007; Cortez 1991).

The need for the management of chronic stress is required when long-term exposure could lead to a various of stress-related diseases such as hypertension, anxiety, depression, cognitive dysfunction, and chronic fatigue syndrome (Ajala

2017). The individual response to the same type of stress may be varied due to complex responses exhibited by the cells and tissues and genetic factors. Therefore, the management of stress cannot be straightforward or universal. Every day, organism deals with stress in one or the other way. The first step for effective management of stress is recognizing stress and the source of stress and identifying whether it's acute or chronic stress. If stress leads to uneasiness in life and disturbances in physiological or psychological functions, identifying the stage of stress would be beneficial in stress management.

The management of acute stress, though not easy, can be regulated through relaxation, meditation, yoga, avoiding stressful environments, etc. However, in case of long-term adaptation, pharmacological interventions may be required. Numerous herbs are well documented to alleviate stress and extreme stress including *Cordyceps sinensis*, *Ganoderma lucidum*, *Ginkgo biloba*, *Ginseng*, *Hippophae rhamnoides*, *Rhodiola imbricata*, etc.

In this chapter, types of stress and the management of stress through adaptogenic plants and medicinal mushrooms have been elaborated with special reference to extreme stress environments, i.e., high altitude. These adaptogenic plants synthesize a class of phyto-compounds that help in enhancing the body's resistance to various stressors including emotional, environmental, and physical stress.

1.2 Classification of Stress

Stress can be classified according to the nature of the stressor, whereby the stressor can be physiological or psychological (Kogler et al. 2015). Physiological stress is induced by an unpleasant sensory or an emotional experience that is usually associated with potential bodily harm or damage to body physiology, for example, pain, hunger, oxidative stress, etc. (Zubieta and Stohler 2009). On the contrary, psychosocial stress is triggered by social situations of threat such as social assessment, rejection, and even by incidents that highlight goal-oriented accomplishments (Pruessner et al. 2010) Table 1.1.

Another classification of stress is based on the type of influence a stressor has that is reflected on an individual. Good stress or stress that can be described as a positive response to a stressor is known as eustress. Simultaneously, a negative response to a stressor is known as distress, or bad stress (Gong and Geertshuis 2023). The type of response to distress is severely high and may lead to psychological or physical damage (Table 1.1). Distress can also be further classified into acute stress and chronic stress, according to the duration of the stress response. Acute stress is when the stress response is high in intensity, but remains short-term, whereas, chronic stress is when the stress response may not be highly intense, but lasts for a prolonged interval of time (Lu et al. 2021) Table 1.1.

Table 1.1 Classification of stress

Classification of Stress	
A. Based on the nature of the stressor	
Physiological stress	Psychological stress
B. Based on the influence of stress on an individual	
Eustress	Distress
C. Based on the duration of stress response	
Acute stress (short-term)	Chronic stress (long-term)

1.3 What Are Adaptogens?

The original definition of adaptogens stated that these are agents that "increase resistance to a broad spectrum of harmful factors (stressors) of different physical, chemical, and biological natures" (Brekhman and Dardymov 1969). These are substances that improve the state of nonspecific tolerance to stress (Wagner et al. 1994). Drugs are classified as adaptogens if they exhibit anti-stress effects including anti-fatigue, antidepressant, and anti-infectious; stimulating effects that are responsible for increasing work and mental performance; however, they do not affect the normal functioning of the anabolic system and exhibit no adverse effect such as withdrawal syndrome (Brekhman and Dardymov 1969).

Adaptogens are known to have complex chemical structures of phenolic or tetracyclic triterpenoids/ steroids. Some of the phenolic adaptogens such as rhodioloside, rosavin, syringin, triandrin, and tyrosol are derivatives of phenylpropanoids and phenylethane. Since adaptogens are responsible to mediate the stress response system, they structurally mimic catecholamines that initiate the regulation of sympathoadrenal system (SAS) (Panossian and Wikman 2010). In addition, compounds such as ginsenosides, eleutheroside, cucurbitacin R diglucoside, sitoindosides, etc. are tetracyclic triterpenoids that chemically resemble the structure of corticosteroids, also known as the stress hormone (Munck et al. 1984). Furthermore, extracts from *Rhodiola* sp., such as rosiridin, are monoterpene glucosides that act as inhibitors of monoamine oxidases A and B (Van Diermen et al. 2009).

1.4 Possible Mechanism of Adaptogens in Stress Management

Adaptogens, as the name portrays, are regulators of metabolic pathways that increase an organism's potential to adapt to any environmental or any other external factors. According to the available literature, several mediators such as stress-activated protein kinases, FOXO transcription factor, heat shock proteins (HSP), and nitric oxide play an important role in the regulation of apoptogenic potential of various plant extracts (Alexander Panossian 2017).

The plant extracts provide a safe treatment as they stimulate the release of HSP70 without being cytotoxic in nature. Inhibition of HSF1 and HSP70 expression has been associated with age-related diseases and neurodegenerative disorders, including Alzheimer's disease, and other metabolic diseases such as diabetes (Asea et al. 2013). Adaptogens also act as inhibitors of stress-activated kinases and have been reported to significantly decrease the levels of phosphorylated forms of SAPK/JNK (Wiegant et al. 2009). In addition to this, adaptogens are also observed to be responsible for the inducing the translocation of FOXO (DAF-16) into the nucleus, thereby favoring a transcriptome that is longevity-biased and stress-resistant (Perez et al. 2012).

1.5 Role of Adaptogens: Few Examples

The role of adaptogens from medicinal plants and mushrooms such as *Rhodiola imbricata*, *Hippophae rhamnoides*, *Ganoderma lucidum*, and *Cordyceps sinensis* has been discussed below and their probable mechanism has been specified in Table 1.2.

1.5.1 Rhodiola imbricata

Rhodiola imbricata grows at high altitude and is from the family Crassulaceae and subfamily Sedoideae which is widely distributed throughout Europe and Asia. Komarov (1939) has defined 20 species and it is estimated that there are about 200 species of Rhodiola worldwide. Rhodiola rhizome is widely used in China, Soviet Republics, Tibet, and Mongolia to alleviate stress, extreme high-altitude stress, longevity, and endurance (Kelly 2001). Further, it has been reported to prevent fatigue and gastrointestinal illness (Kelly 2001; Petkov et al. 1986). *Rhodiola species* have been extensively investigated in animal and human studies indicating adaptogenic and anti-stress properties and found to be effective against lifestyle disorders and diseases (Panossian et al. 2007; Gupta et al. 2008, 2009, 2010; Ajala 2017; Li et al. 2017a, b; Shieh et al. 2018; Zheng et al. 2019; Agapouda et al. 2022; Ivanova Stojcheva and Quintela 2022). The medicinal properties are attributed to *Rhodiola rosea* owing to the presence of certain bioactive substances like rosavins (rosavin, rosarian, and rosin), phenylethanolderivatives (salidroside, tyrosol), organic acids, flavonoids, tannins, and phenolic glycosides (Khanum et al. 2005; Mao et al. 2007). Its protective efficacy against mental and physical stress, toxins, and cold in experimental animals and humans is well documented (Khanum et al. 2005; Saratikov and Krasnov 1987; Kurkin and Zapesochnaya 1986; Petkov et al. 1986, 1990; Saratikov et al. 1978). Rhodiola acts as an antidepressant and psychostimulant (Saratikov and Krasnov 1987; Konstantinos and Heun 2020) and is useful in alleviating neurological disorders (Table 1.2).

Table 1.2 Mechanism of action of few adaptogens along with their intended pharmacological use

Name of Adaptogen	Mechanism of Action	Intentional use	References
Rhodiola imbricata	• Decreases production of ROS • Inhibits H_2O_2 • Stimulates interleukin-6 (IL-6) and tumor necrosis factor (TNFα) • Increases production of NO • Induces apoptosis and cell death	o Anti-cytotoxic o Hepatoprotective o Antioxidant o Anti-stress	(Zhuang et al. 2019) (Bhardwaj et al. 2018)
Hippophae rhamnoides	• Inhibits NMDA receptors • Antagonises Serotonin (5HT) receptor • Reduces corticosterone levels • Reduces the levels of dopamine and norepinephrine • Reduces oxidative stress • Upregulates endogenous antioxidants	o Anti-stress o Antidepressant o Antioxidant and immunomodulatory o Cardioprotective o Antibacterial and antiviral	(Suryakumar and Gupta 2011) (Bae et al. 2016)
Ganoderma lucidum	• Enhances levels of neurotransmitters (catecholamines, serotonin, glutamate) • Prevents increase in glucocorticoid and α-synuclein levels • Improves neuroplasticity by upregulating CREB/p-CREB/BDNF expression via ERK1/ERK2 induction • Attenuates pro-inflammatory cytokines (NFκB, IL6, TNFα) • Maintains oxidative status by normalizing antioxidant enzymes (MDA, GSH, SOD)	o Neuroprotective o Antioxidant, anti-inflammatory, and immunomodulatory o Anti-stress	(Sharma et al. 2019) (Sharma and Tulsawani 2020)
Cordyceps sinensis	• Improves the activity of anti-oxidative enzymes such as SOD, catalase, GSH-Px and lowers the activity of lipid peroxidation and monoamine oxidase B • Increases production of pro-inflammatory cytokines from activated microphages • Regulation of the *foxo* signaling pathway (hsa04068) and HIF-1 signaling pathway (hsa04066)	o Antidepression o Anti-oxidative stress o Antiaging/ senescence o Anticancer o Anti-fatigue/ asthenia	(Chen et al. 2013) (Jordan et al. 2008) (Zhang et al. 2022)

1.5.2 Hippophae rhamnoides

Hippophae rhamnoides, commonly known as seabuckthorn, belongs to family Elaeagnaceae and genus *Hippophae*. It is a wild thorny deciduous shrub, grows at high altitude (2500–4000 m) in adverse climatic conditions, and is native to Europe and Asia (Rousi 1970). In India, it grows naturally in Ladakh, Uttarkhand, Himachal Pradesh, Jammu and Kashmir, Sikkim, and Arunachal Pradesh regions (Dhyani et al. 2010). There are six species of seabuckthorn found in the world, of which three species are reported in India, viz., *Hippophae rhamnoides subspecies Turkestania Rousi*, *Hippophae salicifolia*, and *Hippophae tibitana*. Among these species,

H. rhamnoides has been extensively studied and many of the reports are from Russia and China (Li and Beveridge 2003). *Hippophae rhamnoides* has been reported to possess adaptogenic potential and initiate adaptive response to stress (Saggu et al. 2007; Tulsawani et al. 2010; Patel et al. 2012). It is reported to possess antioxidant and anti-stress activity (Suryakumar et al. 2002; Narayanan et al. 2005; Liao et al. 2018; Ji et al. 2020). The medicinal properties of extracts of *Hippophae rhamnoides* leaves are attributed due to the presence of high content of flavonoids, tannins, and triterpenes (Kallio et al. 2002). The major flavonoids found in the leaves of seabuckthorn include quercetin, kaempferol, and isorhamentin, catechin, and rutin (Zu et al. 2006). It is documented that *Hippophae rhamnoides* L. possesses neuroprotective activity and is effective in aging and neurodegenerative diseases including Alzheimer's disease (Xu et al. 2005; Shivapriya et al. 2015; Bala et al. 2017; Cho et al. 2017; Dong et al. 2020; Ladol and Sharma 2021; Veronique et al. 2022; Wang et al. 2022) (Table 1.2).

1.5.3 Ganoderma lucidum

Mushrooms are known to possess antioxidant property; one such mushroom is *Ganoderma lucidum,* commonly found at high altitudes. *Ganoderma lucidum,* also referred as lingzhi or reishi medicinal mushroom, belongs to family ganodermataceae and has been traditionally used in Chinese Medicine and Asian countries for centuries due to its immunomodulatory and adaptogenic properties (Gao et al. 2004; Mizuno et al. 1991; Wang et al. 2002). The pharmacological properties of *Ganoderma lucidum* are assigned due to the presence of array of bioactive compounds including adenosine, coumarin, ergosterols, ganoderic acids, lactones, mannitol, organic germanium, polysaccharides, triterpenoids, and rare minerals (Boh et al. 2007). These phytoconstituents help in improving blood circulation, eliminate fatigue, enhance energy, strengthen the immune system, and help in detoxification and many aliments and lifestyle disorders (Liu et al. 2009, 2017; Min et al. 1998; Sliva 2003; Wihastuti and Heriansyah 2017; Wu et al. 2016; Xu et al. 2017; Zhao et al. 2018). *Ganoderma lucidum* possesses anti-stress (Tulsawani et al. 2014, 2015) and adaptogenic potential (Sharma et al. 2019). *Ganoderma lucidum* prevented extreme hypobaric hypoxia-induced oxidative stress and memory impairment and has been found to be effective in oxidative stress and neurodegenerative diseases including Parkinson's diseases in experimental animals (Zhou et al. 2012; Zhang et al. 2014; Xuan et al. 2015; Sun et al. 2017; Koganti et al. 2018; Ren et al. 2019; Abate et al. 2020; Sharma and Tulsawani 2020; Jatwani and Tulsawani 2021) (Table 1.2).

1.5.4 Cordyceps sinensis

In general, mushrooms have been widely used as alternative remedies to cure various ailments. *Cordyceps sinensis* (*C. sinensis*) (Berk.) Sacc, kingdom Fungi, division

Ascomycota, family Ophiocordycipitaceae, commonly known as caterpillar fungus, has been used as a stimulant for endurance, longevity, and vitality in traditional medicine system (Zhu et al. 1998). It naturally occurs at an altitude of 3600–5000 m in Northern Himalayan regions due to its adaptogenic potential (Li et al. 2011). Phytoconstituents identified in *Cordyceps sinensis* such as cordycepin, polysaccharides, phenols, and flavonoids are known to exhibit pharmacological effects including antioxidant potential and immunomodulatory properties (Wasser and Weis 1999; Li et al. 2001; Kuo et al. 2001). It has been demonstrated that aqueous extract of *C. sinensis* induced dose-dependent tolerance in human lung epithelial cells by reducing oxidative stress via NRF2 induction under low oxygen environment (Singh et al. 2013). It is well documented that *Cordyceps sinensis* possesses anti-stress and adaptogenic activity against severe and extreme stressors (Koh et al. 2003; Sharma and Tulsawani 2022). Cordyceps sinensis has been shown to prevent oxidative stress and neurodegenerative in many experimental conditions (Jin et al. 2004; Li et al. 2012; Olatunji et al. 2016; He et al. 2018, 2019, 2020, 2021; Liu et al. 2022; Wu et al. 2022; Zhang et al. 2022) (Table 1.2).

1.6 Conclusion

Here we presented a basic overview of stress and role of few adaptogens. Stress is a state in which the bodily homeostasis is compromised. The occurrence of stress systematically involves the activation of HPA axis that is triggered in the presence of a stressor. According to the type of stressful stimuli, the conception of stress can be summarized into different types of stress responses such as psychological stress, oxidative stress, metabolic stress, emotional stress, etc. The stress response system is a complex process that revolves around multiple mediators and includes both genomic and nongenomic interacting pathway systems. Acute stress responses are adaptable and beneficial; however, exposure to a chronic stressor can cause bodily harm and lead to several life-threatening diseases in extreme cases. Management of such chronic stress may require external force, and thereby, provides a need for the development of compounds such as adaptogens.

Adaptogens are responsible for improving stress-induced weakness and fatigue, enhancing mental well-being, and reducing pathophysiological impairments caused by stress-induced diseases. The anti-stress potential of adaptogens such as *Rhodiola imbricata, Hippophae rhamnoides, Ganoderma lucidum,* and *Cordyceps sinensis* as mentioned in the chapter is regulated by maintaining homeostasis through multi-mechanistic approach including the mediation of key factors involved in the stress response system, such as glucocorticoids, corticosteroids, nitric oxide, and protein kinases. Besides their anti-stress activity, adaptogens also exhibit beneficial potential against various pharmacological defects including anti-cytotoxic, antidepression, cognitive enhancement, cardioprotective, etc. As the beneficial effects of adaptogens are being investigated, a combination of different plant herbs along with other dietary supplements may provide a potential synergistic effect against the physical and psychological response to chronic stress. Therefore, further investigations are required to fully explore the potential of adaptogens against the management of chronic stress and to provide long-term stress relief.

References

Abate M, Pepe G, Randino R, Pisanti S, Basilicata MG, Covelli V et al (2020) Ganoderma lucidum ethanol extracts enhance re-epithelialization and prevent keratinocytes from free-radical injury. Pharmaceuticals (Basel) 13(9):224. https://doi.org/10.3390/ph13090224. PMID 32872510, PMCID PMC7557611

Agapouda A, Grimm A, Lejri I, Eckert A (2022) Rhodiola rosea extract counteracts stress in an adaptogenic response curve manner via elimination of ROS and induction of neurite outgrowth. Oxidative Med Cell Longev 2022:5647599. https://doi.org/10.1155/2022/5647599. PMID 3 5602107. PMCID PMC9122715

Ajala TO (2017) The effects of adaptogens on the physical and psychological symptoms of chronic stress. Discovery 4:2. https://doi.org/10.31922/disc4.2

Arnsten AF (2009) Stress signalling pathways that impair prefrontal cortex structure and function. Nat Rev Neurosci 10(6):410–422. https://doi.org/10.1038/nrn2648. PMID 19455173

Asea A, Kaur P, Panossian A, Wikman KG (2013) Evaluation of molecular chaperons Hsp72 and neuropeptide Y as characteristic markers of adaptogenic activity of plant extracts. Phytomedicine 20(14):1323–1329. https://doi.org/10.1016/j.phymed.2013.07.001. PMID 23 920279

Bae D, Kim J, Oh D-R, Kim Y, Choi E-j, Lee H et al (2016) Multifunctional antistress effects of standardized aqueous extracts from Hippophae rhamnoides L. Anim Cells Syst 20(6):369–383. https://doi.org/10.1080/19768354.2016.1250816

Bala M, Gupta V, Prasad J (2017) A standardized Hippophae extract (SBL-1) counters neuronal tissue injuries and changes in neurotransmitters: implications in radiation protection. Pharm Biol 55(1):1833–1842. https://doi.org/10.1080/13880209.2017.1331365. PMID 28552029. PMCID PMC6130468

Bhardwaj P, Bhardwaj G, Raghuvanshi R, Thakur MS, Kumar R, Chaurasia OP (2018) Rhodiola: an overview of phytochemistry and pharmacological applications. In: Singh B, Peter KV (eds) New age herbals: resource, quality and pharmacognosy. Springer Singapore, Singapore, pp 71–113

Boh B, Berovic M, Zhang JS, Zhi-Bin L (2007) Ganoderma lucidum and its pharmaceutically active compounds. Biotechnol Annu Rev 13:265–301. https://doi.org/10.1016/S1387-2656(07)13010-6. PMID 17875480

Brekhman II, Dardymov IV (1969) New substances of plant origin which increase nonspecific resistance. Annu Rev Pharmacol 9(1):419–430. https://doi.org/10.1146/annurev.pa.09.040169.002223. PMID 4892434

Chen PX, Wang S, Nie S, Marcone M (2013) Properties of Cordyceps sinensis: a review. J Funct Foods 5(2):550–569. https://doi.org/10.1016/j.jff.2013.01.034. PMID 32288794

Cho CH, Jang H, Lee M, Kang H, Heo HJ, Kim DO (2017) Sea buckthorn (Hippophae rhamnoides L.) leaf extracts protect neuronal PC-12 cells from oxidative stress. J Microbiol Biotechnol 27(7):1257–1265. https://doi.org/10.4014/jmb.1704.04033. PMID 28535611

Cortez CM (1991) Stress and its physiological inferences. A Folha Médica 103:175–181

Cortez CM, Silva D (2020) Biological stress as a principle of nature: a review of literature. Open J Biophys 10(3):150–173. https://doi.org/10.4236/ojbiphy.2020.103012

Cortez CM, Cruz FAO, Silva D (2008) Psychicological and physiological responses to stress: a review based on results from PET and MRI studies. Braz Arch Biol Technol 51(spe):7–12. https://doi.org/10.1590/S1516-89132008000700003

Dhyani D, Maikhuri RK, Misra S, Rao KS (2010) Endorsing the declining indigenous ethnobotanical knowledge system of seabuckthorn in Central, India Himalaya. J Ethnopharmacol 127(2): 329–334. https://doi.org/10.1016/j.jep.2009.10.037. PMID 19900529

Dong K, Fernando W, Durham R, Stockmann R, Jayatunga W, Jayasena V (2020) A role of sea buckthorn on Alzheimer's disease. Int J Food Sci Technol 55(9):3073–3081. https://doi.org/10.1111/ijfs.14571

Gao Y, Chan E, Zhou S (2004) Immunomodulating activities of Ganoderma, a mushroom with medicinal properties. Food Rev Int 20(2):123–161. https://doi.org/10.1081/FRI-120037158

Suryakumar G, Sai Ram M, Singh V, Ilavazhagan G, Sawhney RC (2002) Anti-oxidant and immunomodulatory properties of seabuckthorn (Hippophae rhamnoides) - an in vitro study. J Ethnopharmacol 79(3):373–378. https://doi.org/10.1016/s0378-8741(01)00406-8. PMID 11849845

Gong W, Geertshuis SA (2023) Distress and eustress: an analysis of the stress experiences of offshore international students. Front Psychol 14:1144767. https://doi.org/10.3389/fpsyg.2023.1144767. PMID 37213394

Gupta V, Saggu S, Tulsawani R, Sawhney RC, Kumar R (2008) A dose dependent adaptogenic and safety evaluation of Rhodiola imbricata Edgew., a high-altitude rhizome. Food Chem Toxicol 46(5):1645–1652. https://doi.org/10.1016/j.fct.2007.12.027. PMID 18280024

Gupta V, Lahiri SS, Sultana S, Tulsawani R, Kumar R (2009) In vitro evaluation of antioxidant and free radical scavenging activities of Rhodiola imbricata. J Complement Integr Med 6(1). https://doi.org/10.2202/1553-3840.1235

Gupta V, Lahiri SS, Sultana S, Tulsawani R, Kumar R (2010) Anti-oxidative effect of Rhodiola imbricata root extract in rats during cold, hypoxia and restraint (C-H-R) exposure and post-stress recovery. Food Chem Toxicol 48(4):1019–1025. https://doi.org/10.1016/j.fct.2010.01.012. PMID 20079793

He MT, Lee AY, Kim JH, Park CH, Shin YS, Cho EJ (2018) Protective role of Cordyceps militaris in Aβ1-42-induced Alzheimer's disease in vivo. Food Sci Biotechnol 28(3):865–872. https://doi.org/10.1007/s10068-018-0521-z. PMID 31093445. PMCID PMC6484065

He MT, Lee AY, Park CH, Cho EJ (2019) Protective effect of Cordyceps militaris against hydrogen peroxide-induced oxidative stress in vitro. Nutr Res Pract 13(4):279–285. https://doi.org/10.4162/nrp.2019.13.4.279. PMID 31388403. PMCID PMC6669067

He MT, Park CH, Shin YS, Choi JM, Cho EJ (2020) Caterpillar medicinal mushroom, Cordyceps militaris (ascomycetes), protects Aβ1-42-induced neurologic damage in C6 glial cells. Int J Med Mushrooms 22(12):1203–1213. https://doi.org/10.1615/IntJMedMushrooms.2020037055. PMID 33463937

He MT, Park CH, Cho EJ (2021) Caterpillar medicinal mushroom, Cordyceps militaris (Ascomycota), attenuates Aβ1-42-induced amyloidogenesis and inflammatory response by suppressing amyloid precursor protein progression and p38 MAPK/JNK activation. Int J Med Mushrooms 23(11):71–83. https://doi.org/10.1615/IntJMedMushrooms.2021040404. PMID 34936310

Ivanova Stojcheva E, Quintela JC (2022) The effectiveness of Rhodiola rosea L. preparations in alleviating various aspects of life-stress symptoms and stress-induced conditions—encouraging clinical evidence. Molecules 27(12):3902. https://doi.org/10.3390/molecules27123902. PMID 35745023

Jatwani A, Tulsawani R (2021) Ganoderma lucidum induces myogenesis markers to avert damage to skeletal muscles in rats exposed to hypobaric hypoxia. High Alt Med Biol. https://doi.org/10.1089/ham.2020.0172. [Epub ahead of print]. PMID 34142874

Ji M, Gong X, Li X, Wang C, Li M (2020) Advanced research on the antioxidant activity and mechanism of polyphenols from Hippophae species-a review. Molecules 25(4):917. https://doi.org/10.3390/molecules25040917. PMID 32092874, PMCID PMC7071004

Jin DQ, Park BC, Lee JS, Choi HD, Lee YS, Yang JH et al (2004) Mycelial extract of Cordyceps ophioglossoides prevents neuronal cell death and ameliorates beta-amyloid peptide-induced memory deficits in rats. Biol Pharm Bull 27(7):1126–1129. https://doi.org/10.1248/bpb.27.1126. PMID 15256753

Jordan JL, Sullivan AM, Lee TDG (2008) Immune activation by a sterile aqueous extract of Cordyceps sinensis: mechanism of action. Immunopharmacol Immunotoxicol 30(1):53–70. https://doi.org/10.1080/08923970701812332. PMID 18306104

Kallio H, Yang B, Peippo P (2002) Effects of different origins and harvesting time on vitamin C, tocopherols, and tocotrienols in sea buckthorn (Hippophaë rhamnoides L.) berries. J Agric Food Chem 50(21):6136–6142. https://doi.org/10.1021/jf020421v. PMID 12358492, Google Scholar

Kelly GS (2001) Rhodiola rosea: a possible plant adaptogen. Altern Med Rev 6(3):293–302. PMID 11410073

Khanum F, Bawa AS, Singh B (2005) Rhodiola rosea: a versatile adaptogen. Compr Rev Food Sci Food Saf 4(3):55–62. https://doi.org/10.1111/j.1541-4337.2005.tb00073.x. PMID 33430554

Koganti P, Tulsawani R, Sharma P, Sharma M, Arora S, Misra K (2018) Role of hydroalcoholic extract of lingzhi or reishi medicinal mushroom, Ganoderma lucidum (Agaricomycetes), in facilitating cellular acclimatization in a low-oxygen microenvironment. Int J Med Mushrooms 20(5):431–444. https://doi.org/10.1615/IntJMedMushrooms.2018026075. PMID 29953358

Kogler L, Müller VI, Chang A, Eickhoff SB, Fox PT, Gur RC et al (2015) Psychosocial versus physiological stress - meta-analyses on deactivations and activations of the neural correlates of stress reactions. NeuroImage 119:235–251. https://doi.org/10.1016/j.neuroimage.2015.06.059. PMID 26123376

Koh JH, Kim KM, Kim JM, Song JC, Suh HJ (2003) Antifatigue and antistress effect of the hot-water fraction from mycelia of Cordyceps sinensis. Biol Pharm Bull 26(5):691–694. https://doi.org/10.1248/bpb.26.691. PMID 12736514

Komarov VL (ed) (1939) Flora of the USSR. Translation. 1971; IX, Rosales and Sarraceniales. Genus 698: Rhodiola L.:20–36

Konstantinos F, Heun R (2020) The effects of Rhodiola rosea supplementation on depression, anxiety and mood – a systematic review. Glob Psychiatry 3(1):72–82. https://doi.org/10.2478/gp-2019-0022

Kuo YC, Tsai WJ, Wang JY, Chang SC, Lin CY, Shiao MS (2001) Regulation of bronchoalveolar lavage fluids cell function by the immunomodulatory agents from Cordyceps sinensis. Life Sci 68(9):1067–1082. https://doi.org/10.1016/s0024-3205(00)01011-0. PMID 11212870

Kurkin VA, Zapesochnaya GG (1986) Chemical composition and pharmacological properties of Rhodiola rosea. Chem Pharm J (Moscow) 20(10):1231–1244

Ladol S, Sharma D (2021) The effects of Hippophae rhamnoides in neuroprotection and behavioral alterations against iron-induced epilepsy. Epilepsy Res 175:106695. https://doi.org/10.1016/j.eplepsyres.2021.106695. PMID 34186382

Li TS, Beveridge HJ (2003) [introduction]. Chap 1. Ottawa, Ontariopp. In: Cavers PB (ed) Seabuckthorn (Hippophae rhamnoides): production and utilization. NRC Research Press, pp 1–6

Li SP, Li P, Dong TT, Tsim KW (2001) Antioxidant activity of different types of natural Cordyceps sinensis and cultured Cordyceps mycelia. Phytomedicine 8(3):207–212. https://doi.org/10.1078/0944-7113-00030. PMID 11417914

Li Y, Wang XL, Jiao L, Jiang Y, Li H, Jiang SP et al (2011) A survey of the geographic distribution of Ophiocordyceps sinensis. J Microbiol 49(6):913–919. https://doi.org/10.1007/s12275-011-1193-z. PMID 22203553

Li Z, Zhang Z, Zhang J, Jia J, Ding J, Luo R et al (2012) Cordyceps militaris extract attenuates D-galactose-induced memory impairment in mice. J Med Food 15(12):1057–1063. https://doi.org/10.1089/jmf.2011.2105. PMID 23216110

Li Y, Pham V, Bui M, Song L, Wu C, Walia A et al (2017a) Rhodiola rosea L.: an herb with anti-stress, anti-aging, and immunostimulating properties for cancer chemoprevention. Curr Pharmacol Rep 3(6):384–395. https://doi.org/10.1007/s40495-017-0106-1. PMID 30393593, PMCID PMC6208354

Li Y, Wu J, Shi R, Li N, Xu Z, Sun M (2017b) Antioxidative effects of Rhodiola Genus: phytochemistry and pharmacological mechanisms against the diseases. Curr Top Med Chem 17(15):1692–1708. https://doi.org/10.2174/1568026617666161116141334. PMID 27848900
Liao LY, He YF, Li L, Meng H, Dong YM, Yi F et al (2018) A preliminary review of studies on adaptogens: comparison of their bioactivity in TCM with that of ginseng-like herbs used worldwide. Chin Med 13:57. https://doi.org/10.1186/s13020-018-0214-9. PMID 30479654. PMCID PMC6240259
Liu YW, Gao JL, Guan J, Qian ZM, Feng K, Li SP (2009) Evaluation of antiproliferative activities and action mechanisms of extracts from two species of Ganoderma on tumor cell lines. J Agric Food Chem 57(8):3087–3093. https://doi.org/10.1021/jf900011f. PMID 19368349
Liu Y, Zhang C, Du J, Jia R, Cao L, Jeney G et al (2017) Protective effect of Ganoderma lucidum polysaccharide against carbon tetrachloride-induced hepatic damage in precision-cut carp liver slices. Fish Physiol Biochem 43(5):1209–1221. https://doi.org/10.1007/s10695-016-0333-0. PMID 28681206
Liu W, Gao Y, Zhou Y, Yu F, Li X, Zhang N (2022) Mechanism of Cordyceps sinensis and its extracts in the treatment of diabetic kidney disease: a review. Front Pharmacol 13:881835. https://doi.org/10.3389/fphar.2022.881835. PMID 35645822. PMCID PMC9136174
Lu S, Wei F, Li G (2021) The evolution of the concept of stress and the framework of the stress system. Cell Stress 5(6):76–85. https://doi.org/10.15698/cst2021.06.250. PMID 34124582
Lupien SJ, Juster RP, Raymond C, Marin MF (2018) The effects of chronic stress on the human brain: from neurotoxicity, to vulnerability, to opportunity. Front Neuroendocrinol 49:91–105. https://doi.org/10.1016/j.yfrne.2018.02.001. PMID 29421159
Mao Y, Li Y, Yao N (2007) Simultaneous determination of salidroside and tyrosol in extracts of Rhodiola L. by microwave assisted extraction and high-performance liquid chromatography. J Pharm Biomed Anal 45(3):510–515. https://doi.org/10.1016/j.jpba.2007.05.031. PMID 176283 86
McEwen BS (1998) Protective and damaging effects of stress mediators. N Engl J Med 338(3): 171–179. https://doi.org/10.1056/NEJM199801153380307. PMID 9428819
McEwen BS, Wingfield JC (2003) The concept of allostasis in biology and biomedicine. Horm Behav 43(1):2–15. https://doi.org/10.1016/s0018-506x(02)00024-7. PMID 12614627
McEwen BS, Bowles NP, Gray JD, Hill MN, Hunter RG, Karatsoreos IN et al (2015) Mechanisms of stress in the brain. Nat Neurosci 18(10):1353–1363. https://doi.org/10.1038/nn.4086. PMID 26404710
McEwen BS, Nasca C, Gray JD (2016) Stress effects on neuronal structure: hippocampus, amygdala, and prefrontal cortex. Neuropsychopharmacology 41(1):3–23. https://doi.org/10.1038/npp.2015.171. PMID 26076834
Min BS, Nakamura N, Miyashiro H, Bae KW, Hattori M (1998) Triterpenes from the spores of Ganoderma lucidum and their inhibitory against HIV-1 protease. Chem Pharm Bull (Tokyo) 46(10):1607–1612. https://doi.org/10.1248/cpb.46.1607. PMID 9810695
Mizuno T, Wang G, Zhang J, Kawagishi H, Nishitoba T, Li J (1991) Reishi, Ganoderma lucidum and Ganoderma tsugae: bioactive substances and medicinal effects. Food Rev Int 11(1): 151–166
Munck A, Guyre PM, Holbrook NJ (1984) Physiological functions of glucocorticoids in stress and their relation to pharmacological actions. Endocr Rev 5(1):25–44. https://doi.org/10.1210/edrv-5-1-25. PMID 6368214
Narayanan S, Ruma D, Gitika B, Sharma SK, Pauline T, Ram MS et al (2005) Antioxidant activities of seabuckthorn (Hippophae rhamnoides) during hypoxia induced oxidative stress in glial cells. Mol Cell Biochem 278(1–2):9–14. https://doi.org/10.1007/s11010-005-7636-2. PMID 161800 83
Olatunji OJ, Feng Y, Olatunji OO, Tang J, Wei Y, Ouyang Z et al (2016) Polysaccharides purified from Cordyceps cicadae protects PC12 cells against glutamate-induced oxidative damage. Carbohydr Polym 153:187–195. https://doi.org/10.1016/j.carbpol.2016.06.108. PMID 275614 86

Panossian A (2017) Understanding adaptogenic activity: specificity of the pharmacological action of adaptogens and other phytochemicals. Ann N Y Acad Sci 1401(1):49–64. https://doi.org/10.1111/nyas.13399. PMID 28640972

Panossian A, Wikman G (2010) Effects of adaptogens on the central nervous system and the molecular mechanisms associated with their stress-protective activity. Pharmaceuticals (Basel) 3(1):188–224. https://doi.org/10.3390/ph3010188. PMID 27713248

Panossian A, Hambardzumyan M, Hovhanissyan A, Wikman G (2007) The adaptogens Rhodiola and Schizandra modify the response to immobilization stress in rabbits by suppressing the increase of phosphorylated stress-activated protein kinase, nitric oxide and cortisol. Drug Target Insights 2:39–54. https://doi.org/10.1177/117739280700200011. PMID 21901061, PMCID PMC3155223

Patel CA, Divakar K, Santani D, Solanki HK, Thakkar JH (2012) Remedial prospective of Hippophae rhamnoides linn. (Sea Buckthorn). ISRN Pharmacol 2012:436857. https://doi.org/10.5402/2012/436857. PMID 22530142, PMCID PMC3317027

Perez FP, Moinuddin SS, ul ain Shamim Q, Joseph DJ, Morisaki J, Zhou X (2012) Longevity pathways: HSF1 and FoxO pathways, a new therapeutic target to prevent age-related diseases. Curr Aging Sci 5(2):87–95. https://doi.org/10.2174/1874609811205020087. PMID 21834787

Petkov VD, Yonkov D, Mosharoff A, Kambourova T, Alova L, Petkov VV et al (1986) Effects of alcohol aqueous extract from Rhodiola rosea L. roots on learning and memory. Acta Physiol Pharmacol Bulg 12(1):3–16. PMID 3751623

Petkov VD, Stancheva SL, Tocuschieva L, Petkov VV (1990) Changes in brain biogenic monoamines induced by the nootropic drugs adafenoxate and meclofenoxate and by citicholine (experiments on rats). Gen Pharmacol 21(1):71–75. https://doi.org/10.1016/0306-3623(90)90598-g. PMID 2105261

Prado CM, Ramos SG, Alves-Filho JC, Elias J Jr, Cunha FQ, Rossi MA (2006) Turbulent flow/low wall shear stress and stretch differentially affect aorta remodeling in rats. J Hypertens 24(3):503–515. https://doi.org/10.1097/01.hjh.0000209987.51606.23. PMID 16467654

Pruessner JC, Dedovic K, Pruessner M, Lord C, Buss C, Collins L et al (2010) Stress regulation in the central nervous system: evidence from structural and functional neuroimaging studies in human populations-2008 Curt Richter Award Winner. Psychoneuroendocrinology 35(1):179–191. https://doi.org/10.1016/j.psyneuen.2009.02.016. PMID 19362426

Ren ZL, Wang CD, Wang T, Ding H, Zhou M, Yang N et al (2019) Ganoderma lucidum extract ameliorates MPTP-induced parkinsonism and protects dopaminergic neurons from oxidative stress via regulating mitochondrial function, autophagy, and apoptosis. Acta Pharmacol Sin 40(4):441–450. https://doi.org/10.1038/s41401-018-0077-8. PMID 29991712. PMCID PMC6461997

Rom O, Reznick AZ (2016) The stress reaction: a historical perspective. Adv Exp Med Biol 905:1–4. https://doi.org/10.1007/5584_2015_195. PMID 26909884

Rousi A (1970) The genus Hippophae L., a taxonomic study. Ann Bot 8:177–227. Google Scholar

Saggu S, Divekar HM, Gupta V, Sawhney RC, Banerjee PK, Kumar R (2007) Adaptogenic and safety evaluation of seabuckthorn (Hippophae rhamnoides) leaf extract: a dose dependent study. Food Chem Toxicol 45(4):609–617. https://doi.org/10.1016/j.fct.2006.10.008. PMID 17141932, Google Scholar

Saratikov AS, Krasnov EA (1987) Rhodiola rosea is a Valuable Medicinal Plant (Golden Root). Tomsk State University Press, Tomsk

Saratikov A, Marina TF, Fisanova LL (1978) Effect of golden root extract on processes of serotonin synthesis in central nervous system. J Biol Sci 6:142

Schneiderman N, Ironson G, Siegel SD (2005) Stress and health: psychological, behavioral, and biological determinants. Annu Rev Clin Psychol 1:607–628. https://doi.org/10.1146/annurev.clinpsy.1.102803.144141., PMID 17716101 (Volume publication date 27 Apr 2005)

Selye H (1956) The stress of life. McGraw-Hill, New York, p 294

Sevillano-García MD, Manso-Calderón R, Cacabelos-Pérez P (2007) Comorbidity in the migraine: depression, anxiety, stress and insomnia. Rev Neurol 45(7):400–405. https://doi.org/10.33588/rn.4507.2007122. PMID 17918105

Sharma P, Tulsawani R (2020) Ganoderma lucidum aqueous extract prevents hypobaric hypoxia induced memory deficit by modulating neurotransmission, neuroplasticity and maintaining redox homeostasis. Sci Rep 10(1):8944. https://doi.org/10.1038/s41598-020-65812-5. PMID 32488040

Sharma P, Tulsawani R (2022) Efficacy of aqueous extract of Chinese caterpillar mushroom Ophiocordyceps sinensis (Ascomycota) against simulated altitude stress and subacute toxicity studies. Int J Med Mushrooms 24(11):21–34. https://doi.org/10.1615/IntJMedMushrooms.2022045194. PMID 36374946

Sharma P, Tulsawani R, Agrawal U (2019) Pharmacological effects of Ganoderma lucidum extract against high-altitude stressors and its subchronic toxicity assessment. J Food Biochem 43(12):e13081. https://doi.org/10.1111/jfbc.13081. PMID 31609024

Shieh JM, Kuo SC, Niu KC, Cheng JT (2018) Herbal remedies as an alternative medication for altitude sickness. J Altern Complement Intern Med 4:57

Shivapriya S, Ilango K, Dubey GP (2015) Evaluation of antioxidant and neuroprotective effect of Hippophae rhamnoides (L.) on oxidative stress induced cytotoxicity in human neural cell line IMR32. Saudi J Biol Sci 22(5):645–650. https://doi.org/10.1016/j.sjbs.2015.04.011. PMID 262 88571. PMCID PMC4537860

Silva JR, Vivanco-Carlevari A, Barrientos M, Martínez C, Salazar LA, Krause M (2017) Biological stress reactivity as an index of the two polarities of the experience model. Psychoneuroendocrinology 84:83–86. https://doi.org/10.1016/j.psyneuen.2017.06.016. PMID 28672276

Singh M, Tulsawani R, Koganti P, Chauhan A, Manickam M, Misra K (2013) Cordyceps sinensis increases hypoxia tolerance by inducing heme oxygenase-1 and metallothionein via Nrf2 activation in human lung epithelial cells. Biomed Res Int 2013:569206. https://doi.org/10.1155/2013/569206. PMID 24063008

Sliva D (2003) Ganoderma lucidum in cancer research. Integr Cancer Ther 2(4):358–364. https://doi.org/10.1177/1534735403259066. PMID 14713328

Sun XZ, Liao Y, Li W, Guo LM (2017) Neuroprotective effects of Ganoderma lucidum polysaccharides against oxidative stress-induced neuronal apoptosis. Neural Regen Res 12(6):953–958. https://doi.org/10.4103/1673-5374.208590. PMID 28761429. PMCID PMC5514871

Suryakumar G, Gupta A (2011) Medicinal and therapeutic potential of Sea buckthorn (Hippophae rhamnoides L.). J Ethnopharmacol 138(2):268–278. https://doi.org/10.1016/j.jep.2011.09.024. PMID 21963559

Tost H, Champagne FA, Meyer-Lindenberg A (2015) Environmental influence in the brain, human welfare and mental health. Nat Neurosci 18(10):1421–1431. https://doi.org/10.1038/nn.4108. PMID 26404717

Tulsawani R, Sharma P, Divekar HM, Meena RN, Singh M, Kumar R (2010) Supplementation of fruit extract of Hippophae rhamnoides speeds adaptation to simulated high altitude stressors in rats. J Complement Integr Med 7(1). https://doi.org/10.2202/1553-3840.1323. Google Scholar

Tulsawani R, Mrinalini S, Manimaran M (2014) Aqueous extract of Ganoderma lucidum facilitates tolerance in rat heart by inducing nuclear factor (erythroid-derived 2)-Like 2 and hypoxia inducible factor 1 dependent signaling. Indian J Pharmacol 46(7):3–6

Tulsawani R, Manickam M, Suryakumar G, Kaur N (2015) Aqueous extract of Ganoderma lucidum prevent transvascular leakage in rat lung and hypothermia following hypobaric hypoxia exposure. Indian J Pharmacol 47(1):S13–S57

Van Diermen D, Marston A, Bravo J, Reist M, Carrupt PA, Hostettmann K (2009) Monoamine oxidase inhibition by Rhodiola rosea L. roots. J Ethnopharmacol 122(2):397–401. https://doi.org/10.1016/j.jep.2009.01.007. PMID 19168123

Veronique T, Muk WY, Christian D (2022) Seabuckthorn polyphenols: characterization, bioactivities and associated health benefits [internet]. In Phenolic compounds – chemistry, synthesis, diversity, non-conventional industrial, pharmaceutical and therapeutic applications. IntechOpen. https://doi.org/10.5772/intechopen.98706

Wagner H, Nörr H, Winterhoff H (1994) Plant adaptogens. Phytomedicine 1(1):63–76. https://doi.org/10.1016/S0944-7113(11)80025-5. PMID 23195818

Wang YY, Khoo KH, Chen ST, Lin CC, Wong CH, Lin CH (2002) Studies on the immunomodulating and antitumor activities of Ganoderma lucidum (Reishi) polysaccharides: functional and proteomic analyses of a fucose-containing glycoprotein fraction responsible for the activities. Bioorg Med Chem 10(4):1057–1062. https://doi.org/10.1016/S0968-0896(01)00377-7

Wang Z, Wang W, Zhu C, Gao X, Chu W (2022) Evaluation of antioxidative and neuroprotective activities of total flavonoids from sea buckthorn (Hippophae rhamnoides L.). Front Nutr 9: 861097. https://doi.org/10.3389/fnut.2022.861097. PMID 35799585. PMCID PMC9253677

Wasser SP, Weis AL (1999) Medicinal properties of substances occurring in higher basidiomycete mushrooms: current perspectives. Int J Med Mushrooms 1(1):31–62. https://doi.org/10.1615/IntJMedMushrooms.v1.i1.30

Wiegant FA, Surinova S, Ytsma E, Langelaar-Makkinje M, Wikman G, Post JA (2009) Plant adaptogens increase lifespan and stress resistance in C. elegans. Biogerontology 10(1):27–42. https://doi.org/10.1007/s10522-008-9151-9. PMID 18536978

Wihastuti TA, Heriansyah T (2017) The inhibitory effects of polysaccharide peptides (PsP) of Ganoderma lucidum against atherosclerosis in rats with dyslipidemia. Heart Int 12(1):e1–e7. https://doi.org/10.5301/heartint.5000234. PMID 29114382

Wu YS, Ho SY, Nan FH, Chen SN (2016) Ganoderma lucidum beta 1,3/1,6 glucan as an immunomodulator in inflammation induced by a high-cholesterol diet. BMC Complement Altern Med 16(1):500. https://doi.org/10.1186/s12906-016-1476-3. PMID 27912745

Wu S, Wu Q, Wang J, Li Y, Chen B, Zhu Z et al (2022) Novel selenium peptides obtained from selenium-enriched Cordyceps militaris alleviate neuroinflammation and gut microbiota dysbacteriosis in LPS-injured mice. J Agric Food Chem 70(10):3194–3206. https://doi.org/10.1021/acs.jafc.1c08393. PMID 35238567

Xu Y, Li G, Han C, Sun L, Zhao R, Cui S (2005) Protective effects of Hippophae rhamnoides L. juice on lead-induced neurotoxicity in mice. Biol Pharm Bull 28(3):490–494. https://doi.org/10.1248/bpb.28.490. PMID 15744075

Xu F, Li X, Xiao X, Liu LF, Zhang L, Lin PP et al (2017) Effects of Ganoderma lucidum polysaccharides against doxorubicine-induced cardiotoxicity. Biomed Pharmacother 95:504–512. https://doi.org/10.1016/j.biopha.2017.08.118. PMID 28866417

Xuan M, Okazaki M, Iwata N, Asano S, Kamiuchi S, Matsuzaki H et al (2015) Chronic treatment with a water-soluble extract from the culture medium of Ganoderma lucidum mycelia prevents apoptosis and necroptosis in hypoxia/ischemia-induced injury of Type 2 diabetic mouse brain. Evid Based Complement Alternat Med 2015:865986. https://doi.org/10.1155/2015/865986. PMID 25945116. PMCID PMC4402482

Zhang W, Zhang Q, Deng W, Li Y, Xing G, Shi X et al (2014) Neuroprotective effect of pretreatment with Ganoderma lucidum in cerebral ischemia/reperfusion injury in rat hippocampus. Neural Regen Res 9(15):1446–1452. https://doi.org/10.4103/1673-5374.139461. PMID 25317156. PMCID PMC4192946

Zhang X, Wang M, Qiao Y, Shan Z, Yang M, Li G et al (2022) Exploring the mechanisms of action of Cordyceps sinensis for the treatment of depression using network pharmacology and molecular docking. Ann Transl Med 10(6):282. https://doi.org/10.21037/atm-22-762. PMID 35434037. PMCID PMC9011256

Zhao X, Zhou D, Liu Y, Li C, Zhao X, Li Y et al (2018) Ganoderma lucidum polysaccharide inhibits prostate cancer cell migration via the protein arginine methyltransferase 6 signaling pathway. Mol Med Rep 17(1):147–157. https://doi.org/10.3892/mmr.2017.7904. PMID 29115463

Zheng T, Bian F, Chen L, Wang Q, Jin S (2019) Beneficial effects of Rhodiola and salidroside in diabetes: potential role of AMP-activated protein kinase. Mol Diagn Ther 23(4):489–498. https://doi.org/10.1007/s40291-019-00402-4. PMID 31069710

Zhou Y, Qu ZQ, Zeng YS, Lin YK, Li Y, Chung P et al (2012) Neuroprotective effect of preadministration with Ganoderma lucidum spore on rat hippocampus. Exp Toxicol Pathol 64(7–8):673–680. 21242065. https://doi.org/10.1016/j.etp.2010.12.011

Zhu JS, Halpern GM, Jones K (1998) The scientific rediscovery of an ancient Chinese herbal medicine: Cordyceps sinensis: part I. J Altern Complement Med 4(3):289–303. https://doi.org/10.1089/acm.1998.4.3-289. PMID 9764768

Zhuang W, Yue L, Dang X, Chen F, Gong Y, Lin X et al (2019) Rosenroot (Rhodiola): potential applications in aging-related diseases. Aging Dis 10(1):134–146. https://doi.org/10.14336/AD.2018.0511. PMID 30705774

Zu Y, Li C, Fu Y, Zhao C (2006) Simultaneous determination of catechin, rutin, quercetin kaempferol and isorhamnetin in the extract of sea buckthorn (Hippophae rhamnoides L.) leaves by RP-HPLC with DAD. J Pharm Biomed Anal 41(3):714–719. https://doi.org/10.1016/j.jpba.2005.04.052. PMID 16520013, Google Scholar

Zubieta JK, Stohler CS (2009) Neurobiological mechanisms of placebo responses. Ann N Y Acad Sci 1156(1):198–210. https://doi.org/10.1111/j.1749-6632.2009.04424.x. PMID 19338509

Molecular Basis of Response to Hypoxia

2

Anuja Krishnan and Hanaan Kasim Ansari

Abstract

Cellular hypoxia is a condition when there is insufficient oxygen available. This can be due to physiological situations, like high altitude environment, extreme exercise, or pathological conditions like cancer, myocardial infarction, inflammation, etc. Under such stressful conditions, the goal of the cell is to adapt and reprogram pathways so that homeostasis is maintained. The principal regulator of a response to hypoxia is called hypoxia-inducible factor (HIF). HIF is a transcription factor that is regulated by various mechanisms including non-oxygen-dependent ones. Metabolic reprogramming and enhanced angiogenesis are two major pathways that are brought on by HIF activation and help to counteract the consequences of hypoxia. In this chapter, we will understand pathways that regulate HIF, target genes, and pathways modulated by HIF and their crosstalk. Such an understanding should help develop therapeutics for various diseases and markers for hypoxia.

Keywords

Hypoxia · Hypoxia-inducible factor · Metabolism · Angiogenesis · Pathways

A. Krishnan (✉) · H. K. Ansari
Department of Molecular Medicine, School of Interdisciplinary Sciences and Technology, New Delhi, India
e-mail: anuja.krishnan@jamiahamdard.ac.in

R. Tulsawani, D. Vohora (eds.), *Adaptation under Stressful Environments through Biological Adjustments and Interventions*,
https://doi.org/10.1007/978-981-99-7652-2_2

2.1 Introduction

Oxygen is the most vital gas for the survival of most vertebrate species. We breathe around 21% of the oxygen in the air, of which a small portion reaches our tissues, with a range of 4.5 to 9.5%. An oxygen level below 2% is considered hypoxic. At high altitudes, where the amount of oxygen in the air decreases, hypoxia is a major factor. Hypoxia is also linked to several clinical disorders, including pulmonary edema, cancer, anemia, stroke, and ischaemic heart disease. Hypobaric hypoxia, a serious stress that has an impact on every part of cellular physiology, is common at high altitudes. The absolute oxygen level and air pressure both fall as the altitude rises. For instance, just 60% of the oxygen present at sea level is available at 4000 m, a typical height for the Tibetan Plateau (Petousi et al. 2014). In order to comprehend the molecular characteristics and signaling involved in hypoxia, several studies have been conducted at the in vitro and in vivo levels. Long-term studies, however, have also been carried out on populations who have inhabited high altitudes for many generations, evaluating phenotypic features and genetic alterations. The hypoxic problem of increasing altitude results in a decrease in arterial partial pressure of oxygen (PO2) and a subsequent decrease in arterial oxygen saturation, lowering ambient oxygen levels (SaO2). The body reacts in several ways, by continuing to increase ventilation in the near term and by boosting red blood cell creation in the long term, in order to counteract this fall in arterial oxygen content and maintain O_2 homeostasis.

Widespread hypoxia causes altitude sickness in healthy persons who go to high altitudes. Acute mountain sickness (AMS) is the most common ailment. In two-thirds of susceptible patients, AMS normally begins after 12 h of climb and manifests within 36 h in the remaining third. It is frequently accompanied by headaches, loss of appetite, disrupted sleep, nausea, exhaustion, and dizziness and sometimes with more severe outcomes, like, periorbital edema, altered mental status, and ataxia. Rapid ascent to high altitude can result in the potentially lethal condition known as high-altitude pulmonary edema (HAPE) which is distinguished by incapacitating exhaustion, dyspnea that worsens with exertion and progresses to dyspnea at rest (Basnyat and Murdoch 2003). Preterm delivery in newborn infants frequently results in hypoxia.

Under pathological conditions, poor blood flow causes hypoxia, which is dangerous for the structure and function of the organs. Stroke (cerebral ischaemia) and cardiac infarction are two examples where this is particularly true (myocardial ischaemia). Similarly in cancer, a tumour's blood supply is quickly exhausted as it expands, leaving some of the tumours with areas where the oxygen level is noticeably lower than in healthy tissues. It has been shown that cancer cells modify their metabolism to allow ongoing growth and proliferation under difficult hypoxic settings. Usually, extremely aggressive tumours that typically do not respond well to therapy are accompanied with tumour hypoxia.

In 2019, William G. Kaelin Jr., Sir Peter J. Ratcliffe, and Gregg L. Semenza received the Nobel Prize in Physiology in appreciation of their discovery of cellular mechanisms to sense and adapt to different oxygen concentrations, providing a foundation for how oxygen levels affect physiological function. Their investigations revealed that hypoxia induces erythropoietin (EPO) mRNA and protein production, thus increasing blood oxygen transport. Detailed analysis of the enhancer region of EPO identified a protein complex that gets bound only in hypoxia conditions. This complex is called the hypoxia-inducible transcription factor (HIF) (Zhang et al. 2019).

Dr. Semenza identified hypoxia-inducible factor (HIF-1α and HIF-1β), the primary transcription factor that stimulates expression of hypoxia-induced proteins. Drs. Kaelin and Ratcliffe discovered the regulation of HIF-1 α involved in the adjustment of the initial physiological response to hypoxia. HIF1α pathway regulates the expression of target genes initiating cellular reprogramming in metabolism, oxygen delivery, angiogenesis, mitochondrial function, ECM production, inflammation, etc. (Samanta and Semenza 2018; Taylor and Scholz 2022).

In this chapter, we describe current understandings regarding the molecular mechanism of response to hypoxia.

2.2 Hypoxia-Inducible Factor (HIF)

2.2.1 Isoforms and Domain Organization

HIF has been established as a master regulator that is involved in adaptation to hypoxic conditions. HIF is a heterodimeric transcription factor consisting of an oxygen-sensitive alpha subunit and the constitutive beta subunit. Both subunits belong to the basic Helix–Loop– Helix PER-ARNT-SIM (bHLH-PAS) family of transcription factors (Wang et al. 1995). Three isoforms exist for HIF-1; 1α, 2α and 3α. HIF1A encodes HIF-1α, EPAS1 (endothelial PAS domain-containing protein 1) encodes HIF-2α, and HIF3A encodes multiple splice variants of HIF-3α. HIF-1α and HIF-2α share a similar domain structure as compared to HIF-3α. Generally, the protein domain structure of the HIF family is conserved. The three alpha isoforms and HIF-1β each has the PAS-A and PAS-B domains necessary for heterodimerization as well as an amino-terminal beta-helix loop helix (bHLH) that is required for DNA binding. The N- and C-terminal transactivation domains (N-TAD and C-TAD) of HIF-1α and HIF-2α are present and necessary for the activation of HIF target genes. Due to the presence of the important asparagine (N) and proline (P) residues that are targeted for hydroxylation under normoxic circumstances, the oxygen-dependent degradation (ODD) domain is a highly conserved domain that regulates the activity and stability of the alpha subunits. The alpha subunits' ODD domain grants oxygen-regulated turnover and overlaps the N-TADs. Whereas HIF1A and HIF2A's PAS domains share about 70% identity, their bHLH domains have roughly 85% identity. As a result, HIF1A and HIF2A can bind DNA sequences that are difficult to differentiate (Smythies et al. 2019; Wenger et al. 2005). Compared to HIF1A and HIF2A, the bHLH and PAS domains of HIF3A only share 74% and 52–58% similarity, respectively, demonstrating the more diverse character of this paralog.

2.2.2 HIF Regulation

2.2.2.1 Post-Translational Modification

HIF- α subunits have a very short half-life in normoxic environments. The HIF α protein is continuously made and destroyed by cells. Under the normoxic condition, prolyl-4-hydroxylases (PHD) add hydroxyl group at two proline residues (Pro402 and Pro564) present within the LXXLAP motifs found in the ODD of the HIF α-subunits. In mammals, there are three isoforms of PHDs (1–3) (Berra et al. 2003). PHD2 plays the most important role in the hydroxylation and degradation of HIF-1α and HIF-2α. PHDs are non-heme Fe(II) and 2-oxoglutarate-dependent dioxygenases that split molecular oxygen. The HIF- α peptide gains one oxygen atom, and the other one interacts with 2-oxoglutarate to create succinate and CO_2. This hydroxylation allows the interaction of HIF α with the von Hippel Lindau (pVHL) E3 complex. HIF-α is targeted for proteolysis via the ubiquitin-proteosome pathway by the recognition component of an E3 ubiquitin ligase complex, pVHL. Under the hypoxic condition, since oxygen is not available, it allows HIF-1α protein to evade hydroxylation and thereby escapes its degradation by proteasome pathway (Berra et al. 2003). HIF-1α moves to the nucleus and forms a dimer with HIF-1β, also known as aryl hydrocarbon receptor nuclear translocators (ARNT). The hypoxia response element (HRE) in the promoter or enhancer sequences of target genes is then bound by the heterodimeric transactivating complex HIF (Fig. 2.1).

In addition to PHD, HIF-1α and HIF-1β are further regulated by another protein called Factor Inhibiting HIF (FIH). This protein hydroxylates asparagine residue present within the C-TAD (Asn-803 for HIF1A, Asn-847 for HIF2A). This hydroxylation disrupts transactivation potential due to its inability to bind to its coactivators like cyclic AMP response element binding protein (CBP) and p300 (Lando et al. 2002). Figure 2.1 illustrates the HIF regulation and signaling.

Other kinds of post-translational regulation of HIF also have been implicated, like phosphorylation, sumoylation, and acetylation. It has been demonstrated that casein kinase II-mediated phosphorylation of HIF-1α and HIF-2α subunits in the C-TAD (Thr796 for HIF1A, Thr844 for HIF2A) enhances transactivation of target genes by either preventing HIF from interacting with VHL and stabilizing it or by increasing the affinity of HIF with transcriptional coactivators (Mottet et al. 2005). Nuclear protein kinases like p42/p44 MAPK enhance the transcriptional activity of HIF-1 and increase its accumulation in the nucleus by phosphorylating it at serine residues 641 and 643 (Richard et al. 1999). On the other hand, phosphorylation by casein kinase I within the PAS-B domain (Ser-247) of HIF 1α disrupts its association with HIF 1β, thus reducing HIF1A-induced gene expression (Knippschild et al. 2005). G9a and G9a-like protein (GLP) are lysine methyltransferases, where mono and dimethylation of K674 were found to suppress transcriptional activity of HIF-1α by inhibiting its transactivation function (Bao et al. 2018).

2.2.2.2 Epigenetic Regulation

An initiation complex made up of general transcription factors and RNA polymerase II must get past the chromatin's physical barrier for transcription to start. Epigenetic

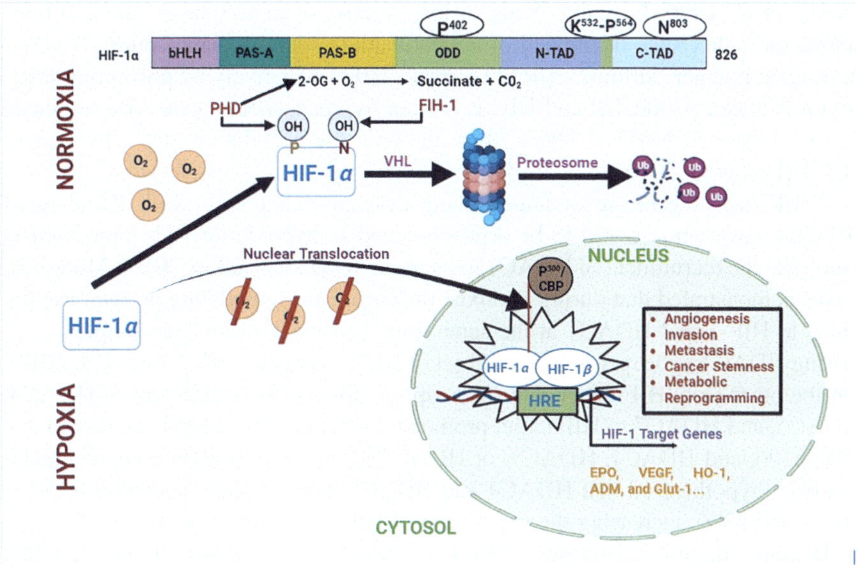

Fig. 2.1 HIF protein domain organization and signaling during hypoxia. HIF protein is comprised of DNA binding domain (bHLH), dimerization domain (PAS), oxygen degradation domain (ODD), and N-terminal and C-terminal transactivation domain (N-TAD and C-TAD). During nomoxia, prolyl hydroxylase (PHD) catalyses hydroxylation of prolinr residue of HIF by using oxygen and oxoglutarate. This hydroxylation of prolines in ODD domain of HIF enables it to bind to VHL targeting HIF to proteasomal degradation. FIH hydroxylates the asparagine residue of HIF and decreases its coactivation potential. During hypoxia, when no oxygen is present, post-translational modification does not occur allowing HIF1α to enter the nucleus and bind to HIF-1β. The HIF1 α β heterodimer binds to the HRE of target genes to enhance the transcription of genes

regulation of genes occurs by post-translational modification of histone tails which can affect the chromatin compaction and accessibility of transcriptional machinery to the DNA. Nucleosome is the basic unit containing 2 histone molecules of each of H2A, H2B, H3, and H4 around which 146 bp of DNA is wrapped.

Histone tails may be modified by acetylation, methylation, ubiquitination, phosphorylation, and other processes. Histone acetylation is controlled by acetylation by histone acetyltransferases (HAT) and deacetylation by histone deacetylase (HDAC). Lysine residues located in the N-terminal tails of histones are acetylated, a process that is closely related to transcriptional activation. By loosening high-order chromatin structures, histone acetylation stimulates transcription. In addition, acetylated histones generate signals for the binding of proteins to bromodomains, which have been demonstrated to either have intrinsic HAT activity or to be a component of chromatin remodelling complexes, thereby stimulating transcription. It is well demonstrated that multiple levels of epigenetic factors regulate HIF transcriptional activity. HIF recruits co-activators: histone acetyltransferases (HAT) like p300 and cAMP response element binding (CREB) protein, CBP, which acetylates and relaxes the higher order structure of chromatin and is associated with transcription activation

(Arany et al. 1996; Luo and Wang 2018). A class of deacetylases called sirtuins senses the redox state of the cell in a nicotinamide adenine dinucleotide (NAD+)-dependent manner. Sirtuin1 (SIRT1) controls HIF-1 negatively by preventing interaction between p300/CBP and HIF-1. During hypoxia, SIRT1 gets downregulated due to decreased NAD(+) levels, thus allowing the acetylation and activation of HIF-1 (Lim et al. 2010).

A HIF target gene, reversion-inducing cysteine-rich protein with Kazal motif (RECK) was demonstrated to be downregulated in hypoxia through a mechanism involving the recruitment of HDAC1 to its promoter (Jeon and Lee 2010). Moreover, it was demonstrated that during hypoxia, the chromatin-remodeling protein, Reptin, binds to HIF-1 and HDAC1 at the same time, causing a restrictive complex comprising HDAC1 to be recruited to a subset of HIF target genes (K. J. Lee et al. 2010). On the other side, HIF-dependent transcription allows the recruitment of HDAC4, HDAC5, and HDAC7 to HIF target promoters. Additionally, it has been shown that HIF, p300, and HDAC4, HDAC5, or HDAC7 form a multiprotein complex, and it has been hypothesized that HDAC4 and HDAC5 increase the association between HIF-1 and p300, increasing the expression of HIF target genes (Kim et al. 2022).

Histone methyltransferases (KMTs) and histone demethylases (KDMs) dynamically control the methylation status of histone lysine. Histone methylation can occur resulting in mono-, di- or trimethylated forms and can result in enhancement or inhibition of transcription. A marker of chromatin active regions is methylation at lysines 4, 36, or 79 of H3, whereas methylation at lysines 9 and 27 of H3 and at lysine 20 of H4 is linked to the formation of heterochromatin and transcriptional repression. (Tsai and Wu 2014). The Jumonji C (JmjC)-domain-containing histone demethylases (JHDM) are Fe(II)/ 2-oxoglutarate- (2OG) dependent oxygenase family of enzymes that use O_2 as a co-substrate in the demethylation process (Islam et al. 2018). JHDMs are known to be not only involved in global histone methylation homeostasis under hypoxia, but certain family members might also play a direct role in specific gene expression. Krieg and colleagues showed that JMJD1A is essential for the hypoxia activation of several genes, of which many are HIF targets like, ADM, EDN1, SERPINE1, PLAUR, and HMOX1 (Krieg et al. 2010). By demethylating H3K9me2 in several hypoxia-responsive promoters, including ADM and GDF15, it has been shown that JMJD1A contributes to hypoxia-dependent gene expression (Beyer et al. 2008). A substantial rise in H3K9ac and H3K4me3 levels and a decline in H3K27me3 levels in the promoters of the HIF target genes VEGF and EGR1 with hypoxic induction— events typically connected with transcriptionally active loci, are observed. The hypoxia-inducible genes EPO, HMOX1, and DAF's promoters have also been found to undergo the same or related epigenetic modifications (Beyer et al. 2008).

In addition to controlling gene expression, hypoxia-induced histone demethylases also alter chromatin architecture such as polycomb complexes and heterochromatin (Thinnes et al. 2014). For instance, lysine-specific demethylase 3A (KDM3A) and HIF-1α interaction in endothelial cells allow hypoxia to alter chromatin conformation and increase the expression of SLC2A3 (GLUT3), which is a glucose transporter (Mimura et al. 2012). In a HIF-1α-dependent manner, KDM3A is recruited at the SLC2A3 gene's promoter, stimulating SLC2A3 expression and subsequently boosting glycolysis.

The methylation status of the hypoxia response element (HRE) is also important for HIF transactivation activity. HRE sites were discovered in the promoter regions of two histone demethylases, KDM4C (JMJD2C) and KDM5B. In hypoxia, both KDMs showed a substantial upregulation. Histone H3's lysine 4 (H3K4me3/2/1) has been tri-, di-, and monomethylated. KDM5B catalyses the removal of methyl groups from this residue, while KDM4C converts few trimethylated histone residues to dimethylated histone residues. KDM4C interacts with HIF-1α, thereby enhancing its binding to HRE of genes involved in metastasis and metabolism (Luo et al. 2012). Further, HRE hypermethylation suppresses the production of erythropoietin (EPO) by preventing HIF from binding to its promoter (Wenger et al. 1998).

In addition to the histone tail modifications that occur post-translationally, chromatin can also be altered by a variety of ATP-dependent chromatin remodelling complexes which alter the accessibility of the transcription machinery to specific loci. One of the regulators is a unique chromatin remodeling complex termed SWI/SNF (SWItch/Sucrose Non-Fermentable) which disrupts interactions between DNA and histones using ATP, thus facilitating access of transcription factors to their target sequences. HIF-1α is a direct target of the SWI/SNF chromatin-remodeling complex and expression levels of HIF and its ability to transactivate shown to be modulated by SWI/SNF levels (Kenneth et al. 2009).

The imitation switch (ISWI) is also chromatin remodeling complex linked to the hypoxic response. However, in contrast to SWI/SNF complex, ISWI downregulation promotes HIF-1α activity without affecting its levels. ISWI has been shown to be essential for expression of FIH mRNA and protein levels (Melvin et al. 2011).

2.2.3 Oxygen-Independent Regulation

HIF stabilization can also result from a variety of oxygen-independent processes. Other factors like cytokines, growth factor, heat shock, nutrient deprivation, etc. can regulate HIF-1α. Studies in cancer have demonstrated that various pro-survival pathways, like, ERK/MAPK, JAK/STAT, and PI3K/Akt/mTOR, increase transcription and translation of HIF1A. MAPK pathway phosphorylates co-activators p300/CPB, thus regulating HIF transactivation. JAK/STAT pathway has been demonstrated to activate Akt-mediated HIF1A transcription involving STAT3. PI3K/Akt/mTOR signaling directly enhances HIF1A transcription and translation, irrespective of oxygen concentrations, whereas conditions like nutrient deprivation inhibits mTOR pathway which may subsequently lead to HIF-1α downregulation. mTOR phosphorylates Mint3 which along with MT1-MMP inhibits FIH-1 and thereby activates HIF1 (Sakamoto et al. 2014). It has been demonstrated that nutrient levels independently control HIF stability. Crucially, growth factor signaling causes HIF to become stabilized as a result of the activation of mTOR and an increase in the rate of HIF translation (Laughner et al. 2001). HIF proteins become stabilized as a result of the number of HIF alpha proteins increasing disproportionately to PHDs

(Pollard et al. 2005; Selak et al. 2005). Different teams have revealed the inverse association between prolyl-hydroxylated HIF α and intracellular reactive oxygen species (ROS) (Brunelle et al. 2005). Finley et al. (2011) found that SIRT3 increases the prolyl-hydroxylase activity of PHDs to their highest level while destabilizing HIF1α by preventing the generation of ROS (Finley et al. 2011). Accumulation of Kreb's cycle intermediates like succinate and fumarate as found in certain tumors is found to stabilize HIF-1α. These metabolites are proposed to suppress PHDs enzymatic activity or compete with them for cofactors like iron and ascorbate, which are necessary for their proper function (Hewitson et al. 2007).

Mouse double minute 2 homolog (MDM2) was demonstrated to ubiquitinate HIF-1α and its destabilization in a hydroxylation-independent manner, despite hypoxic atmosphere (Joshi et al. 2014; Nieminen et al. 2005). This process of HIF-1α degradation seems to occur upon binding with tumour suppressor proteins, such as TAp73 or p53 (Zhou et al. 2015).

2.2.4 HIF Signaling

The HIFαβ heterodimer enters the nucleus under hypoxia and binds to the target genes' hypoxia response elements (HREs), which include the core pentanucleotide sequence (RCGTG). However, genome-wide analysis shows that less than 1% of potential sites are bound by HIFs in response to hypoxia (Schödel et al. 2013).

In addition to the core HRE, the promoters of several HIF target genes possess an inverted, imperfect repeat of the HRE (5'-CAGGT-3′), also called HIF ancillary sequence (HAS) (Kimura et al. 2001). There are more than 100 genes which have been validated as HIF targets containing the HIF binding site (HBS). To ensure that HIF and other transcription factors are properly linked to coactivator complexes and the basic transcriptional machinery, paralogous transcriptional coactivators CBP and p300 are crucial. They are also required for effective transcriptional activation. The CH1 domains of CBP and p300 are recognized by the C-TADs of HIF1A/2A. Moreover, as discussed previously, CPB/p300 has the histone acetyltransferase activity required for chromatin remodeling prior to transcription (Guidez et al. 2005).

The coactivators CBP/p300, SRC-1, and transcription intermediary factor 2 (TIF-2) are recruited by both the NAD and CAD. The coactivation of HIF target genes has also been associated with several additional coactivators with lysine acetyl-transferase activity. It has been demonstrated that three coactivator members of the p160 family, NCOA1, NCOA2, and NCOA3, interact with HIF1A and enhance its transcriptional activity at target genes. The coactivation of HIF1A by NCOA1 and NCOA2 was shown to be synergistic with CBP/p300 and hypoxia-dependent, the latter of which is most likely brought on by their shared capacity to bind to both HIF1A TADs and CBP/p300 (Carrero et al. 2000; Wang et al. 2010)). Subsequent studies suggested that CBP is required for the interaction of NCOA1 with HIF1A. PKM2 is considered an important enzyme that converts phosphoenolpyruvate to pyruvate in glycolysis. Pyruvate kinase M2 isoform (PKM2) functions as a coactivator for HIF target genes (SLC2A1/GLUT1, LDHA, PDK1) whose products help to transition organisms from oxidative to glycolytic metabolism (Luo et al. 2011). This action was supposedly made stronger by PHD3 prolyl hydroxylating PKM2. PKM2 hydroxylation with prolyl hydroxylase 3 (PHD3) enhances PKM2 binding to HIF-1α and PKM2 coactivator function. The expression of PKM2 is regulated by heterogeneous ribonucleoproteins (hnRNPs), which alternatively splice PKM. It was demonstrated that mTOR-mediated upregulation of HIF α leads to transactivation of PKM2. Also, mTOR-mediated upregulation of c-myc leads to upregulation of hnRNPs, which initiate alternate splicing inducing PKM2 isoform over PKM1 (David et al. 2010). Sun and co. experimentally demonstrated that activated mTOR, a key activator of the Warburg effect in cancer cells, induces PKM2 expression. HIF-1 is triggered by mTOR, which also initiates alternative splicing via c-Myc-hnRNPs, which results in aerobic glycolysis in cancer cells (Sun et al. 2011).

HIF target loci in normoxia contain transcriptionally active, but stopped RNAPII in an open chromatin setting. HIF1A is stabilized during hypoxia and joins HIF1B in binding to chromatin. This mainly includes histone acetylation by CBP/p300 and binding of the Bromodomain protein BRD4. Another is the recruitment of CDK8-Mediator together with the SEC results in the release of stalled RNAPII and successful elongation. Acetylation of Lysine-674 on HIF1 α by the p300/CBP-associated factor (PCAF) improves the binding between p300 and HIF1 α. Conversely, SIRT1 deacetylates the same residue, which prevents p300 from being recruited and prevents HIF1a-target genes from being transactivated (Xenaki et al. 2008; Lim et al. 2010). Further it has been demonstrated that CITED2 and CITED4 bind p300/CBP at the same region necessary for binding to HIFα, thus inhibiting HIF-1α activity (Bhattacharya et al. 1999; Fox et al. 2004). Pontin, chromatin remodelling factor, was demonstrated to be a HIF-1α coactivator by mediating the binding between p300 and HIF-1α under hypoxia (J. S. Lee et al. 2011). De-SUMOylation of p300 by sentrin-specific protease (SENP)3 under oxidative stress enhances the binding of p300 to HIF1 α, hence enhancing HIF1α transcriptional activity (Huang et al. 2009).

Main goal of cellular response to hypoxia is to adapt to the new situation, promote cell survival, and restore oxygen homeostasis. The adaptations include strategies to sustain energy levels, activate pathways that require reduced energy need, and promote pathways that would help deliver more oxygen to tissues. However, when these adaptations fail or are not enough, it can lead to the deregulation of cellular organelle and finally result in cell death.

2.3 Adaptations

2.3.1 Metabolic

Most mammals utilize aerobic respiration as the source of metabolic strategy, where glucose is broken down and oxygen is used as the final electron acceptor in the electron transport chain (ETC) to provide energy currency, that is, ATP. This means that a constant supply of oxygen is required to maintain bioenergetic homeostasis. When sufficient levels of oxygen are present, glucose enters glycolysis and gets converted to pyruvate producing ATP. Pyruvate enters the mitochondria and gets converted to acetyl CoA which then enters the TCA cycle to produce NADH and $FADH_2$ which serve as electron sources for the ETC. For ATP synthase to make ATP, the ETC creates a proton gradient across the inner mitochondrial membrane. This process produces further large amounts of ATP (36 molecules of ATP per molecule of glucose). When oxygen is less, ATP gets re-directed towards glycolysis without entering oxidative phosphorylation in the mitochondria (Formenti et al. 2010). With less ATP produced, ATP/AMP ratio allosterically reduces the allosteric inhibition of ATP on glycolytic enzyme phosphofructokinase (PFK). HIF's primary job in hypoxia is to switch the body's energy-producing systems from oxygen-consuming ones to less efficient ones. Various enzymes are regulated by HIF under hypoxia. HIF1 activates pyruvate dehydrogenase kinase 1 (PDK1) which is an inhibitor of the enzyme, pyruvate dehydrogenase (PDH) which is responsible for pyruvate to acetyl CoA conversion. By doing so, pyruvate is prevented from entering the TCA cycle. In addition, with less oxidative phosphorylation, fewer reactive oxygen species are produced thereby reducing damage to tissues. Glycolysis alone is less efficient in ATP production. In order to compensate, HIF1 upregulates transcription of glucose transporters, GLUT1 and GLUT 3 (Formenti et al. 2010). Various enzymes of glycolysis are upregulated under the condition of hypoxia by HIF1. HIF-1α also enhances expression of lactate dehydrogenase A (LDHA), which catalyses the conversion of pyruvate generated from glycolytic metabolism to lactate and concurrently generates NAD^+, a cofactor essential in permitting continued glycolytic activity (Semenza et al. 1994). Main goal of both PDK1 and LDHA is to divert pyruvate away from the TCA cycle by decreasing acetyl Co-A production and instead generate lactic acid. The lactate produced is expelled from the cell by the actions of plasma membrane transporter, monocarboxylate transport 4 (MCT4), to avoid competitive inhibition of LDHA (Ullah et al. 2006). Another strategy to reduce ROS generation under hypoxic conditions is selective mitochondrial autophagy, called mitophagy. HIF increases expression of BCL-2/adenovirus E1B 19-kDa interacting protein 3 (BNIP3) and BNIP3-like (BNIP3L) resulting in increased mitophagy and decreased mitochondrial biogenesis (Bellot et al. 2009). HIF1 also reduces mitochondrial biogenesis by increasing the expression of MAX interactor 1 (MXI1) which negatively regulates MYC which further decreases expression of peroxisome proliferator- activated receptor-γ coactivator 1β (PGC1β) which is crucial in mitochondrial biogenesis (Zhang et al. 2007). HIF1 also modifies the mitochondrial ECT protein complexes, cyclooxygenases (COX)-that transfers electrons received from NADH and $FADH_2$ to generate ATP through oxidative phosphorylation. Under normoxic conditions, COX contains the subunit COX4–1. However, in hypoxic conditions, HIF1 activity enhances the expression of another COX subunit, COX4–2, and of the mitochondrial protease LON, which degrades the COX4–1 subunits, thereby increasing the electron transfer efficiency under hypoxic conditions (Fukuda et al. 2007). HIF-1 also activates the transcription of the microRNA, miR-210. This miRNA is shown to inhibit expression of the

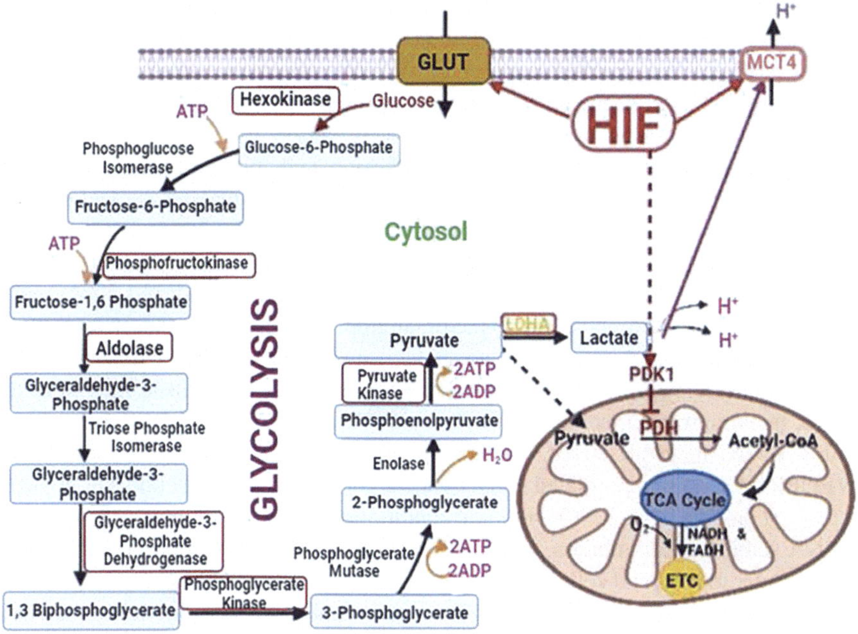

Fig. 2.2 Metabolic reprogramming by HIF during hypoxia. HIF upregulates key enzymes involved in glycolysis. HIF upregulates glucose transporter (GLUT1). Glycolytic enzymes upregulated in glycolysis are circled in RED. HIF activates enzyme, PDK1, which thereby inhibits PDF thus inhibiting pyruvate to be converted to acetyl CoA. HIF also activates LDH which converts pyruvate to lactate. The acidic pH due to lactic enzyme is pumped out by MCT4 which is also activated by HIF

iron-sulfur cluster assembly proteins ISCU1/2, which are important for the function of enzymes in TCA cycle and ETC (Chan et al. 2009) (Fig. 2.2).

2.3.2 Angiogenesis

A balance between O_2 supply, given by the vasculature, and demand, produced by tissues' metabolic outputs, controls the oxygenation of tissues. One way of adaptation is to increase the delivery of oxygen by promoting multiple pro-angiogenic pathways. Angiogenesis is important for tissue or organ regeneration, wound healing, and developmental processes, but it is also involved in pathological processes including cancer and some retinopathies. Angiogenesis consists of multiple steps. Injury in any form due to hypoxia or any other reason upregulates the expression of proangiogenic growth factors that activate their receptors. VEGF increases vascular permeability, thereby allowing the extravasation of plasma proteins that form a primitive scaffold for migrating endothelial cells (Carmeliet 2003). On the other hand, angiopoietin-1 and angiopoietin-2 (Ang-1 and Ang-2) inhibit vascular permeability. Ang-1 prevents plasma leakage and Ang-2 is involved in vessel destabilization via the detachment of smooth muscle cells and the promotion of permeabilization (Maisonpierre et al. 1997). For new vessel development, matrix components need to be degraded which is accomplished by matrix metalloproteinases (MMPs) (Bajbouj et al. 2021; Nelson et al. 2000). Factors such as integrins $\alpha\beta$ promote endothelial cell adhesion and migration, and once new vessels are formed, they are stabilized by Ang-1, PDGF-BB (platelet-derived growth factor-BB), and PDGFR (platelet-derived growth factor receptor) (Eliceiri and Cheresh 1999).

HIF1 α directly activates several genes, like VEGF, PDGF, induced nitric oxide synthases (iNOS), fibroblast growth factor, angiopoietin-2, PDGF, etc. (Krock et al. 2011). Factors like urokinase plasminogen activator receptor uPAR, collagen prolyl-4-hydroxylases, matrix metalloproteinases (like MMP-2), and tissue inhibitors of matrix metalloproteinases have been shown to be influenced by HIF-1 α signaling pathways (Nishi et al. 2016). HIF-1α directly induces VEGF, while VEGFR-1 (VEGFR-1/Flt-1) is upregulated under hypoxic conditions to further boost biological activity (Forsythe et al. 1996). Hypoxic conditions also lead to an increase in VEGF mRNA stability. Vasodilation, enhanced vascular permeability, modification of the extracellular matrix, and cell proliferation are all processes that are directly regulated by the HIF-1 α gene. Many growth hormones and cytokines, including insulin, insulin-like growth factors, and PDGF, can stimulate the production of HIF-1 α protein. HIF-1 α has been linked to P42/44 mitogen-activated kinase (MAPK), and MAPK activation increases HIF's transcriptional response. MAPK signaling may affect HIF activity by promoting the formation of the HIF-p300/CBP complex and by modulating the transactivation activity of p300/CBP (Sang et al. 2003). Most of the studies have been done in cancer model (Nicolas et al. 2019).

Apart from metabolic pathways and angiogenesis, HIF is shown to regulate immune system both innate and adaptive systems. HIF stabilization can occur in absence of oxygen. Toll like receptors (TLRs) in human monocytes have been shown to recognize pathogen-associated molecular patterns (PAMPs) such as lipopolysaccharide (LPS) and activate the mitogen-activated protein kinase (MAPK) pathway and ultimately NF-kB, which in turn transactivates HIF1A and stimulates HIF-1α production under normoxic condition (Frede et al. 2006). In hypoxic condition, PHD-mediated inhibition of IkB kinase-b (IKK-b) has been observed (Cummins et al. 2006). This prevented NF-kB activation and subsequent upregulation of HIF. It has been observed that pro-inflammatory TH17 cells, as opposed to anti-inflammatory Treg cells, are differentiated from naive T cells by mTOR-mediated activation of HIF-1α. It has been suggested that HIF-1α transactivates the RORgt gene which acts on promoters of Th17 genes. By actively inhibiting Foxp3's ability to function, HIF-1α prevents the differentiation of T-regs. Reports suggest that cytokines, TGF-b, and IL-6 are involved in activation of HIF1A in a STAT3-dependent manner, which then causes RORgt to transactivate and, ultimately, TH17 signature genes to transactivate through the HIF1a/RORgt/p300 pathway. In addition, Foxp3 is bound by HIF-1 and targeted for proteasomal degradation, which inhibits the formation of Tregs (Dang et al. 2011) (Table 2.1).

Table 2.1 Metabolic and angiogenesis proteins regulated upon hypoxia

Protein	Function	Reference
Hexokinase	Glucose phosphorylation utilizing ATP	(Mathupala et al. 2001)
Phosphofructokinase	Catalyses the reaction Fructose-6- phosphate to fructose-1,6-bisphosphate using ATP	(Semenza et al. 1994)
Aldolases	Catalyses conversion of fructose-1,6-bisphosphate into dihydroxyacetone phosphate and glyceraldehyde-3- phosphate	(Semenza et al. 1996)
Glyceraldehyde 3-phosphate dehydrogenase (GAPDH)	Catalyses reaction Glyceraldehyde 3- phosphate to 1,3-bisphosphoglycerate and NADH	(Graven et al. 1999)
Phosphoglycerate kinase 1 (PGK1)	Catalyses reaction 1,3- bisphosphoglycerate to 3-phosphoglycerate and ATP	(Semenza et al. 1996)
Pyruvate kinase M (PKM)	Catalyses reaction Phosphoenolpyruvate to pyruvate and 2ATP	(Luo et al. 2011)
Lactate dehydrogenase A (LDHA)	Catalyses reaction Pyruvate and NADH to lactate and NAD+	(Firth et al. 1995)
Monocarboxylate Transporter 4 (MCT4	Cellular lactate efflux	(Ullah et al. 2006)
GLUT1, GLUT 4	Cellular glucose uptake	(Iyer et al. 1998)
BNIP3L	Can induce cell death, autophagy, and mitophagy	(Luo et al. 2012)
MX11	Mitochondria biogenesis	(H. Zhang et al. 2007)
Angiogenesis	Function	Reference
VEGF	Regulates vasculogenesis, the development of blood vessels from precursor cells, and angiogenesis	(Tang et al. 2004)
Angiopoietin 1	Vessel maturation, adhesion, migration, and survival	(Gu et al. 2006)
Angiopoietin 2, 4	Promotes cell death and disrupts vascularization	(Simon et al. 2008)
Angiopoietin-like 4	Angiogenesis and cell migration	(Zhang et al. 2012)
CXCl12	Chemokine that induces migration of hematopoietic progenitor cells, endothelial cells, and leukocytes	(Martin et al. 2010)
FGF 3	Mitogenesis, cellular migration, differentiation, angiogenesis	(Blick et al. 2013)
PDGF	Blood vessel formation and growth of blood vessels from already-existing blood vessel	(Kourembanas et al. 1990)
bFGF	Potent angiogenic factor and endothelial cell mitogen	(Krock et al. 2011)
MMP-2	Cleaves a protein called type IV collagen.	(Ben-Yosef et al. 2005)

(continued)

Table 2.1 (continued)

Angiogenesis	Function	Reference
Erythropoietin	Stimulates red blood cell production.	(Semenza and Wang 1992)
iNOS	Catalyses formation of nitric oxide from L-arginine Involved in vascular tone and angiogenesis	(Kannan et al. 2011)
Heme oxygenase 1	Catalyses the degradation of heme, which produces biliverdin, ferrous iron, and carbon monoxide	(Lee et al. 1997)
Endothelin 1	Vasoconstrictor secreted by endothelial cells regulates vascular tone	(Yamashita et al. 2001)
Thrombospondin 1	Thrombospondin-1 stimulates platelet aggregation and vascular remodeling	(Labrousse-Arias et al. 2016)
VEGF R1	VEGF receptor 1 regulates angiogenesis and remodeling of existing blood vessels	(Okuyama et al. 2006)

2.4 Conclusion

In this chapter, we analysed the role of hypoxia and the intersection of signaling pathways involved. HIF is the major transcription factor in response to hypoxia. We have covered the physiological adaptations and significance of the HIF pathway in this review. Several studies have shown that manipulating the HIF pathway can aid in the treatment of a range of illnesses. In various pathological circumstances, the HIF plays different roles. The fact that HIF(s) are also elevated under inflammatory circumstances suggests that they play a role in preserving homeostasis and preventing cellular inflammation. HIF activators might be used to treat stroke and spinal cord injuries, while HIF inhibitors have been created to treat cancer and ischaemia. HIF regulation and pathways are context-dependent. Identification of new treatment targets and methods will need a complete understanding of how hypoxia affects an ever-growing number of pathways in various cell types. Therefore, while translating results from in vitro research to practical applications, care should always be exercised.

References

Arany Z, Huang LE, Eckner R, Bhattacharya S, Jiang C, Goldberg MA et al (1996) An essential role for p300/CBP in the cellular response to hypoxia. Proc Natl Acad Sci U S A 93(23): 12969–12973. https://doi.org/10.1073/pnas.93.23.12969. PMID 8917528

Bajbouj K, Ramakrishnan RK, Hamid Q (2021) Role of matrix metalloproteinases in angiogenesis and its implications in asthma. J Immunol Res 2021:6645072. https://doi.org/10.1155/2021/6645072. PMID 33628848

Bao L, Chen Y, Lai HT, Wu SY, Wang JE, Hatanpaa KJ et al (2018) Methylation of hypoxia-inducible factor (HIF)-1α by G9a/GLP inhibits HIF-1 transcriptional activity and cell migration. Nucleic Acids Res 46(13):6576–6591. https://doi.org/10.1093/nar/gky449. PMID 29860315

Basnyat B, Murdoch DR (2003) High-altitude illness. Lancet 361(9373):1967–1974. https://doi.org/10.1016/S0140-6736(03)13591-X

Bellot G, Garcia-Medina R, Gounon P, Chiche J, Roux D, Pouysségur J et al (2009) Hypoxia-induced autophagy is mediated through hypoxia-inducible factor induction of BNIP3 and BNIP3L via their BH3 domains. Mol Cell Biol 29(10):2570–2581. https://doi.org/10.1128/MCB.00166-09. PMID 19273585

Ben-Yosef Y, Miller A, Shapiro S, Lahat N (2005) Hypoxia of endothelial cells leads to MMP-2-dependent survival and death. Am J Physiol Cell Physiol 289(5):C1321–C1331. https://doi.org/10.1152/ajpcell.00079.2005. PMID 16210427

Berra E, Benizri E, Ginouvès A, Volmat V, Roux D, Pouysségur J (2003) HIF prolyl-hydroxylase 2 is the key oxygen sensor setting low steady-state levels of HIF-1 in normoxia. EMBO J 22(16):4082–4090. https://doi.org/10.1093/emboj/cdg392. PMID 12912907

Beyer S, Kristensen MM, Jensen KS, Johansen JV, Staller P (2008) The histone demethylases JMJD1A and JMJD2B are transcriptional targets of hypoxia-inducible factor HIF. J Biol Chem 283(52):36542–36552. https://doi.org/10.1074/jbc.M804578200. PMID 18984585

Bhattacharya S, Michels CL, Leung MK, Arany ZP, Kung AL, Livingston DM (1999) Functional role of p35srj, a novel p300/CBP binding protein, during transactivation by HIF-1. Genes Dev 13(1):64–75. https://doi.org/10.1101/gad.13.1.64. PMID 9887100

Blick C, Ramachandran A, Wigfield S, McCormick R, Jubb A, Buffa FM et al (2013) Hypoxia regulates FGFR3 expression via HIF-1α and miR-100 and contributes to cell survival in non-muscle invasive bladder cancer. Br J Cancer 109(1):50–59. https://doi.org/10.1038/bjc.2013.240. PMID 23778527

Brunelle JK, Bell EL, Quesada NM, Vercauteren K, Tiranti V, Zeviani M et al (2005) Oxygen sensing requires mitochondrial ROS but not oxidative phosphorylation. Cell Metab 1(6):409–414. https://doi.org/10.1016/j.cmet.2005.05.002. PMID 16054090

Carmeliet P (2003) Angiogenesis in health and disease. Nat Med 9(6):653–660. https://doi.org/10.1038/nm0603-653. PMID 12778163

Carrero P, Okamoto K, Coumailleau P, O'brien S, Tanaka H, Poellinger L (2000) Redox-regulated recruitment of the transcriptional coactivators CREB-binding protein and SRC-1 to hypoxia-inducible factor 1. Mol Cell Biol 20(1):402–415

Chan SY, Zhang YY, Hemann C, Mahoney CE, Zweier JL, Loscalzo J (2009) MicroRNA-210 controls mitochondrial metabolism during hypoxia by repressing the iron-sulfur cluster assembly proteins ISCU1/2. Cell Metab 10(4):273–284. https://doi.org/10.1016/j.cmet.2009.08.015. PMID 19808020

Cummins EP, Berra E, Comerford KM, Ginouves A, Fitzgerald KT, Seeballuck F et al (2006) Prolyl hydroxylase-1 negatively regulates IκB kinase-β, giving insight into hypoxia-induced NF-κB activity. Proc Natl Acad Sci U S A 103(48):18154–18159. https://doi.org/10.1073/pnas.0602235103. PMID 17114296

Dang EV, Barbi J, Yang HY, Jinasena D, Yu H, Zheng Y et al (2011) Control of TH17/Treg balance by hypoxia-inducible factor 1. Cell 146(5):772–784. https://doi.org/10.1016/j.cell.2011.07.033. PMID 21871655

David CJ, Chen M, Assanah M, Canoll P, Manley JL (2010) HnRNP proteins controlled by c-Myc deregulate pyruvate kinase mRNA splicing in cancer. Nature 463(7279):364–368. https://doi.org/10.1038/nature08697. PMID 20010808

Eliceiri BP, Cheresh DA (1999) The role of αv integrins during angiogenesis: insights into potential mechanisms of action and clinical development. J Clin Invest 103(9):1227–1230. https://doi.org/10.1172/JCI6869. PMID 10225964

Finley LWS, Carracedo A, Lee J, Souza A, Egia A, Zhang J et al (2011) SIRT3 opposes reprogramming of cancer cell metabolism through HIF1α destabilization. Cancer Cell 19(3):416–428. https://doi.org/10.1016/j.ccr.2011.02.014. PMID 21397863

Firth JD, Ebert BL, Ratcliffe PJ (1995) Hypoxic regulation of lactate dehydrogenase a. interaction between hypoxia-inducible factor 1 and cAMP response elements. J Biol Chem 270(36): 21021–21027. https://doi.org/10.1074/jbc.270.36.21021. PMID 7673128

Formenti F, Constantin-Teodosiu D, Emmanuel Y, Cheeseman J, Dorrington KL, Edwards LM et al (2010) Regulation of human metabolism by hypoxia-inducible factor. Proc Natl Acad Sci U S A 107(28):12722–12727. https://doi.org/10.1073/pnas.1002339107. PMID 20616028

Forsythe JA, Jiang BH, Iyer NV, Agani F, Leung SW, Koos RD et al (1996) Activation of vascular endothelial growth factor gene transcription by hypoxia-inducible factor 1. Mol Cell Biol 16(9): 4604–4613. https://doi.org/10.1128/MCB.16.9.4604. PMID 8756616

Fox SB, Bragança J, Turley H, Campo L, Han C, Gatter KC et al (2004) Inhibits hypoxia-activated transcription in cancer cells, and its cytoplasmic location in breast cancer is associated with elevated expression of tumor cell hypoxia-inducible factor 1α. Cancer Res 64(17):6075–6081. https://doi.org/10.1158/0008-5472.CAN-04-0708. PMID 15342390

Frede S, Stockmann C, Freitag P, Fandrey J (2006) Bacterial lipopolysaccharide induces HIF-1 activation in human monocytes via p44/42 MAPK and NF-κB. Biochem J 396(3):517–527. https://doi.org/10.1042/BJ20051839. PMID 16533170

Fukuda R, Zhang H, Kim JW, Shimoda L, Dang CV, Semenza GL (2007) HIF-1 regulates cytochrome oxidase subunits to optimize efficiency of respiration in hypoxic cells. Cell 129(1):111–122. https://doi.org/10.1016/j.cell.2007.01.047. PMID 17418790

Graven KK, Yu Q, Pan D, Roncarati JS, Farber HW (1999) Identification of an oxygen responsive enhancer element in the glyceraldehyde-3-phosphate dehydrogenase gene. Biochim Biophys Acta 1447(2–3):208–218. https://doi.org/10.1016/s0167-4781(99)00118-9. PMID 10542317

Gu J, Yamamoto H, Ogawa M, Ngan CY, Danno K, Hemmi H et al (2006) Hypoxia-induced up-regulation of angiopoietin-2 in colorectal cancer. Oncol Rep 15(4):779–783. https://doi.org/10.3892/or.15.4.779. PMID 16525658

Guidez F, Howell L, Isalan M, Cebrat M, Alani RM, Ivins S et al (2005) Histone acetyltransferase activity of p300 is required for transcriptional repression by the promyelocytic leukemia zinc finger protein. Mol Cell Biol 25(13):5552–5566. https://doi.org/10.1128/MCB.25.13.5552-5566.2005. PMID 15964811

Hewitson KS, Liénard BMR, McDonough MA, Clifton IJ, Butler D, Soares AS et al (2007) Structural and mechanistic studies on the inhibition of the hypoxia-inducible transcription factor hydroxylases by tricarboxylic acid cycle intermediates. J Biol Chem 282(5):3293–3301. https://doi.org/10.1074/jbc.M608337200. PMID 17135241

Huang C, Han Y, Wang Y, Sun X, Yan S, Yeh ETH et al (2009) SENP3 is responsible for HIF-1 transactivation under mild oxidative stress via p300 de-SUMOylation. EMBO J 28(18): 2748–2762. https://doi.org/10.1038/emboj.2009.210. PMID 19680224

Islam MDS, Leissing TM, Chowdhury R, Hopkinson RJ, Schofield CJ (2018) 2-oxoglutarate-dependent oxygenases. Annu Rev Biochem 87(1):585–620. 29494239. https://doi.org/10.1146/annurev-biochem-061516-044724

Iyer NV, Kotch LE, Agani F, Leung SW, Laughner E, Wenger RH et al (1998) Cellular and developmental control of O 2 homeostasis by hypoxia-inducible factor 1α. Genes Dev 12(2): 149–162. https://doi.org/10.1101/gad.12.2.149. PMID 9436976

Jeon HW, Lee YM (2010) Inhibition of histone deacetylase attenuates hypoxia-induced migration and invasion of cancer cells via the restoration of RECK expression. Mol Cancer Ther 9(5): 1361–1370. https://doi.org/10.1158/1535-7163.MCT-09-0717. PMID 20442303

Joshi S, Singh AR, Durden DL (2014) MDM2 regulates hypoxic hypoxia-inducible factor 1α stability in an E3 ligase, proteasome, and PTEN-phosphatidylinositol 3-kinase-AKT-dependent manner. J Biol Chem 289(33):22785–22797. https://doi.org/10.1074/jbc.M114.587493. PMID 24982421

Kannan KB, Colorado I, Reino D, Palange D, Lu Q, Qin X et al (2011) Hypoxia-inducible factor plays a gut-injurious role in intestinal ischemia reperfusion injury. Am J Physiol Gastrointest Liver Physiol 300(5):G853–G861. https://doi.org/10.1152/ajpgi.00459.2010. PMID 21183660

Kenneth NS, Mudie S, van Uden P, Rocha S (2009) SWI/SNF regulates the cellular response to hypoxia. J Biol Chem 284(7):4123–4131. https://doi.org/10.1074/jbc.M808491200. PMID 19097995

Kim J, Lee H, Yi SJ, Kim K (2022) Gene regulation by histone-modifying enzymes under hypoxic conditions: a focus on histone methylation and acetylation. Exp Mol Med 54(7):878–889. https://doi.org/10.1038/s12276-022-00812-1. PMID 35869366

Kimura H, Weisz A, Ogura T, Hitomi Y, Kurashima Y, Hashimoto K et al (2001) Identification of hypoxia-inducible factor 1 ancillary sequence and its function in vascular endothelial growth factor gene induction by hypoxia and nitric oxide. J Biol Chem 276(3):2292–2298. https://doi.org/10.1074/jbc.M008398200. PMID 11056166

Knippschild U, Gocht A, Wolff S, Huber N, Löhler J, Stöter M (2005) The casein kinase 1 family: participation in multiple cellular processes in eukaryotes. Cell Signal 17(6):675–689. https://doi.org/10.1016/j.cellsig.2004.12.011. PMID 15722192

Kourembanas S, Hannan RL, Faller DV (1990) Oxygen tension regulates the expression of the platelet-derived growth factor-B chain gene in human endothelial cells. J Clin Invest 86(2):670–674. https://doi.org/10.1172/JCI114759. PMID 2384608

Krieg AJ, Rankin EB, Chan D, Razorenova O, Fernandez S, Giaccia AJ (2010) Regulation of the histone demethylase JMJD1A by hypoxia-inducible factor 1α enhances hypoxic gene expression and tumor growth. Mol Cell Biol 30(1):344–353. https://doi.org/10.1128/MCB.00444-09. PMID 19858293

Krock BL, Skuli N, Simon MC (2011) Hypoxia-induced angiogenesis: good and evil. Genes Cancer 2(12):1117–1133. https://doi.org/10.1177/1947601911423654. PMID 22866203

Labrousse-Arias D, Castillo-González R, Rogers NM, Torres-Capelli M, Barreira B, Aragonés J et al (2016) HIF-2α-mediated induction of pulmonary thrombospondin-1 contributes to hypoxia-driven vascular remodelling and vasoconstriction. Cardiovasc Res 109(1):115–130. https://doi.org/10.1093/cvr/cvv243. PMID 26503986

Lando D, Peet DJ, Gorman JJ, Whelan DA, Whitelaw ML, Bruick RK (2002) FIH-1 is an asparaginyl hydroxylase enzyme that regulates the transcriptional activity of hypoxia-inducible factor. Genes Dev 16(12):1466–1471. https://doi.org/10.1101/gad.991402. PMID 12080085

Laughner E, Taghavi P, Chiles K, Mahon PC, Semenza GL (2001) HER2 (neu) signaling increases the rate of hypoxia-inducible factor 1α (HIF-1α) synthesis: novel mechanism for HIF-1-mediated vascular endothelial growth factor expression. Mol Cell Biol 21(12):3995–4004. https://doi.org/10.1128/mcb.21.12.3995-4004.2001

Lee PJ, Jiang BH, Chin BY, Iyer NV, Alam J, Semenza GL et al (1997) Hypoxia-inducible factor-1 mediates transcriptional activation of the heme oxygenase-1 gene in response to hypoxia. J Biol Chem 272(9):5375–5381. https://doi.org/10.1074/jbc.272.9.5375. PMID 9038135

Lee KJ, Lee KY, Lee YM (2010) Downregulation of a tumor suppressor RECK by hypoxia through recruitment of HDAC1 and HIF-1α to reverse HRE site in the promoter. Biochim Biophys Acta 1803(5):608–616. https://doi.org/10.1016/j.bbamcr.2010.01.004. PMID 20080132

Lee JS, Kim Y, Bhin J, Shin HJ, Nam HJ, Lee SH et al (2011) Hypoxia-induced methylation of a pontin chromatin remodeling factor. Proc Natl Acad Sci U S A 108(33):13510–13515. https://doi.org/10.1073/pnas.1106106108. PMID 21825155

Lim JH, Lee YM, Chun YS, Chen J, Kim JE, Park JW (2010) Sirtuin 1 modulates cellular responses to hypoxia by deacetylating hypoxia-inducible factor 1α. Mol Cell 38(6):864–878. https://doi.org/10.1016/j.molcel.2010.05.023. PMID 20620956

Luo W, Wang Y (2018) Epigenetic regulators: multifunctional proteins modulating hypoxia-inducible factor-α protein stability and activity. Cell Mol Life Sci 75(6):1043–1056. https://doi.org/10.1007/s00018-017-2684-9. PMID 29032501

Luo W, Hu H, Chang R, Zhong J, Knabel M, O'Meally R et al (2011) Pyruvate kinase M2 is a PHD3-stimulated coactivator for hypoxia-inducible factor 1. Cell 145(5):732–744. https://doi.org/10.1016/j.cell.2011.03.054. PMID 21620138

Luo W, Chang R, Zhong J, Pandey A, Semenza GL (2012) Histone demethylase JMJD2C is a coactivator for hypoxia-inducible factor 1 that is required for breast cancer progression. Proc

Natl Acad Sci U S A 109(49):E3367–E3376. https://doi.org/10.1073/pnas.1217394109. PMID 23129632

Maisonpierre PC, Suri C, Jones PF, Bartunkova S, Wiegand SJ, Radziejewski C et al (1997) Angiopoietin-2, a natural antagonist for Tie2 that disrupts in vivo angiogenesis. Science 277(5322):55–60. https://doi.org/10.1126/science.277.5322.55. PMID 9204896

Martin SK, Diamond P, Williams SA, To LB, Peet DJ, Fujii N et al (2010) Hypoxia-inducible factor-2 is a novel regulator of aberrant CXCL12 expression in multiple myeloma plasma cells. Haematologica 95(5):776–784. https://doi.org/10.3324/haematol.2009.015628. PMID 2001 5878

Mathupala SP, Rempel A, Pedersen PL (2001) Glucose catabolism in cancer cells: identification and characterization of a marked activation response of the type II hexokinase gene to hypoxic conditions. J Biol Chem 276(46):43407–43412. https://doi.org/10.1074/jbc.M108181200. PMID 11557773

Melvin A, Mudie S, Rocha S (2011) The chromatin remodeler ISWI regulates the cellular response to hypoxia: role of FIH. Mol Biol Cell 22(21):4171–4181. https://doi.org/10.1091/mbc.E11-02-0163. PMID 21900490

Mimura I, Nangaku M, Kanki Y, Tsutsumi S, Inoue T, Kohro T et al (2012) Dynamic change of chromatin conformation in response to hypoxia enhances the expression of GLUT3 (SLC2A3) by cooperative interaction of hypoxia-inducible factor 1 and KDM3A. Mol Cell Biol 32(15): 3018–3032. https://doi.org/10.1128/MCB.06643-11. PMID 22645302

Mottet D, Ruys SPD, Demazy C, Raes M, Michiels C (2005) Role for casein kinase 2 in the regulation of HIF-1 activity. Int J Cancer 117(5):764–774. https://doi.org/10.1002/ijc.21268. PMID 15957168

Nelson AR, Fingleton B, Rothenberg ML, Matrisian LM (2000) Matrix metalloproteinases: biologic activity and clinical implications. J Clin Oncol 18(5):1135. https://doi.org/10.1200/JCO.2000.18.5.1135. PMID 10694567

Nicolas S, Abdellatef S, Al Haddad MA, Fakhoury I, El-Sibai M (2019) Hypoxia and EGF stimulation regulate VEGF expression in human glioblastoma multiforme (GBM) cells by differential regulation of the PI3K/rho-GTPase and MAPK pathways. Cell 8(11):1397. https://doi.org/10.3390/cells8111397. PMID 31698752

Nieminen AL, Qanungo S, Schneider EA, Jiang BH, Agani FH (2005) Mdm2 and HIF-1α interaction in tumor cells during hypoxia. J Cell Physiol 204(2):364–369. https://doi.org/10.1002/jcp.20406. PMID 15880652

Nishi H, Sasaki T, Nagamitsu Y, Terauchi F, Nagai T, Nagao T et al (2016) Hypoxia inducible factor-1 mediates upregulation of urokinase-type plasminogen activator receptor gene transcription during hypoxia in cervical cancer cells. Oncol Rep 35(2):992–998. https://doi.org/10.3892/or.2015.4449. PMID 26718775

Okuyama H, Krishnamachary B, Zhou YF, Nagasawa H, Bosch-Marce M, Semenza GL (2006) Expression of vascular endothelial growth factor receptor 1 in bone marrow-derived mesenchymal cells is dependent on hypoxia-inducible factor 1. J Biol Chem 281(22):15554–15563. https://doi.org/10.1074/jbc.M602003200. PMID 16574650

Petousi N, Croft QPP, Cavalleri GL, Cheng HY, Formenti F, Ishida K et al (2014) Tibetans living at sea level have a hyporesponsive hypoxia-inducible factor system and blunted physiological responses to hypoxia. J Appl Physiol (1985) 116(7):893–904. https://doi.org/10.1152/japplphysiol.00535.2013. PMID 24030663

Pollard PJ, Brière JJ, Alam NA, Barwell J, Barclay E, Wortham NC et al (2005) Accumulation of Krebs cycle intermediates and over-expression of HIF1α in tumours which result from germline FH and SDH mutations. Hum Mol Genet 14(15):2231–2239. https://doi.org/10.1093/hmg/ddi227. PMID 15987702

Richard DE, Berra E, Gothié E, Roux D, Pouysségur J (1999) p42/p44 mitogen-activated protein kinases phosphorylate hypoxia-inducible factor 1α (HIF-1α) and enhance the transcriptional activity of HIF-1. J Biol Chem 274(46):32631–32637. https://doi.org/10.1074/jbc.274.46.32631. PMID 10551817

Sakamoto T, Weng JS, Hara T, Yoshino S, Kozuka-Hata H, Oyama M et al (2014) Hypoxia-inducible factor 1 regulation through cross talk between mTOR and MT1-MMP. Mol Cell Biol 34(1):30–42. https://doi.org/10.1128/MCB.01169-13. PMID 24164895

Samanta D, Semenza GL (2018) Metabolic adaptation of cancer and immune cells mediated by hypoxia-inducible factors. Biochim Biophys Acta Rev Cancer 1870(1):15–22. https://doi.org/10.1016/j.bbcan.2018.07.002. PMID 30006019

Sang N, Stiehl DP, Bohensky J, Leshchinsky I, Srinivas V, Caro J (2003) MAPK signaling up-regulates the activity of hypoxia-inducible factors by its effects on p300. J Biol Chem 278(16):14013–14019. https://doi.org/10.1074/jbc.M209702200. PMID 12588875

Schödel J, Mole DR, Ratcliffe PJ (2013) Pan-genomic binding of hypoxia-inducible transcription factors. Biol Chem 394(4):507–517. https://doi.org/10.1515/hsz-2012-0351. PMID 23324384

Selak MA, Armour SM, MacKenzie ED, Boulahbel H, Watson DG, Mansfield KD et al (2005) Succinate links TCA cycle dysfunction to oncogenesis by inhibiting HIF-α prolyl hydroxylase. Cancer Cell 7(1):77–85. https://doi.org/10.1016/j.ccr.2004.11.022. PMID 15652751

Semenza GL, Wang GL (1992) A nuclear factor induced by hypoxia via de novo protein synthesis binds to the human erythropoietin gene enhancer at a site required for transcriptional activation. Mol Cell Biol 12(12):5447–5454. https://doi.org/10.1128/mcb.12.12.5447-5454.1992. PMID 1448077

Semenza GL, Roth PH, Fang HM, Wang GL (1994) Transcriptional regulation of genes encoding glycolytic enzymes by hypoxia-inducible factor 1. J Biol Chem 269(38):23757–23763. https://doi.org/10.1016/S0021-9258(17)31580-6. PMID 8089148

Semenza GL, Jiang BH, Leung SW, Passantino R, Concordet JP, Maire P et al (1996) Hypoxia response elements in the aldolase a, enolase 1, and lactate dehydrogenase a gene promoters contain essential binding sites for hypoxia-inducible factor 1. J Biol Chem 271(51): 32529–32537. https://doi.org/10.1074/jbc.271.51.32529. PMID 8955077

Simon MP, Tournaire R, Pouyssegur J (2008) The angiopoietin-2 gene of endothelial cells is up-regulated in hypoxia by a HIF binding site located in its first intron and by the central factors GATA-2 and Ets-1. J Cell Physiol 217(3):809–818. https://doi.org/10.1002/jcp.21558. PMID 1 8720385

Smythies JA, Sun M, Masson N, Salama R, Simpson PD, Murray E et al (2019) Inherent DNA -binding specificities of the HIF -1α and HIF -2α transcription factors in chromatin. EMBO Rep 20(1). https://doi.org/10.15252/embr.201846401. PMID 30429208

Sun Q, Chen X, Ma J, Peng H, Wang F, Zha X et al (2011) Mammalian target of rapamycin up-regulation of pyruvate kinase isoenzyme type M2 is critical for aerobic glycolysis and tumor growth. Proc Natl Acad Sci U S A 108(10):4129–4134. https://doi.org/10.1073/pnas.1014769108. PMID 21325052

Tang N, Wang L, Esko J, Giordano FJ, Huang Y, Gerber HP et al (2004) Loss of HIF-1α in endothelial cells disrupts a hypoxia-driven VEGF autocrine loop necessary for tumorigenesis. Cancer Cell 6(5):485–495. https://doi.org/10.1016/j.ccr.2004.09.026. PMID 15542432

Taylor CT, Scholz CC (2022) The effect of HIF on metabolism and immunity. Nat Rev Nephrol 18(9):573–587. https://doi.org/10.1038/s41581-022-00587-8. PMID 35726016

Thinnes CC, England KS, Kawamura A, Chowdhury R, Schofield CJ, Hopkinson RJ (2014) Targeting histone lysine demethylases - progress, challenges, and the future. Biochim Biophys Acta 1839(12):1416–1432. https://doi.org/10.1016/j.bbagrm.2014.05.009. PMID 24859458

Tsai YP, Wu KJ (2014) Epigenetic regulation of hypoxia-responsive gene expression: focusing on chromatin and DNA modifications. Int J Cancer 134(2):249–256. https://doi.org/10.1002/ijc.28190. PMID 23564219

Ullah MS, Davies AJ, Halestrap AP (2006) The plasma membrane lactate transporter MCT4, but not MCT1, is up-regulated by hypoxia through a HIF-1alpha-dependent mechanism. J Biol Chem 281(14):9030–9037. https://doi.org/10.1074/jbc.M511397200. PMID 16452478

Wang GL, Jiang BH, Rue EA, Semenza GL (1995) Hypoxia-inducible factor 1 is a basic-helix-loop-helix-PAS heterodimer regulated by cellular O2 tension. Proc Natl Acad Sci U S A 92(12): 5510–5514. https://doi.org/10.1073/pnas.92.12.5510. PMID 7539918

Wang F, Zhang R, Wu X, Hankinson O (2010) Roles of coactivators in hypoxic induction of the erythropoietin gene. PLoS One 5(4):e10002. https://doi.org/10.1371/journal.pone.0010002. PMID 20368990

Wenger RH, Kvietikova I, Rolfs A, Camenisch G, Gassmann M (1998) Oxygen-regulated erythropoietin gene expression is dependent on a CpG methylation-free hypoxia-inducible factor-1 DNA-binding site. Eur J Biochem 253(3):771–777. https://doi.org/10.1046/j.1432-1327.1998.2530771.x. PMID 9654078

Wenger RH, Stiehl DP, Camenisch G (2005) Integration of oxygen signaling at the consensus HRE. Sci STKE 2005(306):re12. https://doi.org/10.1126/stke.3062005re12. PMID 16234508

Xenaki G, Ontikatze T, Rajendran R, Stratford IJ, Dive C, Krstic-Demonacos M, Demonacos C (2008) PCAF is an HIF-1α cofactor that regulates p53 transcriptional activity in hypoxia. Oncogene 27(44):5785–5796. https://doi.org/10.1038/onc.2008.192

Yamashita K, Discher DJ, Hu J, Bishopric NH, Webster KA (2001) Molecular regulation of the endothelin-1 gene by hypoxia. Contributions of hypoxia-inducible factor-1, activator protein-1, GATA-2, AND p300/CBP. J Biol Chem 276(16):12645–12653. https://doi.org/10.1074/jbc.M011344200. PMID 11278891

Zhang H, Gao P, Fukuda R, Kumar G, Krishnamachary B, Zeller KI et al (2007) HIF-1 inhibits mitochondrial biogenesis and cellular respiration in VHL-deficient renal cell carcinoma by repression of C-MYC activity. Cancer Cell 11(5):407–420. https://doi.org/10.1016/j.ccr.2007.04.001. PMID 17482131

Zhang H, Wong CCL, Wei H, Gilkes DM, Korangath P, Chaturvedi P et al (2012) HIF-1-dependent expression of angiopoietin-like 4 and L1CAM mediates vascular metastasis of hypoxic breast cancer cells to the lungs. Oncogene 31(14):1757–1770. https://doi.org/10.1038/onc.2011.365. PMID 21860410

Zhang Q, Yan Q, Yang H, Wei W (2019) Oxygen sensing and adaptability won the 2019 Nobel prize in physiology or medicine. Genes Dis 6(4):328–332. https://doi.org/10.1016/j.gendis.2019.10.006. PMID 31832511

Zhou CH, Zhang XP, Liu F, Wang W (2015) Modeling the interplay between the HIF-1 and p53 pathways in hypoxia. Sci Rep 5(1):13834. https://doi.org/10.1038/srep13834. PMID 26346319

Hypoxia-Inducible Factor Activation in High-Altitude Stress

3

Subhodip Majumdar, Ratnika Sharma, Rekha Maithani, and Rajkumar Tulsawani

Abstract

As altitude increases, the tropospheric barometric pressure falls. A proportionate fall in the partial pressure of oxygen happens along with a decrease in barometric pressure since the oxygen content of the air stays constant. Investigations into the longer-term impacts of acclimatization and adaptation as well as short-term changes brought on by exposure to hypobaric hypoxia are the two primary areas of research in high-altitude physiology. In order to maintain correct homeostasis, the body must have adequate access to oxygen at the tissue level. Hypoxia is a state when this is not the case. This condition can be brought on by either inadequate blood supply or thus low oxygen transport or by lower levels of oxygen in the blood. In hypoxic condition, hypoxia-inducible factor-1 controls the expression of genes in response to oxygen levels. In this chapter, role of the hypoxia-inducible factor-1 (HIF-1) in hypoxic situation is reviewed.

Keywords

Hypoxia · HIF-1 · Hypobaric hypoxia · Hypoxia-inducible factor · Erythropoiesis · Angiogenesis · Lung injury

S. Majumdar
Defence Institute of Physiology and Allied Sciences (DIPAS), Defence Research and Development Organization (DRDO), Delhi, India

Amity Institute of Biotechnology, Amity University, Noida, Uttar Pradesh, India

R. Sharma · R. Maithani · R. Tulsawani (✉)
Defence Institute of Physiology and Allied Sciences (DIPAS), Defence Research and Development Organization (DRDO), Delhi, India
e-mail: rktulsawani@yahoo.com

R. Tulsawani, D. Vohora (eds.), *Adaptation under Stressful Environments through Biological Adjustments and Interventions*,
https://doi.org/10.1007/978-981-99-7652-2_3

3.1 Introduction

The whole surface of the planet is covered by the troposphere, which is the lowest layer of the atmosphere. As altitude (vertical height above sea level) rises, the barometric pressure within the troposphere decreases. Since the oxygen content of the air is constant, a proportional decrease in the partial pressure of oxygen occurs along with a drop in barometric pressure. Hypobaric hypoxia is the term used to describe this situation. At 5300 m above sea level, partial pressure of oxygen is roughly half than that at sea level. At 8850 m above sea level, Mount Everest's peak is the highest point on Earth's surface above sea level, and the partial pressure of oxygen there is roughly one-third that of sea level (Grocott et al. 2007). The two main areas of research in high-altitude physiology are investigations of the longer-term effects of acclimatization and adaption as well as the short-term alterations that result from exposure to hypobaric hypoxia (the acute response to hypoxia). Acute exposure to the surrounding environment at a very high altitude is immediately lethal. At altitudes, the body's ability to acclimatize helps in sustaining in lower inspired partial pressure of oxygen (Grocott et al. 2007).

Natural selection under hypobaric hypoxia environment may help develop modifications in population of high-altitude people over several generations. Acute syndromes, for instance, acute mountain sickness, high-altitude pulmonary, and cerebral edema, are observed in people living in lowlands or highland dwellers who may experience altitudes higher than they are used to (Grocott et al. 2007).

Hypoxia is a condition arising from poor blood supply or low oxygen concentration in the blood (hypoxemia) that further causes insufficient supply of oxygen at tissue level and thus prevents proper homeostasis in the body. Hypoxia can manifest as acute, chronic based on its severity from moderate to severe. Different tissues respond differently to low oxygen concentrations where certain tissues can tolerate hypoxia/ischemia for a longer duration, while other tissues may be severely injured (Bhutta et al. 2022).

The hypoxemic type of hypoxia is caused by one of two mechanisms: a decrease in the amount of breathable oxygen caused by a drop in barometric pressure, a condition commonly experienced by pilots, mountaineers, and people living at high altitudes (see altitude sickness); or cardiopulmonary failure, in which the lungs are unable to effectively transfer oxygen from the alveoli to the blood (Britannica 2023). When there is anemic hypoxia, either there is insufficient hemoglobin to supply the body's oxygen requirements, as in anemia, or the hemoglobin that is there is rendered inactive. Carbon monoxide poisoning and acquired methemoglobinemia are two examples of the latter, in which the hemoglobin is altered by toxic chemicals to the point that it is no longer functional for transporting oxygen and provides no respiratory benefit (Britannica 2023).

Stagnant hypoxia, which happens when there is insufficient blood flowing through the capillaries to feed the tissues, can be general or local. If it is widespread, it might be caused by a heart issue that limits blood flow, a difficulty with the capacity of veins to return blood, or a shock-causing injury. Any condition that restricts or prevents blood flow to any area of the body may result in local stagnant hypoxia. Tourniquet usage to stop bleeding, ergot poisoning, exposure to cold, and overwhelming systemic disease with shock are all examples.

3.2 HIF-1 Domain Structure

The hypoxia-inducible transcription factor, i.e., HIF-1, is a heterodimeric transcription factor. It controls expression of genes based on the oxygen levels and composed of α and β subunits. HIF-1α is an oxygen-regulated subunit, whereas HIF-1β subunit is independent of oxygen sensing machinery. Under hypoxic conditions, HIF-1α binds to hypoxia response element (HRE) in promoter regions of HIF-responsive genes (Ziello et al. 2007).

Under normoxic conditions, two conserved proline residues of oxygen-dependent degradation domain (ODD) of HIF-1 are hydroxylated by prolyl hydroxylases. This hydroxylation allows the von Hippel Lindau (VHL) protein, a component of E3 ubiquitin ligase complex, to recognize and bind to hydroxylated HIF-1. HIF-1α hydroxylation at proline residue is essential for binding of VHL with HIF-1α. Several reports investigating at ODD modification provides more evidence of the relevance of proline residues. HIF-1α through interaction with VHL degrades itself via *ubiquitin*-dependent 26S proteasomal pathway. However, during hypoxic conditions in cells and tissues, hydroxylation of HIF-1 is inhibited which in turn reduces the interaction between HIF-1 and VHL. Thus, under hypoxia, HIF-1 protein is able to dimerize with ARNT and further activates transcription of target genes (Chowdhury et al. 2016). Genes involved in glucose-energy metabolism and oxygen balance are among those that HIF-1 in general regulates. While certain HIF-regulated genes contribute to increasing tissue oxygen supply, others promote tolerance to severe oxygen deprivation (Ziello et al. 2007).

3.3 Role of Hypoxia-Inducible Factor 1 in Various Organs

3.3.1 HIF-1 Mediated Angiogenesis in Cardio Protection

The ischemic heart's substrate preference switches from fatty acid to glucose usage. HIF-1 plays a critical function in the early stages of stress response by controlling components of glucose metabolism, when aerobic metabolism gives way to anaerobic metabolism. HIF-1 has also been reported to regulate mitochondrial metabolism via limiting the entry of pyruvate by activating pyruvate dehydrogenase, by changing cytochrome oxidase complex, and by reducing mitochondrial respiration (Li et al. 2020). A hybrid constitutively stable HIF-1, produced by adenovirus, induces survival genes against ischemia-reperfusion injury in cultured new-born cardiomyocytes. Furthermore, when individuals suffer from acute coronary artery blockage, levels of HIF-1 and VEGF rise in the myocardium. Thus, coronary blood flow and oxygen delivery to heart muscle may be enhanced during myocardial hypoxia due to VEGF production induced by HIF-1 (Tekin et al. 2010).

In CKD patients, hypoxia promotes myocardial infarction, cardiac remodelling, atherosclerosis, and peripheral arterial disease. HIF signalling influences the development, metabolic response, ischemia, and atherosclerosis of heart disease in a variety of ways under hypoxia. HIF-regulated transcription is required for critical heart developmental processes and it has been demonstrated that HIF-1 deletion may result in cardiac defects such as septation and trabeculation which in turn may be influenced by genetic makeup of mice. As a result, HIF-regulated transcription is required for critical heart developmental processes. Cellular metabolism and cardiac contractility are closely intertwined, and HIF plays an important role in defining the metabolic response in the heart. Cardiometabolic function is also maintained by HIF-1 during hypoxia by escalating glycolysis and regulating mitochondrial metabolism. Further, HIF-1 has also been reported to boost NO which is a vasodilator and thus helps in heart conditions such as ischemia (Pan et al. 2021). It also promotes angiogenesis by targetting VEGF, endothelin-1, and matrix metalloproteinases (MMPs) in various cardiac cell types (Li et al. 2020).

3.3.2 Adverse Role of HIF-1 Activation in Myocardium

In contrast to the substantial data supporting HIF-1's cardioprotective activities, many researchers have proposed the myocardial damage in response to HIF-1 activation. Tang et al. (2008) for example, revealed that during ischemia-reperfusion, polyol pathway activation raises transferrin receptor levels by activating HIF-1, resulting in enhanced cellular absorption of transferrin and iron and exacerbating oxidative tissue damage caused by lipid peroxidation. Also, it has been demonstrated that during heart failure due to pressure or volume overload in rat model system, there is an elevated HIF-1 α and VEGF mRNA and protein expression. It is recorded that aberrant regulation of HIF-1 and VEGF in deteriorating ventricular myocardium can be reversed with the treatment of adrenergic β receptor blocker, carvedilol. Prolonged stimulation of HIF-1 in transgenic mice missing prolyl hydroxylase (PHD2) results in early death with significant venous congestion and dilated cardiomyopathy. It was reasoned that cardiomyopathy might be caused by the blood hyperviscosity and volume overload due to the prolonged HIF-1 activation along with direct deleterious effect on cardiac myocytes (Shimoda and Semenza 2011).

3.3.3 HIF-1 Action in Hypoxic Lung Injury

Hypoxic acute lung injury has been known to be caused due to residency at high altitude or chronic lung disease, resulting in abnormalities in lung function and repair. Early events in acute lung injury are characterized by lung edema, destruction of alveolar lining layer, and death of alveolar epithelial cells, whereas later stages are distinguished by hyperplasia of alveolar type II cells leading to fibrosis. Hypoxia, ischemia, and/or inflammatory stimulation can all lead to increased vascular permeability after acute lung damage (Shimoda and Semenza 2011). HIF-1 has been reported to play both a barrier-protective and barrier-disruptive functions during lung damage. The study on hypoxic/ischemic model by Becker et al. (2000) demonstrated an increase in VEGF level and augmented barrier disruption due to an increase in HIF-1 protein levels, whereas Morote-Garcia et al. (2008) described the HIF-1-dependent inhibition of adenosine kinase, the enzyme that is responsible for converting adenosine to AMP and reduces hypoxia-induced pulmonary vascular leak.

Additionally, HIF-1α causes alveolar type II cell death during epithelial cell injury and consequent fibrotic lung disease. Furthermore, inflammatory levels of NO may cause overexpression of HIF-1α, which may hinder the effective healing response during epithelial wound repair (Bove et al. 2008). Lately, individuals with acute lung damage have been reported to suffer from pulmonary fibrosis due to the effect of epithelial-mesenchymal transition (EMT) (Kim et al. 2006; Shimoda and Semenza 2011; Hill et al. 2019).

Reactive oxygen species produced during aerobic respiration in mitochondria play an important role in sustaining HIF-1α in alveolar epithelial cells (Zepeda et al. 2013; Zuo and Wijegunawardana 2021). In comparison to these secondary evidences, splanchnic ischemia-reperfusion and intestinal inflammation models of lung injury provide direct evidence about the contribution of HIF in the development of acute lung injury. There is a fast and protracted rise in nuclear HIF-1 protein levels in mouse models of traumatic haemorrhagic shock caused by gut ischemia-reperfusion, with animals partly defective in HIF-1 displaying decreased intestinal mucosal damage, bacterial translocation, and subsequent lung injury (Shimoda and Semenza 2011).

3.3.4 Transcriptional Activity in Kidney

Oxygen supply plays a key role in regulating the erythropoietin (EPO) production in the kidney and local oxygen tension drives the transcription of EPO gene. The hypoxia-inducible factor regulates EPO synthesis (HIF). As local oxygen tension falls, accumulating HIF binds to the hypoxia-responsive element (HRE), a critical region in the EPO gene, and triggers EPO transcription. Erythropoietin synthesis has been reported to be influenced by the interaction of three oxygen-regulated HIF-1 subunits (HIF-1α, HIF-2α, and HIF-3α). Yet, the kidney's extreme sensitivity to variations in oxygen tension makes it vulnerable to hypoxic damage. In hypoxic circumstances, severe energy depletion occurs, followed by the activation of a variety of essential metabolic changes. Hypoxia is a pro-fibrogenic trigger as well (Nangaku and Eckardt 2007; Shu et al. 2019). Besides ischemic acute renal failure, hypoxia is a contributing factor in many kidney disorders such as acute glomerulonephritis, nephrotoxic acute kidney damage, and radiocontrast nephropathy. Moreover, hypoxia has also been reported to be involved in end-stage renal failure in chronic kidney disease. Therefore, hypoxia may be considered as a viable therapeutic target against acute and chronic kidney diseases. The activation of HIF results in the coordinated production of a number of adaptive genes (Nangaku and Eckardt 2007; Tanaka et al. 2016).

3.3.5 Erythropoietic Action of HIF-1 in Brain

In animals, oxygen supply determines the normal tissue function and a specialized adaptive response may be triggered at the cellular, local, and systemic levels due to a mismatch between oxygen supply and its consumption. These processes are mediated by HIF-1 and HIF-2 which in turn regulate genes encoding VEGF and EPO under hypoxic conditions. EPO is a glycoprotein which is synthesized in fetal hepatocytes and in adult kidneys by interstitial fibroblasts. It then travels to bone marrow via blood where it inhibits the death of erythroid progenitor cells and thus controls red cell synthesis. Recently, EPO has also been reported to play a substantial role in the neurological system (Marti 2004; Ureña-Guerrero et al. 2020).

EPO and its receptor are found in endothelial cells, glial cells and neurons, and throughout the brain and is regulated by transcription factors induced by hypoxia and ischemia. In in vivo and in vitro, EPO has powerful neuroprotective capabilities and appears to function in two ways: directly shielding neurons from ischemic damage and activating endothelial cells, thereby enhancing the angiogenic impact of VEGF in the nervous system. Hence, during restricted blood supply (chronic ischemia), self-regulated physiological system involving hypoxia-induced gene products such as VEGF and EPO may help to avoid neuronal harm (Acker and Acker 2004; Rey et al. 2019; Ureña-Guerrero et al. 2020).

References

Acker T, Acker H (2004) Cellular oxygen sensing need in CNS function: physiological and pathological implications. J Exp Biol 207(18):3171–3188. https://doi.org/10.1242/jeb.01075, PMID 15299039

Becker PM, Alcasabas A, Yu AY, Semenza GL, Bunton TE (2000) Oxygen-independent upregulation of vascular endothelial growth factor and vascular barrier dysfunction during ventilated pulmonary ischemia in isolated ferret lungs. Am J Respir Cell Mol Biol 22:272–279

Bhutta BS, Alghoula F, Berim I (2022) Hypoxia. In: Stat Pearls. StatPearls Publishing

Bove PF, Hristova M, Wesley UV, Olson N, Lounsbury KM, van der Vliet A (2008) Inflammatory levels of nitric oxide inhibit airway epithelial cell migration by inhibition of the kinase ERK1/2 and activation of hypoxia-inducible factor-1α. J Biol Chem 283:17919–17928

Britannica T (2023) Editors of encyclopaedia. Hypoxia. Encyclopedia Britannica

Chowdhury R, Leung IK, Tian YM, Abboud MI, Ge W, Domene C, Cantrelle FX, Landrieu I, Hardy AP, Pugh CW, Ratcliffe PJ (2016) Structural basis for oxygen degradation domain selectivity of the HIF prolyl hydroxylases. Nat Commun 7(1):12673

Grocott M, Montgomery H, Vercueil A (2007) High-altitude physiology and pathophysiology: implications and relevance for intensive care medicine. Crit Care 11(1):203. https://doi.org/10.1186/cc5142, PMID 17291330

Hill C, Li J, Liu D, Conforti F, Brereton CJ, Yao L, Zhou Y, Alzetani A, Chee SJ, Marshall BG, Fletcher SV, Hancock D, Ottensmeier CH, Steele AJ, Downward J, Richeldi L, Lu X, Davies DE, Jones MG, Wang Y (2019) Autophagy inhibition-mediated epithelial-mesenchymal transition augments local myofibroblast differentiation in pulmonary fibrosis. Cell Death Dis 10 (8):591. https://doi.org/10.1038/s41419-019-1820-x. PMID: 31391462; PMCID: PMC6685977

Kim KK, Kugler MC, Wolters PJ, Robillard L, Galvez MG, Brumwell AN, Sheppard D, Chapman HA (2006) Alveolar epithelial cell mesenchymal transition develops in vivo during pulmonary fibrosis and is regulated by the extracellular matrix. Proc Natl Acad Sci U S A 103:13180–13185

Li X, Zhang Q, Nasser MI, Xu L, Zhang X, Zhu P, He Q, Zhao M (2020) Oxygen homeostasis and cardiovascular disease: a role for HIF? Biomed Pharmacother 128:110338

Marti HH (2004) Erythropoietin and the hypoxic brain. J Exp Biol 207(18):3233–3242. https://doi.org/10.1242/jeb.01049, PMID 15299044

Morote-Garcia JC, Rosenberger P, Kuhlicke J, Eltzschig HK (2008) HIF-1-dependent repression of adenosine kinase attenuates hypoxia-induced vascular leak. Blood 111:5571–5580

Nangaku M, Eckardt KU (2007) Hypoxia and the HIF system in kidney disease. J Mol Med 85(12): 1325–1330. https://doi.org/10.1007/s00109-007-0278-y

Pan Z, Ma G, Kong L, Du G (2021) Hypoxia-inducible factor-1: regulatory mechanisms and drug development in stroke. Pharmacol Res 170:105742

Rey F, Balsari A, Giallongo T, Ottolenghi S, Di Giulio AM, Samaja M, Carelli S (2019) Erythropoietin as a neuroprotective molecule: an overview of its therapeutic potential in neurodegenerative diseases. ASN Neuro 11:1759091419871420

Shimoda LA, Semenza GL (2011) HIF and the lung: role of hypoxia-inducible factors in pulmonary development and disease. Am J Respir Crit Care Med 183(2):152–156. https://doi.org/10.1164/rccm.201009-1393PP, PMID 21242594

Shu S, Wang Y, Zheng M, Liu Z, Cai J, Tang C, Dong Z (2019) Hypoxia and hypoxia-inducible factors in kidney injury and repair. Cell 8(3):207

Tanaka S, Tanaka T, Nangaku M (2016) Hypoxia and hypoxia-inducible factors in chronic kidney disease. Renal Replace Ther 2(1):1–8

Tang WH, Wu S, Wong TM, Chung SK, Chung SS (2008) Polyol pathway mediates iron-induced oxidative injury in ischemic–reperfused rat heart. Free Radic Biol Med 45(5):602–610

Tekin D, Dursun AD, Xi L (2010) Hypoxia inducible factor 1 (HIF-1) and cardioprotection. Acta Pharmacol Sin 31(9):1085–1094. https://doi.org/10.1038/aps.2010.132, PMID 20711226

Ureña-Guerrero ME, Castañeda-Cabral JL, Rivera-Cervantes MC, Macias-Velez RJ, Jarero-Basulto JJ, Gudiño-Cabrera G, Beas-Zárate C (2020) Neuroprotective and neurorestorative effects of Epo and VEGF: perspectives for new therapeutic approaches to neurological diseases. Curr Pharm Des 26(12):1263–1276

Zepeda AB, Pessoa A Jr, Castillo RL, Figueroa CA, Pulgar VM, Farías JG (2013) Cellular and molecular mechanisms in the hypoxic tissue: role of HIF-1 and ROS. Cell Biochem Funct 6: 451–459

Ziello JE, Jovin IS, Huang Y (2007) Hypoxia-inducible factor (HIF)-1 regulatory pathway and its potential for therapeutic intervention in malignancy and ischemia. Yale J Biol Med 80(2):51–60, PMID 18160990

Zuo L, Wijegunawardana D (2021) Redox role of ROS and inflammation in pulmonary diseases. In: Lung inflammation in health and disease, Volume II, vol 22, pp 187–204

Regulation of Vascular Endothelial Growth Factor by SUMOylation in Hypoxic Cardiomyocytes

4

Dolly Sharma, Chhavi Rai, Rajeev Varshney, and Ramesh C. Meena

Abstract

Cardiovascular diseases are the foremost cause of death worldwide due to a variety of physiological and environmental conditions in humans. Hypoxia affects the cardiovascular homeostasis and compensates by delivering more blood to tissues to counteract the lower oxygen delivery. As the severity of hypoxia increases, the acute central cardiovascular response causes an increase in heart rate with an unchanged stroke volume, which is mediated primarily by increased sympathetic neural discharge. Stroke, coronary artery disease, and thrombosis are among the most common cardiovascular diseases. The VEGF family regulates functions such as angiogenesis, lymphopoiesis, inflammation, oxidative stress resistance, and lipid metabolism which have potential therapeutic targets for cardiovascular diseases. Cardiomyocytes are the most common cell types in the heart that are both the source and the target of VEGF-A and their cell surface receptors, e.g., VEGFR1 and VEGFR2. In the present chapter, the authors highlight the cardiovascular responses during hypoxia in human cardiomyocytes and the role of SUMOylation in the regulation of VEGF activity and its functions.

Keywords

Cardiomyocytes · Cardiovascular · Hypoxia · Inflammation · SUMOylation · VEGF

D. Sharma · C. Rai · R. Varshney · R. C. Meena (✉)
Department of Pathophysiology and Disruptive Technologies, Defence Institute of Physiology and Allied Sciences (DIPAS), Defence Research and Development Organization (DRDO), Delhi, India
e-mail: rcmeena.dipas@gov.in

R. Tulsawani, D. Vohora (eds.), *Adaptation under Stressful Environments through Biological Adjustments and Interventions*,
https://doi.org/10.1007/978-981-99-7652-2_4

4.1 Introduction

Cardiovascular disease (CVD) is a general term for diseases concerning the heart and the blood vessels. CVDs can be caused by a variety of environmental and physiological conditions and are one of the leading causes of morbidity and mortality in humans. CVDs are a severe hazard to human life and have sparked a lot of research in recent years (Du et al. 2021). The physiological function of each tissue depends on peripheral circulation. Circulation disturbances reduce oxygen supply which leads to tissue hypoxia. Tissue hypoxia occurs in a variety of cardiovascular illnesses such as atherosclerosis, pulmonary arterial hypertension, and heart failure. Hypoxia, in general, is defined as a drop-in oxygen concentration below the tissue-specific physiological threshold and manifests as suboptimal levels of molecular oxygen at the tissue level. The energy metabolism is inhibited due to decreased oxygen availability in the cells. While cardiovascular illnesses may or may not involve in tissue hypoxia and its signaling has the capacity to regulate tissue remodeling processes and/or enhance the severity of cardiovascular complications (Abe et al. 2017). HIF-1 and HIF-2 are important transcription factors that mediate most of the physiological responses to hypoxia at the transcriptional level. In a hypoxic environment, differential expression of genes is severely inhibited; however, transcript levels of hypoxia-inducible genes such as vascular endothelial growth factor (VEGF) are increased.

The VEGF is a homodimeric vasoactive glycoprotein with a major role as a mediator of angiogenesis. Angiogenesis or the growth of new blood vessels is involved in several physiological and pathological processes including cardiovascular diseases. VEGF-A plays a major role in the regulation of myocardial angiogenesis in cardiomyocytes (CMs). Lack of VEGF-A results in circumstances that compromise heart functions. Furthermore, VEGF-A is a target of CMs because it affects cardiac biology, such as mobilization. CMs are the most common cell types in the heart that are both the source and the target of VEGF-A and their cell surface receptors, e.g., VEGFR1 and VEGFR2. VEGF-A stimulates morphogenesis, contractility, and wound healing via activating CMs. Hypoxia causes inflammation, mechanical stress, and cytokine activation that results in the generation of VEGF-A in CMs. Furthermore, elevated levels of VEGF-A have been discovered in individuals with various CVDs and are frequently linked to a poor prognosis and severity of illness (Braile et al. 2020). Each cell responds to hypoxia in a variety of ways at transcriptional, translational, and posttranslational levels.

Posttranslational modification (PTM) is an essential regulatory mechanism in a cell, which works by modifying cellular proteins that serve in a variety of biological activities. SUMOylation is a PTM method that affects a wide range of biological functions including cell cycle control, gene transcription, differentiation, and cellular localization. The protein localization, protein-protein interactions, protein-DNA interactions, and/or protein activity, all can be regulated by SUMO modification. SUMOylation has a substrate-specific influence on protein activity. In the heart tissue, SUMO modifies a vast variety of target proteins that impact cell proliferation, apoptosis, and metabolism. SUMOylation has been uncovered to play a critical role in maintaining normal cellular activities, and dysregulation of SUMOylation has been linked to the pathogenesis of various cardiac disorders (Hannoun et al. 2010). SUMO can directly interact with and modify a set of proteins that participate in the hypoxic signaling cascade and their downstream functions. An imbalance in SUMOylation can result in aberrant heart functions. Therefore, in the present chapter, we emphasize the role of SUMOylation in the regulation of VEGF activity and its functions during hypoxia in human cardiomyocytes.

4.1.1 Oxygen (O_2): A Fundamental Element of Human Body

One of the most important and abundant elements on the planet is oxygen. Oxygen makes up about 46% of the Earth's crust and 20.95% of the atmosphere by volume. It is necessary for the survival of higher organisms and is found in almost every biological molecule in the body. Oxygen is a vital element found in abundance in humans, accounting for 65 and 90% of the adult human body's mass and energy, respectively, which can only be produced using O_2. It is required for multicellular organism growth and development and is required for cell survival. All cells and tissue types require it for growth, repair, and maintenance. It is important in the breathing and metabolic processes of living organisms. Oxygen is necessary for cells to generate adenosine triphosphate (ATP) through oxidative phosphorylation as the final electron acceptor in the electron transport chain. Mammalian evolution has resulted in the development of a physiological network that ensures O_2 capture, binding, transport, and delivery to every organ of the body, as well as tissue-level O_2 homeostasis (Michiels 2004). The circulatory system delivers oxygen to every cell in the body, and breathing ensures that there is enough oxygen for all the cellular processes. The cardiopulmonary system is responsible for mediating tissue oxygen requirement and its distribution to every cell in the higher organisms. O_2 bound with hemoglobin is the main source of O_2 supply in the body, with a small amount of dissolved O_2 in plasma. The oxygen then diffuses from the bloodstream to the peripheral tissues, where it is used for aerobic cellular metabolism. Advanced multilevel systems within an organism's body extract O_2 from the atmosphere and distribute it throughout the body. The partial pressure of oxygen (pO_2) decreases from 159 mmHg in the air to 3 or 4 mmHg in the respiratory tract, alveoli, arterial blood, capillaries, and cells, and finally, the mitochondria where O_2 is used to form and/or produce ATP. This oxygen cascade can be broken down into four steps and four key resistances in order. The exchange of air in and out of alveolar spaces, diffusive exchange of O_2 between alveoli and blood in pulmonary capillaries, diffusion from the pulmonary to the systemic circulation, and diffusive exchange of O_2 between blood and tissues are the sequential steps in the oxygen cascade. Respiratory, circulatory, tissue, and mitochondrial resistance are the different types of challenges. These steps work together to keep the body's O_2 homeostasis by keeping O_2 concentration within a narrow physiological range. In normal circumstances, oxygen delivery to body cells exceeds oxygen consumption. However, various physiological and pathological conditions such as reduced barometric pressure, impaired respiratory function, cardiac dysfunction, hypoventilation, deficiency of functional hemoglobin, and diffusion abnormality lead to low cardiac output, thus causing a further drop in pO_2, in the subsequent steps of the oxygen cascade, and resulting in hypoxia (Dunn et al. 2016).

4.2 Lack of Oxygen: Hypoxia

Hypoxia is a prevalent physiological stimulus. When the body is deprived of an adequate O_2 supply, even for a brief period, it leads to the impairment of various functions of organs in the body. Hampering of physiological functions of tissues due to a reduction in oxygen availability/supply below the threshold requirement is known as “hypoxia”. In other words, hypoxia results when the tissue demand exceeds its O_2 supply (Höpfl et al. 2003) at both local and systemic levels. Hypoxia can be classified as follows:

(a) Hypoxic/hypobaric hypoxia: This occurs when there is a deficiency in O_2 exchange in the lungs due to low pO_2 in the arterial blood. Its causes include decreased pO_2 in environments at high altitudes, sleep apnea, pneumonia, pulmonary edema, and/or asthma.
(b) Anemic/hypemic hypoxia: This condition arises when the body is unable to transport the available O_2 to the target tissue due to reduced hemoglobin or red blood corpuscles in the blood. This is caused due to anemia from blood loss, carbon monoxide (CO) poisoning, methemoglobinemia, and sickle cell disease.
(c) Histologic/histotoxic hypoxia: This type of hypoxia takes place when the cells fail to effectively utilize O_2 despite its normal availability. The tissue oxygenation levels may be at or above normal and therefore are not treated as ‘true hypoxia’. Its causes include cyanide poisoning and consumption of alcohol or narcotics.
(d) Stagnant hypoxia/ischemia: The local disruptions in blood supply result in inadequate availability of oxygen in tissues. This type of hypoxia is caused by heart failure, decreased circulating blood volume, and vasodilation.

4.3 General Cardiovascular Responses to Hypoxia

Hypoxia causes the cardiovascular system to compensate by delivering more blood to tissues for counteracting the lower oxygen delivery, which is detected by oxygen-sensing mechanisms such as carotid bodies. As the severity of hypoxia increases, the acute central cardiovascular response causes an increase in heart rate with an unchanged stroke volume, which is mediated primarily by increased sympathetic neural discharge. Stroke, coronary artery disease, and thrombosis are among the most common cardiovascular diseases. Understanding how these diseases work allows for the development of prophylactics and therapeutics. At the cellular level, it is well understood that the pathophysiology of most cardiovascular diseases involves complex yet coordinated signaling networks triggered in response to a hypoxic milieu at either the cellular or tissue level (Gupta and Ashraf 2018). Lower levels of hypoxia may cause some systemic vasodilation at rest, but as the severity of hypoxia increases, the peripheral vasculature constricts to redistribute oxygen delivery to the most critically dependent organs, such as the heart and brain (Heinonen et al. 2016). This regulation is exacerbated in obstructive sleep apnea, which causes chronic intermittent hypoxia, a physiological condition that can compromise important physiological functions. Similarly, when exercising in hypoxia, skeletal muscle perfusion increases to match oxygen demand, creating a circulatory competition between locomotor skeletal muscles and other organs, resulting in decreased exercise capacity as hypoxia severity increases. Hypoxia affects the body's energy metabolism in addition to causing cardiovascular stress. Although hypoxia may theoretically reduce oxygen requirements at the local tissue level due to reduced oxygen supply, whole-body metabolism often increases in response to hypoxia due to increased sympathetic neural activation and the release of various stress hormones. The circulatory system distributes blood flow and transports oxygen and nutrients to each tissue. The physiological function of each tissue depends on peripheral circulation. Circulation disturbances reduce oxygen delivery, resulting in tissue hypoxia, which occurs in a variety of cardiovascular disorders such as atherosclerosis, pulmonary arterial hypertension, and heart failure. While cardiovascular disorders can cause tissue hypoxia, hypoxia signaling has the potential to modulate tissue remodeling processes and the severity of cardiovascular disorders. Each cell responds to hypoxia in a variety of ways at the transcriptional, translational, and posttranslational levels. Most of the genes that are expressed at the transcriptional level are significantly suppressed in the hypoxic environment. In contrast, the expression of a group of genes is significantly enhanced in hypoxia called hypoxia-inducible genes. Representatives of hypoxia-inducible genes are genes related to angiogenesis (vascular endothelial growth factor-A, VEGF-A), erythropoiesis (erythropoietin, EPO), cellular metabolism (pyruvate dehydrogenase kinase, isoform 1, Pdk 1, or lactate dehydrogenase-A, LDH-A), and inflammation (inducible nitric oxide synthase, i-NOS) (Kumar and Prabhakar 2008).

4.3.1 Vascular Endothelial Growth Factor

The key mediator of angiogenesis is vascular endothelial growth factor (VEGF), a homodimer vasoactive glycoprotein. The VEGF family regulates functions such as angiogenesis, lymphopoiesis, inflammation, oxidative stress resistance, and lipid metabolism which have the potential therapeutic targets for cardiovascular diseases. VEGF-A, VEGF-B, VEGF-C, VEGF-D, VEGF-E (source: virus), VEGF-F (source: snake venom), placenta growth factor (PGF), and EG-VEGF are members of the VEGF family. VEGFR-1, VEGFR-2, and VEGFR-3 are VEGF receptors (VEGFRs) (Fig. 4.1). VEGFR-1 and VEGFR-2 are mostly found in vascular endothelial cells, whereas VEGFR-3 is mostly found in lymphatic endothelial cells (Zhou et al. 2021). Angiogenesis or the growth of new blood vessels is involved in several

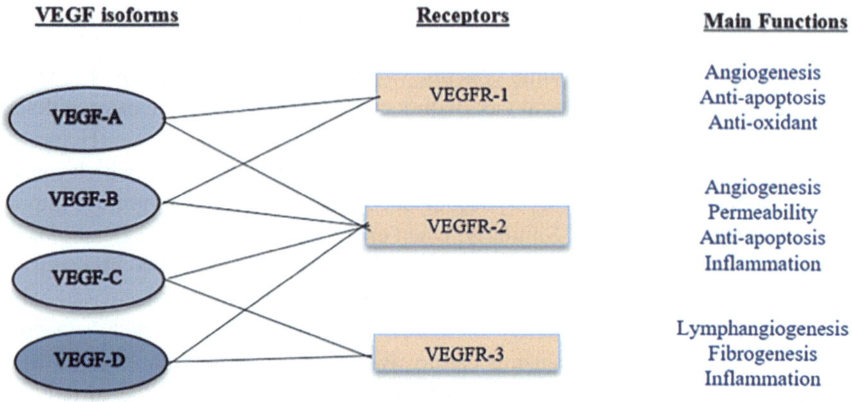

Fig. 4.1 VEGF family and their functions

physiological and pathological processes, including CVDs. CMs, the most common cell type in the heart, are both the source and the target of VEGF-A and their cell surfaces express the receptors, VEGFR-1 and VEGFR-2. The heart and VEGF-A have a complicated relationship. On the one hand, VEGF-A stimulates morphogenesis, contractility, and wound healing by activating CM. Inflammation, mechanical stress, and cytokine stimulation, on the other hand, cause CM to produce VEGF-A. Furthermore, high levels of VEGF-A have been found in patients with various CVDs and are frequently linked to a poor prognosis and disease severity. Based on current clinical studies on angiogenesis therapy, it is reasonable to believe that controlling angiogenesis and VEGF-A can improve the quality and length of life of patients with heart diseases (Braile et al. 2020).

4.3.2 Cardiomyocytes as a Producer and Target of VEGF-A

CMs produce VEGF-A, and it is also known to regulate vascular permeability and angiogenesis in CMs (Abhinand et al. 2016). The deletion of VEGF-A in CMs causes negative vasculogenesis/angiogenesis and a thinner ventricular wall in the mouse model, confirming mutual signaling from CM to EC during the heart development process (Giordano et al. 2001). The myocardial microvasculature is underdeveloped in mice lacking VEGF-A, but the coronary artery structure is preserved, suggesting a different signaling pathway for vasculogenesis/angiogenesis in the myocardium and epicardial coronary arteries. Mice with CM-specific deletion of VEGF-A164 and VEGF-A188 isoforms develop ischemic cardiomyopathy and heart failure (Carmeliet et al. 1999). In these mice, heart capillaries are also more irregular, tortuous, and dilated indicating incomplete vessel remodeling. VEGF-A stimulates myocardial angiogenesis by increasing vascular permeability and EC proliferation. In a mouse model, CM-produced VEGF-A inhibits the transformation of endocardial to mesenchymal cells. This mechanism is critical for cardiac cushion formation and necessitates careful VEGF-A concentration control. Low levels of VEGF-A cause endocardial cells to become mesenchymal in the same animal model, whereas high levels of VEGF-A prevent the transformation. This signal for endocardial-mesenchymal transformation induced by CM VEGF-A production might be regulated by an endothelial-derived feedback process via the calcineurin/NFAT pathway, demonstrating the importance of EC-CM interactions for cardiac morphogenesis.

CMs are both VEGF-A producers and targets (Pearlman et al. 1995; Neufeld et al. 1999; Chen et al. 2006) and later exhibit a variety of actions in myocardial reparative wound healing, including vasculogenesis, stem cell recruitment and homing, decreased apoptosis, and increased vasodilation and autonomic response modulation (Conway et al. 2001). VEGF-A exerts its biological effects by interacting with two major tyrosine kinase receptors, VEGFR1 and VEGFR2, with VEGFR1 having a higher affinity than VEGFR2. Although both the receptors are expressed on the CM surface, VEGFR1 is upregulated in response to hypoxia and oxidative stress (Zentilin et al. 2010). The heart uses VEGF-A phospholipase C1 signaling to control the strength of the heartbeat; therefore, VEGFR1 is playing important roles in cardiac contractility. VEGF-A has been shown to inhibit CM apoptosis and activate the expression of genes involved in myocardial contractility and metabolism in a rat model. VEGF-A promotes cardiac stem cell migration via PI3K/Akt pathway and plays a critical role in heart repairing process. The VEGF-A pathway, in combination with myocardial stromal cell-derived factor-1 (SDF-1), promotes cardiac stem cell mobilization and myocardial repair in infarcted hearts. Following an acute infarcted myocardium, VEGF-A plays a critical role in triggering the cardiac angiogenic response. VEGF-A and VEGFR expression are increased on the day first and then suppressed during the first week of angiogenesis in myocardial infarcted rats (Fig. 4.2).

4.3.3 Mechanisms Involved in the Regulation of VEGF-A Functions

Cardiomyocytes and endothelial cells (ECs) are frequently exposed to hypoxia and inflammation in cardiovascular disorders, which leads to an increase in hypoxia-inducible factors (HIFs). HIFs can promote angiogenesis by upregulating several pro-angiogenic factors, particularly VEGF, via the HIF-1 pathway (Fig. 4.2). Increased VEGF-A promotes VEC proliferation, increases vascular permeability, and restores endothelial integrity and function. As a result, it compensates for ischemia and hypoxia while also safeguarding the wounded myocardium.

4.3.3.1 Upregulation Mechanisms of VEGF-A

Recently, researchers have discovered new methods for upregulating VEGF-A in addition to the traditional pathway. According to a study published in Nature, a shortage of oxygen and nutrients activated the transcription coactivator PGC-1, a key regulator of mitochondrial function, and increased angiogenesis by increasing VEGF expression. Jin et al. discovered that hypoxia caused ECs to generate TNF,

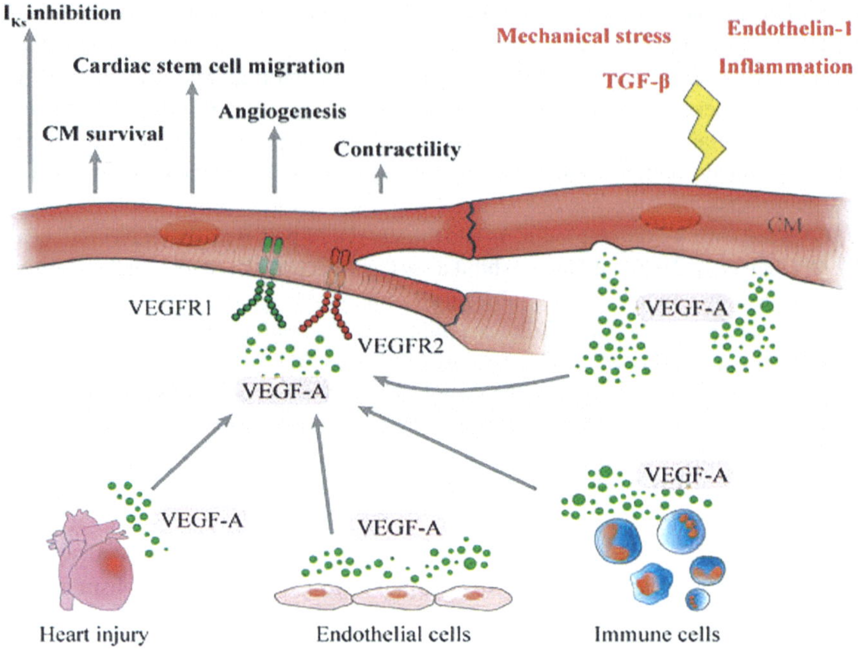

Fig. 4.2 Schematic representation of cardiomyocytes as source and target of vascular endothelial growth factor-A. A plethora of stimuli including inflammation, mechanical stress, endothelin-1, and transforming growth factor-β induce CMs to produce and release VEGF-A, whose function is to promote angiogenesis in myocardial tissue. CMs are also a target of VEGF-A, produced by several cells during heart injury through binding with VEGFR1 and VEGFR2 receptors expressed on their surface. CMs activation induced by VEGF-A enhances CM survival, contractility, cardiac stem cell recruitment, cardiac angiogenesis, and reduction of potassium current (IKs) (Braile et al. 2020)

implying that the autocrine cycle may activate HIFs through an NF-kB-dependent mechanism to upregulate VEGFs and promote angiogenesis. In ECs, this results in the TNF-α/NF-kB/HIF/VEGF signaling cascade (Ferraro et al. 2019). Annexin A1 (AnxA1) induced angiogenesis and cardiac repair by causing cardiac macrophages to polarize to angiogenic and repair phenotypes and stimulating them to generate substantial amounts of VEGF-A (Fig. 4.3). Mild hypoxia (14% O_2), on the other hand, did not stimulate coronary angiogenesis or VEGF-A expression in growing mouse hearts, according to the study done by Zhou et al. 2021.

4.3.3.2 Downregulation Mechanisms of VEGF-A

LDL inhibits the TNF-α/NF-κB/HIF/VEGF signaling cascade in ECs, as it can suppress TNFR1, prevent the expression of HIFs in ECs exposed to hypoxia or TNF-α, and thus decrease VEGF expression (Fig. 4.3). Similarly, knocking down HIF-1 or HIF-2 inhibited angiogenesis by drastically reducing the production of VEGFs caused by hypoxia (Jin et al. 2019). Furthermore, through p53-dependent and independent methods, the E2F1 transcription factor blocked the overexpression

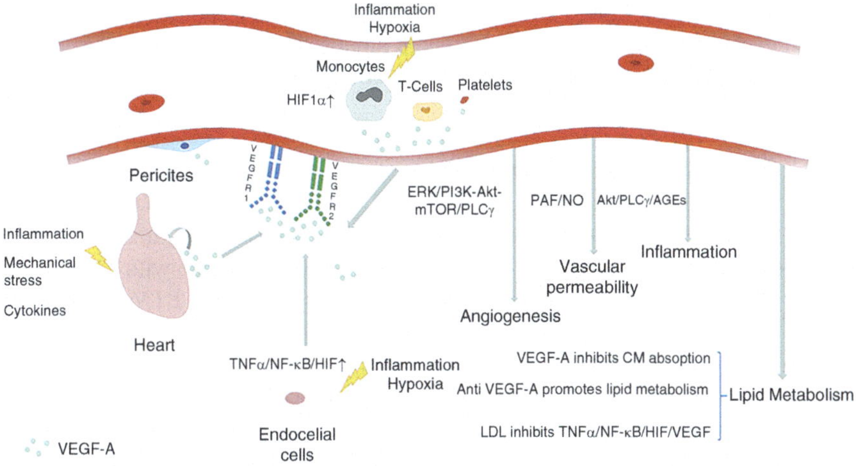

Fig. 4.3 The important functions and regulatory mechanism of VEGF-A

of VEGFs and PIGF, respectively, restricting heart neovascularization and functional recovery after MI. Overexpression of ANXA1 suppressed inflammatory markers as well as the infiltration and death of polymorphonuclear neutrophils (PMNs) in a rat model of cardiac reperfusion damage, which further downregulated the STAT3 signaling pathway and hampered VEGF expression. Lipid-lowering medications have been found to downregulate VEGF in many investigations.

4.4 SUMOylation: A Posttranslational Protein Modification

Posttranslational modification (PTM) is an essential regulatory mechanism in a cell, which works by modifying cellular proteins that serve in a variety of biological activities. PTMs can alter any protein after or during its translation. Different types of modifications affect protein's charge state, hydrophobicity, conformation, and stability, thus its functions. Protein modification is reversible and different organelles have different functions (Zamaraev et al. 2017). All the physiological processes including immune functions, location, duration, and intensity of the exact physiological process are controlled by specific protein modifications to ensure the cells' rapid and dynamic response to extracellular and intracellular stimulation (Liu et al. 2016). Phosphorylation, acetylation, ubiquitination, and SUMOylation are examples of posttranslational protein modifications that can affect target protein activity, intracellular distribution, protein-protein interactions, and protein longevity (Choudhary et al. 2009; Johnson 2004). As widely used PTM, SUMOylation is gaining popularity since it is found in almost all eukaryotes and is required for genomic integrity, transcriptional regulation, gene expression, and intracellular signal transduction regulation. One of the most pressing issues is the relationship between SUMOylation and other PTMs such as phosphorylation, ubiquitination, methylation, and acetylation. DNA damage repair, immune responses, carcinogenesis, cell cycle progression, and apoptosis are among the most important biological processes regulated by SUMOylation (Han et al. 2018).

4.4.1 The Family of SUMO Proteins

Sentrin was the first SUMO protein isoform discovered in 1996 (Okura et al. 1996) which regulates protein activities by forming a covalent conjugate with proteins by binding SUMO to particular lysine residues, implying a unique PTM mechanism (Chang and Yeh 2020; Hay 2005). As a result, SUMOylation is recommended as a term to represent the process of diverse proteins being modified by reversibly binding to lysine residues via a multistep enzymatic reaction cascade. A total five SUMO isoforms have been found so far (SUMO1, 2, 3, 4, and 5), in which SUMO1 and 2/3 are found throughout the body, whereas SUMO4 and SUMO5 appear to be tissue-specific and have received less attention (Liang et al. 2016). The kidney, spleen, and lymph nodes have SUMO4 only. Protein localization, protein-protein interactions, protein-DNA interactions, and/or protein activity are all regulated by SUMO modification. SUMOylation, on the other hand, has a substrate-specific influence on protein activity. Activating (E1), conjugating (E2), and ligating (E3) enzymes dynamically regulate protein SUMOylation, although a class of SUMO-specific proteases can reverse it (SENPs). Six SENPs have been discovered, each with its own expression profile, substrate specificity, and cellular location. SENP1 and SENP2 recognize all three mammalian SUMOs (SUMO1–3), but SENP3, SENP5, SENP6, and SENP7 choose SUMO2/3 as a substrate. SENP1 is a protease that deconjugates a large number of SUMOylated proteins and appears to be localized in various compartments. SUMO1 shares 48% sequence commonality with both SUMO2 and SUMO3, while the subsequent two isoforms are highly similar with 97% sequence identity (Guo et al. 2004; Lee et al. 2016). Although no evidence indicates conjugation with substrates, SUMO4 is 85% identical to SUMO2/3 (Saitoh and Hinchey 2000). SUMO5 and SUMO1 are extremely similar (Hammoudi et al. 2016). SUMOylation, on the other hand, happens at a specific lysine residue within the canonical/consensus pattern of -K-X-E/D, known as the SUMO-interacting motif (SIM), where there is a hydrophobic residue, K is lysine for SUMO conjugation, X is any amino acid residue, and E/D is an acidic residue (Maejima and Sadoshima 2014).

4.4.2 SUMOylation in Response to Hypoxia

Multicellular organisms have evolved homeostatic mechanisms in order to maintain cellular and tissue functions upon changes in oxygen tension. Exposure of human tissues or cells to hypoxia is encountered both during physiological (high altitude, intense exercise, and embryogenesis) and pathological processes. SUMOylation is a reversible posttranslational modification that plays a role in DNA damage and repair, immune responses, carcinogenesis, cell cycle progression, and apoptosis. The small ubiquitin-like modifier (SUMO) pathway is found in all eukaryotes and is involved in gene regulation, cellular signaling, and the preservation of genomic integrity (Han et al. 2018). The SUMO family mediated SUMOylation with SUMO1, being one of the most important members. SUMO1 is important for early heart development and reduced SUMOylation activity causes cardiac pathology and SUMO1 played a key role in normal cardiac development. The altered SUMOylation levels have an impact on several transcription factors and enzymes that keep the heart healthy. SUMO1 has also been discovered to improve the oxidative stress response in patients with increased cardiac hypertrophy. Hypoxia causes a dramatic reprogramming of gene expression, which is followed by a series of events such as switching to anaerobic energy production that results in cardiovascular diseases and so on, all of which help cells adapt and survive in a hypoxic environment. In addition to SUMO, the hypoxia-inducible transcription factors HIF-1 and HIF-2 (EPAS1) are also involved in angiogenesis which are critical for hypoxic responses.

4.4.3 SUMOylation in Cardiovascular Diseases

The cell division, DNA repair, genetic and epigenetic transcriptional regulation, and cellular metabolism are regulated by SUMOylation (Lee et al. 2016; Oh et al. 2018; Chang and Yeh 2020). As a result, SUMOylation dysregulation is linked to the pathogenesis of a variety of illnesses, including cancer, diabetes, epilepsy, and cardiovascular diseases (Heun 2007; Johnson 2004). SUMOylation appears to be a therapeutic target in the cardiovascular system, according to several recent investigations. SUMOylation has a significant impact on cardiac function and development, as well as the heart's ability to respond to diverse pathological stressors (Lee et al. 2014; Kim et al. 2015; Mendler et al. 2016). Specific SUMO isoforms (SUMO1 and 2/3) are required in the heart which are playing distinctive roles. Congenital heart disorders such as atrial and septal abnormalities, progressive cardiac dysfunction, and sudden death plagued SUMO1 knockout mice (Wang et al. 2011). Overexpression of SUMO1 improves cardiac function and increases contractility in isolated cardiomyocytes in mice with heart failure (Kho et al. 2011). SUMO2 KO mice perished at embryonic stage E10.5 with severe developmental delay and no distinct cardiac abnormality, but SUMO3 KO mice survived (Wang et al. 2014). The SUMOylation system is thought to regulate the heart's response to hypoxic stress (Hotz et al. 2020).

4.5 VEGF Regulation by SUMOylation During Hypoxic Stress

The wide-ranging SUMO modification was increased significantly in hypoxic cardiomyocytes both in vivo and in vitro. It leads to increased levels of HIF1-α uninterruptedly encouraged downstream VEGF expression (Liu et al. 2021). The VEGF-A/VEGFR2 signaling majorly activates in hypoxic conditions (Fig. 4.4).

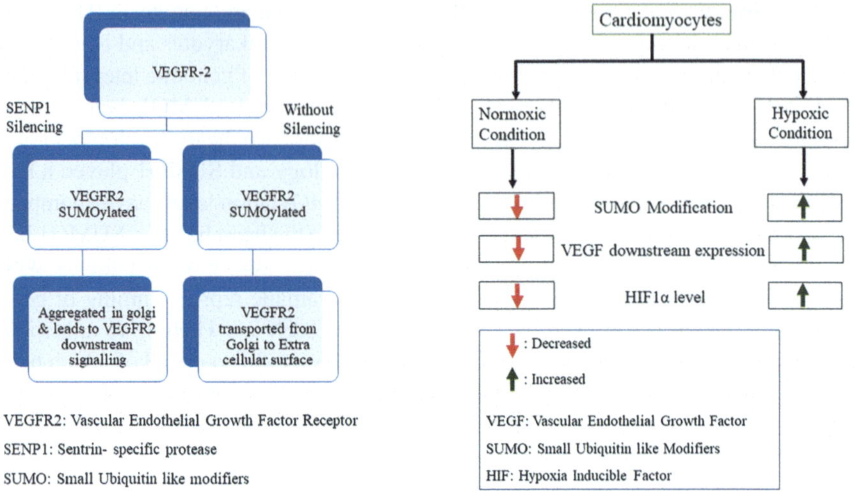

Fig. 4.4 VEGF regulation by SUMOylation during hypoxic stress

Studies showed that in the absence of SENP1, VEGFR2 remains in SUMOylated form and congregated in Golgi bodies leading to alteration in VEGFR2 surface expression and its downstream signaling. Zhou et al. also reveal that SENP1 works as an endogenous activator in the signaling of VEGFR2 by regulating VEGFR2 transportation from the Golgi to ECs surface (Fig. 4.4). The VEGFR2 remains in a deSUMOylated state by SENP1 to ensure VEGFR2 membrane targeting in ECs (Zhou et al. 2018). Furthermore, SUMO conjugation to FGFR1 preserves stability in the competitive enrolment of the adaptor protein FRS2-α between FGFR1 and VEGFR2 receptor complexes. Hypoxia-enhanced FGFR1 SUMOylation alters the dimerization of FGFR1 and FGFR1 binding with its phosphatase, leading to limited tyrosine kinase activation of FGFR1. Therefore, it supported the stimulation of VEGF-A/VEGFR2 signaling in endothelial cells by facilitating the binding of FRS2 to VEGFR2, but restricting further recruitment of FRS2 to FGFR1. Zhu et al. identify that SENP1-regulated FGFR1 SUMOylation and deSUMOylation control the competitive recruitment of FRS2-α by FGFR1 and VEGFR2 to switch receptor-complex formation responding to hypoxic and normoxic angiogenic environments (Zhu et al. 2022).

In summary, hypoxia is associated with many cardiovascular conditions including atherosclerosis, pulmonary arterial hypertension, and heart failure. VEGF-A and heart share a very complicated relationship where cardiomyocytes are both the source and the target of VEGF-A, and its cell surfaces express both VEGFR1 and VEGFR2 receptors. It also stimulates morphogenesis, contractility, and wound healing by activating cardiomyocytes. Lack of VEGF-A may result in circumstances that compromise heart functions. SUMOylation has a substrate-specific influence on protein activity and it regulates the expression and signaling associated with

VEGF-A. SUMOylation dysregulation is linked to the pathogenesis of a variety of illnesses, including cancer, diabetes, epilepsy, and cardiovascular diseases.

References

Abe H, Semba H, Takeda N (2017) The roles of hypoxia signaling in the pathogenesis of cardiovascular diseases. J Atheroscler Thromb 24(9):884–894. https://doi.org/10.5551/jat.RV17009, PMID 28757538

Abhinand CS, Raju R, Soumya SJ, Arya PS, Sudhakaran PR, VEGF-A (2016) VEGF-A/VEGFR2 signaling network in endothelial cells relevant to angiogenesis. J Cell Commun Signal 10(4): 347–354. https://doi.org/10.1007/s12079-016-0352-8, PMID 27619687

Braile M, Marcella S, Cristinziano L, Galdiero MR, Modestino L, Ferrara AL et al (2020) VEGF-A in cardiomyocytes and heart diseases. Int J Mol Sci 21(15):5294. https://doi.org/10.3390/ijms21155294, PMID 32722551

Carmeliet P, Ng YS, Nuyens D, Theilmeier G, Brusselmans K, Cornelissen I et al (1999) Impaired myocardial angiogenesis and ischemic cardiomyopathy in mice lacking the vascular endothelial growth factor isoforms VEGF164 and VEGF188. Nat Med 5(5):495–502. https://doi.org/10.1038/8379, PMID 10229225

Chang HM, Yeh ETH (2020) SUMO: from bench to bedside. Physiol Rev 100(4):1599–1619. https://doi.org/10.1152/physrev.00025.2019, PMID 32666886

Chen Y, Amende I, Hampton TG, Yang Y, Ke Q, Min JY et al (2006) Vascular endothelial growth factor promotes cardiomyocyte differentiation of embryonic stem cells. Am J Physiol Heart Circ Physiol 291(4):H1653–H1658. https://doi.org/10.1152/ajpheart.00363.2005, PMID 16699073

Choudhary C, Kumar C, Gnad F, Nielsen ML, Rehman M, Walther TC et al (2009) Lysine acetylation targets protein complexes and co-regulates major cellular functions. Science 325(5942):834–840. https://doi.org/10.1126/science.1175371, PMID 19608861

Conway EM, Collen D, Carmeliet P (2001) Molecular mechanisms of blood vessel growth. Cardiovasc Res 49(3):507–521. https://doi.org/10.1016/s0008-6363(00)00281-9, PMID 11166264

Du C, Chen X, Su Q, Lu W, Wang Q, Yuan H et al (2021) The function of SUMOylation and its critical roles in cardiovascular diseases and potential clinical implications. Int J Mol Sci 22(19): 10618. https://doi.org/10.3390/ijms221910618, PMID 34638970

Dunn J-OC, Mythen MG, Grocott MP (2016) Physiology of oxygen transport. BJA Education 16 (10):341–348. https://doi.org/10.1093/bjaed/mkw012

Ferraro B, Leoni G, Hinkel R, Ormanns S, Paulin N, Ortega-Gomez A et al (2019) Pro-angiogenic macrophage phenotype to promote myocardial repair. J Am Coll Cardiol 73(23):2990–3002. https://doi.org/10.1016/j.jacc.2019.03.503, PMID 31196457

Giordano FJ, Gerber HP, Williams SP, VanBruggen N, Bunting S, Ruiz-Lozano P et al (2001) A cardiac myocyte vascular endothelial growth factor paracrine pathway is required to maintain cardiac function. Proc Natl Acad Sci U S A 98(10):5780–5785. https://doi.org/10.1073/pnas.091415198, PMID 11331753

Guo D, Li M, Zhang Y, Yang P, Eckenrode S, Hopkins D et al (2004) A functional variant of SUMO4, a new IκBα modifier, is associated with type 1 diabetes. Nat Genet 36(8):837–841. https://doi.org/10.1038/ng1391, PMID 15247916

Gupta N, Ashraf ZM (2018) Hypoxia signaling in cardiovascular diseases. IntechOpen. https://doi.org/10.5772/intechopen.80456

Hammoudi V, Vlachakis G, Schranz ME, van den Burg HA (2016) Whole-genome duplications followed by tandem duplications drive diversification of the protein modifier SUMO in angiosperms. New Phytol 211(1):172–185. https://doi.org/10.1111/nph.13911, PMID 26934536

Han ZJ, Feng YH, Gu BH, Li YM, Chen H (2018) The post-translational modification, SUMOylation, and cancer (review). Int J Oncol 52(4):1081–1094. https://doi.org/10.3892/ijo.2018.4280, PMID 29484374

Hannoun Z, Greenhough S, Jaffray E, Hay RT, Hay DC (2010) Post-translational modification by SUMO. Toxicology 278(3):288–293. https://doi.org/10.1016/j.tox.2010.07.013, PMID 20674646

Hay RT (2005) SUMO: a history of modification. Mol Cell 18(1):1–12. https://doi.org/10.1016/j.molcel.2005.03.012, PMID 15808504, 18

Heinonen IH, Boushel R, Kalliokoski KK (2016) The circulatory and metabolic responses to hypoxia in humans–with special reference to adipose tissue physiology and obesity. Front Endocrinol 7:116. https://doi.org/10.3389/fendo.2016.00116, PMID 27621722

Heun PSUM (2007) SUMOrganization of the nucleus. Curr Opin Cell Biol 19(3):350–355. https://doi.org/10.1016/j.ceb.2007.04.014, PMID 17467254

Höpfl G, Ogunshola O, Gassmann M (2003) Hypoxia and high altitude: the molecular response. Adv Exp Med Biol 543:89–115

Hotz PW, Wiesnet M, Tascher G, Braun T, Müller S, Mendler L (2020) Profiling the murine SUMO proteome in response to cardiac ischemia and reperfusion injury. Molecules 25(23):5571. https://doi.org/10.3390/molecules25235571, PMID 33260959

Jin F, Zheng X, Yang Y, Yao G, Ye L, Doeppner TR et al (2019) Impairment of hypoxia-induced angiogenesis by LDL involves a HIF-centered signaling network linking inflammatory TNFα and angiogenic VEGF. Aging (Albany NY) 11(2):328–349. https://doi.org/10.18632/aging.101726, PMID 30659163

Johnson ES (2004) Protein modification by SUMO. Annu Rev Biochem 73(1):355–382. https://doi.org/10.1146/annurev.biochem.73.011303.074118

Kho C, Lee A, Jeong D, Oh JG, Chaanine AH, Kizana E et al (2011) SUMO1-dependent modulation of SERCA2a in heart failure. Nature 477(7366):601–605. https://doi.org/10.1038/nature10407, PMID 21900893

Kim EY, Zhang Y, Beketaev I, Segura AM, Yu W, Xi Y et al (2015) SENP5, a SUMO isopeptidase, induces apoptosis and cardiomyopathy. J Mol Cell Cardiol 78:154–164. https://doi.org/10.1016/j.yjmcc.2014.08.003, PMID 25128087

Kumar GK, Prabhakar NR (2008) Post-translational modification of proteins during intermittent hypoxia. Respir Physiol Neurobiol 164(1–2):272–276. https://doi.org/10.1016/j.resp.2008.05.017, PMID 18602876

Lee A, Jeong D, Mitsuyama S, Oh JG, Liang L, Ikeda Y et al (2014) The role of SUMO-1 in cardiac oxidative stress and hypertrophy. Antioxid Redox Signal 21(14):1986–2001. https://doi.org/10.1089/ars.2014.5983, PMID 24893265

Lee A, Oh JG, Gorski PA, Hajjar RJ, Kho C (2016) Post-translational modifications in heart failure: small changes, big impact. Heart Lung Circ 25(4):319–324. https://doi.org/10.1016/j.hlc.2015.11.008, PMID 26795636

Liang YC, Lee CC, Yao YL, Lai CC, Schmitz ML, Yang WM (2016) SUMO5, a novel poly-SUMO isoform, regulates PML nuclear bodies. Sci Rep 6(1):26509. https://doi.org/10.1038/srep26509, PMID 27211601

Liu J, Qian C, Cao X (2016) Post-translational modification control of innate immunity. Immunity 45(1):15–30. https://doi.org/10.1016/j.immuni.2016.06.020, PMID 27438764

Liu H, Bian X, Xu M, Ma X, Zhang C, Jiang J, Fu J, Liu X (2021) HIF-1α protein SUMOylation is an important protective mechanism of action of hypothermia in hypoxic cardiomyocytes. Mol Med Rep 23(6):476

Maejima Y, Sadoshima J (2014) SUMOylation: a novel protein quality control modifier in the heart. Circ Res 115(8):686–689. https://doi.org/10.1161/CIRCRESAHA.114.304989, PMID 25258400

Mendler L, Braun T, Müller S (2016) The ubiquitin-like SUMO system and heart function: from development to disease. Circ Res 118(1):132–144. https://doi.org/10.1161/CIRCRESAHA.115.307730, PMID 26837744

Michiels C (2004) Physiological and pathological responses to hypoxia. Am J Pathol 164(6): 1875–1882. https://doi.org/10.1016/S0002-9440(10)63747-9, PMID 15161623

Neufeld G, Cohen T, Gengrinovitch S, Poltorak Z (1999) Vascular endothelial growth factor (VEGF) and its receptors. FASEB J 13(1):9–22. https://doi.org/10.1096/fasebj.13.1.9, PMID 9872925

Oh JG, Watanabe S, Lee A, Gorski PA, Lee P, Jeong D et al (2018) miR-146a suppresses SUMO1 expression and induces cardiac dysfunction in maladaptive hypertrophy. Circ Res 123(6): 673–685. https://doi.org/10.1161/CIRCRESAHA.118.312751, PMID 30355233

Okura T, Gong L, Kamitani T, Wada T, Okura I, Wei CF et al (1996) Protection against Fas/APO-1- and tumor necrosis factor-mediated cell death by a novel protein, sentrin. J Immunol 157(10): 4277–4281

Pearlman JD, Hibberd MG, Chuang ML, Harada K, Lopez JJ, Gladstone SR et al (1995) Magnetic resonance mapping demonstrates benefits of VEGF–induced myocardial angiogenesis. Nat Med 1(10):1085–1089. https://doi.org/10.1038/nm1095-1085, PMID 7489368

Saitoh H, Hinchey J (2000) Functional heterogeneity of small ubiquitin-related protein modifiers SUMO-1 versus SUMO-2/3. J Biol Chem 275(9):6252–6258. https://doi.org/10.1074/jbc.275.9.6252, PMID 10692421

Wang J, Chen L, Wen S, Zhu H, Yu W, Moskowitz IP et al (2011) Defective SUMOylation pathway directs congenital heart disease. Birth Defects Res A Clin Mol Teratol 91(6):468–476. https://doi.org/10.1002/bdra.20816, PMID 21563299

Wang L, Wansleeben C, Zhao S, Miao P, Paschen W, Yang W et al (2014) 2 is essential while SUMO 3 is dispensable for mouse embryonic development. EMBO Rep 15(8):878–885

Zamaraev AV, Kopeina GS, Prokhorova EA, Zhivotovsky B, Lavrik IN (2017) Post-translational modification of caspases: the other side of apoptosis regulation. Trends Cell Biol 27(5): 322–339. https://doi.org/10.1016/j.tcb.2017.01.003, PMID 28188028

Zentilin L, Puligadda U, Lionetti V, Zacchigna S, Collesi C, Pattarini L et al (2010) Cardiomyocyte VEGFR-1 activation by VEGF-B induces compensatory hypertrophy and preserves cardiac function after myocardial infarction. FASEB J 24(5):1467–1478. https://doi.org/10.1096/fj.09-143180, PMID 20019242

Zhou HJ, Xu Z, Wang Z, Zhang H, Zhuang ZW, Simons M, Min W (2018) SUMOylation of VEGFR2 regulates its intracellular trafficking and pathological angiogenesis. Nat Commun 9 (3303). https://doi.org/10.1038/s41467-018-05812-2

Zhou Y, Zhu X, Cui H, Shi J, Yuan G, Shi S et al (2021) The role of the VEGF family in coronary heart disease. Front Cardiovasc Med 8:738325. https://doi.org/10.3389/fcvm.2021.738325, PMID 34504884

Zhu X, Qiu C, Wang Y, Jiang Y, Chen Y, Fan L, Ren R, Wang Y, Chen Y, Feng Y, Zhou X, Zhu Y, Ge Z, Lai D, Qin L, Simons M, Yu L (2022) FGFR1 SUMOylation coordinates endothelial angiogenic signaling in angiogenesis. Proc Natl Acad Sci U S A 119(26):e2202631119. https://doi.org/10.1073/pnas.2202631119

Expression of Vascular Endothelial Growth Factor at High Altitude

5

Ritika Jain, Ratnika Sharma, Rekha Maithani, and Rajkumar Tulsawani

Abstract

Oxygen plays an important role in aerobic respiration in mammals. However, there is subsequent decrease in partial pressure of oxygen with increase in altitude, thus resulting in hypobaric hypoxia. Hypoxia being a state of low oxygen levels in the body tissues, at high altitude, causes acute mountain sickness, high altitude pulmonary edema, and high altitude cerebral edema. Altitude sickness results in manifestations such as shortness of breath, insomnia, dehydration, severe headache, hypoglycemia, and hypothermia. In severe cases, failure in ability of successful acclimatization may cause death. Vascular endothelial growth factor is regulated by hypoxia-inducible factor-1. It is a potent angiogenic factor, secreted by tumor cells, platelets, and macrophages and many cells play major physiological role in hematopoiesis, bone formation, wound healing, and development. Vascular endothelial growth factor comprises of five classes, plays an essential and differential role in angiogenesis in retina and brain, and maintains skeletal muscle capillarity.

Keywords

High Altitude · Hypobaric hypoxia · VEGF · HACE · Angiogenesis

R. Jain
Defence Institute of Physiology and Allied Sciences (DIPAS), Defence Research & Development Organization (DRDO), Delhi, India

Faculty of Life Sciences, Mandsaur University, Mandsaur, Madhya Pradesh, India

R. Sharma · R. Maithani · R. Tulsawani (✉)
Defence Institute of Physiology and Allied Sciences (DIPAS), Defence Research & Development Organization (DRDO), Delhi, India
e-mail: rktulsawani@yahoo.com

R. Tulsawani, D. Vohora (eds.), *Adaptation under Stressful Environments through Biological Adjustments and Interventions*,
https://doi.org/10.1007/978-981-99-7652-2_5

5.1 Introduction

The altitude regions are categorized into larger three zones; high altitude (1500–3500 m), very high altitude (3500–5500 m), and extreme altitude (5500 m and above sea level). The common syndromes observed at these altitude regions are acute mountain sickness (AMS), high altitude pulmonary edema (HAPE), and high altitude cerebral edema (HACE), which are characterized by insomnia, shortness of breath, dehydration, exhaustion, severe headache, hypoglycemia, hypothermia, hyponatremia, carbon monoxide poisoning, and neurologic problems including migraine (Hackett 1999; Mehta et al. 2008; Luks et al. 2017; Basnyat and Tabin 2018; Thapa and Basnyat 2018; Aksel et al. 2019; Li et al. 2022; Zubieta-Calleja and Zubieta-DeUrioste 2022).

Ascend to altitude causes barometric pressure fall and thus drop in partial pressure of oxygen, leading to AMS, HAPE, and HACE affecting biological functions (Table 5.1).

In mammals, presence of oxygen drives the very key basic reaction for aerobic metabolism that takes place in mitochondria of cells, i.e., synthesis of ATP from ADP that generates energy, carbon dioxide, and water. Hypoxia is characterized by lower availability of oxygen at cellular level; this can be caused by various reasons, which are described in the following portion.

5.1.1 Hypoxemia

Hypoxemia refers to lowering of partial pressure of oxygen (Po_2) from ~150 mmHg (at sea level) to 43 mmHg in the arterial blood with increase in altitude. One of the reasons behind it may be the lower oxygen content of the inspired gas at high altitudes.

Table 5.1 Impact of altitude in biological functions and symptoms

Altitude	High altitude: Altitude of 1500–3500 m above sea level	Very high altitude: Altitude of 3500–5500 m above sea level	Extreme altitude: Altitude of 5500 m or more above sea level
Syndrome	Acute Mountain Sickness (AMS)	High Altitude Pulmonary Edema (HAPE)	High Altitude Cerebral Edema (HACE)
Symptoms	• Rapid heart rate • Dizziness • Headache • Lethargy • Irritability • Nausea and vomiting • Insomnia • Anorexia • Shortness of breath during exertion	• Shortness of breath even during rest • Cyanosis • Chest congestion • Persistent cough • Tiredness and weakness • Frothy sputum	• Severe headache • Weakness • Loss of coordination • Disorientation/ confusion, • Psychosis

5.1.2 Oxygen Delivery

It corresponds to taking up of oxygenated blood from the alveolar capillaries to the tissues. It is mainly dependent on two factors: the first is oxygen concentration and the second is cardiac output. Oxygen delivery can be expressed as.

$$Do_2 = Ca_{O2} \times Q_T,$$

where Do_2 is oxygen delivered, Ca_{O2} is content of blood oxygen, and Q_T is cardiac output.

5.1.3 Oxygen Extraction or Utilization

This is a process for oxidative metabolism, i.e., it involves the movement of oxygen molecule from Hemoglobin (Hb) by dissociation into the cytoplasm of cells and finally into the mitochondria. At rest condition, only 25% of oxygen enters into the tissues; it means that only 250 mL is utilized or extracted by the cells and the rest 750 mL remains in the blood for every 1000 mL of oxygen that is delivered per minute. Under the non-resting stage such as exercising conditions, around 50% or even to 60–75% (in elite athletes) oxygen can be extracted into the cells. Thus, the oxygen extraction differs with the blood flow and oxygen delivery (MacIntyre 2014).

Vascular Endothelial Growth Factor (VEGF): VEGF was originally termed as endothelial cell-specific mitogen and secreted by different cell types such as keratinocytes, macrophages, platelets, and renal mesangial cells (Ferrara et al. 1992; Iijima et al. 1993; Sunderkötter et al. 1994; Frank et al. 1995; Verheul et al. 1997; Itakura et al. 2000; Miatmoko et al. 2023). VEGF significantly contributes to physiological functions including many developmental aspects, angiogenesis, hematopoiesis, osteogenesis, and wound healing (Reichardt and Tomaselli 1991; Ferrara et al. 1998; Gerber et al. 1999; Chintalgattu et al. 2003; Duffy et al. 2004). The five homodimeric family members of the VEGF include VEGF-A, VEGF-B, VEGF-C, VEGF-D, and PIGF (placental growth factor) (Takahashi 2011). Such peptides are individually encoded by genes. VEGF-A, generally referred as VEGF, has regulatory functions during developmental angiogenesis, vasculogenesis, and progenitor endothelial cell differentiation (Podar and Anderson 2005; Shibuya 2013; Apte et al. 2019). VEGF-A also comes in four different isoforms: VEGF121, VEGF165, VEGF189, and VEGF206, which are created through alternative mRNA splicing (Park et al. 1993). VEGF165 is the main VEGF-A isoform in terms of quantity and biological activity (Takahashi and Shibuya 2005). Many human malignancies overexpress VEGF165, and this overexpression is linked to the growth, invasion, and metastasis of cancers. Gene targeting experiments proposed role of VEGF-A role in the developmental vasculogenesis and even single VEGF-A allele defect causes embryonic lethality Ferrara et al. 1996). In healthy tissues, VEGF-B and PlGF also promote angiogenesis, though considerably less powerfully than

Table 5.2 VEGF(s) binding to the receptors and functions (West et al. 2005; Takahashi 2011; Apte et al. 2019)

VEGFs alone or in combination	Receptor	Function(s)
VEGFA	sVEGFR1	• VEGFA inhibition
VEGFA	VEGFR1	• Release of growth factors
PIGC	VEGFR1	• Tumor vascularization
VEGFB and PIGC	VEGFR1	• Normal development and angiogenesis
VEGFA	VEGFR2	• Physiological and pathological angiogenesis • Cell survival • Proliferation • Migration
VEGFC and VEGFD	VEGFR2	• Angiogenesis
VEGFC and VEGFD	VEGFR3	• Vasculo-angiogenesis • Hematopoiesis

VEGF-A. In cancers and inflammatory lesions, a rise in PIGF, however, encourages pathological angiogenesis. By means of proteolytic cleavage, VEGF-C and VEGF-D are produced from their precursors. In homozygotes and heterozygotes of VEGF-C gene knockout mice, the absence of lymph vessels causes fetal death as a result of fluid accumulation in organs, demonstrating that VEGF-C plays a significant role in embryonic lymphangiogenesis (Karkkainen et al. 2004; Haiko et al. 2008; Apte et al. 2019) These VEGF family members engage with receptors on cell membranes. The three subtypes of VEGFRs, VEGFR-1, VEGFR-2, and VEGFR-3, share structural similarities with platelet-derived growth factor (PDGF) receptors (Takahashi 2011). Structurally, all subtypes consist of extracellular space with seven immunoglobulin-like domains and an intracellular space with a tyrosine kinase domain. However, functionally, hematopoietic stem cells, tumor cells, and vascular endothelial cells express VEGFR-1 and VEGFR-2, while macrophages and monocytes express only VEGFR-1. Contrary to this, VEGFR is represented in lymphatic endothelial cells. These VEGFR subtypes respond differently to these VEGF members. VEGF-B and PIGF have affinity to receptor VEGFsR-1, whereas VEGF-A activates receptors, VEGFR-1 and VEGFR-2. In addition, receptor VEGFR-3 has only affinity to ligand; VEGF-C and VEGF-D. Heparin and neuropilin-1 may also bind VEGF-A165 and are involved in the activation of VEGFRs by VEGF-A (Table 5.2).

Vasculogenesis and angiogenesis during development and physiological balance are now well understood, because of the identification of vascular endothelial-derived growth factor (VEGF) (Takahashi 2011; Apte et al. 2019). Our knowledge of the molecular processes by which VEGF regulates neurovascular homeostasis has advanced during the past two decades. Involvement of VEGF in pathogenesis of numerous malignancies (Boocock et al. 1995; Podar and Anderson 2005) and blinding eye disorders has been emerging.

5.2 Role of VEGF in Maintaining Skeletal Muscle Capillarity

Many cellular mechanisms in skeletal muscles are activated as a result of prolonged exposure to hypoxia in humans and many other mammals. These systems help to make up for the lack of cellular oxygen. Among these, one of the biological mechanisms is the increased production of a VEGF, a subset of hypoxia-induced genes which is activated by the HIF-1 binding to hypoxic response element of VEGF promoter. For skeletal muscle angiogenesis, VEGF gene expression is crucial, and deletion of the VEGF gene in mice has been found to significantly impair skeletal muscle capillarity (Breen et al. 2008). Yet, there are additional signalling pathways that contribute to elevated or maintained VEGF levels during acute or protracted hypoxia in addition to HIF1-dependent transcriptional activation of the VEGF gene. Also, AMP kinase activity or inflammation may lead to fluctuations in energy status due to ROS generation, which may further induce VEGF in skeletal muscles. These routes could offer VEGF overexpression in the context of muscle activity when exposed for an extended period of time to a hypoxic environment, such as that found at high altitudes. A significant decrease in body mass and muscle mass was recorded in mountaineers at 5000 m for 8-week journey (Harvey et al. 1979; Rose et al. 1988) which was accompanied with reduced oxidative capability and total muscle mitochondrial volume density (Hoppeler et al. 1990a). However, another noteworthy finding about high altitude participants showed increased capillary density per area of muscle in the skeletal muscles compared to unaltered capillary to fiber ratio (C:F) (Hoppeler et al. 1990b). This shows that, despite acute hypoxia, the rise in capillary density was just a result of muscle atrophy and not the development of new capillaries or angiogenesis. This finding was supported by another controlled simulated trial Operation Everest II in a hypobaric chamber simulating high altitude (Green et al. 1989; MacDougall et al. 1991). Hence, it seems that following continuous exposure to a hypoxic environment, the number of capillaries having low oxidative capacity supplying oxygen to myocytes may remain stable.

Nonetheless, this experiment has now undergone further review. Deveci et al. (2001) reported an increase in C:F ratio in rats exposed to hypoxia that corresponds to angiogenesis in active muscles (diaphragm and soleus). These data were further corroborated by significant elevated C:F ration of active muscles (soleus and diaphragm) with bigger fibers. Despite the fact that these studies show that long-term hypoxia can trigger the establishment of new capillaries, it is still unclear whether skeletal muscle can simultaneously express more VEGF and generate new blood vessels. In fact, it was discovered that rats trained in a hypoxic environment for 8 weeks had lower mRNA levels of VEGF, VEGFR-1, and VEGFR-2 (Olfert et al. 2001). According to Milkiewicz et al. (2006), higher VEGF expressions are represented in biphasic manner during capillary expansion in non-inflammatory skeletal muscle ischemia. It is evident that VEGF is necessary to keep capillaries in mature skeletal muscle (Tang et al. 2004). It was also observed that number of skeletal capillaries did not decrease in proportion to reduction in cross-sectional area of fiber for the mountaineers at high altitude which further suggests that VEGF plays key role in preventing apoptosis during prolonged exposure to hypoxia (Gerber et al.

1998). However, angiogenesis will occur in the skeletal muscles when the right signals and cellular environment are present while negative feedback loop will control the levels of VEGF mRNA and protein, as the number of vessels is adjusted to match the needs of each muscle fiber (Breen et al. 2008).

5.3 Role of VEGF in Retina and Brain Angiogenesis

An intricately planned process called vascular development creates artery, vein, and capillary patterns and densities tuning to satisfy the needs of the local tissue. VEGF, a secreted polypeptide, is produced under hypoxia conditions by target tissues and is the principal regulator of vascular development and homeostasis (Usui et al. 2015). Vascular endothelial cells (ECs) are required for the development and maintenance of blood vessels, and VEGF stimulates EC survival, proliferation, permeability, and taxis towards the VEGF source. HIFs, α and β heterodimeric transcription factors that are stabilized by tissue hypoxia and operate as intracellular oxygen sensors, regulate the transcription of the Vegf gene. Since VEGF production is under HIF control, vascular expansion takes place where and when it is required. Vascular expansion, stabilization, or regression can occur as a result, depending on the HIF/VEGF system's level of activation and the surrounding developmental environment (West et al. 2005; Usui et al. 2015).

Retina has a highly vascularized structure and has been used to study the vascular development, especially CNS. Therefore, it plays a central role in development of therapeutics. Most mammalian retinas, including those of mice and people, have three layers of capillaries to meet the inner retina's metabolic requirements. The intraretinal blood vessels enter the body at the optic nerve, where it branches as it travels radially along the retina's inner surface near the vitreous. The surface vascular plexus and two tiers of intraretinal capillaries are connected by smaller vessels that extend perpendicularly into the retina. The choroidal circulation supplies the outer retina as well as the retinal pigment epithelium.

The role of VEGF in retinal vasculature's radial developmental process has been defined and involves; (a) retinal astrocytes that are present in vascular front that expands along the radius have high levels of Vegf mRNA, whereas (b) retinal astrocytes in the center of developing vasculature have low levels of Vegf mRNA (Stone et al. 1995). Newly formed capillaries respond to VEGF due to the presence of endothelial tip cells which are the specialized exploratory cells and help in capillary outgrowth (Gerhardt et al. 2003; Zarkada et al. 2021). It is proposed that physiological hypoxia in astrocytes release VEGF and help in vascular network development (Stone et al. 1995). The vascular plexes are developed in astrocytes and later development of intraretinal capillaries occurs (Usui et al. 2015; Rattner et al. 2019).

5.4 Role of Hypoxia Responsive Genes; HIF1-α and VEGF in Sickle Cell Anemia

Hypoxia-inducible factor regulates vascular endothelial growth factor as primary target gene in vasculature and inflammation and casually interconnected in various physiological and pathophysiological processes. Vascular inflammation is associated with pathogenesis of vaso-occlusive events in sickle cell anemia (SCA) (Gürkan et al. 2005). Polymerization of hemoglobin S (HbS) and hypoxia play central role in microvascular erythrocyte sickling (SCA). The degree and intensity of hemoglobin deoxygenation are proportional to the level of HbS polymerization, which is a key factor in determining the severity of SCA (Caboot and Allen 2014). In addition to modifications in the biophysical and rheological properties of erythrocytes, there is compromise in delivery of blood and oxygen to the tissues and organs (Li et al. 2017), and under such hypoxic environment, erythrocytes undergo sickling altering cell dynamics and result in vaso-occlusive crises, hemolysis, extreme pain episodes, and multi-organ failure (Li et al. 2017; Machogu and Machado 2018). Pedrosa and Lemes (2020) demonstrated that SCA patients have higher HIF-1α and VEGF genes expressions and hydroxyurea therapy reduced expression of HIF-1α and VEGF in doses-dependent manner. It is believed that the hypoxia/HIF-1a signalling pathway, particularly through VEGF transcription, is the primary regulating mechanism of altering cell dynamics and may be potential target for drug development for sickle cell disease (Pedrosa and Lemes 2020).

5.5 Concluding Remarks

This chapter discusses about the importance of oxygen in tissues and differential role of VEGFs under normal oxygen and hypoxic conditions including developmental processes and angiogenesis.

References

Aksel G, Çorbacıoğlu ŞK, Özen C (2019) High-altitude illness: management approach. Turk J Emerg Med 19(4):121–126. https://doi.org/10.1016/j.tjem.2019.09.002. PMID 31687609, PMCID PMC6819752

Apte RS, Chen DS, Ferrara N (2019) VEGF in signaling and disease: beyond discovery and development. Cell 176(6):1248–1264. https://doi.org/10.1016/j.cell.2019.01.021. PMID 3084 9371

Basnyat B, Tabin G (2018) Altitude illness. In: Jameson J, Fauci AS, Kasper DL, Hauser SL, Longo DL, Loscalzo J (eds) Harrison's principles of internal medicine, 20e. McGraw-Hill

Boocock CA, Charnock-Jones DS, Sharkey AM et al (1995) Expression of vascular endothelial growth factor and its receptors flt and KDR in ovarian carcinoma. J Natl Cancer Inst 87:506–516. [PubMed]

Breen E, 2008. Tang K, Olfert M, Knapp A, Wagner P. Skeletal muscle capillarity during hypoxia: VEGF and its activation. High Alt Med Biol. 9(2):158–166. doi: https://doi.org/10.1089/ham.2008.1010, PMID 18578647

Caboot JB, Allen JL (2014) Hypoxemia in sickle cell disease: significance and management. Paediatr Respir Rev 15(1):17–23. https://doi.org/10.1016/j.prrv.2013.12.004. PMID 24461342

Chintalgattu V, Nair DM, Katwa LC (2003) Cardiac myofibroblasts: a novel source of vascular endothelial growth factor (VEGF) and its receptors Flt-1 and KDR. J Mol Cell Cardiol 35(3): 277–286. https://doi.org/10.1016/s0022-2828(03)00006-3. PMID 12676542

Deveci D, Marshall JM, Egginton S (2001) Relationship between capillary angiogenesis, fiber type, and fiber size in chronic systemic hypoxia. Am J Physiol Heart Circ Physiol 281(1):H241–H252. https://doi.org/10.1152/ajpheart.2001.281.1.H241. PMID 11406491

Duffy AM, Bouchier-Hayes DJ, Harmey JH (2004) Vascular endothelial growth factor (VEGF) and its role in non-endothelial cells: autocrine signalling by VEGF. VEGF Cancer 1:133–144

Ferrara N, Houck K, Jakeman L et al (1992) Molecular and biological properties of the vascular endothelial growth factor family of proteins. Endocr Rev. 13:18

Ferrara N, Carver-Moore K, Chen H, Dowd M, Lu L, O'Shea KS et al (1996) Heterozygous embryonic lethality induced by targeted inactivation of the VEGF gene. Nature 380(6573): 439–442. https://doi.org/10.1038/380439a0. PMID 8602242

Ferrara N, Chen H, Davis-Smyth T, Gerber H-P, Nguyen T-N, Peers D, Chisholm V, Hillan KJ, Schwall RH (1998) Vascular endothelial growth factor is essential for corpus luteum angiogenesis. Nat Med 4(3):336–340. https://doi.org/10.1038/nm0398-336

Frank S, Hübner G, Breier G, Longaker MT, Greenhalgh DG, Werner S (1995) Regulation of vascular endothelial growth factor expression in cultured keratinocytes. Implications for normal and impaired wound healing. J Biol Chem 270(21):12607–12613. https://doi.org/10.1074/jbc.270.21.12607. PMID 7759509

Gerber HP, McMurtrey A, Kowalski J, Yan M, Keyt BA, Dixit V et al (1998) Vascular endothelial growth factor regulates endothelial cell survival through the phosphatidylinositol 3′-kinase/Akt signal transduction pathway: requirement for Flk-1/KDR activation. J Biol Chem 273(46): 30336–30343. https://doi.org/10.1074/jbc.273.46.30336. PMID 9804796

Gerber HP, Vu TH, Ryan AM, Kowalski J, Werb Z, Ferrara N (1999) VEGF couples hypertrophic cartilage remodeling, ossification and angiogenesis during endochondral bone formation. Nat Med 5(6):623–628. https://doi.org/10.1038/9467. PMID 10371499

Gerhardt H, Golding M, Fruttiger M, Ruhrberg C, Lundkvist A, Abramsson A et al (2003) VEGF guides angiogenic sprouting utilizing endothelial tip cell filopodia. J Cell Biol 161 (6):1163–1177. https://doi.org/10.1083/jcb.200302047. (PMC Free article). PMID 12810700 [CrossRef], Google Scholar

Green HJ, Sutton JR, Cymerman A, Young PM, Houston CS (1989) Operation EVEREST II: adaptations in human skeletal muscle. J Appl Physiol (1985) 66(5):2454–2461. https://doi.org/10.1152/jappl.1989.66.5.2454. PMID 2745306

Gürkan E, Tanriverdi K, Başlamişli F (2005) Clinical relevance of vascular endothelial growth factor levels in sickle cell disease. Ann Hematol 84(2):71–75. https://doi.org/10.1007/s00277-004-0935-y. PMID 15322765

Hackett PH (1999) High altitude cerebral Edema and acute mountain sickness. A pathophysiology update. In: Roach RC, Wagner PD, Hackett PH (eds) Hypoxia. Advances in experimental medicine and biology, vol 474. Springer, Boston, pp 23–45. https://doi.org/10.1007/978-1-4615-4711-2_2. PMID 10634991

Haiko P, Makinen T, Keskitalo S, Taipale J, Karkkainen MJ, Baldwin ME et al (2008) Deletion of vascular endothelial growth factor C (VEGF-C) and VEGF-D is not equivalent to VEGF receptor 3 deletion in mouse embyos. Mol Cell Biol 28(15):4843–4850. https://doi.org/10.1128/MCB.02214-07. PMID 18519586, PMCID PMC2493372

Harvey TC, James HM, Chettle DR, Birmingham Medical Research Expeditionary Society (1979) Birmingham Medical Research Expeditionary Society 1977 Expedition: effect of a Himalayan trek on whole body composition, nitrogen and potassium. Postgrad Med J 55(645):475–477. https://doi.org/10.1136/pgmj.55.645.475. PMID 493200

Hoppeler H, Howald H, Cerretelli P (1990a) Human muscle structure after exposure to extreme altitude. Experientia 46(11–12):1185–1187. https://doi.org/10.1007/BF01936933. PMID 2253 720

Hoppeler H, Kleinert E, Schlegel C, Claassen H, Howald H, Kayar SR et al (1990b) Morphological adaptations of human skeletal muscle to chronic hypoxia. Int J Sports Med 11(Suppl 1):S3–S9. https://doi.org/10.1055/s-2007-1024846. PMID 2323861

Iijima K, Yoshikawa N, Connolly DT, Nakamura H (1993) Human mesangial cells and peripheral blood mononuclear cells produce vascular permeability factor. Kidney Int 44(5):959–966. https://doi.org/10.1038/ki.1993.337. PMID 8264155

Itakura J, Ishiwata T, Shen B, Kornmann M, Korc M (2000) Concomitant over-expression of vascular endothelial growth factor and its receptors in pancreatic cancer. Int J Cancer 85(1): 27–34. https://doi.org/10.1002/(sici)1097-0215(20000101)85:1<27::aid-ijc5>3.0.co;2-8. PMID 10585578

Karkkainen MJ, Haiko P, Sainio K, Partanen J, Taipale J, Petrova TV et al (2004) Vascular endothelial growth factor C is required for sprouting of the first lymphatic vessels from embryonic veins. Nat Immunol 5(1):74–80. https://doi.org/10.1038/ni1013. PMID 14634646

Li X, Du E, Dao M, Suresh S, Karniadakis GE (2017) Patient-specific modeling of individual sickle cell behavior under transient hypoxia. PLoS Comput Biol 13(3):e1005426. https://doi.org/10.1371/journal.pcbi.1005426. PMID 28288152, PMCID PMC5367819

Li H, Li A, Gao R, Zhou H, Si P (2022) An efficient breathing zone with targeted oxygenation for improving the sleep environment in hypoxic areas. Build Environ 222:109371; ISSN 0360-1323. https://doi.org/10.1016/j.buildenv.2022.109371

Luks AM, Swenson ER, Bärtsch P (2017) Acute high-altitude sickness. Eur Respir Rev 26(143): 160096. https://doi.org/10.1183/16000617.0096-2016. PMID 28143879

MacDougall JD, Green HJ, Sutton JR, Coates G, Cymerman A, Young P et al (1991) Operation EVEREST II: structural adaptations in skeletal muscle in response to extreme simulated altitude. Acta Physiol Scand 142(3):421–427. https://doi.org/10.1111/j.1748-1716.1991.tb09176.x. PMID 1927554

Machogu EM, Machado RF (2018) How I treat hypoxia in adults with hemoglobinopathies and hemolytic disorders. Blood 132(17):1770–1780. https://doi.org/10.1182/blood-2018-03-818195. PMID 30206115

MacIntyre NR (2014) Tissue hypoxia: implications for the respiratory clinician. Respir Care 59(10): 1590–1596. https://doi.org/10.4187/respcare.03357. PMID 25161296

Mehta SR, Chawla A, Kashyap AS (2008) Acute mountain sickness, high altitude cerebral oedema, high altitude pulmonary oedema: the current concepts. Med J Armed Forces India 64(2): 149–153. https://doi.org/10.1016/S0377-1237(08)80062-7. PMID 27408119, PMCID PMC4921562

Miatmoko A, Hariawan BS, Cahyani DM, Dewangga SS, Handoko KK, Purwati et al (2023) Prospective use of amniotic mesenchymal stem cell metabolite products for tissue regeneration. J Biol Eng 17(1):11. https://doi.org/10.1186/s13036-023-00331-1. PMID 36759827

Milkiewicz M, Hudlicka O, Shiner R, Egginton S, Brown MD (2006) Vascular endothelial growth factor mRNA and protein do not change in parallel during non-inflammatory skeletal muscle ischaemia in rat. J Physiol 577(2):671–678. https://doi.org/10.1113/jphysiol.2006.113357. PMID 16990404

Olfert IM, Breen EC, Mathieu-Costello O, Wagner PD (2001) Skeletal muscle capillarity and angiogenic mRNA levels after exercise training in normoxia and chronic hypoxia. J Appl Physiol 91(3):1176–1184. https://doi.org/10.1152/jappl.2001.91.3.1176. PMID 11509513

Park JE, Keller GA, Ferrara N (1993) The vascular endothelial growth factor (VEGF) isoforms: differential deposition into the subepithelial extracellular matrix and bioactivity of extracellular matrix-bound VEGF. Mol Biol Cell 4(12):1317–1326. https://doi.org/10.1091/mbc.4.12.1317. PMID 8167412, PMCID PMC275767

Pedrosa AM, Lemes RPG (2020) Gene expression of HIF-1α and VEGF in response to hypoxia in sickle cell anaemia: influence of hydroxycarbamide. Br J Haematol 190(1):e39–e42. https://doi.org/10.1111/bjh.16693. PMID 32352161

Podar K, Anderson KC (2005) The pathophysiologic role of VEGF in hematologic malignancies: therapeutic implications. Blood 105(4):1383–1395. https://doi.org/10.1182/blood-2004-07-2909. PMID 15471951

Rattner A, Williams J, Nathans J (2019) Roles of HIFs and VEGF in angiogenesis in the retina and brain. J Clin Invest 129(9):3807–3820. https://doi.org/10.1172/JCI126655. PMID 31403471

Reichardt LF, Tomaselli KJ (1991) Extracellular matrix molecules and their receptors: functions in neural development. Annu Rev Neurosci 14:531–570. https://doi.org/10.1146/annurev.ne.14.030191.002531. (PMC Free article). PMID 1851608

Rose MS, Houston CS, Fulco CS, Coates G, Sutton JR, Cymerman A (1988) Operation Everest. II: nutrition and body composition. J Appl Physiol 65(6):2545–2551. https://doi.org/10.1152/jappl.1988.65.6.2545. PMID 3215854

Shibuya M (2013) Vascular endothelial growth factor and its receptor system: physiological functions in angiogenesis and pathological roles in various diseases. J Biochem 153(1, Jan):13–19. https://doi.org/10.1093/jb/mvs136. PMID 23172303

Stone J, Itin A, Alon T, Pe'er J, Gnessin H, Chan-Ling T et al (1995) Development of retinal vasculature is mediated by hypoxia-induced vascular endothelial growth factor (VEGF) expression by neuroglia. J Neurosci 15(7 pt 1):4738–4747. https://doi.org/10.1523/JNEUROSCI.15-07-04738.1995. (PMC Free article), PMID 7623107

Sunderkötter C, Steinbrink K, Goebeler M, Bhardwaj R, Sorg C (1994) Macrophages and angiogenesis. J Leukoc Biol 55(3):410–422. https://doi.org/10.1002/jlb.55.3.410. PMID 7509844

Takahashi S (2011) Vascular endothelial growth factor (VEGF), VEGF receptors and their inhibitors for antiangiogenic tumor therapy. Biol Pharm Bull 34(12):1785–1788. https://doi.org/10.1248/bpb.34.1785. PMID 22130231

Takahashi H, Shibuya M (2005) The vascular endothelial growth factor (VEGF)/VEGF receptor system and its role under physiological and pathological conditions. Clin Sci (Lond) 109(3):227–241. https://doi.org/10.1042/CS20040370. PMID 16104843

Tang K, Breen EC, Gerber HP, Ferrara NM, Wagner PD (2004) Capillary regression in vascular endothelial growth factor-deficient skeletal muscle. Physiol Genomics 18(1):63–69. https://doi.org/10.1152/physiolgenomics.00023.2004. PMID 15084712

Thapa SS, Basnyat B (2018) Acute mountain sickness (AMS) in a Nepali pilgrim after rapid ascent to a sacred lake (4380 m) in the Himalayas. BMJ Case Rep 11(1):bcr2017222888. https://doi.org/10.1136/bcr-2017-222888. PMID 30567154, PMCID PMC6301503

Usui Y, Westenskow PD, Kurihara T, Aguilar E, Sakimoto S, Paris LP et al (2015) Neurovascular crosstalk between interneurons capillaries is required for vision. J Clin Invest 125(6):2335–2346. https://doi.org/10.1172/JCI80297. (PMC Free article). PMID 25915585 [CrossRef], Google Scholar

Verheul HM, Hoekman K, Luykx-de Bakker S, Eekman CA, Folman CC, Broxterman HJ et al (1997) Platelet: transporter of vascular endothelial growth factor. Clin Cancer Res 3(12 Pt 1):2187–2190. PMID 9815613

West H, Richardson WD, Fruttiger M (2005) Stabilization of the retinal vascular network by reciprocal feedback between blood vessels and astrocytes. Development 132(8):1855–1862. https://doi.org/10.1242/dev.01732. PMID 15790963 [CrossRef], Google Scholar

Zarkada G, Howard JP, Xiao X, Park H, Bizou M, Leclerc S et al (2021) Specialized endothelial tip cells guide neuroretina vascularization and blood-retina-barrier formation. Dev Cell 56(15):2237–2251.e6. https://doi.org/10.1016/j.devcel.2021.06.021. PMID 34273276, PMCID PMC9951594

Zubieta-Calleja GR, Zubieta-DeUrioste N (2022) High Altitude Pulmonary Edema, High Altitude Cerebral Edema, and acute mountain sickness: an enhanced opinion from the High Andes – la Paz, Bolivia 3,500 m. Rev Environ Health. https://doi.org/10.1515/reveh-2021-0172. PMID 3 5487499

Thrombogenetic Risk at High Altitude: Early Diagnosis and Treatment Modalities

6

Swati Srivastava and Iti Garg

Abstract

Venous thrombosis (VT) is a complex preventable medical condition. Its manifestation depends on several acquired and inherited genetic risk factors. Venous thromboembolism (VTE) is clinically represented by deep vein thrombosis (DVT) and life-threatening pulmonary embolism (PE). Although VTE can affect both men and women of all ages and ethnicities, those with advanced age, recent surgery, prolonged immobilization, malignancy, injury, or trauma are at higher risk. Also, individuals with mutations in genetic factors leading to increased production of procoagulants or decreased synthesis of natural anticoagulants are likely to develop VT. Also, exposure to hypoxic conditions at high altitudes (HA) is regarded as an independent risk factor for developing VT. Environmental challenges at HA such as hypoxia, dehydration, hemoconcentration, reduced mobility, and use of constricted clothing contribute to a prothrombotic milieu. Moreover, HA hypoxia triggers platelet adhesion, endothelial activation, and inflammation, thus creating an environment more conducive to VTE occurrence. Early and accurate diagnosis of DVT is extremely important to prevent PE events. Several treatment options are available for VTE; however, the dose and duration of anticoagulants prescribed depend on clinical condition and risk of recurrence. This chapter consolidates the existing knowledge on venous thromboembolism with an emphasis on high altitude-induced VTE.

Keywords

Venous thrombosis · Thromboembolism · DVT · High altitude · VTE

S. Srivastava (✉) · I. Garg
Division of Pathophysiology and Disruptive Technology, Defence Institute of Physiology and Allied Sciences (DIPAS), Defence Research and Development Organization (DRDO), Delhi, India
e-mail: sri_swati@rediffmail.com

R. Tulsawani, D. Vohora (eds.), *Adaptation under Stressful Environments through Biological Adjustments and Interventions*,
https://doi.org/10.1007/978-981-99-7652-2_6

6.1 Introduction

The term 'thrombosis' refers to a medical condition, wherein a blood clot or 'thrombus' causes restriction of blood flow inside a vessel. Timely treatment of these blood clots is essential to avoid serious complications. An 'embolus' refers to a piece of thrombus that may dislodge itself from the site of formation and travel to other parts of the body. Thrombosis is categorized into arterial and venous thrombosis, based on whether the thrombus develops into arteries and veins. Depending on the location and size of the thrombus, thrombosis could be minor or severe. For instance, if the thrombus develops in the blood vessel that supplies blood to the brain or heart, it may result in a stroke or heart attack, respectively. Thrombus in arteries (arterial thrombosis) typically comprises atherosclerotic plaques that rupture in left heart chambers or arteries leading to Acute coronary syndrome or Ischemic stroke. This type of thrombus is made up of mainly platelets and thus can be treated with antiplatelet drugs.

On the other hand, the formation of thrombus in the veins is relatively sudden, mainly occurring in venous sinusoids of muscles and valves in veins. Venous thromboembolism (VTE) is considered the third most common vascular disease after myocardial infarction (MI) and stroke. The two most common clinical implications of VTE include deep vein thrombosis (DVT) and pulmonary embolism (PE). PE is usually followed by DVT and it is a potentially fatal condition. Venous thrombosis of the lower limb may involve superficial leg veins, deep calf veins, or more proximal veins. The occurrence of DVT restricts oxygen supply to the affected tissue. Venous thrombi are formed by intravascular deposition of mainly red blood cells and fibrin along with variable components of platelets and leukocytes (Freiman 1987). The occurrence of VTE is attributed to at least one of the three underlying risk factors; vascular endothelial damage, stasis of blood flow, and hypercoagulability of blood as proposed by Rudolph Virchow in 1884 (Kumar et al. 2010). Up to one-third of DVT cases lead to its dreaded complication, PE, which is the prime cause of mortality (Kearon 2003). One percent to eight percent of PE patients die despite adequate therapy (Hirsh and Bates 1999; Fedullo et al. 2001).

A complex interaction of multiple pro-thrombotic and antithrombotic components exists in the body to maintain normal blood flow. Any external or internal factor that disturbs this balance may lead to blood hemostasis. VTE is a complex multifactorial disease, and its etiology is dependent upon several genetic risk factors along with age, comorbidity, and lifestyle habits. Besides several inherited and acquired risk factors contributing to VTE pathogenesis, exposure to extreme conditions of high altitude (HA) could also provoke thrombotic complications. Hypoxic conditions of HA trigger the formation and propagation of blood clots in the vasculature leading to thrombosis (Jha et al. 2002; Kumar 2006). HA altitude ascent should be planned with prominent checkpoints such as prior exposure to HA, preexisting medical conditions, recent surgery/trauma, family or self-history of blood clots, etc.

This chapter reviews the pathophysiology of venous thrombosis, its risk factors, symptoms, diagnosis, and treatment, with special emphasis on high altitude environment being an independent triggering factor for the development of thrombus. This compilation of existing knowledge on venous thrombosis increases scientific awareness for researchers working in related areas and even the general population as a whole.

6.1.1 Epidemiology of Venous Thromboembolism (VTE)

Venous thromboembolism has been categorized as one of the major health concerns, especially in Western countries. The annual incidence rate of overall VTE ranges from 104 to 183 cases per lac per year in the European population (Anderson Jr et al. 1991; Silverstein et al. 1998; Oger 2000; Spencer et al. 2006; Tagalakis et al. 2013). Incidences of VTE increase with age, both for men and women (Silverstein et al. 1998; Naess et al. 2007). The age-adjusted overall annual incidence of the rate of VTE is 130 for men and 110 for women per 100,000 (Silverstein et al. 1998; Naess et al. 2007). However, incidence rates in women become higher during childbearing age compared to men of similar age groups.

Several studies have shown ethnic differences in the occurrence of VTE. Incidences of VTE are reportedly higher in African-Americans (White et al. 1998; Zakai et al. 2014) compared to Asians (Cheuk et al. 2004) and Native Americans (Hooper et al. 2002).Lower incidence rate is reported in Asian than in European populations (Lee et al. 2011). Incidence rate of VTE is 16.5 per 100,000 per year in Taiwan, which is ten-folds lower than reported from Europe and the USA (Lee et al. 2011). Annual incidences of VTE are estimated to range from 104 to 183 per 100,000 people in European population (Spencer et al. 2006; Tagalakis et al. 2013; Huang et al. 2014). Higher percentage of idiopathic VTE events have been reported in European and African populations (25% to 40%), compared to those from Asia and the Pacific Islands (19%) (Heit et al. 2002; Hooper et al. 2002; White et al. 2005).

Definite epidemiological data for VTE occurrence rates at HA are lacking. However, in a recent follow-up study between January 2013 and June 2021, medical records of 1,112,014 patients admitted to the hospital were screened. Out of these, HA-VTE events occurred in 1.2% (13,843) patients, which increased over time and were found to be associated with greater adverse outcomes (Neeman et al. 2022). Another population-based study was conducted between the years 2002 and 2009 in over 100 public and private hospitals on 3,331,677 patients. Researchers found that admitted patients showed incidence rates of HA-VTE 11.45 per 1000 patients; however, this incidence rate varied among medical and surgical patients, public and private hospitals, and different hospital groups (Assareh et al. 2016). These studies provided evidence that incidence rates of high altitude-induced VTE could be managed with targeted interventions and improved facilities.

6.1.2 Risk Factors

VTE can affect both men and women of all ages, ethnicities, and races. However, there are several acquired and inherited risk factors for venous thrombosis that contribute to the development of thrombus by enhancing blood coagulability (Fig. 6.1) (as reviewed by McLendon et al. 2022). It has been reported that more than 50 percent of the individuals who suffer from DVT have at least one or more acquired risk factors (Spencer et al. 2006). One of the strongest risk factors for VTE is advancing age. The risk of developing VTE in older age is several folds higher than in younger age (Oger 2000; Rosendaal 1997a, b). This could be due to decreased mobility and muscular tone. Classical risk factors of VTE such as prolonged immobilization, fractures, use of oral contraceptives, puerperium, paralysis, and the antiphospholipid antibody syndrome, not only predispose patients to thrombosis, but also trigger VTE in people with inherited thrombophilic abnormalities (Rosendaal 1997a, b; Anderson and Spencer 2003). The combination of these acquired risk factors with one or more genetic defects in the coagulation/anticoagulation system results in an increased risk of VTE, wherein the risk exceeds the sum of the separate effects of the single factors (Rosendaal 1997a, b).

6.1.2.1 Acquired

Venous thrombosis is a multifactorial disease. A single factor is seldom enough for the development of VTE. An inactive lifestyle, obesity, and diabetes are associated with increased coagulation and decreased fibrinolytic activity and thus contribute to an increased risk of venous thrombosis (MacCallum and Meade 1999). Several other established risk factors for VTE include trauma, surgery, and cancer (Goldhaber 1998). All major surgeries including orthopedic and neurovascular surgeries are

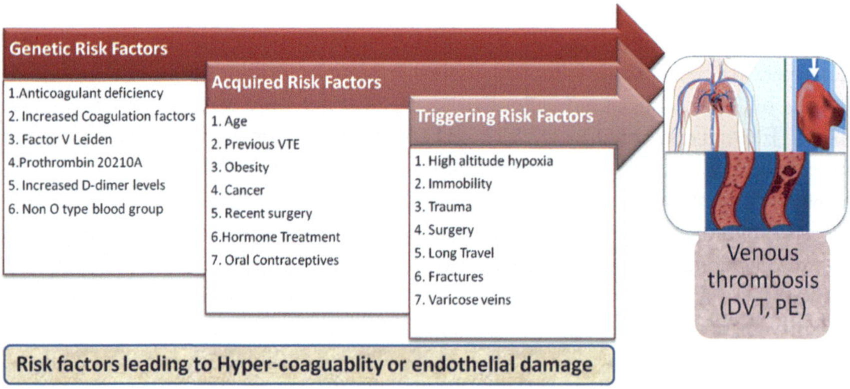

Fig. 6.1 Risk factors associated with the pathophysiology of venous thromboembolism

associated with a higher risk of DVT, especially in individuals with advancing age and prior history of DVT. During hip and knee replacement surgery, the risk of thrombosis increases up to 30–50% (Cohen et al. 1973; Hull and Raskob 1986). Significant risk of DVT is conferred due to major or minor trauma (McLaughlin et al. 2009).

As stasis is one of the major causes of VTE, immobilization of the extremities in conditions like paralysis, bed rest, plaster casts, and prolonged travel is associated with an increased risk of thrombosis (Homans 1954; Gibbs 1957; Warlow et al. 1976). Immobilization may also be associated with prolonged air or ground travel, which increases DVT risk by 2–4 folds (Lapostolle et al. 2001). Obesity is also one of the major risk factors for the development of thrombus. Increased levels of coagulation factor VIII and IX in obese individuals could be one of the reasons; however, even after adjustment of these clotting factors, the risk of VTE remains (Abdollahi et al. 2003).

Another risk factor for having thrombogenic potential is malignancy. The presence of tumors in less differentiated cell lines has higher tendencies of hypercoagulation, and sometimes malignancy associated VTEs may precede the diagnosis of malignancy (Sørensen et al. 1998). In a cancer patient, the use of chemotherapy agents during treatment and surgical procedures as well as myeloma treatment drugs also increases DVT risk (Al-Asadi et al. 2019).

Some of the risk factors of VTE in women include the use of oral contraceptives (Sidney et al. 2004; Rosendaal 1999), pregnancy (Lim et al. 2007), and hormone replacement therapy.Increased risk of VTE remains during all trimesters of pregnancy and even during the postpartum period.

6.1.2.2 Inherited/Genetic Risk Factors

Genetic risk factors play a key role in predisposition to thrombophilia. Several studies have been conducted to establish that venous thrombosis is a consequence of genetic and environmental risk factors. The presence of genetic risk factors when combined with acquired risk factors such as advancing age, obesity, surgery, trauma, cancer, immobilization, pregnancy, and use of exogenous hormones enhances the risk of VTE several times (Cushman 2005). Idiopathic VTE (VTE in the absence of environmental factors) may be divided into "loss-of-coagulation function" and "gain-of-coagulation function" disorders (Crowther and Kelton 2003). Loss of coagulation function disorders causes deficiencies in natural endogenous anticoagulants such as protein C (PC), protein S (PS), and antithrombin (ATIII). These are potent risk factors for VTE, although less commonly found (Crowther and Kelton 2003). On the other hand, a gain of coagulation function disorders includes factor V Leiden, prothrombin 20210A variant, and increased production of other pro-coagulants such as factor VIII, V, VII, IX, XI, and von Willebrand factor (Crowther and Kelton 2003).

Inherited thrombophilia has been associated with ATIII deficiency to a greater extent than compared deficiency of other natural anticoagulants like protein C and protein S. ATIII deficiency occurs due to mutation in the SERPINC1 gene. In one of the study reports, 85% of patients with antithrombin deficiency suffered from VTE by 50 years of age (Thaler and Lechner 1981).Though ATIII deficiency is rare in the general population, it accounts for the relative risk of VTE around 10 (Bucciarelli et al. 2012).

Protein C is another anticoagulant that plays a crucial role in inhibiting thrombus formation. The inactive form of PC circulating in plasma gets activated at the surface of endothelium by the membrane-bound thrombin–thrombomodulin complex. Upon activation, it binds to its cofactor PS and inactivates coagulation factors FVa and FVIIIa, thus limiting the coagulation cascade (Dahlbäck 2008). Low levels of PC in blood and mutations in the PC gene have been widely associated with VT risk (Folsom et al. 2002; Reiner et al. 2008).

Two mutations namely Factor V Leiden and prothrombin G20210A gene mutations are most commonly associated with VTE pathology causing increased morbidity and mortality; however, their prevalence is distinctive worldwide (White 2003; Frutos 2007; Naess et al. 2007). These mutations are mostly found in Caucasians and Middle Eastern populations; however, their frequency of occurrence is moderate to rare among other populations including Africans. The presence of Factor V Leiden mutations renders Factor Va resistant to activated protein C (APC) cleavage resulting in increased production of thrombin and higher levels of prothrombin fragment 1 + 2 (Simioni et al. 1996). This particular mutation confers high risk of developing VTE (5 to 10 folds higher) in heterozygous individuals and a 50- to 100-fold higher risk in homozygotes, compared to individuals who do not have this mutation (Dahlbäck 1995). Besides this, second most common genetic mutation associated with VTE susceptibility is substitution of G to A at nucleotide at position 20,210 in prothrombin gene. This single nucleotide polymorphism change results in increased prothrombin production and confers 3-fold greater risk of venous thrombosis (Poort et al. 1996; EGAPP Working Group 2011).

Elevation of D-dimer levels in a healthy individual is considered to be an independent risk factor for VTE (Cushman et al. 2003). Elevated D-dimer levels may also be co-regulated with other coagulation factors resulting in thrombosis (Van Hylckama et al. 2003). The relative risk of venous thrombosis in individuals with higher D-dimer levels is increased up to 2.5 to 3-fold and even more in the case of idiopathic thrombosis (Cushman et al. 2003).

Besides these, non-O-type blood groups have been associated with increased incidences of VTE. It has been shown that A and B-type of blood groups are at a higher risk of thrombosis (Wu et al. 2008, Trégouët et al. 2009).

6.1.3 Environmental Conditions at High Altitudes (HA)

High altitude (HA) is defined as heights 9000 ft. (2700 m) above from sea level. Heights above 18,000–19,000 ft. (5500–5800 m) are defined as extreme altitude. To overcome harsh climatic conditions at these altitudes such as hypoxia (oxygen deprivation), low humidity, solar and ultraviolet radiation, etc., individual's body needs to undergo several physiological changes during the process of acclimatization. Failure during the acclimatization may lead to development of various high altitude maladies.

Development of venous thrombosis has been widely studied in relation to exposure of an individual to high altitude conditions. Hypoxic condition prevailing at high altitude pertubates molecules associated with vascular integrity and thus alters hemostatic system (Prabhakar et al. 2019). Higher incidences of thrombosis at HA may be attributed to stasis of blood flow, increased viscosity, and a procoagulant state with diminished fibrinolysis (Singh and Chohan 1972; Mannucci et al. 2002; Tyagi et al. 2014). There are few clinical case studies available that reported death of trekkers in Himalyan region due to VTE (Dickinson et al. 1983; Schreijer et al. 2005; Cheng et al. 2009; Hull et al. 2016). Naik and coworkers found higher incidences of chronic venous thrombosis in combatants who remained at high altitude for a longer duration (Naik et al. 2021).

A recent meta-analysis study summarized that overall incidences of VTE (DVT and PE combined) are significantly higher at high altitude compared to low altitudes. Also, VTE incidences are significantly higher in 30 days and 90 days postoperative patients group at HA than low altitudes (Yang et al. 2022). In a recent report by Nair and coworkers, higher incidences of thrombosis at HA have been attributed to inflammation, endothelial dysfunction, a prothrombotic state, and dampened fibrinolysis (Nair et al. 2022). Risk of developing a rare cerebrovascular disorder, cerebral venous thrombosis (CVT), is much higher at high altitude areas compared to low altitudes (Hassan et al. 2019; Khattar et al. 2019).

6.2 Risk of Recurrent Thrombosis

Recurrent VTE incidences increase risk of morbidity and mortality. Absolute risk of VTE increases with advancing age. The risk of VTE recurrence per year is about 5–7%; however, the risk increases to 50 times more in patients with previous history of thrombosis (White et al. 2005). Anticoagulants continued after first incidence of VTE reduces the chances of recurrent VTE; however, its dose and duration should be carefully monitored to avoid anticoagulant-related bleeding. Thus, it is very important to assess the risk of recurrent VTE to decide optimal duration of treatment (Fahrni et al. 2015). VTE treatment is divided into two important phases, i.e., acute phase treatment and secondary prophylaxis. During acute phase, a minimal duration of anticoagulation is essential to dissolve thrombus and prevent embolism. This continues till the time when activated inflammatory and coagulation processes return to normal baseline levels (van Aken et al. 2000).

Continuation of anticoagulation treatment after three to 6 months of thrombotic episode reduces chances of recurrent VTE (Castellucci et al. 2013). It has been reported that, after when anticoagulation treatment is discontinued, patients with a first unprovoked VTE possess much higher risk of VTE recurrence, in comparison to patients with provoked VTE (Iorio et al. 2010; Boutitie et al. 2011). In a recent meta-analysis study involving 18 studies involving 7515 patients, researchers found that in patients who complete 3 months of anticoagulation treatment after first episode of unprovoked VTE, the risk of VTE recurrence was 10% after first year, 16% after 2 years, 25%after 5 years, and 36% after 10 years of VTE episode. They further reported that 4% of recurrent VTE events resulted in death of patients (Khan et al. 2019). Anticoagulation therapy in patients with unprovoked proximal VTE and PE may continue for indefinite period in patients with non-high bleeding risk (Kearon et al. 2016; Mazzolai et al. 2018).

6.3 Early Symptoms and Diagnosis of VTE

Clumping of blood cells together in deep veins leads to DVT and this usually occurs in the leg veins. However, the clot may dislodge itself and travel to lungs through the blood stream leading to deadly condition of PE. Thus, recognizing early symptoms of DVT is extremely essential to avoid life-threatening complications. Symptoms of DVT include swelling, pain and tenderness in legs, and increased warmth in the swollen area. The affected area often has red or discolored skin. Clinical presentation of DVT depends on the extent and location of thrombus. The symptoms of pulmonary embolism include chest pain, irregular or rapid heart beat, difficulty in breathing, and coughing. There are other clinical complications whose symptoms resemble those of VTE, such as muscle injury, cellulitis (a bacterial skin infection), and inflammation (swelling) of veins, etc. Also, symptoms of pulmonary embolism are similar to that of heart attack and pneumonia. Thus, diagnosis of VTE needs to be done by special imaging techniques which can observe blood clots in veins and lungs.

Scoring systems have been designed to estimate the pretest probability of DVT. One of the most popular scoring systems to assess DVT is Wells Scoring Criteria (Wells et al. 1997). This scoring system defines three categories of patients with high, intermediate, or low risk depending on presence or absence of 9 clinical criteria. Based on this score, Wells and coworkers estimated DVT prevalence to be 5% in low risk and 53% in high risk groups (Wells et al. 2006). Few years later, the original scoring system was revised to include "previously documented DVT" criteria and risk categories were defined as "unlikely" or "likely" (Wells et al. 2003). Sensitivity and specificity of Wells scoring criteria to evaluate DVT risk were estimated by the National Institute for Health and Care Excellence (NICE) guidelines in 2012 and were found to be 77–98% and 38–58%, respectively (Langford et al. 2012).

Most commonly used clinical biomarker for DVT testing is plasma levels of D-dimer, which is degradation product of a cross-linked fibrin blood clot. Levels of D-dimer elevate significantly during venous thromboembolism. However, D-dimer levels are also elevated in other conditions such as major surgery, hemorrhage, trauma, pregnancy, or cancer (Kelly et al. 2002). In addition to these, higher levels of D-dimer are also present during postoperative period and after trauma. As estimated by the NICE guidelines, D-dimer assay has a high sensitivity (75–100%) and relatively lower specificity (26–83%) (Langford et al. 2012). Thus,

D-dimer as a clinical biomarker for VTE is sensitive to the event but nonspecific in nature. Negative test result of D-dimer elevation rules out likelihood of DVT.

To confirm the presence of DVT, diagnostic imaging test is essential. Ultrasound imaging determines size, chronicity, and degree of occlusion of a thrombus in proximal DVT. This method is safe, reliable, and cost-effective (Kearon et al. 1998; Min et al. 2016). Inability to compress the vein during an ultrasound test confirms DVT diagnosis; however, real-time imaging such as duplex and color-flow Doppler can be used to characterize the blood clot further. Conventional contrast venography, computed tomography (CT) venography, and magnetic resonance (MR) venography are among the other diagnostic imaging modalities. One of the most popular imaging tests to diagnose DVT is Compression ultrasonography (Scarvelis and Wells 2006).

6.4 Treatment and Management of VTE

Early treatment of deep vein thrombosis is essential to prevent the risk of pulmonary embolism. Treatment strategies primarily focus on prevention of thrombus extension and development of PE and also to prevent long-term recurrent events. Most common and effective treatment strategy for VTE is with anticoagulants (Fig. 6.2). VTE therapy usually starts with therapeutic doses of unfractionated heparin or LMW heparin. This is a preferred anticoagulant treatment due to its ease of administration and efficacy. The time and duration of anticoagulation treatment have to be decided with utmost care. Recurrent episodes of VTE are so common that guidelines have

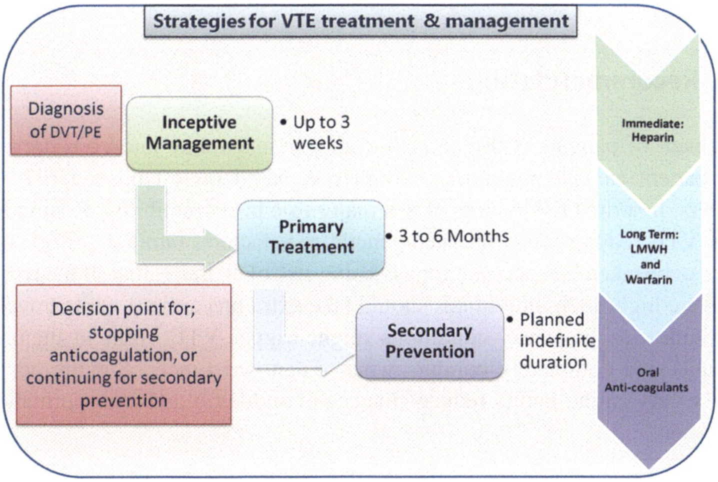

Fig. 6.2 Treatment and management plan for venous thromboembolism

been established for the duration of anticoagulation therapy. On the other hand, people taking anticoagulation treatment have an increased risk of bleeding or hemorrhage. Treatment may itself have some side effects such as severe bruising, bleeding gums or nose bleeds, vomiting, or excessive menstrual bleeding.

Therapy usually starts with unfractionated heparin or low molecular weight (LMW) heparin (Brandjes et al. 1992). Among the two, LMW heparin is preferred over unfractionated heparin for its ease of administration and efficacy resulting in lesser number of deaths and recurrent thrombosis cases (Van Dongen et al. 2004). In general, VTE treatment for initial 1 week is done with weight-adjusted LMW heparin dose, once or twice daily. LMW heparin is also considered as a safe and easily accessible treatment option for CVT at high altitude (Khattar et al. 2019). An alternative agent for treatment of DVT is Fondaparinux, a synthetic pentasaccharide, which is as effective and safe as LMW heparin (Matisse Investigators 2004).

Effective treatment of DVT and long-term prevention of its recurrence includes oral anticoagulants such as vitamin K antagonists (e.g., warfarin) (Hull et al. 1982). Direct oral anticoagulants (s) (DOACs) are good alternative treatment options which are very effective and safe. DOACs are broadly classified into two types (i) factor Xa inhibitors which include rivaroxaban, apixaban, edoxaban, and betrixaban and (ii) direct thrombin inhibitors like dabigatran. This new class of therapeutic agents is at par or even better than vitamin K antagonists such as warfarin or LMWHs in reducing thrombotic complications and risk of excessive bleeding (Ruff et al. 2014; van der Hulle et al. 2014; Kapoor et al. 2017). Efficacy and safety of DOACs give an added advantage of their use as they require less monitoring and follow-up of patients and fewer drug-food interactions (Rose and Bar 2018). Clinicians should carefully examine patients' comorbidities, prior medication, and history of thrombus formation before prescribing the new oral anticoagulant agents.

6.5 Recommendations

Management of patients with venous thrombosis must take into consideration the risk of patient as low, medium, or high risk based on established risk scoring parameters. Inherited risk factors play a major role in susceptibility of an individual towards VTE. Along with inherited/genetic risk factors, familial history of VTE should also be taken into account for assessing risk of an individual. If the risk scores for VTE are high, such individuals should take extra precautions while travelling to high altitudes, as 'hypoxic' conditions might trigger VTE event in them. VT is preventable and completely curable, when diagnosed timely. Staying active and having healthy eating habits reduce chances of sudden blood clots formation to a greater extent.

6.6 Summary

Venous thromboembolism is a fairly common and complex disease. It has high rate of recurrence. Early diagnosis and treatment of VTE reduces health care costs and also increases survival rates. Incidences of VTE including CVT are significantly higher at high altitudes. Current diagnostics for VTE in high altitude areas mostly includes D-dimer test, which is a sensitive and nonspecific marker. Other confirmatory tests such as venography and venous ultrasonography are not readily available in remote areas of HA. Data of VTE events at HA are very limited, though it has been a persistent problem at HA and is associated with adverse outcomes. Delay in diagnosis of VTE can lead to serious life-threatening complications. Treatment modalities include usage of anticoagulants such as unfractioned heparin or low molecular weight heparin followed by use of direct oral anticoagulants such as dabigatran, apixaban, rivaroxaban, and edoxaban. Though anticoagulant treatment strategy is effective in reducing thrombus, the dosage and duration of treatment vary case to case depending upon severity of the complication. More research and newer approaches are required to mitigate VTE occurrence at HA. Better VTE management techniques will reduce the economic burden associated with it.

Acknowledgments The authors are extremely thankful to Director DIPAS for his guidance and support in writing this manuscript.

References

Abdollahi M, Cushman M, Rosendaal FR (2003) Obesity: risk of venous thrombosis and the interaction with coagulation factor levels and oral contraceptive use. Thromb Haemost 89 (3):493–498. PMID: 12624633

Al-Asadi O, Almusarhed M, Eldeeb H (2019) Predictive risk factors of venous thromboembolism (VTE) associated with peripherally inserted central catheters (PICC) in ambulant solid cancer patients: retrospective single Centre cohort study. Thromb J 17:2. https://doi.org/10.1186/s12959-019-0191-y. PMID 30697126

Anderson FA Jr, Wheeler HB, Goldberg RJ, Hosmer DW, Patwardhan NA, Jovanovic B et al (1991) A population-based perspective of the hospital incidence and case-fatality rates of deep vein thrombosis and pulmonary embolism. The Worcester DVT study. Arch Intern Med 151(5): 933–938. https://doi.org/10.1001/archinte.1991.00400050081016. PMID 2025141

Anderson FA, Spencer FA (2003) Risk factors for venous thromboembolism. Circulation 107(23; Suppl 1):I9–I16. https://doi.org/10.1161/01.CIR.0000078469.07362.E6. PMID 12814980

Assareh H, Chen J, Ou L, Hillman K, Flabouris A (2016) Incidences and variations of hospital acquired venous thromboembolism in Australian hospitals: a population-based study. BMC Health Serv Res 16(1):511. https://doi.org/10.1186/s12913-016-1766-y,. PMID 27659903

Boutitie F, Pinede L, Schulman S, Agnelli G, Raskob G, Julian J et al (2011) Influence of preceding length of anticoagulant treatment and initial presentation of venous thromboembolism on risk of recurrence after stopping treatment: analysis of individual participants' data from seven trials. BMJ 342:d3036. https://doi.org/10.1136/bmj.d3036. PMID 21610040

Brandjes DPM, Heijboer H, Büller HR, de Rijk M, Jagt H, ten Cate JW (1992) Acenocoumarol and heparin compared with acenocoumarol alone in the initial treatment of proximal-vein thrombosis. N Engl J Med 327(21):1485–1489. https://doi.org/10.1056/NEJM199211193272103. PMID 1406880

Bucciarelli P, Passamonti SM, Biguzzi E, Gianniello F, Franchi F, Mannucci PM et al (2012) Low borderline plasma levels ofantithrombin, protein C and protein S are risk factors for venous thromboembolism. J Thromb Haemost 10(9):1783–1791. https://doi.org/10.1111/j.1538-7836.2012.04858.x. PMID 22812604

Castellucci LA, Cameron C, Le Gal G et al (2013) Efficacy and safety outcomes of oral anticoagulants and anti-platelet drugs in the secondary prevention of venous thromboembolism: systematic review and network meta-analysis. BMJ 347:f5133

Cheng S, Chng SM, Singh R (2009) Cerebral venous infarction during a high altitude expedition. Singap Med J 50(8):e306–e308. PMID 19710966

Cheuk BL, Cheung GC, Cheng SW (2004) Epidemiology of venous thromboembolism in a Chinese population. Br J Surg 91(4):424–428. https://doi.org/10.1002/bjs.4454. PMID 15048741

Cohen SH, Ehrlich GE, Kauffman MS, Cope C (1973) Thrombophlebitis following knee surgery. J Bone Joint Surg Am 55(1):106–112. https://doi.org/10.2106/00004623-197355010-00010. PMID 4691649

Crowther MA, Kelton JG (2003) Congenital thrombophilic states associated with venous thrombosis: a qualitative overview and proposed classification system. Ann Intern Med 138(2):128–134. https://doi.org/10.7326/0003-4819-138-2-200301210-00014. PMID 12529095

Cushman M (2005) Inherited risk factors for venous thrombosis. Hematology Am Soc Hematol Educ Program 1:452–457. https://doi.org/10.1182/asheducation-2005.1.452. PMID 16304419

Cushman M, Folsom AR, Wang L, Aleksic N, Rosamond WD, Tracy RP et al (2003) Fibrin fragment D-dimer and the risk of future venous thrombosis. Blood 101(4):1243–1248. https://doi.org/10.1182/blood-2002-05-1416. PMID 12393393

Dahlbäck B (1995) New molecular insights into the genetics of thrombophilia. Resistance to activated protein C caused by Arg506 to Gln mutation in factor V as a pathogenic risk factor for venous thrombosis. Thromb Haemost 74(1):139–148. https://doi.org/10.1055/s-0038-1642667. PMID 8578447

Dahlbäck B (2008) Advances in understanding pathogenic mechanisms of thrombophilic disorders. Blood 112(1):19–27. https://doi.org/10.1182/blood-2008-01-077909. PMID 18574041

Dickinson J, Heath D, Gosney J, Williams D (1983) Altitude-related deaths in seven trekkers in the Himalayas. Thorax 38(9):646–656. https://doi.org/10.1136/thx.38.9.646. PMID 6623417

Evaluation of Genomic Applications in Practice and Prevention (EGAPP) Working Group (2011) Recommendations from the EGAPP Working Group: routine testing for Factor v Leiden (R506Q) and prothrombin (20210G>A) mutations in adults with a history of idiopathic venous thromboembolism and their adult family members. Genet Med 13(1):67–76. https://doi.org/10.1097/GIM.0b013e3181fbe46f. PMID 21150787

Fahrni J, Husmann M, Gretener SB, Keo HH (2015) Assessing the risk of recurrent venous thromboembolism--A practical approach. Vasc Health Risk Manag 11(11):451–459. https://doi.org/10.2147/VHRM.S83718. PMID 26316770

Fedullo PF, Auger WR, Channick RN, Kerr KM, Rubin LJ (2001) Chronic thromboembolic pulmonary hypertension. Clin Chest Med 22(3):561–581. https://doi.org/10.1016/S0272-5231(05)70292-6

Folsom AR, Aleksic N, Wang L, Cushman M, Wu KK, White RH (2002) Protein C, antithrombin, and venous thromboembolism incidence: a prospective population-based study. Arterioscler Thromb Vasc Biol 22(6):1018–1022. https://doi.org/10.1161/01.atv.0000017470.08363.ab. PMID 12067914

Freiman DG (1987) The structure of thrombi. In: Colman RW, Hirsh J, Marder VJ, Salzman EW (eds) Thrombosis and hemostasis: basic principles and clinical practice. J B Lippincott Publishers, Philadelphia, pp 1123–1135

García de Frutos PG (2007) Mechanisms of thrombophilia. Thromb Haemost 98(3):485–487. https://doi.org/10.1160/TH07-08-0504. PMID 17849035

Gibbs NM (1957) Venous thrombosis of the lower limbs with particular reference to bed-rest. Br J Surg 45(191):209–236. https://doi.org/10.1002/bjs.18004519102. PMID 13510687

Goldhaber SZ (1998) Pulmonary embolism. N Engl J Med 339(2):93–104. https://doi.org/10.1056/NEJM199807093390207

Hassan WU, Syed MJ, Alamgir W, Awan S, Bell SM, Majid A et al (2019) Cerebral venous thrombosis at high altitude: analysis of 28 cases. Cerebrovasc Dis 48(3–6):184–192. https://doi.org/10.1159/000504504. PMID 31775151

Heit JA, O'Fallon WM, Petterson TM, Lohse CM, Silverstein MD, Mohr DN et al (2002) Relative impact of risk factors for deep vein thrombosis and pulmonary embolism: a population-based study. Arch Intern Med 162(11):1245–1248. https://doi.org/10.1001/archinte.162.11.1245. PMID 12038942

Hirsh J, Bates SM (1999) Prognosis in acute pulmonary embolism. Lancet 353(9162):1375–1376. https://doi.org/10.1016/S0140-6736(99)90053-3. PMID 10227212

Homans J (1954) Thrombosis of the leg veins due to prolonged sitting. N Engl J Med 250(4): 148–149. https://doi.org/10.1056/NEJM195401282500404. PMID 13119864

Hooper WC, Holman RC, Heit JA, Cobb N (2002) Venous thromboembolism hospitalizations among American Indians and Alaska Natives. Thromb Res 108(5–6):273–278. https://doi.org/10.1016/s0049-3848(03)00058-6. PMID 12676185

Huang W, Goldberg RJ, Anderson FA, Kiefe CI, Spencer FA (2014) Secular trends in occurrence of acute venous thromboembolism: the Worcester VTE study (1985-2009). Am J Med 127(9): 829–39.e5. https://doi.org/10.1016/j.amjmed.2014.03.041. PMID 24813864

Hull CM, Rajendran D, Fernandez BA (2016) Deep vein thrombosis and pulmonary embolism in a mountain guide: awareness, diagnostic challenges, and management considerations at altitude. Wilderness Environ Med 27(1):100–106. https://doi.org/10.1016/j.wem.2015.10.010. PMID 2 6723546

Hull RD, Raskob GE (1986) Prophylaxis of venous thromboembolic disease following hip and knee surgery. J Bone Joint Surg Am 68(1):146–150. https://doi.org/10.2106/00004623-198668010-00022. PMID 2416763

Hull R, Hirsh J, Jay R, Carter C, England C, Gent M et al (1982) Different intensities or oral anticoagulant therapy in the treatment of proximal-vein thrombosis. N Engl J Med 307(27): 1676–1681. https://doi.org/10.1056/NEJM198212303072704. PMID 6755255

Iorio A, Kearon C, Filippucci E, Marcucci M, Macura A, Pengo V et al (2010) Risk of recurrence after a first episode of symptomatic venous thromboembolism provoked by a transient risk factor: a systematic review. Arch Intern Med 170(19):1710–1716. https://doi.org/10.1001/archinternmed.2010.367. PMID 20975016

Jha SK, Anand AC, Sharma V, Kumar N, Adya CM (2002) Stroke at high-altitude: Indian experience. High Alt Med Biol 3(1):21–27. https://doi.org/10.1089/15270290275363951 3. PMID 12006161

Kapoor A, Ellis A, Shaffer N, Gurwitz J, Chandramohan A, Saulino J et al (2017) Comparative effectiveness of venous thromboembolism prophylaxis options for the patient undergoing total hip and knee replacement: a network meta-analysis. J Thromb Haemost 15(2):284–294. https://doi.org/10.1111/jth.13566. PMID 28102615

Kearon C (2003) Natural history of venous thromboembolism. Circulation 107(23; Suppl 1):I22–I30. https://doi.org/10.1161/01.CIR.0000078464.82671.78. PMID 12814982

Kearon C, Julian JA, Newman TE, Ginsberg JS (1998) Non-invasive diagnosis of deep venous thrombosis. McMaster diagnostic imaging practice guidelines initiative. Ann Intern Med 128(8):663–677. https://doi.org/10.7326/0003-4819-128-8-199804150-00011. PMID 9537941

Kearon C, Akl EA, Ornelas J, Blaivas A, Jimenez D, Bounameaux H et al (2016) Antithrombotic therapy for VTE disease: CHEST guideline and expert panel report. Chest 149(2):315–352. https://doi.org/10.1016/j.chest.2015.11.026. PMID 26867832

Kelly J, Rudd A, Lewis RR, Hunt BJ (2002) Plasma D-dimers in the diagnosis of venous thromboembolism. Arch Intern Med 162(7):747–756. https://doi.org/10.1001/archinte.162.7.747. PMID 11926847

Khan F, Rahman A, Carrier M, Kearon C, Weitz JI, Schulman S et al (2019) Long term risk of symptomatic recurrent venous thromboembolism after discontinuation of anticoagulant

treatment for first unprovoked venous thromboembolism event: systematic review and meta-analysis. BMJ 366:l4363. https://doi.org/10.1136/bmj.l4363. PMID 31340984
Khattar NK, Sumardi F, Zemmar A, Liang Q, Li H, Xing Y et al (2019) Cerebral venous thrombosis at high altitude: a retrospective cohort of twenty-one consecutive patients. Cureus 11(6):e4940. https://doi.org/10.7759/cureus.4940. PMID 31431845
Kumar S (2006) High-altitude induced deep venous thrombosis: a study of 28 cases. Indian J Surg 68(2):84–88
Kumar DR, Hanlin E, Glurich I, Mazza JJ, Yale SH (2010) Virchow's contribution to the understanding of thrombosis and cellular biology. Clin Med Res 8(3–4):168–172. https://doi.org/10.3121/cmr.2009.866. PMID 20739582
Langford NJ, Stansby G, Avital L (2012) LThe management of venous thromboembolic diseases and the role of thrombophilia testing: summary of NICE Guideline cg144. Acute Med 11(3): 138–142. PMID 22993743
Lapostolle F, Surget V, Borron SW, Desmaizières M, Sordelet D, Lapandry C et al (2001) Severe pulmonary embolism associated with air travel. N Engl J Med 345(11):779–783. https://doi.org/10.1056/NEJMoa010378. PMID 11556296
Lee CH, Cheng CL, Lin LJ, Tsai LM, Yang YH (2011) Epidemiology and predictors of short-term mortality in symptomatic venous thromboembolism. Circ J 75(8):1998–2004. https://doi.org/10.1253/circj.cj-10-0992. PMID 21697611
Lim W, Eikelboom JW, Ginsberg JS (2007) Inherited thrombophilia and pregnancy associated venous thromboembolism. BMJ 334(7607):1318–1321
MacCallum PK, Meade TW (1999) Haemostatic function, arterial disease and the prevention of arterial thrombosis. Baillieres Best Pract Res Clin Haematol 12(3):577–599. https://doi.org/10.1053/beha.1999.0041. PMID 10856986
Mannucci PM, Gringeri A, Peyvandi F, Di Paolantonio T, Mariani G (2002) Short-term exposure to high altitude causes coagulation activation and inhibits fibrinolysis. Thromb Haemost 87(2): 342–343. https://doi.org/10.1055/s-0037-1612997. PMID 11858498
Matisse Investigators (2004) Fondaparinux or enoxaparin for the initial treatment of symptomatic deep vein thrombosis. A randomized trial. Ann Intern Med 140:867–873
Mazzolai L, Aboyans V, Ageno W, Agnelli G, Alatri A, Bauersachs R et al (2018) Diagnosis and management of acute deep vein thrombosis: a joint consensus document from the European Society of Cardiology working groups of aorta and peripheral vascular diseases and pulmonary circulation and right ventricular function. Eur Heart J 39(47):4208–4218. https://doi.org/10.1093/eurheartj/ehx003
McLaughlin DF, Wade CE, Champion HR, Salinas J, Holcomb JB (2009) Thromboembolic complications following trauma. Transfusion 49(Suppl 5):256S–263S. https://doi.org/10.1111/j.1537-2995.2008.01989.x. PMID 19954488
McLendon K, Goyal A, Attia M. Deep venous thrombosis risk factors. StatPearls. Updated 2022 Apr 21:2022 Jan-
Min SK, Kim YH, Joh JH, Kang JM, Park UJ, Kim HK et al (2016) Diagnosis and treatment of lower extremity deep vein thrombosis: Korean practice guidelines. Vasc Spec Int 32(3):77–104. https://doi.org/10.5758/vsi.2016.32.3.77. PMID 27699156
Naess IA, Christiansen SC, Romundstad P, Cannegieter SC, Rosendaal FR, Hammerstrøm J (2007) Incidence and mortality of venous thrombosis: a population-based study. J Thromb Haemost 5(4):692–699. https://doi.org/10.1111/j.1538-7836.2007.02450.x. PMID 17367492
Naik AK, Prasad JP, Singh V, Singh SK, Chakrabarty BK, Sharma J et al (2021) Study of clinical spectrum, laboratory parameters, and radiological patterns in cerebral venous thrombosis cases occurring at high altitude. J Mar Med Soc 1(23):33–38
Nair V, Singh S, Ashraf MZ, Yanamandra U, Sharma V, Prabhakar A et al (2022) Epidemiology and pathophysiology of vascular thrombosis in acclimatized lowlanders at high altitude: a prospective longitudinal study. The Lancet Regional Health – Southeast Asia 3:100016. ISSN 2772-3682

Neeman E, Liu V, Mishra P, Thai KK, Xu J, Clancy HA et al (2022) Trends and risk factors for venous thromboembolism among hospitalized medical patients. JAMA Netw Open 5(11): e2240373. https://doi.org/10.1001/jamanetworkopen.2022.40373. PMID 36409498

Oger E (2000) Incidence of venous thromboembolism: a community-based study in western France. EPI-GETBP Study Group. Groupe d'Etude de la Thrombose de Bretagne Occidentale. Thromb Haemost 83(5):657–660. https://doi.org/10.1055/s-0037-1613887. PMID 10823257

Poort SR, Rosendaal FR, Reitsma PH, Bertina RM (1996) A common genetic variation in the 3′-untranslated region of the prothrombin gene is associated with elevated plasma prothrombin levels and an increase in venous thrombosis. Blood 88(10):3698–3703. https://doi.org/10.1182/blood.V88.10.3698.bloodjournal88103698. PMID 8916933

Prabhakar A, Chatterjee T, Bajaj N, Tyagi T, Sahu A, Gupta N et al (2019) Venous thrombosis at altitude presents with distinct biochemical profiles: a comparative study from the Himalayas to the plains. Blood Adv 3(22):3713–3723. https://doi.org/10.1182/bloodadvances.2018024554. PMID 31765479

Reiner AP, Carty CL, Jenny NS, Nievergelt C, Cushman M, Stearns-Kurosawa DJ et al (2008) PROC, PROCR and PROS1 polymorphisms, plasma anticoagulant phenotypes, and risk of cardiovascular disease and mortality in older adults: the cardiovascular Health Study. J Thromb Haemost 6(10):1625–1632. https://doi.org/10.1111/j.1538-7836.2008.03118.x. PMID 18680534

Rose DK, Bar B (2018) Direct oral anticoagulant agents: pharmacologic profile, indications, coagulation monitoring, and reversal agents. J Stroke Cerebrovasc Dis 27(8):2049–2058. https://doi.org/10.1016/j.jstrokecerebrovasdis.2018.04.004. PMID 29753603

Rosendaal FR (1997a) Risk factors for venous thrombosis: prevalence, risk, and interaction. Semin Hematol 34(3):171–187. PMID 9241704

Rosendaal FR (1997b) Thrombosis in the young: epidemiology and risk factors, a focus on venous thrombosis. Thromb Haemost 78(1):1–6. https://doi.org/10.1055/s-0038-1657492. PMID 9198119

Rosendaal FR (1999) Venous thrombosis: a multicausal disease. Lancet 353(9159):1167–1173. https://doi.org/10.1016/s0140-6736(98)10266-0. PMID 10209995

Ruff CT, Giugliano RP, Braunwald E, Hoffman EB, Deenadayalu N, Ezekowitz MD et al (2014) Comparison of the efficacy and safety of new oral anticoagulants with warfarin in patients with atrial fibrillation: a meta-analysis of randomised trials. Lancet 383(9921):955–962. https://doi.org/10.1016/S0140-6736(13)62343-0. PMID 24315724

Scarvelis D, Wells PS (2006) Diagnosis and treatment of deep-vein thrombosis. CMAJ 175(9): 1087–1092. https://doi.org/10.1503/cmaj.060366. PMID 17060659

Schreijer AJ, Cannegieter SC, Rosendaal FR, Helmerhorst FM (2005) A case of thrombosis at high altitude. Thromb Haemost 94(5):1104–1105. https://doi.org/10.1160/TH05-05-1104. PMID 16363256

Sidney S, Petitti DB, Soff GA, Cundiff DL, Tolan KK, Quesenberry CP Jr (2004) Venous thromboembolic disease in users of low-estrogen combined estrogen-progestin oral contraceptives. Contraception 70(1):3–10. https://doi.org/10.1016/j.contraception.2004.02.010. PMID 15208046

Silverstein MD, Heit JA, Mohr DN, Petterson TM, O'Fallon WM, Melton LJ 3rd. (1998) Trends in the incidence of deep vein thrombosis and pulmonary embolism: a 25-year population-based study. Arch Intern Med 158(6):585–593. https://doi.org/10.1001/archinte.158.6.585. PMID 9521222

Simioni P, Scarano L, Gavasso S, Sardella C, Girolami B, Scudeller A et al (1996) Prothrombin fragment 1+2 and thrombin–antithrombin complex levels in patients with inherited APC resistance due to factor V Leiden mutation. Br J Haematol 92(2):435–441. https://doi.org/10.1046/j.1365-2141.1996.d01-1500.x. PMID 8603014

Singh I, Chohan IS (1972) Blood coagulation changes at high altitude predisposing to pulmonary hypertension. Br Heart J 34(6):611–617. https://doi.org/10.1136/hrt.34.6.611. PMID 5040260

Sørensen HT, Mellemkjaer L, Steffensen FH, Olsen JH, Nielsen GL (1998) The risk of a diagnosis of cancer after primary deep venous thrombosis or pulmonary embolism. N Engl J Med 23;338 (17):1169-73. https://doi.org/10.1056/NEJM199804233381701. PMID 9554856

Spencer FA, Emery C, Lessard D, Anderson F, Emani S, Aragam J et al (2006) The Worcester venous thromboembolism study: a population-based study of the clinical epidemiology of venous thromboembolism. J Gen Intern Med 21(7):722–727. https://doi.org/10.1111/j.1525-1497.2006.00458.x. PMID 16808773

Tagalakis V, Patenaude V, Kahn SR, Suissa S (2013) Incidence of and mortality from venous thromboembolism in a real-world population: the Q-VTE Study Cohort. Am J Med 126(9):832.e13–832.e21. https://doi.org/10.1016/j.amjmed.2013.02.024. PMID 23830539

Thaler E, Lechner K (1981) Antithrombin III deficiency and thromboembolism. Clin Haematol 10(2):369–390. https://doi.org/10.1016/S0308-2261(21)00229-0. PMID 7032781

Trégouët DA, Heath S, Saut N, Biron-Andreani C, Schved JF, Pernod G et al (2009) Common susceptibility alleles are unlikely to contribute as strongly as the FV and ABO loci to VTE risk: results from a GWAS approach. Blood 113(21):5298–5303. https://doi.org/10.1182/blood-2008-11-190389. PMID 19278955

Tyagi T, Ahmad S, Gupta N, Sahu A, Ahmad Y, Nair V et al (2014) Altered expression of platelet proteins and calpain activity mediate hypoxia-induced prothrombotic phenotype. Blood 123(8):1250–1260. https://doi.org/10.1182/blood-2013-05-501924. PMID 24297866

van Aken BE, den Heijer M, Bos GM, van Deventer SJ, Reitsma PH (2000) Recurrent venous thrombosis and markers of inflammation. Thromb Haemost 83(4):536–539. https://doi.org/10.1055/s-0037-1613858. PMID 10780312

van der Hulle T, Kooiman J, den Exter PL, Dekkers OM, Klok FA, Huisman MV (2014) Effectiveness and safety of novel oral anticoagulants as compared with vitamin K antagonists in the treatment of acute symptomatic venous thromboembolism: a systematic review and meta-analysis. J Thromb Haemost 12(3):320–328. https://doi.org/10.1111/jth.12485. PMID 24330006

Van Dongen CJ, van den Belt AG, Prins MH, Lensing AW (2004) Fixed dose subcutaneous low molecular weight heparins versus adjusted dose unfractionated heparin for venous thromboembolism [review]. Cochrane Database Syst Rev 18(4):CD001100. https://doi.org/10.1002/14651858.CD001100.pub2. PMID 15495007

Van Hylckama VA, Callas PW, Cushman M, Bertina RM, Rosendaal FR (2003) Inter-relation of coagulation factors and d-dimer levels in healthy individuals. J Thromb Haemost 1(3):516–522. https://doi.org/10.1046/j.1538-7836.2003.00080.x. PMID 12871460

Warlow C, Ogston D, Douglas AS (1976) Deep venous thrombosis of the legs after strokes. Part I–incidence and predisposing factors. Br Med J 1(6019):1178–1181. https://doi.org/10.1136/bmj.1.6019.1178. PMID 1268614

Wells PS, Anderson DR, Bormanis J, Guy F, Mitchell M, Gray L et al (1997) Value of assessment of pretest probability of deep-vein thrombosis in clinical management. Lancet 350(9094):1795–1798. https://doi.org/10.1016/S0140-6736(97)08140-3. PMID 9428249

Wells PS, Anderson DR, Rodger M, Forgie M, Kearon C, Dreyer J et al (2003) Evaluation of D-dimer in the diagnosis of suspected deep-vein thrombosis. N Engl J Med 349(13):1227–1235. https://doi.org/10.1056/NEJMoa023153. PMID 14507948

Wells PS, Owen C, Doucette S, Fergusson D, Tran H (2006) Does this patient have deep vein thrombosis? JAMA 295(2):199–207. https://doi.org/10.1001/jama.295.2.199. PMID 16403932

White RH (2003) The epidemiology of venous thromboembolism. Circulation 107(23; Suppl 1):I4–I8. https://doi.org/10.1161/01.CIR.0000078468.11849.66. PMID 12814979

White RH, Zhou H, Romano PS (1998) Incidence of idiopathic deep venous thrombosis and secondary thromboembolism among ethnic groups in California. Ann Intern Med 128(9):737–740. https://doi.org/10.7326/0003-4819-128-9-199805010-00006. PMID 9556467

White RH, Zhou H, Murin S, Harvey D (2005) Effect of ethnicity and gender on the incidence of venous thromboembolism in a diverse population in California in 1996. Thromb Haemost 93(2):298–305. https://doi.org/10.1160/TH04-08-0506. PMID 15711746

Wu O, Bayoumi N, Vickers MA, Clark P (2008) ABO(H) blood groups and vascular disease: a systematic review and meta-analysis. J Thromb Haemost 6(1):62–69. https://doi.org/10.1111/j.1538-7836.2007.02818.x. PMID 17973651

Yang H, Yang G, Xu M, Zhao Y, He S, Wang Q et al (2022) Impact of high altitude on the incidence of postoperative venous thromboembolism and its genetic susceptibility: a meta-analysis and systematic review. Sci Total Environ 4(4):156632. https://doi.org/10.1016/j.scitotenv.2022.156632. PMID 35691358

Zakai NA, McClure LA, Judd SE, Safford MM, Folsom AR, Lutsey PL et al (2014) Racial and regional differences in venous thromboembolism in the United States in 3 cohorts. Circulation 129(14):1502–1509. https://doi.org/10.1161/CIRCULATIONAHA.113.006472. PMID 24508826

Leaky Gut at High Altitude: Synbiotics May Be a Possible Solution

7

Kamla Prasad Mishra, Jyotsana Bakshi, and Rajeev Varshney

Abstract

Sojourn to high altitude may lead to gastrointestinal (GI) discomfort due to acute hypoxia. GI discomfort is one of the problems observed in people traveling to high altitudes. The increased intestinal permeability gives rise to a physiological condition known as leaky gut. We have reported that increased intestinal permeability, inflammation, and microbial dysbiosis are the major causes of GI discomfort. Under this condition, the tight junctions guarding the anti-lumen space of body become leaky or loose and allow an uncontrolled exchange of harmful molecules in the bloodstream causing local or systemic inflammation. In our previous study, we reported an increased TH17 cell population in Peyer's patches, increased the zonulin level in peripheral blood, and microbial dysbiosis in the gut. Furthermore, we reported that synbiotics (a mixture of prebiotics and probiotics) treatment decreased inflammation and microbial dysbiosis and strengthened the gastrointestinal epithelial barrier. This review aims to elucidate the role of synbiotics in intestinal mucosal barrier injury and inflammation under hypobaric hypoxia. In addition to this, we also reviewed various potential peripheral blood biomarkers that may be associated with intestinal mucosal barrier injury.

Keywords

High altitude · Gastrointestinal tract · Inflammation · Synbiotics: Microbialdysbiosis

K. P. Mishra (✉) · J. Bakshi · R. Varshney
Defence Institute of Physiology and Allied Sciences (DIPAS), Defence Research and Development Organization (DRDO), Delhi, India
e-mail: kpmpgi@rediffmail.com

R. Tulsawani, D. Vohora (eds.), *Adaptation under Stressful Environments through Biological Adjustments and Interventions*,
https://doi.org/10.1007/978-981-99-7652-2_7

7.1 Introduction

India has a long border with neighbors in high-altitude ranges. Mountaineers, trekkers, pilgrims, and defense personnel regularly visit some of the extreme high-altitude areas including the Siachen Glacier. At high altitudes, mountaineers are exposed to the most hazardous and life-threatening environmental conditions such as low barometric pressures and rarity of oxygen (hypobaric hypoxia), sub-zero temperature, high winds, and intense ultraviolet radiation exposure. Personnel who are acutely sojourn to those places may develop some of the environmentally related illnesses known as acute mountain sickness, high altitude pulmonary edema, high altitude cerebral edema, frostbite, ultraviolet radiation injuries to skin and eye, and gut bleeding (Berré et al. 1999). Out of these life-threatening diseases, gastrointestinal bleeding at high altitudes is very less studied. Our research on the animal model in a simulated high-altitude environment reported some aspects of GI-related problems (Khanna et al. 2018; 2019; 2021a, b) which include intestinal mucosal barrier injury, inflammation, increased sIgA level, increased Th17 cell types, and microbial dysbiosis (Khanna et al. 2021a, b).

Increased intestinal permeability at high altitude hypoxia was observed in animals which could happen when tight junction proteins are less expressed leading to uncontrolled release of substances into the bloodstream (Luo et al. 2017). This phenomenon is known as leaky gut, in some cases, even bleeding occurs in the intestine and it can be seen in the stool (Camilleri 2019). Exposure to heights above 5000 m increased the risk of gastrointestinal bleeding and it further increased when sojourners use drugs to treat high altitude-related illnesses such as aspirin and dexamethasone (Khanna et al. 2018, 2019; Wu et al. 2007). Kashyap et al. (2005) have also reported cases of gastrointestinal bleeding due to the high incidence of peptic and duodenal ulcers from moderated altitude.

During the construction of the Qinghai-Tibet railroad from 2001 to 2003, about 0.5% of gastrointestinal bleeding cases were reported and interpreted as low oxygen levels and low temperature leading to the acute gastric mucosal lesion (AGML) that might be resulted in gut bleeding. If the disease is early diagnosed, it can help in early recovery from life-threatening disease condition of gut bleeding (Wu et al. 2007). Wu (2001) had reported gut bleeding incidences in labor and staff working between 4767 m and 5232 m to construct and repair the Qinghai-Tibetan Highway as 0.95% and 1.93%, respectively. Similarly, in another study, 0.8% of the soldiers working from 3700 m to 5380 m reported gut bleeding.

7.2 Intestinal Barrier Injury/Dysfunction with High Altitude Exposure

High-altitude environment could result in intestinal barrier injury/dysfunction (Figs. 7.1a, b). Our study performed on animal models under simulated high-altitude conditions reported the effects of acute exposure to hypobaric hypoxia for 14 days at 7620 m on GI immune function have been studied and we found that hypobaric

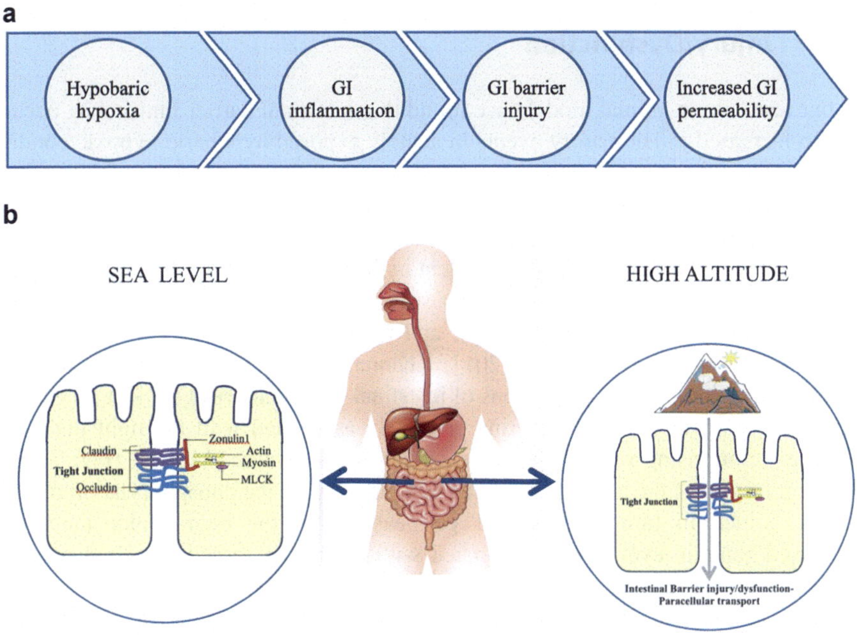

Fig. 7.1 (**a**) Progress sequence for gastrointestinal barrier injury. (**b**) A comparative representation of healthy and damaged epithelial tight junctions at sea level and high altitude

hypoxia exposure of rats for 7 days significantly increased TH17 cell types in Peyer's patches, increased interleukin-17 (IL-17) and dendritic cell populations compared with unexposed control rats. We have also reported the decreased expression of tight junction proteins and increased level of zonulin in peripheral blood. This study highlighted how hypobaric hypoxia may lead to intestinal barrier dysfunction and injury in the gut wall (Khanna et al. 2018, 2019, 2021a, b). Zhang et al. (2013) have reported that hypobaric hypoxia exposure of 3 days to altitudes of 3842 m and 4767 m severely injures intestinal mucosa. They also found increased expressions of HIF-1$\alpha$and iNOS with increasing hypobaric hypoxia exposure which may play a critical role in the intestinal mucosa damage.

In a Genome-wide transcriptional analysis, Li et al. (2016) showed that hypoxia-induced oxidative stress causes intestinal injury in Tibetans via the inhibition of growth factor receptor-bound protein 2 (GRB2), epidermal growth factor receptor (EGFR), and tyrosine-protein phosphatase nonreceptor type 11 (PTPN11) {GRB2/EGFR/PTPN11} pathways (Li et al. 2016). In another study, Luo et al. (2017) reported the downregulation of tight junction proteins in intestinal epithelial cells of rats exposed to simulated hypobaric hypoxia of 4000 m for 3 days (Luo et al. 2017).

7.3 Potential Peripheral Blood Markers for Intestinal Barrier Injury/Dysfunction

In our research on animal models, we found that intestinal barrier injury may occur due to increased inflammatory events in acutely exposed hypobaric hypoxic condition. Furthermore, the manifestation of increased intestinal permeability was observed by the mannitol test. We were interested in knowing the level of zonulin protein as it is an endogenous modulator of tight junctions of the intestinal epithelial lining. We found a significant enhancement in the zonulin level (Khanna et al. 2019, 2021a, b). The emerging literature also suggests that fatty acid binding proteins (FABP), lipid binding proteins (LBP), and zonulin levels increased in intestinal barrier injury/dysfunction (Fasano 2011; Khanna et al. 2019, 2021a, b). Zonulin is known for modulating tight junctions of intestinal epithelial cells (Tajik et al. 2020). The perijunctional actin myosin ring is closely associated with the tight junction complex, comprising zonulin and myosin light chain kinase (MLCK) (Fig. 7.1b). The contraction of the actin myosin ring is controlled by the phosphorylation of the MLCK which in turn increase the permeability of the intercellular junction. Increased zonulin level is well correlated with in vivo increased intestinal permeability as measured using the lactulose/mannitol test in subjects with type 1 diabetes and their relatives (Sapone et al. 2006). There is a need to correlate these peripheral blood markers with intestinal permeability or mucosal barrier injury/dysfunction in a larger cohort of acutely exposed high-altitude sojourners. It will save tens of lives if diagnosed early as gut bleeding becomes life-threatening if not treated early. It is also important to mention here that apart from high altitude exposure, dietary habits may also affect intestinal permeability and mucosal barrier function (Brun et al. 2007). Other factors such as stress and antibiotics may affect the gut barrier function (Lobionda et al. 2019).

Dietary habits and lifestyle factors have a clear impact on intestinal permeability and barrier function with known detrimental effects for a Western diet, high alcohol intake, stress, and certain medications.

7.4 High Altitude-Induced Alterations in Gut Microbiota

Human gut microbiota research is enormously increased in recent years due to advancements in technology and reduction in cost. A huge amount of data is being generated through these studies that have made a significant understanding of the interaction and effects on the host. Trillions of microbial cells reside in the human gut as commensal symbionts to maintain the normal homeostasis of the body (Chung and Kasper 2010; Lin and Zhang 2017). It has been shown that dysbiosis of gut bacteria affects the immune system (Dupaul-Chicoine et al. 2013; Cianci et al. 2018), and changes in the commensal bacteria in the gut could result in several diseases including gut (Guinane and Cotter 2013; Bhushan et al. 2021; Khanna et al. 2021a, b) (Fig. 7.2).

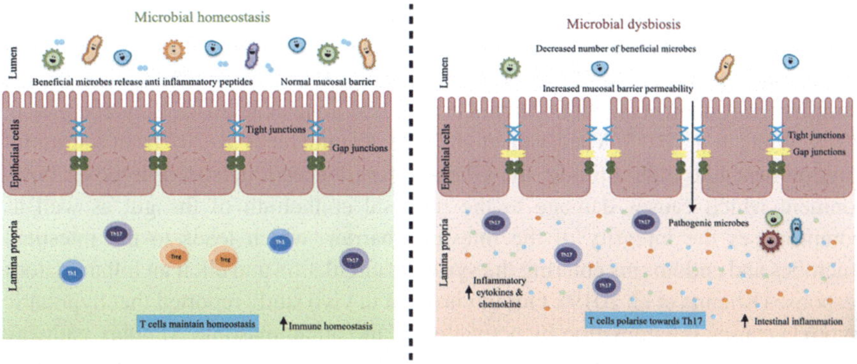

Fig. 7.2 A comparative representation of the intestinal epithelia during microbial dysbiosis and microbial homeostasis

Microbial dysbiosis is associated with diseases like diabetes, obesity and metabolic syndrome, autism, rheumatic disease, inflammatory bowel disease, Alzheimer's disease, cancer, cardiovascular disease, and autoimmune diseases (Yang et al. 2015; Boursier et al. 2016; Thevaranjan et al. 2017; Saffouri et al. 2019; Kesika et al. 2021).

In our previous study, we reported the effect of synbiotics on hypobaric hypoxia exposed animals in terms of microbial dysbiosis and functional analysis. We reported significantly decreased bacteria of the phylum *Deferribacteres* and *Firmicutes* in synbiotics treated and hypobaric hypoxia exposed animals. Deferribacteres phylum bacteria are associated with gut mucus in rodents and play a role in the pathogenesis of inflammatory diseases (Aidy et al. 2014). Therefore, a decreased population of bacteria of the phylum Deferribacteres shows improvement in inflammation under hypobaric hypoxia exposure. We have also reported the restoring of the amino acid metabolism in synbiotics treatment.

7.5 Hypoxia-Induced Inflammation in the Gut

Over the past few years, inflammation in the gut caused due to high altitude hypoxia has gradually gained the attention of researchers. A hypoxic environment at high altitude is known to cause detrimental effects on various organs including the gut (McKenna et al. 2022). Several studies on animal models exposed to hypoxia have pointed out its role in causing inflammation (Grenz et al. 2012; Ramakrishnan et al. 2014; Pham et al. 2021). However, it remains unclear how high altitude leads to immunological disturbances in the GI tract. The homeostasis of the immune system as well as the immunoregulatory mechanism of the body is affected by the change in geographical attitude (Cristofori et al. 2021). The inflammatory responses are activated by the body for a short time during stress as an adaptive mechanism (Walmsley et al. 2014). However, prolonged exposure to stress like hypoxia can induce chronic inflammation which may lead to various diseases. It is proved that healthy individual sojourning to 3400 m has increased levels of pro-inflammatory marker and cytokines like IL-1β, IL-6, IL-8, and C-Reactive protein in their plasma samples (Hartmann et al. 2000; Mishra et al. 2015; Song et al. 2016; Lundeberg et al. 2018; Wang et al. 2018; Malacrida et al. 2019; Kammerer et al. 2020). In a stressful environment like high altitude, the GI tract experiences a persistent hypoxic environment which causes damage to the mucosal epithelium of the gut as well as compromises the integrity of the intestinal barrier, which leads to the intestinal microbes and endotoxins entering the systemic circulation and elicit an inflammatory response (Khanna et al. 2019). Our in vitro and in vivo study reported that hypobaric hypoxia causes inflammation by activation of the inflammasome complex pathway and treatment with andrographolide reduced the hypoxia-induced inflammation in the gut (Khanna et al. 2018).

A study on Wistar rats exposed to a simulated hypobaric hypoxia environment for 48 h has revealed that the hypoxia treatment group was more susceptible to intestinal mucosal barrier injury as well as there was a considerable increase in the expression of markers associated with autophagy like LC3B and Beclin1 (Zhang et al. 2015).

Latest research has pointed out that dysbiosis of gut microbiota due to environmental factors can also cause inflammation and increase gut mucosal barrier permeability (Pellegrini et al. 2018). The role of prebiotics which is food for good bacteria and probiotics, i.e., beneficial bacteria have been examined closely in overcoming problems like inflammatory diseases of the gut (De Greef et al. 2013; de Oliveira et al. 2017; Yoon et al. 2021). A study aimed at identifying the role of prebiotics on tight junction protein occludin and the expression of cytokines in rats exposed to high altitude reported that prebiotic treatment was able to restore the reduced expression of occludin during high-altitude exposure (Zhang et al. 2013). Along with this, the prebiotic treatment was also able to suppress the production of inflammatory cytokines like TNF-α (Zhang et al. 2013). A combination prepared by combining the pre- and probiotics has been developed earlier known as synbiotics which consists of the benefits of both (Ford et al. 2014; Frei et al. 2015). Our lab has also reported that the GI barrier strengthened in rat models exposed to hypobaric hypoxia after synbiotic treatment. The gut barrier function improved after synbiotic treatment as the levels of Th17 decreased as well as reduced the chances of microbial dysbiosis (Khanna et al. 2021a, b).

7.6 Conclusions

Acute exposure to hypobaric hypoxia may lead to gastrointestinal injury resulting in damage to the mucosal epithelium of the intestine and its integrity. Hypoxic stress causes increased microbial dysbiosis and inflammatory molecules leading to a leaky gut which can be prevented by synbiotics. The role of synbiotics in gut-associated illnesses is slowly gaining the attention of researchers. The analysis of biomarkers in peripheral blood such as zonulin and lipid binding proteins should be explored to find out the association between hypoxia-induced mucosal barrier injury and their levels. It will save the lives of many individuals visiting at high altitudes.

Acknowledgement JB thanks DRDO for providing financial support in the form of a Junior Research Fellowship.

Conflict of Interest The authors report no conflict of interest.

Funding Information This work was supported by the Defence Research and Development Organization in the form of project DIP-265.

References

Berré J, Vachiéry JL, Moraine JJ, Naeije R (1999) Cerebral blood flow velocity responses to hypoxia in subjects who are susceptible to high-altitude pulmonary oedema. Eur J Appl Physiol Occup Physiol 80(4):260–263. https://doi.org/10.1007/s004210050591. PMID 10483794

Bhushan B, Eslavath MR, Yadav AP, Srivastava AK, Reddy MPK, Norboo T et al (2021) Metagenomic sequencing reveals altered gut microbiota of sojourners at high altitude: a longitudinal study. J ProteinsProteomics 12(4):271–288. https://doi.org/10.1007/s42485-021-00077-8

Boursier J, Mueller O, Barret M, Machado M, Fizanne L, Araujo-Perez F et al (2016) The severity of nonalcoholic fatty liver disease is associated with gut dysbiosis and shift in the metabolic function of the gut microbiota. Hepatology 63(3):764–775. https://doi.org/10.1002/hep.28356. PMID 26600078

Brun P, Castagliuolo I, Di Leo VD, Buda A, Pinzani M, Palù G et al (2007) Increased intestinal permeability in obese mice: new evidence in the pathogenesis of nonalcoholic steatohepatitis. Am J Physiol Gastrointest Liver Physiol 292(2):G518–G525. https://doi.org/10.1152/ajpgi.00024.2006. PMID 17023554

Camilleri M (2019) Leaky gut: mechanisms, measurement and clinical implications in humans. Gut 68(8):1516–1526. https://doi.org/10.1136/gutjnl-2019-318427. PMID 31076401

Chung H, Kasper DL (2010) Microbiota-stimulated immune mechanisms to maintain gut homeostasis. Curr Opin Immunol 22(4):455–460. https://doi.org/10.1016/j.coi.2010.06.008. PMID 20 656465

Cianci R, Pagliari D, Piccirillo CA, Fritz JH, Gambassi G (2018) The microbiota and immune system crosstalk in health and disease. Mediat Inflam 2018:2912539. https://doi.org/10.1155/2018/2912539. PMID 29849485

Cristofori F, Dargenio VN, Dargenio C, Miniello VL, Barone M, Francavilla R (2021) Anti-inflammatory and immunomodulatory effects of probiotics in gut inflammation: a door to the body. Front Immunol 12:578386. https://doi.org/10.3389/fimmu.2021.578386. PMID 337170 63

De Greef E, Vandenplas Y, Hauser B, Devreker T, Veereman-Wauters G (2013) Probiotics and IBD. Acta Gastroenterol Belg 76(1):15–19. PMID 23650777

de Oliveira GLV, Leite AZ, Higuchi BS, Gonzaga MI, Mariano VS (2017) Intestinal dysbiosis and probiotic applications in autoimmune diseases. Immunology 152(1):1–12. https://doi.org/10.1111/imm.12765. PMID 28556916

Dupaul-Chicoine J, Dagenais M, Saleh M (2013) Crosstalk between the intestinal microbiota and the innate immune system in intestinal homeostasis and inflammatory bowel disease. Inflam Bowel Dis 19(10):2227–2237. https://doi.org/10.1097/MIB.0b013e31828dcac7. PMID 23 669404

El Aidy S, Derrien M, Aardema R, Hooiveld G, Richards SE, Dane A et al (2014) Transient inflammatory-like state and microbial dysbiosis are pivotal in establishment of mucosal

homeostasis during colonisation of germ-free mice. Benef Microbes 5(1):67–77. https://doi.org/10.3920/BM2013.0018. PMID 24322881
Fasano A (2011) Zonulin and its regulation of intestinal barrier function: the biological door to inflammation, autoimmunity, and cancer. Physiol Rev 91(1):151–175. https://doi.org/10.1152/physrev.00003.2008. PMID 21248165
Ford AC, Quigley EM, Lacy BE, Lembo AJ, Saito YA, Schiller LR et al (2014) Efficacy of prebiotics, probiotics, and Synbiotics in irritable bowel syndrome and chronic idiopathic constipation: systematic review and meta-analysis. Am J Gastroenterol 109(10):1547–1561. https://doi.org/10.1038/ajg.2014.202. PMID 25070051
Frei R, Akdis M, O'Mahony L (2015) Prebiotics, probiotics, Synbiotics, and the immune system: experimental data and clinical evidence. Curr Opin Gastroenterol 31(2):153–158. https://doi.org/10.1097/MOG.0000000000000151. PMID 25594887
Grenz A, Clambey E, Eltzschig HK (2012) Hypoxia signaling during intestinal ischemia and inflammation. Curr Opin Crit Care 18(2):178–185. https://doi.org/10.1097/MCC.0b013e3283514bd0. PMID 22322265
Guinane CM, Cotter PD (2013) Role of the gut microbiota in health and chronic gastrointestinal disease: understanding a hidden metabolic organ. Ther Adv Gastroenterol 6(4):295–308. https://doi.org/10.1177/1756283X13482996. PMID 23814609
Hartmann G, Tschöp M, Fischer R, Bidlingmaier C, Riepl R, Tschöp K, Hautmann H, Endres S, Toepfer M (2000) High altitude increases circulating interleukin-6, interleukin-1 receptor antagonist and C-reactive protein. Cytokine 12(3):246–252. https://doi.org/10.1006/cyto.1999.0533. PMID: 10704252
Kammerer T, Faihs V, Hulde N, Stangl M, Brettner F, Rehm M, Horstmann M, Kröpfl J, Spengler C, Kreth S, Schäfer S (2020) Hypoxic-inflammatory responses under acute hypoxia: In Vitro experiments and prospective observational expedition trial. Int J Mol Sci 21(3):1034. doi: https://doi.org/10.3390/ijms21031034. PMID: 32033172
Kashyap R, Mahajan S, Sharma B, Jaret P, Patial RK, Rana S, Pal LS (2005) A clinical profile of acute upper gastrointestinal bleeding at moderate altitude. J Indian Acad Clin Med 6(3):224–228
Kesika P, Suganthy N, Sivamaruthi BS, Chaiyasut C (2021) Role of gut-brain axis, gut microbial composition, and probiotic intervention in Alzheimer's disease. Life Sci 264:118627. https://doi.org/10.1016/j.lfs.2020.118627. PMID 33169684
Khanna K, Mishra KP, Ganju L, Kumar B, Singh SB (2018) High-altitude-induced alterations in gut-immune axis: a review. Int Rev Immunol 37(2):119–126. https://doi.org/10.1080/08830185.2017.1407763. PMID 29231767
Khanna K, Mishra KP, Chanda S, Eslavath MR, Ganju L, Kumar B et al (2019) Effects of acute exposure to hypobaric hypoxia on mucosal barrier injury and the gastrointestinal immune axis in rats. High Alt Med Biol 20(1):35–44. https://doi.org/10.1089/ham.2018.0031. PMID 30484710
Khanna K, Mishra KP, Chanda S, Ganju L, Singh SB, Kumar B (2021a) Effect of Synbiotics on amelioration of intestinal inflammation under hypobaric hypoxia. High Alt Med Biol 22(1):32–44. https://doi.org/10.1089/ham.2020.0062. PMID 33185493
Khanna K, Eslavath MR, Ganju L, Mishra KP, Singh SB, Kumar B (2021b) Metagenomic insights into taxonomic and functional shifts under chronic exposure to hypobaric hypoxia induced microbial communities. J Proteins Proteomics 12(2):105–114. https://doi.org/10.1007/s42485-021-00062-1
Li K, Gesang L, Dan Z, Gusang L (2016) Genome-wide transcriptional analysis reveals the protection against hypoxia-induced oxidative injury in the intestine of Tibetans via the inhibition of GRB2/EGFR/PTPN11 pathways. Oxidative Med Cell Longev 6967396. https://doi.org/10.1155/2016/6967396. Epub 2016 Aug 9. PMID: 27594973
Lin L, Zhang J (2017) Role of intestinal microbiota and metabolites on gut homeostasis and human diseases. BMC Immunol 18(1):2. https://doi.org/10.1186/s12865-016-0187-3. PMID 28061847
Lobionda S, Sittipo P, Kwon HY, Lee YK (2019) The role of gut microbiota in intestinal inflammation with respect to diet and extrinsic stressors. Microorganisms 7(8):271. https://doi.org/10.3390/microorganisms7080271. PMID 31430948

Lundeberg J, Feiner JR, Schober A, Sall JW, Eilers H, Bickler PE (2018) Increased cytokines at high altitude: lack of effect of ibuprofen on acute mountain sickness, physiological variables, or cytokine levels. High Alt Med Biol 19(3):249–258. https://doi.org/10.1089/ham.2017.0144. Epub 2018 Jun 20. PMID: 29924642

Luo H, Zhou DJ, Chen Z, Zhou QQ, Wu K, Tian K et al (2017) Establishment and evaluation of an experimental rat model for high altitude intestinal barrier injury. Exp Ther Med 13(2):475–482. https://doi.org/10.3892/etm.2016.4012. PMID 28352318

Malacrida S, Giannella A, Ceolotto G, Reggiani C, Vezzoli A, Mrakic-Sposta S, Moretti S, Turner R, Falla M, Brugger H, Strapazzon G (2019) Transcription factors regulation in human peripheral white blood cells during hypobaric hypoxia exposure: an in-vivo experimental study. Sci Rep 9(1):9901. https://doi.org/10.1038/s41598-019-46391-6. PMID: 31289332

McKenna ZJ, Gorini Pereira F, Gillum TL, Amorim FT, Deyhle MR, Mermier CM (2022) High-altitude exposures and intestinal barrier dysfunction. Am J Physiol Regul Integr Comp Physiol 322(3):R192–R203. https://doi.org/10.1152/ajpregu.00270.2021. PMID 35043679

Mishra KP, Ganju L, Singh SB (2015) Hypoxia modulates innate immune factors: A review. Int Immunopharmacol 28(1):425–428. https://doi.org/10.1016/j.intimp.2015.07.008. Epub 2015 Jul 14. PMID: 26184693

Pellegrini C, Antonioli L, Colucci R, Blandizzi C, Fornai M (2018) Interplay among gut microbiota, intestinal mucosal barrier and enteric neuro-immune system: a common path to neurodegenerative diseases? Acta Neuropathol 136(3):345–361. https://doi.org/10.1007/s00401-018-1856-5. PMID 29797112

Pham K, Parikh K, Heinrich EC (2021) Hypoxia and inflammation: insights from high-altitude physiology. Front Physiol 12:676782. https://doi.org/10.3389/fphys.2021.676782. PMID 34122145

Ramakrishnan S, Anand V, Roy S (2014) Vascular endothelial growth factor signaling in hypoxia and inflammation. J Neuroimmune Pharmacol 9(2):142–160. https://doi.org/10.1007/s11481-014-9531-7. PMID 24610033

Saffouri GB, Shields-Cutler RR, Chen J, Yang Y, Lekatz HR, Hale VL et al (2019) Small intestinal microbial dysbiosis underlies symptoms associated with functional gastrointestinal disorders. Nat Commun 10(1):2012. https://doi.org/10.1038/s41467-019-09964-7. PMID 31043597

Sapone A, De Magistris L, Pietzak M, Clemente MG, Tripathi A, Cucca F et al (2006) Zonulin upregulation is associated with increased gut permeability in subjects with type 1 diabetes and their relatives. Diabetes 55(5):1443–1449. https://doi.org/10.2337/db05-1593. PMID 16644703

Song TT, Bi YH, Gao YQ, Huang R, Hao K, Xu G, Tang JW, Ma ZQ, Kong FP, Coote JH, Chen XQ (2016) Systemic proinflammatory response facilitates the development of cerebral edema during short hypoxia. J Neuroinflammation 13:1–4. https://doi.org/10.1186/s12974-016-0528-4. PMID: 26968975

Tajik N, Frech M, Schulz O, Schälter F, Lucas S, Azizov V et al (2020) Targeting zonulin and intestinal epithelial barrier function to prevent onset of arthritis. Nat Commun 11(1):1995. https://doi.org/10.1038/s41467-020-15831-7. PMID 32332732

Thevaranjan N, Puchta A, Schulz C, Naidoo A, Szamosi JC, Verschoor CP et al (2017) Age-associated microbial dysbiosis promotes intestinal permeability, systemic inflammation, and macrophage dysfunction. Cell Host Microbe 21(4):455–466.e4. https://doi.org/10.1016/j.chom.2017.03.002

Walmsley SR, Harris AJ, Roger Thompson AA, KB Whyte M (2014) HIF-mediated innate immune responses: cell signaling and therapeutic implications. Hypoxia (Auckl) 2:47–58. https://doi.org/10.2147/HP.S50269. PMID: 27774466

Wang C, Jiang H, Duan J, Chen J, Wang Q, Liu X, Wang C (2018) Exploration of acute phase proteins and inflammatory cytokines in early stage diagnosis of acute mountain sickness. High Alt Med Biol 19(2):170–177. https://doi.org/10.1089/ham.2017.0126. Epub 2018 Apr 2. PMID: 29608374

Wu TY (2001) Take note of altitude gastrointestinal bleeding. Newsletter Int Soc Mountain Med 10:9–11

Wu TY, Ding SQ, Liu JL, Jia JH, Dai RC, Zhu DC, Liang BZ, Qi DT, Sun YF (2007) High-altitude gastrointestinal bleeding: an observation in Qinghai-Tibetan railroad construction workers on Mountain Tanggula. World J Gastroenterol 13(5):774. https://doi.org/10.3748/wjg.v13.i5.774. PMID: 17278202

Yang T, Santisteban MM, Rodriguez V, Li E, Ahmari N, Carvajal JM et al (2015) Gut dysbiosis is linked to hypertension. Hypertension 65(6):1331–1340. https://doi.org/10.1161/HYPERTENSIONAHA.115.05315. PMID 25870193

Yoon W, Park SH, Lee JS, Byeon JH, Kim SH, Lim J et al (2021) Probiotic mixture reduces gut inflammation and microbial dysbiosis in children with atopic dermatitis. Australas J Dermatol 62(3):e386–e392. https://doi.org/10.1111/ajd.13644. PMID 34110005

Zhang F, Wu W, Deng Z, Zheng X, Zhang J, Deng S et al (2015) High altitude increases the expression of hypoxia-inducible factor 1α and inducible nitric oxide synthase with intestinal mucosal barrier failure in rats. Int. J Clin Exp Pathol 8(5):5189–5195. PMID 26191216

Zhang FX, Deng ZY, Dong JZ, Duan Y, Deng SX, Chen JY et al (2013) Effects of prebiotics on the expression of the integral membrane protein occludin and cytokines in rats after acute high altitude exposure. Int J Pharmacol 9(6):385–389. https://doi.org/10.3923/ijp.2013.385.389

Human Gastrointestinal Tract: Impact of Hypobaric Hypoxia

8

Manisha Kumari, Chhavi Rai, Anu Kumari, Rajeev Varshney, and Ramesh C. Meena

Abstract

Sojourning to high-altitude regions often leads to gastrointestinal distress, with symptoms such as anorexia, nausea, diarrhea, and vomiting. The infectious agents that contribute to this distress include microorganisms. Hypoxic stress is a central factor affecting human behavior in high-altitude areas. Hypoxia can cause bacterial displacement and dysfunction in the gastrointestinal barrier leading to a "leaky gut." The intestinal barrier is a semi-permeable structure made up of single-layered intestinal cells that help to preserve structural integrity and maintain selective permeability. Hypoxic exposures induce physiological alterations to the intestinal barrier, resulting in increased permeability and immunological dysregulation. Recent studies demonstrated an increase in intestinal permeability while ascending to high altitudes. In this chapter, we have discussed the importance of the human microbiome, which plays a crucial role in maintaining gut health and is also affected by environmental stressors such as hypoxia. The chapter includes human and animal studies demonstrating the effects of hypoxia on intestinal barrier integrity and immune functions and offers insights into the pathogenesis of gastrointestinal symptoms at high altitudes. The study of the effect of hypobaric hypoxia on the GIT may help to explain common clinical observations at high altitudes and provide insights into the pathogenesis of GI symptoms.

M. Kumari · C. Rai · A. Kumari · R. Varshney · R. C. Meena (✉)
Department of Pathophysiology and Disruptive Technologies, Defence Institute of Physiology and Allied Sciences (DIPAS), Defence Research and Development Organization (DRDO), Delhi, India
e-mail: rcmeena.dipas@gov.in

R. Tulsawani, D. Vohora (eds.), *Adaptation under Stressful Environments through Biological Adjustments and Interventions*,
https://doi.org/10.1007/978-981-99-7652-2_8

Keywords

Dysbiosis · Hypobaric Hypoxia · High Altitude · Microbiome · Mucosal Immunity

8.1 Introduction

Travelling to mountainous regions usually gives a sense of weird feeling in our gut. Tourists generally complain of having at least one symptom of gastrointestinal distress such as anorexia, nausea, diarrhea, and vomiting. Pilgrims of Shri Amarnath Yatra (3800 m) suffered from problems of gastroenteritis (Mir et al. 2008). A few infectious agents responsible for this distress are *Escherichia coli, Salmonella, Campylobacter, Giardia, Entamoeba, and Cyclospora*. It has been found that *Helicobacter pylori* infection may be implicated with an increased incidence of upper gastrointestinal tract (GI) cancer, the frequency of which increases with the rise in altitude from plains to the mountains. The manifestations differ considerably in short-term visitors, long-term residents, and native highlanders. Earlier studies have reported the prevalence of diarrhea in 30–36% of mountaineers who attempted to ascend Mount Everest (Murdoch 1995). In addition, gastroenteritis has also been reported about 23% of trekkers ascending to very high altitudes (5100 m), despite access to proper hygiene. Furthermore, a group of manual laborers working at an altitude of 4905 m reported a potentially fatal condition, "gut bleeding" (Wu et al. 2007).

Oxygen is the central factor affecting human behavior and physical activities under hypoxic stress. Low air pressure and oxygen content in plateau areas cause changes in the digestive system, thus destroying the gastrointestinal barrier and causing dysfunction resulting in bacterial displacement that is called a "leaky gut." It is established that a steep oxygen gradient exists along the length of the intestine from the lumen to the serosa, suggesting that the human intestine is already in a state of intense hypoxic stress (Zheng et al. 2015). This low basal oxygen tension stabilizes hypoxia-inducible factors which elicit several intracellular adaptations that collectively promote intestinal barrier integrity (Fagundes and Taylor 2017). However, exposure to hypoxic stress "physiological hypoxia" might exacerbate and bring severe implications on the physical functioning and working ability of populations living in high-altitude areas (Cheng et al. 2022).

The human intestinal microenvironment is relatively complex as it is home to an estimated 38 trillion bacterial cells. Close to 99% of these microorganism are obligate anaerobes. The human microbiome represents a second genome: a set of genes and gene products that enhance the activity of our own human genome. The normal residents, "autochthonous" microbes live in symbiotic relationships with the host, whereas the transient ones, "allochthonous" microbes colonize only on a temporary basis. This microbial community performs several functions that are beneficial to the host, including metabolic functions (fermenting indigestible food, yielding short chain fatty acids and vitamins), stimulation of intestinal angiogenesis.

Interestingly, the microbiome is also inextricably linked to our immunity, with progressively increasing research agreeing that huge proportion of the entire human immune system is found in the gut. Humans and their gut microbiota maintain a dynamic coexistence and a mutually beneficial relationship, but this intricate interplay between the intestinal microbiota, the intestinal epithelial layer, and the local mucosal immune system, changes when exposed to environmental stressors.

The chapter further presents a detailed discussion about understanding the effect of hypobaric hypoxia on the integrity of the intestinal tract, which may help to explain common clinical observations that occur at high altitude. It will also provide insights into the pathogenesis of these symptoms. The following sections will help us to better understand the impact of hypobaric hypoxia on the human GI tract.

8.1.1 The Intestinal Barrier Injury

The intestinal barrier is a semi-permeable structure that allows the uptake of essential nutrients and helps in immune sensing while being restrictive against pathogenic molecules and bacteria. It is made up of single-layered intestinal cells, such as enterocytes, goblet, paneth, and enteroendocrine cells, which are connected by junction complexes such as desmosomes, tight junctions, and adherens junctions. Desmosomes and adherens junctions help to preserve structural integrity, while tight junction complexes are important for maintaining selective permeability, barrier integrity, and closing paracellular gaps between colonocytes that are in charge of carrying tiny molecules. In fact, it has been demonstrated that several physiological alterations to the intestinal barrier are induced by hypoxic exposures (Fig. 8.1). For

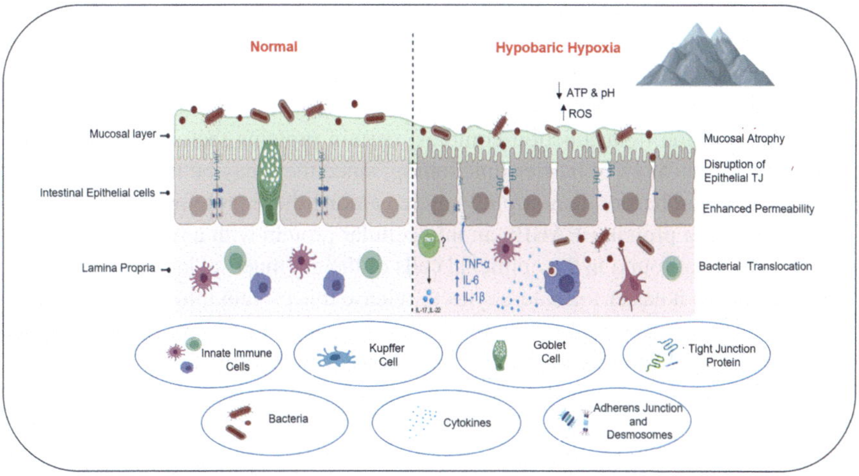

Fig. 8.1 Study of effect of hypobaric hypoxia on intestinal barrier integrity

instance, when subjected to hypobaric and normobaric hypoxia, animal studies showed significant epithelial damage and weakened intestinal barriers that resulted in increased permeability. Additionally, animals exposed to hypobaric hypoxia for 7 and 14 days showed immunological dysregulation, and increased IL-17 expression in tissue mRNA and serum, which is indicative of a shift towards Th17 polarisation (Khanna et al. 2019). Another study found that hypoxia can promote Th17 cell differentiation and suppress Treg differentiation by degrading the Treg master regulator FOXP3 and stabilizing RORt, suggesting that hypoxia has a pro-inflammatory and potentially harmful effect on the intestinal barrier (Dang et al. 2011).

Humans have also demonstrated an increase in intestinal permeability while ascending to high altitudes, either on an expedition or for scientific purposes. For instance, prior research by Dinmore et al. (1994) showed that a group of males ($n = 10$) and females ($n = 4$) on an excursion to 6300 m had reduced intestinal permeability. It is interesting to note that over the first several days when they visited from 60 m to 1050 m, there was a two-fold increase in intestinal permeability. On day 11, at an altitude of 5730 m, there was no appreciable increase in permeability, which would indicate partial acclimatization (Dinmore et al. 1994). Another recent study by Karl et al. (2018) reported increased intestinal permeability in healthy, physically active but un-acclimatized males ($n = 17$) following a rapid 22 days exposure to 4300 m (Karl et al. 2018). Nonetheless, the findings of both studies should be interpreted properly given the limitations within each study design. For example, neither study included a control group nor baseline samples which makes it difficult to determine the direct effects of hypobaric hypoxia without the influence of diet, physical activity, and/or other confounding factors (e.g., circadian disruption and psychological stress).

The research listed above makes it clear that if intestinal permeability can be quantified, serious consequences for gut health can be delayed. The assessment of intestinal permeability has previously been used to assess disease severity and forecast the course of various illnesses. It has been utilized as a disease activity marker for ulcerative colitis and Crohn's disease. Due to their placement in enterocytes and the tight junction complex, certain proteins seem to be intriguing candidates as indicators of barrier functioning, since they may be more mechanistically related to gut barrier disruption. Included among these proteins is the intestinal fatty acid binding protein (iFABP), an intracellular protein with a modest molecular mass (14–15 kDa) found in the epithelial cells of the gut mucosal layer. The position of iFABP makes it easier for enterocytes to release this protein into the bloodstream when there is intestinal injury, which causes this protein to be upregulated. Higher serum levels of iFABP were found in celiac disease patients with increased intestinal permeability, compared to healthy controls, suggesting that this protein may be used as an indicator of an ongoing inflammatory process. Plasma levels of iFABP were found to be higher in patients with a severe form of ulcerative colitis, compared to those with a mild form of the disease (Wells et al. 2017). Another protein involved in tight junction control that has been suggested as a direct biomarker of gut barrier integrity is zonulin (Prehaptoglobin-2). As compared to controls, Crohn's patients had higher blood and fecal levels of zonulin (Fasano 2012). In addition, mice gavage with zonulin displayed an increase in both the small intestine and gastroduodenal permeability in comparison to controls treated with bovine serum albumin (Tripathi et al. 2009).

8.1.2 Alterations in Mucosal Immunity

The mucosal immune system makes up the majority of the total immune system in terms of immunoglobulin synthesis, and immune cell mobilization. It is the first barrier that microorganisms in the gut come into contact with; covering the epithelial cells, it is made primarily of glycoproteins. It protects against invasive pathogens and other chemical or mechanical dangers. According to where the organ's lining is located, the mucosal immune system, also known as mucosa-associated lymphoid tissue (MALT), is separated into additional lymphoid organs. For example, MALT lining the bronchioles are referred to as bronchioles-associated lymphoid tissue (BALT) (Sminia et al. 1989) and lymphoid cells lining the gut are referred to as gut-associated lymphoid tissue (GALT) (Zmora et al. 2017). The Peyer's patches (PPs), intestinal lamina propria, intestinal intra epithelium, crypto patches, lymphoid follicles, and mesenteric lymph nodes are major inductive sites of the GALT (McGhee and Fujihashi 2012).

M cells are specialized epithelial cells that are present in Peyer's patches. By transporting antigens from the lumen to immune cells, they are in charge of starting the immunological responses (Jang et al. 2004). A prominent antibody isotype of the mucosal immune system, immunoglobulin A (IgA), has demonstrated a crucial function in preventing infections from entering the circulation by increasing its concentration when exposed to high altitude hypoxic environments. T cells are among the several diverse kinds of immune cells. The gut also contains macrophages, mast cells, NK cells, plasma cells, and dendritic cells. The number of circulating plasmacytoid dendritic cells (pDCs), in the peripheral blood of 17 healthy individuals subjected to hypobaric hypoxia (3000 m, 3 days), have been demonstrated to drastically decrease at high altitudes (Rohm et al. 2016). Follow-up in a normoxic setting revealed yet another considerable reconstitution of the circulating pDCs. IgA and IgG levels in the blood of high altitude and sea level inhabitants who lived at high altitudes for 2 years were shown to be upregulated in another study related to human volunteers. The peripheral blood lymphocytes of healthy females who spent 21 days at an altitude of 5050 m revealed alterations in T cells and NK cells after acute and chronic exposure to hypobaric hypoxia, but not in the population of B cells. Another study also suggested impairment of homeostatic regulation of the Th1/Th2 immune balance on exposure to HA, which could potentially increase the risk of infections and long-term immunological abnormalities (Facco et al. 2005).

8.1.3 Alteration in Gut Microbiota

The major functions of an immune system have been demonstrated to be impacted by microbial dysbiosis, and changes to the gut microbiota may contribute to a variety of gut-related problems. Numerous types of research have investigated how altitude affects the human intestine (Fig. 8.1). Twelve healthy Tibetans (T group, permanently lived in Tibet and had never left the place), 12 healthy Chinese Han (HH group, 3100 m from sea level), and 11 healthy Chinese Han (LH group, altitude of 450 m above sea level) who had never lived in Tibet were studied. Significant differences were found in the genera *Lactobacillus, Prevotella, Clostridium, Desulfovibrio, and Bacteroides* between the three groups as revealed in the analysis of 454 pyrosequencing (Li and Zhao 2015). It was found that high altitude may have an impact on how the human gut flora develops. In research simulating hypobaric hypoxia at 4872.9 m, the density of total aerobes was found to be significantly decreased, while the density of total anaerobes, grew dramatically after 30 days (8 h per day) of exposure (Adak et al. 2014a). Another study revealed that hypobaric hypoxia for 30 days (8 h per day) at 55 kPa air pressure (4872.9 m approximately) damages the large intestine's epithelial barrier. Hypobaric hypoxia exposure was used to regulate large gut microbial populations, which were thought to be the root of systemic infection and consequent epithelium damage (Adak et al. 2014b). The examination of feces from seven mountaineers who spent 47 days climbing over 5000 m revealed an increase in potentially dangerous *Proteobacteria* and *Enterobacteriaceae* such as *Escherichia coli*. During hypoxic exposure at high altitude environment, the bacteria including *Bifidobacteria* and species from the *Atopobium*, *Coriobacterium*, and *Eggerthella lenta* groups were reduced. Therefore, high altitude-induced immunological challenges might be caused by altered gut microbial composition (Kleessen et al. 2005). In a recent study, intestinal microbial communities of mice housed under normal conditions (2200 m, 16% oxygen) and those subjected to hypobaric hypoxia (5000 m, 11% oxygen) were sequenced using the HiSeq Illumina platform. While no appreciable variations were detected in alpha diversity, significant differences were found in beta diversity, the relative abundance of the class Epsilonproteobacteria, phylum *Actinobacteria*, class Erysipelotrichia, genus *Helicobacter*, and *Alistipes*. Between the two groups, there were no appreciable differences in the populations of the aerobic, facultative anaerobic, possibly pathogenic, stress-tolerant, mobile element, biofilm-forming, Gram-negative, and Gram-positive bacteria. Functional analysis also revealed substantial variations between the two groups, in 34 genes in functional metabolic pathways, including those for energy metabolism, arachidonic acid metabolism, flavonoid biosynthesis, and the RIG-I-like receptor signaling pathway (Zhang et al. 2018). Three Indian groups, including urban and rural people living in the Ballabhgarh area and rural high-altitude regions of Leh, Ladakh, India, participated in research on human subjects. The dominance of *Firmicutes*, *Bacteroidetes*, *Actinobacteria,* and *Proteobacteria* communities was observed in these Indian communities. A high representation of *Bacteroidetes* and a low abundance of *Proteobacteria* was observed in people residing at Leh while individuals living in Ballabhgarh areas harbored a higher number of *Firmicutes* and *Proteobacteria*. The rural Ballabhgarh inhabitants demonstrated a distinctive microbiome with a greater level of inter-individual homogeneity (Das et al. 2018).

8.2 Summary and Future Perspective

As humans continue to venture to high-altitude regions for various purposes such as tourism, research, and occupation, the impact of these environments on human health is becoming an increasingly important topic of research. Studies have already shown that exposure to high altitudes leads to alterations in the composition of gut microbiota, with a decrease in diversity and an increase in the abundance of certain bacterial species. These alterations have been linked to a variety of health outcomes, including gastrointestinal distress, altered nutrient absorption, and immune dysfunction.

One possible future perspective is that as more research is conducted in these areas, we may discover ways to modulate the gut microbiota to mitigate the negative effects of high-altitude exposure. For example, prebiotics or probiotics could be used to promote a more diverse and stable gut microbiota, which could in turn improve nutrient absorption and immune functions. Overall, the study of alterations in human gut microbiota and mucosal immunity at high altitude is a promising area of research with potential applications for improving the health and well-being of those who venture into high altitude environment.

References

Adak A, Ghosh K, Mondal KC (2014a) Modulation of small intestinal homeostasis along with its microflora during acclimatization at simulated hypobaric hypoxia. Indian J Exp Biol 52(11): 1098–1105. PMID 25434105

Adak A, Maity C, Ghosh K, Mondal KC (2014b) Alteration of predominant gastrointestinal flora and oxidative damage of large intestine under simulated hypobaric hypoxia. Z Gastroenterol 52(2):180–186. https://doi.org/10.1055/s-0033-1336007. PMID 24526402

Cheng J, Sun Y, He J, Wang Z, Li W, Wang R (2022) The mechanism of colon tissue damage mediated by HIF-1α/NF-κB/STAT1 in high-altitude environment. Front Physiol 13:933659. https://doi.org/10.3389/fphys.2022.933659. PMID 36164339

Dang EV, Barbi J, Yang HY, Jinasena D, Yu H, Zheng Y et al (2011) Control of TH17/Treg balance by hypoxia-inducible factor 1. Cell 146(5):772–784

Das B, Ghosh TS, Kedia S, Rampal R, Saxena S, Bag S et al (2018) Analysis of the gut microbiome of rural and urban healthy Indians living in sea level and high altitude areas. Sci Rep 8(1):1

Dinmore AJ, Edwards JS, Menzies IS, Travis SP (1994) Intestinal carbohydrate absorption and permeability at high altitude (5,730 m). J Appl Physiol (1985) 76(5):1903–1907. https://doi.org/10.1152/jappl.1994.76.5.1903. PMID 8063648

Facco M, Zilli C, Siviero M, Ermolao A, Travain G, Baesso I et al (2005) Modulation of immune response by the acute and chronic exposure to high altitude. Med Sci Sports Exer 37(5): 768–774. https://doi.org/10.1249/01.MSS.0000162688.54089.CE

Fagundes RR, Taylor CT (2017) Determinants of hypoxia-inducible factor activity in the intestinal mucosa. J Appl Physiol (1985) 123(5):1328–1334. https://doi.org/10.1152/japplphysiol.00203. PMID 28408694

Fasano A (2012) Zonulin, regulation of tight junctions, and autoimmune diseases. Ann N Y Acad Sci 1258(1):25–33. https://doi.org/10.1111/j.1749-6632.2012.06538.x

Jang MH, Kweon MN, Iwatani K, Yamamoto M, Terahara K, Sasakawa C et al (2004) Intestinal villous M cells: an antigen entry site in the mucosal Yokota epithelium. Proc Natl Acad Sci U S A 101(16):6110–6115

Karl JP, Berryman CE, Young AJ, Radcliffe PN, Branck TA, Pantoja-Feliciano IG et al (2018) Associations between the gut microbiota and host responses to high altitude. Am J Physiol Gastrointest Liver Physiol 315(6):G1003–G1015. https://doi.org/10.1152/ajpgi.00253.2018. PMID 30212253

Khanna K, Mishra KP, Chanda S, Eslavath MR, Ganju L, Kumar B et al (2019) Effects of acute exposure to hypobaric hypoxia on mucosal barrier injury and the gastrointestinal immune axis in rats. High Alt Med Biol 20(1):35–44. https://doi.org/10.1089/ham.2018.0031. PMID 30484710

Kleessen BR, Schroedl WI, Stueck MA, Richter A, Rieck O, Krueger MO (2005) Microbial and immunological responses relative to high-altitude exposure in mountaineers. Med Sci Sports Exerc 37(8):1313–1318. https://doi.org/10.1249/01.mss.0000174888.22930.e0. PMID 1611 8577

Li L, Zhao X (2015) Comparative analyses of fecal microbiota in Tibetan and Chinese Han living at low or high altitude by barcoded 454 Pyrosequencing. Sci Rep 5(1):14682. https://doi.org/10. 1038/srep14682. PMID 26443005

McGhee JR, Fujihashi K (2012) Inside the mucosal immune system. PLoS Biol 10(9):e1001397. https://doi.org/10.1371/journal.pbio.1001397. PMID 23049482

Mir IS, Mir M, Ahmed M (2008) Profile of non-traumatic surgical disorders found in the pilgrims/ trekkers travelling to Shri Amarnath Ji cave. Indian J Med Res 128(6):740–743. PMID 1924 6798

Murdoch DR (1995) Symptoms of infection and altitude illness among hikers in the Mount Everest region of Nepal. Aviat Space Environ Med 66(2):148–151. PMID 7726779

Rohm I, Aderhold N, Ratka J, Goebel B, Franz M, Pistulli R et al (2016) Hypobaric hypoxia in 3000 m altitude leads to a significant decrease in circulating plasmacytoid dendritic cells in humans. Clin Hemorheol Microcirc 63(3):257–265. https://doi.org/10.3233/CH-152035. PMID 2689010 7

Sminia T, Van der Brugge-Gamelkoorn GJ, Jeurissen SH (1989) Structure and function of bronchus-associated lymphoid tissue (BALT). Crit Rev Immunol 9(2):119–150. PMID 2 663024

Tripathi A, Lammers KM, Goldblum S, Shea-Donohue T, Netzel-Arnett S, Buzza MS et al (2009) Identification of human zonulin, a physiological modulator of tight junctions, as prehaptoglobin-2. Proc Natl Acad Sci U S A 106(39):16799–16804. https://doi.org/10.1073/ pnas.0906773106. PMID 19805376

Wells JM, Brummer RJ, Derrien M, MacDonald TT, Troost F, Cani PD et al (2017) Homeostasis of the gut barrier and potential biomarkers. Am J Physiol Gastrointest Liver Physiol 312(3):G171–G193. https://doi.org/10.1152/ajpgi.00048.2015. PMID 27908847

Wu TY, Ding SQ, Liu JL, Jia JH, Dai RC, Zhu DC et al (2007) High-altitude gastrointestinal bleeding: an observation in Qinghai-Tibetan railroad construction workers on Mountain Tanggula. World J Gastroenterol 13(5):774–780. https://doi.org/10.3748/wjg.v13.i5.774. PMID 17278202

Zhang W, Jiao L, Liu R, Zhang Y, Ji Q, Zhang H et al (2018) The effect of exposure to high altitude and low oxygen on intestinal microbial communities in mice. PLoS One 13(9):e0203701. https://doi.org/10.1371/journal.pone.0203701. PMID 30208088

Zheng L, Kelly CJ, Colgan SP (2015) Physiologic hypoxia and oxygen homeostasis in the healthy intestine. A review in the theme: cellular responses to hypoxia. Am J Physiol Cell Physiol 309(6):C350–C360. https://doi.org/10.1152/ajpcell.00191.2015. PMID 26179603

Zmora N, Bashiardes S, Levy M, Elinav E (2017) The role of the immune system in metabolic health and disease. Cell Metab 25(3):506–521. https://doi.org/10.1016/j.cmet.2017.02.006. PMID 28273474

Mechanism of Skeletal Muscle Atrophy at High Altitude: Role of Herbals and Nutraceuticals

9

Richa Rathor and Geetha Suryakumar

Abstract

Exposure to high altitude leads to numerous adaptive and maladaptive responses in our body which accentuates a decline in the physical performance. Skeletal muscle atrophy is a well-reported phenomenon during prolonged stays at high altitudes. Hypobaric hypoxia, the major stressor at high altitudes, disturbs the redox homeostasis and activates a plethora of intracellular signaling pathways, including ER stress, altered anabolic pathways, disturbed calcium homeostasis, enhanced protein degradation, and apoptosis in the skeletal muscle. Our recent research has elucidated the multiple signaling pathways responsible for high altitude-induced skeletal muscle atrophy. Understanding the key mechanisms of the atrophic phenomena paves the way for the development of strategies and interventions to ameliorate this protein loss. Dietary phytochemicals and nutraceuticals are gaining attraction for the amelioration of skeletal muscle atrophy and enhancing physical performance. The present chapter described a few phytochemicals that not merely ameliorate skeletal muscle protein loss but also manage active lifestyles under stressful conditions.

Keywords

Hypoxia · High altitude · Skeletal muscle · Nutraceuticals

R. Rathor · G. Suryakumar (✉)
Defence Institute of Physiology and Allied Sciences (DIPAS), Defence Research and Development Organization (DRDO), Delhi, India
e-mail: sgeetha.dipas@gov.in

R. Tulsawani, D. Vohora (eds.), *Adaptation under Stressful Environments through Biological Adjustments and Interventions*,
https://doi.org/10.1007/978-981-99-7652-2_9

9.1 Introduction

Millions of the population permanently reside at high altitudes above 3000 m (9843 ft.) in Asia and South America and are exposed to innumerable stressors like low temperature, increased UV radiations, intense wind velocity, and less oxygen availability because of a decline in atmospheric pressure. Moreover, nowadays, people visiting high altitudes for recreational, professional, and pilgrimage purposes are constantly increasing. Several physiological, biochemical, and metabolic adaptative responses are activated under hypobaric hypoxic conditions to compensate for the low oxygen availability under the extreme environment. Low oxygen availability becomes steadily extra awful as the duration of stay and altitude increase. Thus, high altitude experience put the body and all the vital organs in a serious situation because of low oxygen accessibility.

On the basis of height, high altitude can categorize into three parts such as high altitude, very high altitude, and extreme altitude (Fig. 9.1).

9.1.1 High Altitude (1500–3500 M)

A decline in partial pressure of inspired oxygen (PIO_2) at altitude leads to a hyperventilation response resulting in lower arterial carbon dioxide partial pressure ($PaCO_2$). The arterial partial pressure of oxygen (PaO_2) is significantly declined, and the arterial oxygen transport SaO_2 of around 90% is present.

9.1.2 Very High Altitude (3500–5500 M)

Exposure to this altitude leads to a fall in SaO_2 below 90% that leads to extreme hypoxemia which causes Acute mountain sickness, cerebral and pulmonary edema,

Fig. 9.1 Altitude categories

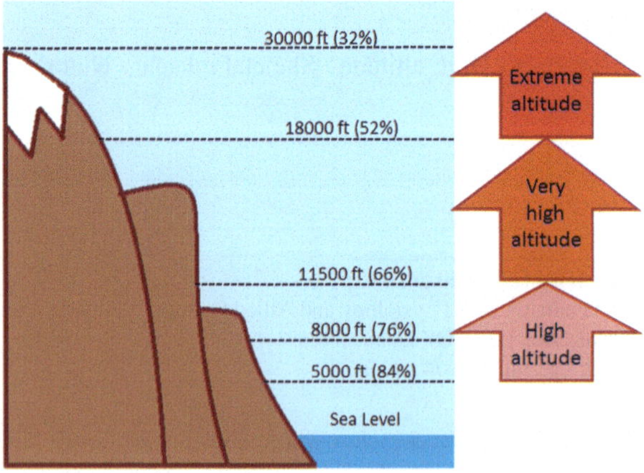

sleep disturbances, immune failure, decrease in appetite, muscle loss, etc. (Gallagher and Hackett 2004; Fuehrer et al. 2022; Hackett and Roach 2004).

9.1.3 Extreme Altitude (>5500 M)

Extreme altitude exposes marked hypoxemia, hypocapnia, and alkalosis environment. Physiological functions are progressively decreased at this height which leads to casualty.

9.2 Maladies Associated with High altitude

9.2.1 Hyperventilation, Hematological Changes, Renal Function, and Fluid Balance

Acute hypoxia exposure causes hyperventilatory response and respiratory alkalosis. After arrival at 3500–4000 m altitude, plasma volume is reduced by around 5% and this deficiency persists for 4 months. The water content is also reduced to 5% and numerous reasons are responsible for the same as a decline in thirst regulation, decrease water intake, or sometimes loss of senses. Furthermore, plasma volume reduction results in an increase in hemoglobin concentration. Moreover, hypoxia exposure also led to enhancing the erythropoiesis process and red cell mass.

9.2.2 Cardiovascular Changes and Oxygen Consumption

Acute hypoxia exposure led to an increase in heart rate and cardiac output. Furthermore, oxygen consumption (VO_2 max) also decreased as the height increased.

9.2.3 Acute Mountain Sickness

Acute exposure to hypoxia detailed symptoms like nausea, loss of appetite, vomiting, fatigue, weakness, dizziness, and light-headedness as well as sleep disturbances.

9.2.4 High Altitude Pulmonary Oedema (HAPE) and Cerebral Oedema (HACE)

Acute mountain sickness results in fluid retention and anti-diuresis, and once these symptoms are retained for a long time, it causes edema in the lungs and heart. Although the incidence of this illness is very rare, it could be dangerous for life (affects 1–2% population who speedily climb to 4500 m) (Hackett and Roach 2004).

Early symptoms of HAPE consist of dry cough but if persists it converted into bubbly, blood-stained wet sputum. HACE warning signs include ataxia with altered consciousness. Human beings can be hallucinated due to HACE.

9.2.5 Sleep Disorder

Disturbance in sleep was also observed during early-stage hypoxia exposure. Furthermore, unpleasant dreams, awakenings, and poor-quality sleep used to be faced by individuals as the altitude increased (Zieliński et al. 2000; Weil 2004).

9.2.6 Skeletal Muscle Atrophy at High altitude

Skeletal muscles are the metabolic organ that is also considered the major reservoir of body proteins. This organ plays a pivotal role in many body functions including mechanical as well as metabolic processes. Besides this, it plays a highly important role in mobility and physical movement. Fat-free mass contributes to 30–40% of the body weight and constitutes around 50% of body proteins. High altitude-linked loss of body weight is one of the most inevitable changes (McDonald et al. 2022), and altitude exposure of more than 5000 m accounts for 60%–70% of body weight loss which includes a significant loss of fat-free mass (FFM) or skeletal muscle mass (Rose et al. 1988; Hamad and Travis 2006; Dünnwald et al. 2019). Loss of skeletal muscle mass and atrophy has been well-documented in several studies on prolonged exposure to high altitude (McKinnell and Rudnicki 2004; Chaudhary et al. 2012; Rathor and Suryakumar 2016). Rose et al. (1988) reported a 10–20% reduction in muscle mass on mountaineering expeditions. Bonaldo and Sandri (2013) described decreased muscle volume and less cross-sectional area (CSA) of myofibers. Earlier reports from our lab suggested the fact that gastrocnemius muscles are more prone to high altitudes with reference to the soleus muscle (Chaudhary et al. 2012). Bharadwaj et al. (2000) investigated the effects of a 10-month stay at high altitude (HA) on body composition of Indian soldiers with reference to body musculature and described muscle mass degradation at high altitude. Soldiers deployed at height 4100 m degraded 1.74 kg muscle mass; while, soldiers deployed at height 3750 m degraded 1.38 kg muscle mass.

Furthermore, few reports suggested the differential response of skeletal muscles during short-term hypoxia whereas muscle mass loss and negative nitrogen balance were observed during chronic stay (Shukla et al. 2005; Vats et al. 2007).

9.2.6.1 Skeletal Muscle Fibers

The number, as well as the size of the muscle fiber, is considered for depicting the structure of skeletal muscle. Skeletal muscles are composed of myofibrils with actin and myosin filaments which play a prime role in muscle contraction. They are multinucleated in which the nuclei are placed along the periphery and the muscle cells are post-mitotic. Each muscle is covered by a connective tissue sheath which is termed an epimysium. Furthermore, a single muscle fiber is made of more than 80% proteins which include contractile, regulatory, and cytoskeletal. The plasma membrane of muscle which is sarcolemma comprises a massive network of proteins and dysfunction or mutation in one of these proteins leads to compromised contractile activity and sometimes results in muscle weakness, damage, wasting, and atrophy (Frontera and Ochala 2015).

On the basis of differences in contractile and metabolic properties, skeletal muscles can be categorized as slow-twitch and fast-twitch fibre types. Slow-twitch/oxidative muscle fibers (SO) are characterized by a high content of myoglobin, mitochondria, and wide capillary density whereas fast-twitch/glycolytic fibres (FG) have reduced myoglobin, less mitochondrial as well as capillary density with bigger cross-sectional areas. Furthermore, there are fast-twitch oxidative glycolytic fibres (FOG) that manifest intermediate properties. Recent research works have now classified skeletal muscle fibre type, based on the isoforms of myosin heavy chain: type-I fibres, type-IIA, and type-IIX/IIB fibres which can be grouped into the groups of SO, FOG, as well as FG, respectively. Soleus which are red muscles have type I fibre while gastrocnemius which are white muscles have a mixture of type I and II fibres composition. Both gastrocnemius and soleus muscles play a major role in physical movements, locomotion, and performance.

The various compartments of proteins such as structural, contractile, and regulatory which particularly form the musculoskeletal system have received much scientific attention due to muscle's contribution to mobility, helps in bone movement, protecting internal organs, exercise capacity, functioning, and health. Skeletal muscle has been classified into three groups based on its location in the muscle tissue: (1) myofibrillar proteins, (2) sarcoplasmic proteins, and (3) stroma proteins. The most crucial proteins responsible for contractile properties are myofibrillar proteins which constitute 55% to 60% of the total proteins in the muscle. Furthermore, sarcoplasmic proteins contribute to about 30% to 35% of the total proteins and are vital for the cytoskeleton structure, excitation-contraction (E-C) coupling, energy release, and force generation (Ottenheijm and Granzier 2010). Approximately, 10–15% is constituted by the stroma proteins which majorly consist of collagen and extracellular matrix proteins. Other regulatory proteins comprise tropomyosin and calcium-dependent troponin complex viz. troponins C, I, and T which actively participates in myofilaments sliding during E-C coupling and force generation. In addition, there are anchoring proteins such as titin and nebulin which anchorage Z-disc to the myofilaments (Ottenheijm and Granzier 2010; Monroy et al. 2012). This massive complex network of multiple proteins constitutes the muscle sarcomere and consequently leads to the initiation of the process of E-C coupling and helps in the generation of muscle force and contractility.

9.3 Energy Metabolism

Oxygen supply plays a key role in regulating skeletal muscle functions like maintaining energetics and redox homeostasis through oxidative phosphorylation (Mason and Johnson 2007). Hypobaric hypoxia affects the energy generation by aerobic metabolism due to decrease in mitochondrial density (Green et al. 2000), increased amino acid and glucose metabolism (Horscroft and Murray 2014), prevention of β-oxidation, oxidative phosphorylation (Jacobs et al. 2012), increased anaerobic capacity as well as respiration (Edwards et al. 2010; Horscroft and Murray 2014). The metabolism in the muscle is affected by the altitude as well as the tenure at high altitude (Levett et al. 2012).

9.4 Altered Redox Homeostasis

Redox status and its homeostasis are very important for the cell that maintains the balance between the levels of oxidants and antioxidants. The alteration of disturbance in the redox signaling leads to oxidative stress. High altitude-induced hypobaric hypoxia due to a decrease in barometric pressure leads to enhanced generation of free radicals and a decline in the antioxidant defense response of the body (Bakonyi and Radak 2004). Mohanraj et al. (1998) reported that due to the disturbance in electron transport chain and aerobic metabolism, there were excess ROS levels under hypoxia Mohanraj et al. (1998). Furthermore, the decreased oxygen availability at high altitudes also leads to restricted ATP generation and this condition is also known as reductive stress (Khan and O'Brien 1995).

Acute and severe hypobaric hypoxia exposure leads to enhance oxidative damage and compromised respiratory function in skeletal muscle in mice (Debevec et al. 2017). Many studies have also supported the direct correlation between the degree of hypoxia, oxidative damage, and its effect on skeletal muscle performance under low oxygen conditions (Singh et al. 2001; Hoppeler et al. 2003). Proteins constitute a major part of the muscle mass, and they are made of many thiol groups containing amino acids. As the sulfydryl group of thiols is susceptible to oxidants, amino acids get oxidized and result in oxidative protein modification (Agrawal et al. 2017). Elevated ROS levels also led to lipid peroxidation that causes dysregulated calcium ion homeostasis in the cell and the progression of muscular atrophy (Barhwal et al. 2009; Semenov et al. 2009).

9.5 Inflammation

Inflammation plays a paramount role in high altitude-induced illnesses and maladies. Mainly, inflammation is regulated by transcription factors nuclear factor-kappa B (NF-κB) and hypoxia-inducible factor (HIF-1α) during exposure to high altitude conditions. NF-κB is a transcription factor that mainly regulates the proinflammatory response; while, HIF-1α, a heterodimeric transcription factor, is considered the master regulator for eliciting the cellular and molecular response against hypoxia (Semenza 2012). Evidence from earlier research demonstrated that the expression of HIF-1α is enhanced significantly during exposure to chronic hypoxia (Chaudhary et al. 2012; Pham et al. 2021). Moreover, other studies have also described the significant role of increased pro-inflammatory cytokines levels (TNF-α, IFN-γ, and IL-1β) in skeletal muscle during hypobaric hypoxia exposure that further has a prominent role in the pathogenesis of altitude (Hackett and Roach 2001; Agrawal et al. 2018; Gangwar et al. 2020). Enhanced NF-κB and TNF- α levels ultimately result in skeletal muscle atrophy via activating muscle-specific E3 ligases, Mafbx-1, and MuRF-1 (Agrawal et al. 2018). Furthermore, human studies also supported the fact as a significant increase in plasma TNF-α, IL-1β, IL-6, and cytokine levels was recorded in the volunteers at 3860 m altitude (Song et al. 2011; Mishra et al. 2016).

9.6 Endoplasmic Reticulum (ER) Stress

ER regulates several cellular functions in maintaining proper protein homeostasis such as protein synthesis, post-translational modifications, protein folding especially of the secretory proteins, and calcium homeostasis (McCaffrey and Braakman 2016; Schwarz and Blower 2016). Hypobaric hypoxia induced at high altitudes causes enhanced oxidative protein modifications and disturbances in the folding machinery of the proteins which further lead to the accumulation of misfolded and oxidized proteins in ER and thus eliciting the ER stress response (Pu et al. 2020). Furthermore, our lab also reported enhancement in ER stress-related protein expressions, GRP78, PERK, IRE1α, ATF6, ATF4, CHOP, and caspase-12 in cardiac tissue during chronic hypobaric hypoxia Jain et al. (2016a, b); while, the expression of chaperones such as GRP78, HSP70, and HSP60 was also found enhanced in skeletal muscles during acute hypobaric hypoxia exposure (Barhwal et al. 2009) and the persistent and severe ER stress resulted in apoptosis and promoted cell death (Agrawal et al. 2018; Iurlaro and Muñoz-Pinedo 2016).

9.7 Remodelling of Ryanodine Receptor-1 (RyR1) and Dysregulated Ca2+ Homeostasis

Primarily, RyR1 is situated at the terminal cisternae of the SR that is engaged in regulating Ca^{2+} ion channeling from SR to the cytoplasm. RyR1 is four homotetramer proteins that consist of FKBP12, a phosphodiesterase (PDE4D3), a phosphatase (PP1), and calmodulin Bellinger et al. (2008). Redox remodeling of this receptor has been reported in many diseases, including Duchenne muscular dystrophy Meyer et al. (2021), intense exercise (Place et al. 2015), heart failure (Kushnir et al. 2018), and aging (Gaboardi et al. 2018). It is a well-known fact that hypobaric hypoxia altered redox homeostasis and enhances oxidative-nitrosative stress. Our recent research has described oxidative damage-promoted remodelling of the ryanodine receptor that alters the cytosolic calcium concentration. The redox modification of RyR1 promoted leakage of calcium (Ca^{2+}) into the cytosol. This increase in cytosolic Ca^{2+} further activated the degradation of proteins via enhancing calpain and caspase activity (Agrawal et al. 2020).

9.8 Protein Homeostasis

Maintenance of protein homeostasis or proteostasis is a fundamental cellular mechanism for sustaining the quality control and equilibrium of proteins in the human body (Hutt and Balch 2010; Cheng et al. 2018). The balance between the rate of protein synthesis and degradation process is mandatory for managing proteostasis. Inequity between antioxidants and ROS levels triggers the modifications of amino acid moieties due to oxidation, enzyme inactivation, and misfolded proteins. This further leads to post-translational modifications in the critical proteins of the skeletal muscle which disrupts the proteostasis. The delicate equilibrium between protein synthesis and degradation is critical for maintaining skeletal muscle mass and both crucial processes are sensitive to intrinsic and extrinsic factors such as hormonal balance, nutritional status, physical activity, exercise and burn or injury, disease, etc.

9.9 Protein Synthesis at High altitude

The intricate pathway of protein synthesis is categorized into three steps: gene transcription, translation, and protein breakdown. Among the three steps, translational control of proteins is considered the main step which commences via the IGF1-Akt-mTOR pathway. Our earlier observations by Chaudhary et al. (2012) observed the altered protein turnover rate in the skeletal muscle under prolonged exposure to hypoxia by an imbalance in the process of protein synthesis and degradation pathway. The results indicate that exposure to hypoxia-activated the process of protein synthesis as well as degradation. However, the rate of protein synthesis was increased by 1.5-fold; however, the degradation pathway was enhanced by 5.0 fold. This imbalance became a prime culprit for hypobaric hypoxia-induced muscle atrophy. Furthermore, a comprehensive study by us also described that the expression of Akt, p-Akt, and p70S6K progressively enhanced during hypobaric hypoxia exposure (Agrawal et al. 2018).

9.10 Protein Degradation

Researchers have established the four main proteolytic pathways which contributed towards skeletal muscle atrophy (Powers et al. 2007): (1) calpains, (2) the ubiquitin proteasome system (UPS), (3) caspases, and (4) autophagy/lysosomal. Prolonged exposure to hypobaric hypoxia leads to activation of ER stress response and further is responsible for upregulation of C/EBP homologous protein (CHOP) expression. The increased expression of CHOP enhances the apoptotic cellular death pathways in the skeletal muscle. In addition, enhanced levels of cytosolic Ca^{2+} activated apoptosis regulators such as caspase-3, caspase-9, and annexin V (Agrawal et al. 2018). Our earlier study by Chaudhary et al. (2012) also reported the upregulation of the ubiquitin-proteasome pathway and calpains during chronic hypoxia exposure. The enhanced protein degradation due to these catabolic processes is primarily accountable for skeletal muscle protein loss under hypobaric hypoxia. Another study too reported that LC3B (puncta) and beclin, the autophagy markers also amplified significantly in hypobaric hypoxia-exposed rats (Agrawal et al. 2022). Hence, the literature evidenced that these multiple pathways are interconnected and aggregately accountable for hypobaric hypoxia-induced muscle degradation (Lee et al. 2004; Bowen et al. 2015).

9.11 Molecular Mechanism of Hypobaric Hypoxia-Induced Muscle Atrophy

The prime adverse environmental stress at high altitudes is hypobaric hypoxia or low oxygen condition that is responsible for disturbance in redox homeostasis of cells. This redox imbalance activated the inflammatory response through NF-kB and TNF-α pathways and also leads to remodelling of ryanodine receptor-1 (RyR1). In addition, the enhanced ROS production also causes an increase in protein oxidation and accumulation of oxidized, unfolded, and misfolded proteins (ER stress). In the initial stage, ER stress is an adaptive response to cope with this enhanced misfolded protein load with the help of chaperones, HSP70, HSP60, GRP78, and GRP94. But if hypobaric hypoxia is chronic or persistent, the cells activate the degradation and apoptosis. In addition, the redox modelling of RyR1 promotes leakage of Ca^{2+} ion to cytosol that ultimately enhances calpain activity and apoptosis process. Hence, low oxygen condition commences with dysregulated redox and protein homeostasis in the skeletal muscle which ends up with cell death and atrophy (Fig. 9.2).

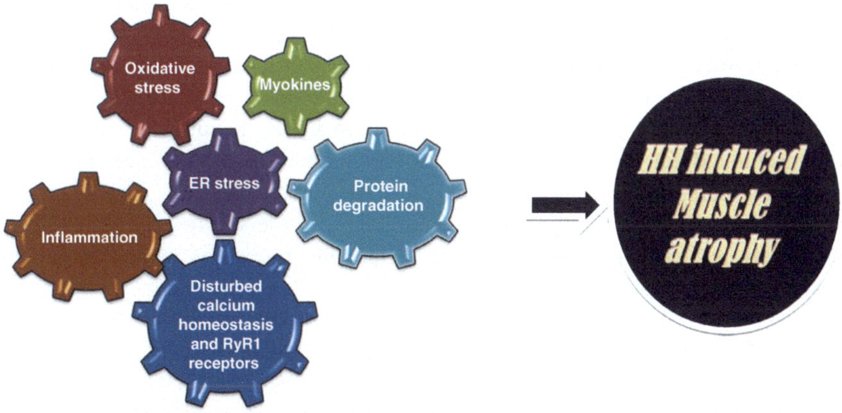

Fig. 9.2 A glimpse of hypobaric hypoxia induced muscle atrophy

9.12 Herbals and Nutraceuticals to Ameliorate Hypobaric Hypoxia Induced Muscle Atrophy

The maintenance of proper skeletal muscle function and strength is vital for the optimal physical performance of soldiers who are deployed in high-altitude regions. Our extensive research has elucidated the mechanism of hypobaric hypoxia-induced skeletal muscle loss that results in compromised physical performance at high altitudes. Therefore, strategies and interventions which can regulate the key pathways in the muscle to maintain protein homeostasis are a prerequisite for soldiers who are serving at strategic borders of our country. Here, we have described a few dietary herbals and nutraceuticals that could be a probable solution to ameliorate hypobaric hypoxia-induced muscle atrophy.

Nutraceuticals are defined as food-based sources which can provide physiological benefits in addition to their nutritional value (Kalra 2003). Nutraceuticals may be used for many purposes, including delay in aging, prevention of chronic diseases, and improve overall health (Zhao 2007).

9.12.1 Sea Buckthorn (SBT)

Scientific classification		
Kingdom	: Plantae	
Order	: Rosales	
Family	: *Elaeagnaceae*	
Genus	: *Hippophae*	
Species	: *rhamnoides*	

A Himalayan medicinal plant, sea buckthorn (SBT), is well recognized all over the world for its nutritional and medicinal value. The scientific name of this plant is *Hippophae rhamnoides L.* Elaeagnaceae family. It is a thorny cold-resistant nitrogen-fixing deciduous shrub that is basically found in Europe and Asia. Many bioactive components, including vitamins, phytosterols, flavonoids, carotenoids, polyunsaturated fatty acids, and some essential amino acids are present in the plant. Traditionally, the SBT has been used for numerous health issues such as asthma, skin diseases, gastric ulcers, and lung disorders. Previously, we have extensively reviewed the medicinal and therapeutic properties of sea buckthorn (Suryakumar and Gupta 2011).

Narayanan et al. (2005) reported the protective efficacy of SBT extract against hypoxia. The SBT extract showed antistress and adaptogenic activity against multiple stress, cold-hypoxia-restraint (CHR). Saggu and Kumar (2007) reported the antistress activity of SBT on oral administration with a dose of 100 mg/kg body weight. The leaf and seed oil of SBT is also able to inhibit hypoxia-associated transvascular leakage in the lungs and brain (Purushothaman et al. 2008). Oral administration of SBT acts as an adaptogen by boosting hypoxic tolerance and survival.

Furthermore, our results also observed the hypoxia protective mechanism of SBT using in vivo studies which indicated that SBT extracts enhanced the expression of hypoxia-responsive proteins such as HIF-1α and HO-1 (Rathor et al. 2015). SBT also exacerbated the production of NO via enhancing iNOS and eNOS expression. Furthermore, the extract of SBT also decreased the hypoxia induced ER stress response by downregulating CHOP and PERK. Conclusively, SBT showed anti-inflammatory and antiapoptotic responses against hypoxia-exposed animals via modulating proteostasis signaling mechanisms (Jain et al. 2016a, b).

9.12.2 Curcumin

Scientific classification

Kingdom	: Plantae
Order	: Zingiberales
Family	: Zingiberaceae
Genus	: *Curcuma*
Species	: *longa*

Curcumin is a well-known nutraceutical and a very important component of Indian spice, commonly known as haldi (turmeric). The beneficial activities of curcumin have been described in Ayurveda as well as in traditional Chinese medicine. Several researchers have demonstrated the biological activities of curcumin, such as antioxidant, immunomodulatory, cytoprotective, and anti-inflammatory. Furthermore, curcumin showed beneficial efficacy against hypoxia by ameliorating the hypoxia-induced oxidative stress, and pulmonary vascular leakage, and managing inflammation by attenuating NF-kB expression (Sarada et al. 2008). Hypobaric hypoxia-induced vascular leakage in the brain was also ameliorated by the administration of curcumin. Furthermore, the supplementation of curcumin attenuated the upregulation of NF-kB expression under hypoxia through the downregulation of pro-inflammatory cytokine levels (Sarada et al. 2015).

Recently, the prophylactic activity of curcumin was also reported by Mathew and Sarada (2018). This study evidenced that curcumin maintained redox homeostasis and enhanced the prosurvival signaling mechanisms. In detail, the study observed an increased expression of antioxidant enzymes (Phase-II) by increasing Nrf2 and HIF-1α signalling, and regulation of these key pathways maintained the pulmonary surfactant homeostasis.

Our recent study demonstrated that curcumin ameliorated the skeletal muscle loss induced by chronic hypobaric hypoxia by inhibition of NF-kB expression and decreased the protein degradation pathways. Furthermore, curcumin administration enhanced myogenesis-related markers such as Myo D, Myf 5, and myogenin. The activation of myogenic regulatory proteins showed the protective efficacy of curcumin in muscle atrophy that also enhanced the physical performance in hypoxia-exposed animals (Chaudhary et al. 2019).

9.12.3 Ursolic Acid

Identifiers and properties	
CAS number	: 77-52-1
Chemical formula	: $C_{3}0H_{48}O_{3}$
Molar mass	: 456.711 g·Mol^{-1}
Melting point	: 285 to 288 °C

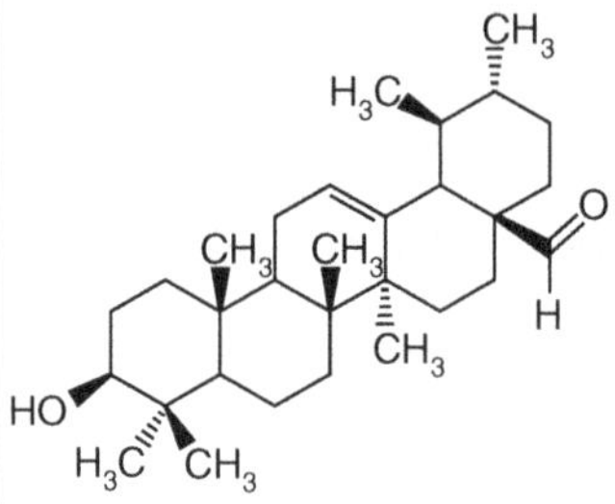

Ursolic acid is a natural pentacyclic triterpene, present in various edible herbs, vegetables, whole grains, dietary fibres and fruits mainly apples, cranberries, pears, star fruit, mahogany, and blueberries (Ikeda et al. 2008; Lin et al. 2011). This compound is lipophilic in nature and present on outer waxy coat of the apple. It is reported that an apple has approximately 50 mg of ursolic acid (Frighetto et al. 2008; Jäger et al. 2009). Ursolic acid have numerous biological activities like anti-oxidant (Saravanan et al. 2006; Senthil et al. 2007), anti-inflammatory (Subbaramaiah et al. 2000; Shishodia et al. 2003), anti-cancer (Harmand et al. 2005; Agarwal et al. 2005), enhance muscle mass (Kunkel et al. 2011; Adams et al. 2015), prevent liver diseases, reduce obesity, and decrease glucose tolerance (Kunkel et al. 2012). Hypobaric hypoxia associated skeletal muscle loss is regulated by multiple signaling pathways. Ursolic acid showed potent protection against hypobaric hypoxia induced muscle mass loss due to its various biological properties. As described in our recent research, ursolic acid has potent antioxidant properties, it showed anti-inflammatory activity and in addition also enhanced protein synthesis pathway by upregulated expression of Akt, pAkt, and p70S6kinase (Rathor et al. 2021).

9.12.4 Carnosine

Identifiers and properties	
CAS number	: 305-84-0
Chemical formula	: $C_9H_{14}N_4O_3$
Molar mass	: 226.236 g·Mol^{-1}
Melting point	: 253 °C

(continued)

Identifiers and properties	
	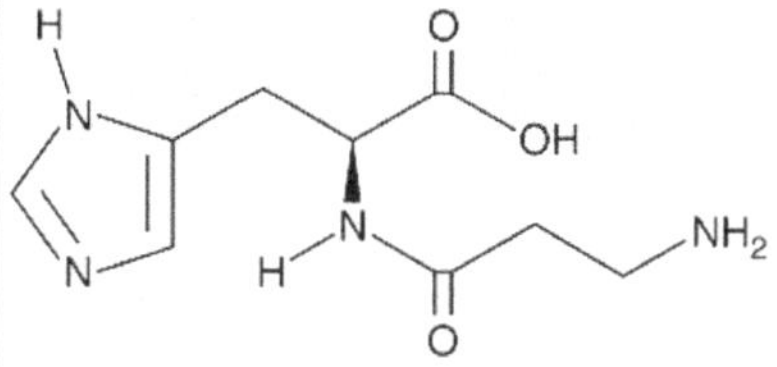

Carnosine is a biologically active dipeptide found in the skeletal muscle brain, and kidney of mammals, fish and birds (Dunnett and Harris 1997). Carnosine is basically a dipeptide with a nonessential amino acid β-alanine and an essential amino acid L-histidine. Carnosine is synthesized in the cell with the help of enzyme carnosine synthase. Carnosine has protective efficacy against various diseases, including cardiovascular diseases, lung injury, ocular diseases, obesity, Alzheimer's disease, diabetes, dementia, ischemia-reperfusion injury, atherosclerosis, Parkinson's disease, wound healing, and cancer (Stuerenburg 2000; Tanaka et al. 2017; Kim et al. 2019; Ansurudeen et al. 2012; Regazzoni et al. 2016; Peters et al. 2012; Zhao et al. 2019; Keskin et al. 2017; Bermúdez et al. 2019). At present, carnosine is being used by athletes and sprinters for improving performance and muscle strength. Furthermore, β-alanine supplementation also synthesizes carnosine that further assist in maintaining the pH homeostasis in the muscle (Peeling et al. 2019). Recently, our group have reported the potential role of carnosine in augmenting hypobaric hypoxia associated muscle protein loss via managing muscle proteostasis (Agrawal et al. 2022). Supplementation with carnosine ameliorated the protein oxidation induced by hypobaric hypoxia which in turn decreased the misfolded protein load on the ER. Further autophage and apoptosis were also decreased in the skeletal muscle with carnosine supplementation. The physical performance was also improved by carnosine administration.

9.13 Conclusion

The molecular mechanism involved in high altitude-associated muscle loss is described in the present chapter. Muscle protein loss due to hypobaric hypoxia is a multifaceted pathophysiological condition that is synchronized by a number of intracellular signaling pathways, including oxidative stress, ER stress, inflammation, distorted anabolic pathways, calcium imbalance, apoptosis, and excessive protein degradation.

Recent scenarios promoted the use of dietary phytochemicals for intervention studies due to their safe nature, easy availability, and effortless supplement. Therefore, numerous dietary interventions were screened and investigated against hypobaric hypoxia-induced muscle atrophy. The detailed in vitro and in vivo analysis established the beneficial role of dietary interventions to ameliorate hypobaric hypoxia-induced muscle protein loss. Now, it is time to translate these studies into clinical studies and establish proof of principle.

Acknowledgements The authors are thankful to Dr. Rajeev Varshney, Director, DIPAS, for his constant support and encouragement.

Conflict of Interest The authors declare that they have no conflicts of interest with the contents of this article.

References

Adams CM, Ebert SM, Dyle MC (2015) Use of mRNA expression signatures to discover small molecule inhibitors of skeletal muscle atrophy. Curr Opin Clin Nutr Metab Care 18(3):263–268. https://doi.org/10.1097/MCO.0000000000000159, PMID 25807353

Agarwal MK, Iqbal M, Athar M (2005) Inhibitory effect of 18beta-glycyrrhetinic acid on 12-O-tetradecanoyl phorbol-13-acetate-induced cutaneous oxidative stress and tumor promotion in mice. Redox Rep 10(3):151–157. https://doi.org/10.1179/135100005X57346, PMID 16156954

Agrawal A, Rathor R, Suryakumar G (2017) Oxidative protein modification alters proteostasis under acute hypobaric hypoxia in skeletal muscles: a comprehensive in vivo study. Cell Stress Chaperones 22(3):429–443. https://doi.org/10.1007/s12192-017-0795-8, PMID 28425050

Agrawal A, Rathor R, Kumar R, Suryakumar G, Ganju L (2018) Role of altered proteostasis network in chronic hypobaric hypoxia induced skeletal muscle atrophy. PLoS One 13(9): e0204283. https://doi.org/10.1371/journal.pone.0204283, PMID 30240405

Agrawal A, Rathor R, Kumar R, Suryakumar G, Singh SN, Kumar B (2020) Redox modification of ryanodine receptor contributes to impaired Ca2+ homeostasis and exacerbates muscle atrophy under high altitude. Free Radic Biol Med 160:643–656. https://doi.org/10.1016/j.freeradbiomed.2020.09.001, PMID 32916280

Agrawal A, Rathor R, Kumar R, Singh SN, Kumar B, Suryakumar G (2022) Endogenous dipeptide-carnosine supplementation ameliorates hypobaric hypoxia-induced skeletal muscle loss via attenuating endoplasmic reticulum stress response and maintaining proteostasis. IUBMB Life 74(1):101–116. https://doi.org/10.1002/iub.2539, PMID 34455667

Ansurudeen I, Sunkari VG, Grünler J, Peters V, Schmitt CP, Catrina SB et al (2012) Carnosine enhances diabetic wound healing in the db/db mouse model of type 2 diabetes. Amino Acids 43(1):127–134. https://doi.org/10.1007/s00726-012-1269-z, PMID 22451275

Bakonyi T, Radak Z (2004) High altitude and free radicals. J Sports Sci Med 3(2):64–69, PMID 24482580

Barhwal K, Hota SK, Baitharu I, Prasad D, Singh SB, Ilavazhagan G (2009) Isradipine antagonizes hypobaric hypoxia induced CA1 damage and memory impairment: complementary roles of L-type calcium channel and NMDA receptors. Neurobiol Dis 34(2):230–244. https://doi.org/10.1016/j.nbd.2009.01.008, PMID 19385055

Bellinger AM, Mongillo M, Marks AR (2008) Stressed out: the skeletal muscle ryanodine receptor as a target of stress. J Clin Invest 118(2):445–453. https://doi.org/10.1172/JCI34006, PMID 18246195

Bermúdez ML, Seroogy KB, Genter MB (2019) Evaluation of carnosine intervention in the Thy1-aSyn mouse model of Parkinson's disease. Neuroscience 411:270–278. https://doi.org/10.1016/j.neuroscience.2019.05.026, PMID 31125602

Bharadwaj H, Prasad J, Pramanik SN, Krishnani S, Zachariah T, Chaudhary KL et al (2000) Effect of prolonged exposure to high altitude on skeletal muscle of Indian soldiers. Def Sci J 50(2): 167–176. https://doi.org/10.14429/dsj.50.3421

Bonaldo P, Sandri M (2013) Cellular and molecular mechanisms of muscle atrophy. Dis Model Mech 6(1):25–39. https://doi.org/10.1242/dmm.010389, PMID 23268536

Bowen TS, Schuler G, Adams V (2015) Skeletal muscle wasting in cachexia and sarcopenia: molecular pathophysiology and impact of exercise training. J Cachexia Sarcopenia Muscle 6(3): 197–207. https://doi.org/10.1002/jcsm.12043, PMID 26401465

Chaudhary P, Suryakumar G, Prasad R, Singh SN, Ali S, Ilavazhagan G (2012) Chronic hypobaric hypoxia mediated skeletal muscle atrophy: role of ubiquitin-proteasome pathway and calpains. Mol Cell Biochem 364(1–2):101–113. https://doi.org/10.1007/s11010-011-1210-x, PMID 22215202

Chaudhary P, Sharma YK, Sharma S, Singh SN, Suryakumar G (2019) High altitude mediated skeletal muscle atrophy: protective role of curcumin. Biochimie 156:138–147. https://doi.org/10.1016/j.biochi.2018.10.012, PMID 30347230

Cheng J, North BJ, Zhang T, Dai X, Tao K, Guo J et al (2018) The emerging roles of protein homeostasis-governing pathways in Alzheimer's disease. Aging Cell 17(5):e12801. https://doi.org/10.1111/acel.12801, PMID 29992725

Debevec T, Millet GP, Pialoux V (2017) Hypoxia-induced oxidative stress modulation with physical activity. Front Physiol 8:84. https://doi.org/10.3389/fphys.2017.00084, PMID 28243207

Dunnett M, Harris RC (1997) High-performance liquid chromatographic determination of imidazole dipeptides, histidine, 1-methylhistidine and 3-methylhistidine in equine and camel muscle and individual muscle fibres. J Chromatogr B Biomed Sci Appl 688(1):47–55. https://doi.org/10.1016/s0378-4347(97)88054-1, PMID 9029312

Dünnwald T, Gatterer H, Faulhaber M, Arvandi M, Schobersberger W (2019) Body composition and body weight changes at different altitude levels: a systematic review and meta-analysis. Front Physiol 10:430. https://doi.org/10.3389/fphys.2019.00430, PMID 31057421

Edwards LM, Murray AJ, Tyler DJ, Kemp GJ, Holloway CJ, Robbins PA et al (2010) The effect of high-altitude on human skeletal muscle energetics: P-MRS results from the Caudwell Xtreme Everest expedition. PLoS One 5(5):e10681. https://doi.org/10.1371/journal.pone.0010681, PMID 20502713

Frighetto RTS, Welendorf RM, Nigro EN, Frighetto N, Siani AC (2008) Isolation of ursolic acid from apple peels by high speed counter-current chromatography. Food Chem 106(2):767–771. https://doi.org/10.1016/j.foodchem.2007.06.003

Frontera WR, Ochala J (2015) Skeletal muscle: a brief review of structure and function. Calcif Tissue Int 96(3):183–195. https://doi.org/10.1007/s00223-014-9915-y, PMID 25294644

Fuehrer J, McGowan J, Huecker MR (2022) High altitude cardiopulmonary diseases. In: StatPearls

Gaboardi AJ, Kressler J, Snow TK, Balog EM (2018) Aging impairs regulation of ryanodine receptors from extensor digitorum longus but not soleus muscles. Muscle Nerve 57(6): 1022–1025. https://doi.org/10.1002/mus.26063, PMID 29315676

Gallagher SA, Hackett PH (2004) High-altitude illness. Emerg Med Clin North Am 22(2):329–viii. https://doi.org/10.1016/j.emc.2004.02.001, PMID 15163571

Gangwar A, Paul S, Ahmad Y, Bhargava K (2020) Intermittent hypoxia modulates redox homeostasis, lipid metabolism associated inflammatory processes and redox post-translational modifications: benefits at high altitude. Sci Rep 10(1):7899. https://doi.org/10.1038/s41598-020-64848-x, PMID 32404929

Green H, Roy B, Grant S, Otto C, Pipe A, McKenzie D et al (2000) Human skeletal muscle exercise metabolism following an expedition to mount Denali. Am J Physiol Regul Integr Comp Physiol 279(5):R1872–R1879. https://doi.org/10.1152/ajpregu.2000.279.5.R1872, PMID 11049873

Hackett PH, Roach RC (2001) High-altitude illness. N Engl J Med 345(2):107–114. https://doi.org/10.1056/NEJM200107123450206, PMID 11450659

Hackett PH, Roach RC (2004) High altitude cerebral edema. High Alt Med Biol 5(2):136–146. https://doi.org/10.1089/1527029041352054, PMID 15265335

Hamad N, Travis SP (2006) Weight loss at high altitude: pathophysiology and practical implications. Eur J Gastroenterol Hepatol 18(1):5–10. https://doi.org/10.1097/00042737-200601000-00002, PMID 16357612

Harmand PO, Duval R, Delage C, Simon A (2005) Ursolic acid induces apoptosis through mitochondrial intrinsic pathway and caspase-3 activation in M4Beu melanoma cells. Int J Cancer 114(1):1–11. https://doi.org/10.1002/ijc.20588, PMID 15523687

Hoppeler H, Vogt M, Weibel ER, Flück M (2003) Response of skeletal muscle mitochondria to hypoxia. Exp Physiol 88(1):109–119. https://doi.org/10.1113/eph8802513, PMID 12525860

Horscroft JA, Murray AJ (2014) Skeletal muscle energy metabolism in environmental hypoxia: climbing towards consensus. Extrem Physiol Med 3(1):19. https://doi.org/10.1186/2046-7648-3-19, PMID 25473486

Hutt D, Balch WE (2010) Cell biology. The proteome in balance. Science 329(5993):766–767. https://doi.org/10.1126/science.1194160, PMID 20705837

Ikeda Y, Murakami A, Ohigashi H (2008) Ursolic acid: an anti- and pro-inflammatory triterpenoid. Mol Nutr Food Res 52(1):26–42. https://doi.org/10.1002/mnfr.200700389, PMID 18203131

Iurlaro R, Muñoz-Pinedo C (2016) Cell death induced by endoplasmic reticulum stress. FEBS J 283(14):2640–2652. https://doi.org/10.1111/febs.13598, PMID 26587781

Jacobs RA, Siebenmann C, Hug M, Toigo M, Meinild AK, Lundby C (2012) Twenty-eight days at 3454-m altitude diminishes respiratory capacity but enhances efficiency in human skeletal muscle mitochondria. FASEB J 26(12):5192–5200. https://doi.org/10.1096/fj.12-218206, PMID 22968913

Jäger S, Trojan H, Kopp T, Laszczyk MN, Scheffler A (2009) Pentacyclic triterpene distribution in various plants—rich sources for a new group of multi-potent plant extracts. Molecules 14(6): 2016–2031. https://doi.org/10.3390/molecules14062016, PMID 19513002

Jain K, Suryakumar G, Ganju L, Singh SB (2016a) Amelioration of ER stress by 4-phenylbutyric acid reduces chronic hypoxia induced cardiac damage and improves hypoxic tolerance through upregulation of HIF-1α. Vasc Pharmacol 83:36–46. https://doi.org/10.1016/j.vph.2016.03.004, PMID 27058435

Jain K, Suryakumar G, Prasad R, Ganju L, Bala Singh SB (2016b) Enhanced hypoxic tolerance by Seabuckthorn is due to upregulation of HIF-1a and attenuation of ER stress. J Appl Biomed 14(1):71–83. https://doi.org/10.1016/j.jab.2015.10.001

Kalra EK (2003) Nutraceutical—definition and introduction. AAPS PharmSci 5(3):E25. https://doi.org/10.1208/ps050325, PMID 14621960

Keskin A, Kanar BG, Kanar RG, Kepez A, Tuneri B (2017) An investigation on the impact of carnosine on the myocardium in lower extremity ischemia-reperfusion injury in rats. Int J Cardiovasc Acad 3:109–113. https://doi.org/10.1016/j.ijcac.2016.11.054

Khan S, O'Brien PJ (1995) Modulating hypoxia-induced hepatocyte injury by affecting intracellular redox state. Biochim Biophys Acta 1269(2):153–161. https://doi.org/10.1016/0167-4889(95)00112-6

Kim Y, Kim E, Kim Y (2019) L-histidine and L-carnosine accelerate wound healing via regulation of corticosterone and PI3K/Akt phosphorylation in d-galactose induced aging models in vitro and in vivo. J Funct Foods 58:227–237. https://doi.org/10.1016/j.jff.2019.04.060

Kunkel SD, Suneja M, Ebert SM, Bongers KS, Fox DK, Malmberg SE et al (2011) mRNA expression signatures of human skeletal muscle atrophy identify a natural compound that increases muscle mass. Cell Metab 13(6):627–638. https://doi.org/10.1016/j.cmet.2011.03.020, PMID 21641545

Kunkel SD, Elmore CJ, Bongers KS, Ebert SM, Fox DK, Dyle MC et al (2012) Ursolic acid increases skeletal muscle and brown fat and decreases diet-induced obesity, glucose intolerance and fatty liver disease. PLoS One 7(6):e39332. https://doi.org/10.1371/journal.pone.0039332, PMID 22745735

Kushnir A, Santulli G, Reiken SR, Coromilas E, Godfrey SJ, Brunjes DL et al (2018) Ryanodine receptor calcium leak in circulating B-lymphocytes as a biomarker in heart failure. Circulation 138(11):1144–1154. https://doi.org/10.1161/CIRCULATIONAHA.117.032703, PMID 29593014

Lee SW, Dai G, Hu Z, Wang X, Du J, Mitch WE (2004) Regulation of muscle protein degradation: coordinated control of apoptotic and ubiquitin-proteasome systems by phosphatidylinositol

3 kinase. J Am Soc Nephrol 15(6):1537–1545. https://doi.org/10.1097/01.asn.0000127211.86206.e1, PMID 15153564

Levett DZ, Radford EJ, Menassa DA, Graber EF, Morash AJ, Hoppeler H et al (2012) Acclimatization of skeletal muscle mitochondria to high-altitude hypoxia during an ascent of Everest. FASEB J 26(4):1431–1441. https://doi.org/10.1096/fj.11-197772, PMID 22186874

Lin Y, Vermeer MA, Trautwein EA (2011) Triterpenic acids present in hawthorn lower plasma cholesterol by inhibiting intestinal ACAT activity in hamsters. Evid Based Complement Alternat Med 2011:801272. https://doi.org/10.1093/ecam/nep007

Mason S, Johnson RS (2007) The role of HIF-1 in hypoxic response in the skeletal muscle. Adv Exp Med Biol 618:229–244. https://doi.org/10.1007/978-0-387-75434-5_18, PMID 18269201

Mathew T, Sarada SKS (2018) Intonation of Nrf2 and HIF1-α pathway by curcumin prophylaxis: a potential strategy to augment survival signaling under hypoxia. Respir Physiol Neurobiol 258: 12–24. https://doi.org/10.1016/j.resp.2018.09.008, PMID 30268739

McCaffrey K, Braakman I (2016) Protein quality control at the endoplasmic reticulum. Essays Biochem 60(2):227–235. https://doi.org/10.1042/EBC20160003, PMID 27744338

McDonald MD, Hunt K, Sivaramakrishnan H, Moullin J, Avenell A, Kerr DA et al (2022) A systematic review examining socioeconomic factors in trials of interventions for men that report weight as an outcome. Obes Rev 23(7):e13436. https://doi.org/10.1111/obr.13436, PMID 35187778

McKinnell IW, Rudnicki MA (2004) Molecular mechanisms of muscle atrophy. Cell 119(7): 907–910. https://doi.org/10.1016/j.cell.2004.12.007, PMID 15620349

Meyer P, Notarnicola C, Meli AC, Matecki S, Hugon G, Salvador J et al (2021) Skeletal ryanodine receptors are involved in impaired myogenic differentiation in Duchenne muscular dystrophy patients. Int J Mol Sci 22(23):12985. https://doi.org/10.3390/ijms222312985, PMID 34884796

Mishra KP, Sharma N, Soree P, Gupta RK, Ganju L, Singh SB (2016) Hypoxia-induced inflammatory chemokines in subjects with a history of high-altitude pulmonary edema. Indian J Clin Biochem 31(1):81–86. https://doi.org/10.1007/s12291-015-0491-3, PMID 26855492

Mohanraj P, Merola AJ, Wright VP, Clanton TL (1998) Antioxidants protect rat diaphragmatic muscle function under hypoxic conditions. J Appl Physiol (1985) 84(6):1960–1966. https://doi.org/10.1152/jappl.1998.84.6.1960

Monroy JA, Powers KL, Gilmore LA, Uyeno TA, Lindstedt SL, Nishikawa KC (2012) What is the role of titin in active muscle? Exerc Sport Sci Rev 40(2):73–78. https://doi.org/10.1097/JES.0b013e31824580c6, PMID 22157359

Narayanan S, Ruma D, Gitika B, Sharma SK, Pauline T, Ram MS et al (2005) Antioxidant activities of seabuckthorn (Hippophae rhamnoides) during hypoxia induced oxidative stress in glial cells. Mol Cell Biochem 278(1–2):9–14. https://doi.org/10.1007/s11010-005-7636-2, PMID 16180083

Ottenheijm CA, Granzier H (2010) Role of titin in skeletal muscle function and disease. Adv Exp Med Biol 682:105–122. https://doi.org/10.1007/978-1-4419-6366-6_6, PMID 20824522

Peeling P, Castell LM, Derave W, de Hon O, Burke LM (2019) Sports foods and dietary supplements for optimal function and performance enhancement in track-and-field athletes. Int J Sport Nutr Exerc Metab 29(2):198–209. https://doi.org/10.1123/ijsnem.2018-0271, PMID 30299192

Peters V, Schmitt CP, Zschocke J, Gross ML, Brismar K, Forsberg E (2012) Carnosine treatment largely prevents alterations of renal carnosine metabolism in diabetic mice. Amino Acids 42(6): 2411–2416. https://doi.org/10.1007/s00726-011-1046-4, PMID 21833769

Pham K, Parikh K, Heinrich EC (2021) Hypoxia and inflammation: insights from high-altitude physiology. Front Physiol 12:676782. https://doi.org/10.3389/fphys.2021.676782, PMID 34122145

Place N, Ivarsson N, Venckunas T, Neyroud D, Brazaitis M, Cheng AJ et al (2015) Ryanodine receptor fragmentation and sarcoplasmic reticulum Ca2+ leak after one session of high-intensity interval exercise. Proc Natl Acad Sci U S A 112(50):15492–15497. https://doi.org/10.1073/pnas.1507176112, PMID 26575622

Powers SK, Kavazis AN, McClung JM (2007) Oxidative stress and disuse muscle atrophy. J Appl Physiol (1985) 102(6):2389–2397. https://doi.org/10.1152/japplphysiol.01202.2006, PMID 17289908

Pu X, Lin X, Duan X, Wang J, Shang J, Yun H et al (2020) Oxidative and endoplasmic reticulum stress responses to chronic high-altitude exposure during the development of high-altitude pulmonary hypertension. High Alt Med Biol 21(4):378–387. https://doi.org/10.1089/ham.2019.0143, PMID 33090046

Purushothaman J, Suryakumar G, Shukla D, Malhotra AS, Kasiganesan H, Kumar R et al (2008) Modulatory effects of seabuckthorn (Hippophae rhamnoides L.) in hypobaric hypoxia induced cerebral vascular injury. Brain Res Bull 77(5):246–252. https://doi.org/10.1016/j.brainresbull.2008.08.026, PMID 18824077

Rathor R, Suryakumar G (2016) Muscle atrophy at high altitude. J Clin Mol Endocrinol 1:18. https://doi.org/10.21767/2572-5432.100016

Rathor R, Sharma P, Suryakumar G, Ganju L (2015) A pharmacological investigation of Hippophae Salicifolia (HS) and Hippophae Rhamnoides Turkestanica (HRT) against multiple stress (C-H-R): an experimental study using rat model. Cell Stress Chaperones 20(5):821–831. https://doi.org/10.1007/s12192-015-0603-2

Rathor R, Agrawal A, Kumar R, Suryakumar G, Singh SN (2021) Ursolic acid ameliorates hypobaric hypoxia-induced skeletal muscle protein loss via upregulating Akt pathway: an experimental study using rat model. IUBMB Life 73(2):375–389. https://doi.org/10.1002/iub.2435, PMID 33368975

Regazzoni L, de Courten B, Garzon D, Altomare A, Marinello C, Jakubova M et al (2016) A carnosine intervention study in overweight human volunteers: bioavailability and reactive carbonyl species sequestering effect. Sci Rep 6:27224. https://doi.org/10.1038/srep27224, PMID 27265207

Rose MS, Houston CS, Fulco CS, Coates G, Sutton JR, Cymerman A (1988) Operation Everest. II: nutrition and body composition. J Appl Physiol (1985) 65(6):2545–2551. https://doi.org/10.1152/jappl.1988.65.6.2545, PMID 3215854

Saggu S, Kumar R (2007) Modulatory effect of seabuckthorn leaf extract on oxidative stress parameters in rats during exposure to cold, hypoxia and restraint (C-H-R) stress and post stress recovery. J Pharm Pharmacol 59(12):1739–1745. https://doi.org/10.1211/jpp.59.12.0018, PMID 18053338

Sarada S, Himadri P, Mishra C, Geetali P, Ram MS, Ilavazhagan G (2008) Role of oxidative stress and NFkB in hypoxia-induced pulmonary edema. Exp Biol Med (Maywood) 233(9):1088–1098. https://doi.org/10.3181/0712-RM-337, PMID 18641050

Sarada SK, Titto M, Himadri P, Saumya S, Vijayalakshmi V (2015) Curcumin prophylaxis mitigates the incidence of hypobaric hypoxia-induced altered ion channels expression and impaired tight junction proteins integrity in rat brain. J Neuroinflammation 12:113. https://doi.org/10.1186/s12974-015-0326-4, PMID 26048285

Saravanan R, Viswanathan P, Pugalendi KV (2006) Protective effect of ursolic acid on ethanol-mediated experimental liver damage in rats. Life Sci 78(7):713–718. https://doi.org/10.1016/j.lfs.2005.05.060, PMID 16137716

Schwarz DS, Blower MD (2016) The endoplasmic reticulum: structure, function and response to cellular signaling. Cell Mol Life Sci 73(1):79–94. https://doi.org/10.1007/s00018-015-2052-6, PMID 26433683

Semenov DG, Beliakov AV, Samoĭlov MO (2009) Comparison of Ca2+ responses to stimulation of glutamate receptors in the rat brain cortex following hypobaric hypoxia of different degree. Ross Fiziol Zh Im I M Sechenova 95(9):977–983, PMID 19899714

Semenza GL (2012) Hypoxia-inducible factors in physiology and medicine. Cell 148(3):399–408. https://doi.org/10.1016/j.cell.2012.01.021, PMID 22304911

Senthil S, Chandramohan G, Pugalendi KV (2007) Isomers (oleanolic and ursolic acids) differ in their protective effect against isoproterenol-induced myocardial ischemia in rats. Int J Cardiol 119(1):131–133. https://doi.org/10.1016/j.ijcard.2006.07.108, PMID 17052790

Shishodia S, Majumdar S, Banerjee S, Aggarwal BB (2003) Ursolic acid inhibits nuclear factor-kappaB activation induced by carcinogenic agents through suppression of IkappaBalpha kinase and p65 phosphorylation: correlation with down-regulation of cyclooxygenase 2, matrix metalloproteinase 9, and cyclin D1. Cancer Res 63(15):4375–4383, PMID 12907607

Shukla V, Singh SN, Vats P, Singh VK, Singh SB, Banerjee PK (2005) Ghrelin and leptin levels of sojourners and acclimatized lowlanders at high altitude. Nutr Neurosci 8(3):161–165. https://doi.org/10.1080/10284150500132823, PMID 16117183

Singh SN, Vats P, Kumria MM, Ranganathan S, Shyam R, Arora MP et al (2001) Effect of high altitude (7,620 m) exposure on glutathione and related metabolism in rats. Eur J Appl Physiol 84(3):233–237. https://doi.org/10.1007/s004210170010, PMID 11320641

Song K, Peng S, Sun Z, Li H, Yang R (2011) Curcumin suppresses TGF-β signaling by inhibition of TGIF degradation in scleroderma fibroblasts. Biochem Biophys Res Commun 411(4):821–825. https://doi.org/10.1016/j.bbrc.2011.07.044, PMID 21798239

Stuerenburg HJ (2000) The roles of carnosine in aging of skeletal muscle and in neuromuscular diseases. Biochemistry (Mosc) 65(7):862–865, PMID 10951106

Subbaramaiah K, Michaluart P, Sporn MB, Dannenberg AJ (2000) Ursolic acid inhibits cyclooxygenase-2 transcription in human mammary epithelial cells. Cancer Res 60(9): 2399–2404, PMID 10811116

Suryakumar G, Gupta A (2011) Medicinal and therapeutic potential of sea buckthorn (Hippophae rhamnoides L.). J Ethnopharmacol 138(2):268–278. https://doi.org/10.1016/j.jep.2011.09.024, PMID 21963559

Tanaka KI, Sugizaki T, Kanda Y, Tamura F, Niino T, Kawahara M (2017) Preventive effects of carnosine on lipopolysaccharide-induced lung injury. Sci Rep 7:42813. https://doi.org/10.1038/srep42813, PMID 28205623

Vats P, Singh VK, Singh SN, Singh SB (2007) High altitude induced anorexia: effect of changes in leptin and oxidative stress levels. Nutr Neurosci 10(5–6):243–249. https://doi.org/10.1080/10284150701722299, PMID 18284032

Weil JV (2004) Sleep at high altitude. High Alt Med Biol 5(2):180–189. https://doi.org/10.1089/1527029041352162, PMID 15265339

Zhao J (2007) Nutraceuticals, nutritional therapy, phytonutrients, and phytotherapy for improvement of human health: a perspective on plant biotechnology application. Bentham Science Publishers, p 75; http://www.benthamscience.com/biot/samples/biot1-1/Zhao

Zhao J, Posa DK, Kumar V, Hoetker D, Kumar A, Ganesan S et al (2019) Carnosine protects cardiac myocytes against lipid peroxidation products. Amino Acids 51(1):123–138. https://doi.org/10.1007/s00726-018-2676-6, PMID 30449006

Zieliński J, Koziej M, Mańkowski M, Sarybaev AS, Tursalieva JS, Sabirov IS et al (2000) The quality of sleep and periodic breathing in healthy subjects at an altitude of 3,200 m. High Alt Med Biol 1(4):331–336. https://doi.org/10.1089/15270290050502408, PMID 11256469

Altitude Acclimatization Via Pharmacological and Non-pharmacological Interventions

10

Poornima Sharma, Swaraj Mohanty, and Yasmin Ahmad

Abstract

The key element of hypobaric hypoxia is declined barometric pressure and lower partial pressure at high altitudes. Hypobaric hypoxia at high altitudes (>2500 m) is correlated with mountain sicknesses due to physiological variations, which on severity lead to high-altitude pulmonary edema and cerebral edema. Pharmacological and non-pharmacological acclimatizations are the core components for conquering physiological changes at high altitudes. Evidence suggests that available non-pharmacological strategies of acclimatization/pre-acclimatization, such as gradual ascent, intermittent hypoxia training, and prior training in simulated chambers before the ascent to high altitude, regulate protective mechanisms for better adaptation at high altitude. The use of conventional acclimatization strategies requires continuous residency, which is a time-dependent process and is counted as a possible disadvantage in terms of the rapid translocation of individuals to high altitudes. In concern to avoid time dependency and altitude residency of traditional acclimatization strategies, there is a need to enhance the use of modern acclimatization strategies in conjunction with pharmacological interventions to decrease the acclimatization time. This chapter is focused on the acclimatization strategies and medical monitoring to conquer the hypoxia responses at high altitudes and the applications of modern acclimatization strategies to reduce acclimatization time with the aim of easing rapid movement operations.

P. Sharma · S. Mohanty · Y. Ahmad (✉)
Defence Institute of Physiology and Allied Sciences (DIPAS), Defence Research and Development Organization (DRDO), New Delhi, India
e-mail: yasminchem@gmail.com

R. Tulsawani, D. Vohora (eds.), *Adaptation under Stressful Environments through Biological Adjustments and Interventions*,
https://doi.org/10.1007/978-981-99-7652-2_10

Keywords

Hypoxia · Hypobaric hypoxia · Normobaric hypoxia · Acclimatization strategies · Medical monitoring

10.1 Introduction

At high altitudes, hypoxia is a significant physiological challenge. Hypoxia is a major physiological challenge at high altitudes. People translocating from plain regions to plateaus typically struggle with hypoxia (Luo et al. 2014). Hypoxia is a medical condition that occurs due to a scarcity of oxygen supply at the cellular level (Bagali et al. 2018). To overcome hypoxia, acclimatization/adaptation to high altitude is required. Individuals sojourning to high altitudes face a lot of struggles to acclimatize and adapt to the high altitudes. High-altitude hypobaric hypoxic conditions are known to cause mountain altitude sicknesses like acute mountain sickness (AMS), chronic mountain sickness, high-altitude cerebral edema (HACE), and high-altitude pulmonary edema (HAPE). It is challenging to forecast the likelihood of high-altitude illnesses since some people are more suspected of high-altitude diseases than others. In this chapter, we will cover the major issues associated with high altitude including hypoxia and diseases caused by it; this chapter also offers an understanding of pharmacological and non-pharmacological approaches to acclimatization for human survival at high altitude (Luo et al. 2014).

10.2 Basic Concepts of High-Altitude Hypoxia

There are multiple destinations at high altitudes that fascinate travelers; for instance, Mount Everest in Asia at 8848 m, Aconcagua in North and South America at 7000 m, Mount Kilimanjaro at 5895 m in Africa, and Mount Elbrus at 5642 m (Netzer et al. 2013). The troposphere is known as the lowest portion of the atmosphere, and it covers the earth's entire surface. The barometric pressure in the troposphere diminishes on the vertical gain of high altitude. On ascending to high altitude, inspired oxygen remains constant while there is a parallel decrease in barometric pressure and partial pressure of oxygen. The clinical definition of this state is hypobaric hypoxia. For instance, pO_2 at the Everest base camp (5300 m) is nearly one-half of the earth's surface, and at the elevation of 8850 m, it is one-third that of sea level. The lower barometric pressure at high altitudes is responsible for causing a decline in arterial oxygen saturation and arterial pressure of oxygen that results in reduced transport of oxygen to tissues. In terms of acclimatization and adaptation, high-altitude physiological research is classified into two groups, including changes caused by acute exposure to hypoxia and changes by chronic exposure to hypoxia. Acute exposure to extreme altitudes (>8000 m) is very lethal. Increased cardiac output and hyperventilation are the primary influences of short-term responses to high-altitude exposure. The peripheral chemoreceptors of carbon dioxide are primarily responsible for initiating HVR (hyperventilatory response) in the hypoxic condition, which leads to a reduction in the alveolar pressure of carbon dioxide and an increase in oxygen partial pressure. Our autonomic nervous system primarily causes an increase in cardiac output, which elevates heart rate and merely serves to maintain the oxygen supply to tissues (Michiels 2004). In acute exposure, high-altitude hypoxia acts as a causative agent for high-altitude illness. The probability of likelihood of developing acute mountain sickness, HACE, or HAPE is dependent on ascent speed. Various studies concluded that inflammation is the contributing factor to developing altitude mountain sickness; however, other experts dispute this. Many experts contended that patients with pre-existing inflammatory disorders, such as upper respiratory tract infection and diarrhoeal illness, are at more risk of developing acute high-altitude illness. Genetic variation accounts for the significant variety in an individual's ability to operate at high altitudes and their susceptibility to high-altitude sickness. For instance, native Tibetans have more vital capacities and maximal oxygen uptakes as compared to the population of Han Chinese as these people were translocated from the plains in the course of the last 60 years. As per the reports, people with lower ACE levels have better performance at high altitudes (Grocott et al. 2007). ACE is a convoluted physiological system, and it is associated with various components and genes that characterize its activity and regulatory actions, such as blood pressure or vascular growth (Zheng and Huang 2020). Following the hypothesis that ACE gene insertion high allele enhances the performance at altitude and ameliorates the childhood outcomes of meningococcal septicemia and cardiovascular response to preterm birth (Harding et al. 2003). Few researchers claimed that gene variants of endothelial nitric oxide synthase might be involved in adaptation to altitude. Nitric oxide is formed in the lungs, and it is involved in the regulation of respiratory blood flow. The level of blowing out nitric oxide is higher in the native populations of elevated areas (Beall et al. 2001). The endothelial nitric oxide synthase gene polymorphisms have been found to reduce the level of nitric oxide synthesis in Japanese people, increasing their susceptibility to HAPE (Droma et al. 2002), while in Caucasians, no variations were recorded for nitric oxide synthase genotype frequencies between the HAPE-resistant and HAPE-susceptible (Weiss et al. 2003). Furthermore, the genotype of nitric oxide synthase and pulmonary artery systolic pressure in acute hypoxia are found to be independent of each other (Smith et al. 2006). In addition to deceleration of exercise performance, the lack of oxygen also causes cephalalgia, sleep disturbances, and high-altitude illnesses like HAPE, HACE, and AMS, as well as pulmonary vasoconstriction, enhanced sympathetic drive, more intracranial pressure, oxygen desaturation, fluid retention, and other factors like oxidative stress and inflammation may also contribute to the pathogenesis of high-altitude illnesses.

10.3 Pathophysiological Consequences at High Altitude

The biggest driver causing high-altitude (>2500 m) diseases in sojourners is high-altitude hypobaric hypoxia. AMS is the first sickness encountered by sojourners on acute exposure to high altitudes. The symptoms of AMS can be resolved with acclimatization. In fatal circumstances, AMS assists in the development of multifactorial illnesses such as HAPE and HACE (Pham et al. 2021). Here, we have discussed the pathophysiological consequences of the sojourners that they have to encounter at high altitudes.

10.3.1 Acute Mountain Sickness

High-altitude hypoxia leads to a range of physiological alterations including acidosis, inflammation, vasogenic edema, abnormalities in cerebral venous blood flow, and inadequate cerebrospinal fluid production, all of which contribute to the development of AMS. Numerous investigations revealed that inflammation is the key factor in the emergence of acute mountain sickness symptoms. Acute exposure to high-altitude hypoxia is reported to increase the levels of pro-inflammatory cytokines and inflammatory markers, such as C-reactive protein (CRP), IL-6, and IL-1β. It has been noted in a few instances that individuals with AMS and without AMS may exhibit fluctuations in expressions of inflammatory mediators. For instance, AMS patients exhibit higher levels of IL-6, IL-1β, and TNF-α and a decline in the level of IL-10 in contrast to non-AMS patients. The blood–brain barrier might be damaged by the production of angiogenic and inflammatory mediators, which may also increase the likelihood of vasogenic edema (Malacrida et al. 2019).

There is a self-report of a questionnaire, which is known as the Lake Louise AMS score. In this clinical assessment, researchers query research participants to record the severity of their symptoms. It is a two-step process, participants evaluate the questionnaire themselves, and researchers verbally confirm their responses. The aggregate of participants' questionnaire responses constitutes the AMS self-score. It was developed in 1991 to establish a diagnosis based on symptom scores. Diagnosis of AMS is based on multifactor such as rapid ascent to altitude, few hours at altitude, headache, emesis, insomnia, dizziness, and Lake Louise Score < 3 (Smedley and Grocott 2013) (Table 10.1).

10.3.2 High-Altitude Cerebral Edema

HACE is the most dangerous fatal condition that is encountered by sojourners at high altitudes. It mainly occurs at the elevated >2000 m. Its severity results in seizures and impaired cognitive functions. Hypoxia mediates key risk factors of HACE including cerebral vasodilation, loss of blood–brain barrier integrity, elevated intracranial pressure, micro-bleeding in entire cerebral in white matter tracts, and macrovascular disruption (Hackett et al. 2019). HACE results in papilledema, brain

Table 10.1 Questionnaire based Lake Louise scoring for the assesment of acute mountain sickness (Smedley and Grocott 2013)

Symptoms	Severity	Scoring
Headache	No headache	0
	Mild headache	1
	Moderate headache	2
	Severe headache, incapacitating	3
Fatigue	Weak	0
	Mild fatigue	1
	Moderate fatigue	2
	Severe fatigue	3
Insomnia	Normal sleep cycle	0
	Slight sleep disturbances	1
	Multiple breaks in sleep	2
	No sleep at all	3
Dizziness	No dizziness	0
	Mild dizziness	1
	Moderate dizziness	2
	Severe dizziness	3
Gastrointestinal symptoms	No symptoms	0
	Nausea	1
	Emesis	2
	Incapacitating	3

edema on neuroimaging, behavioral abnormalities, altered consciousness, and ataxia. If the symptoms are not resolved earlier, it will lead to lethal consequences like coma (Turner et al. 2021).

10.3.3 High-Altitude Pulmonary Edema

It is non-cardiogenic pulmonary edema and its symptoms appear in people who translocate rapidly from plains to altitude (>2500–3000 m) (Paralikar 2012). High-altitude pulmonary edema is primarily caused by hypoxic-mediated pulmonary vasoconstriction. At high altitudes, declined alveolar oxygen pressure is responsible for developing hypoxic pulmonary vasoconstriction (HPV) throughout the entire lungs, and it raises the mean pulmonary artery pressures. Its clinical manifestation is composed of dyspnea, exercise intolerance, and fatigue, and in fatal cases, it involves mental alterations and chest pain; differential diagnoses include myocardial infarction, pneumonia, and bronchospasm. HAPE mostly occurs due to the rapid ascent of sojourners where exaggerated HPV gives rise to acute pulmonary hypertension and elevates the capillary permeability and alveolar fluid buildup (Pham et al. 2021). Visitors to ski resorts in Colorado's Rocky Mountains had a HAPE prevalence of 0.01–0.1% compared to an alpine mountaineering population with a prevalence of 0.2%. The risk of HAPE incidence in people who engage in hiking and mountain climbing in the Himalayas and Alps might reach 15% (Paralikar 2012).

10.4 Molecular Mechanisms of High-Altitude Hypoxia

The transcriptional response of hypoxia is controlled by the hypoxia-inducible factor. It is a protein heterodimer consisting of an oxygen-sensitive α-sub-unit and β-sub-unit. There are three isoforms of HIF, such as HIF-1, HIF-2, and HIF-3, which perform overlapping functions but also exhibit distinct functions in different types of cells. In normoxia, hydroxylation of HIF-α is done by oxygen-dependent prolyl hydroxylases. The von Hippel–Lindau tumor suppressor protein (pVHL) ubiquitinates HIF upon hydroxylation, and it is degraded by proteosomes. The prolyl hydroxylases (PHDs) are oxygen-dependent because its co-substrate activity decreases under hypoxic conditions, which leads to the dimerization of HIF-α to HIF β and migration to the nucleus. Then, this HIF complex may attach to the hypoxia response elements such as 5′-RCGTG-3′ in gene promoters for the regulation of the expression of a minimum of 100 genes to enhance the oxygen supply to hypoxic tissues. The activity of the HIF pathway is composed of the activation of genes associated with the metabolic adaptation such as lactate dehydrogenase A, phosphoglycerate Kinase, vascularization by vascular endothelial growth factor (VEGF), as well as erythropoietin (EPO) for the production of red blood cells, and many other genes involved in enhancing oxygen delivery (Pham et al. 2021). Additionally, HIF is also associated with the escalation of glycolytic metabolism and in reducing oxygen consumption by mitochondria. On the cellular level, the maximum amount of oxygen has been consumed by mitochondria, and it is also known for supplying cellular energy in the form of ATP to various types of cells. Hypoxia mediates the remodeling of the mitochondrial oxidative phosphorylation system to transfer the cellular energy metabolism toward glycolysis. Reduction of oxygen consumption in mitochondria is associated with the upregulation of respiratory complex I inhibitor, NADH dehydrogenase 1α subcomplex 4-like 2, or the production of hypoxia-activated microRNAs like miR-210. It inhibits the induction of the various important components of the electron transport system. Hypoxia-induced modifications of respiratory complex IV are responsible for enhancing oxygen utilization efficiency. In hypoxia, the subunit of complex IV COX4I1 was found to be degraded and replaced by COX4I2, which accelerated the electron transfer efficiency from complex IV to O_2.

Acute exposure to hypoxia for 2 h increases the concentration of plasma erythropoietin, and in the exposure of 48 h, it reaches the maximum production and it decreases subsequently even in unchanged hypoxic conditions. Exposure of hypoxia over an extended period of time, the growth of new capillaries helps to maintain aerobic metabolism via oxygen supply to various tissues like skeletal muscles. Increased muscle capillarity and increased production of inflammatory cytokines and free radicals are the effects of hypoxia-dependent upregulation of VEGF (Fukuda et al. 2007; Lendahl et al. 2009; Tello et al. 2011; Hayashi et al. 2015; Lee et al. 2020; Burtscher et al. 2022).

10.5 Acclimatization Impact on Human Survival at High Altitude

Acclimatization entails those beneficial physiological adjustments that enable individuals to cope with high-altitude hypoxia and maintain the oxygenation of tissues. There are some physiological markers of acclimatization such as increased ventilation, polycythemia, changes in oxidative enzymes, reduction in oxygen affinity of hemoglobin, decrease in the intercapillary distance of some peripheral tissues like skeletal muscle, and respiratory alkalosis that takes place when lowlanders translocate from sea level to high altitude (West 2006). Acclimatization speeds up breathing, which, in turn, increases the alveolar ventilation by five times the values at sea level. It is done by the hypoxic ventilatory response (HVR) of usually carotid bodies (peripheral chemoreceptors). HVR is the reflex response of the carotid body on hypoxia stimulation. The ventilatory response of acclimatization is associated with a time-critical increase in the hypoxic ventilatory response. Preliminary exposure of high-altitude hypobaric hypoxia declines the lung's vital capacity and residual volume, but after 4 weeks of habitation, the values return to equatorial ranges at lower altitudes. Acute exposure to high altitude causes a sudden increase in the erythrocyte concentration due to a decrease in plasma volume. Dehydration is the main cause of a decrease in plasma volume at high altitudes, which is partially caused by insensible fluid loss due to more ventilation of cold dry air. Other factors, such as less fluid intake and probable diuresis, may also be associated with a reduction in plasma volume. Tannheimer et al. (2010) reported that chronic exposure to high altitude caused an increase in hemoglobin and hematocrit due to fluid shift in the interstitium. In response to 1 week of hypoxia, erythropoietin stimulates the bone marrow to accelerate the production of red blood cells. Primary formation of erythropoietin is done in the kidney in response to endurance training/or hypoxia. Athletes are exposed to a natural or artificial hypoxic environment with the purpose of increasing the concentration of erythropoietin, which is known to enhance oxygen uptake and exercise performance. Acute high-altitude hypobaric hypoxia accelerates the pulse rate from 70 beats per minute to 105 beats with the purpose of compensating for less oxygen content in the blood. These changes either may occur due to an increase in sympathetic nerve activity or due to partial parasympathetic withdrawal. In acclimatization instead of an increase in sympathetic nerve activity, there is a decrease in cardiac output, which is associated with a decrease in heart rate. It is either due to the rise in vagal input and down-regulation of adrenergic receptors or due to the time-dependent decrease in stroke volume. Chronic exposure to high altitude stabilizes the stroke volume (Krzywicki et al. 1971; Hoppeler and Vogt 2001; Hopkins et al. 2003; Calbet and Lundby 2009; Palmer 2010; Tannheimer et al. 2010; Paula de and Niebauer 2012; San et al. 2013). Acute exposure to high altitude for 12 h boosts the level of cerebral blood flow, which becomes normal after 3–5 days of acclimatization. The increase in cerebral blood flow declines the arterial oxygen content for the purpose of conserving cerebral oxygen supply during acclimatization (Burtscher et al. 2022).

10.5.1 Pharmacological and Non–pharmacological Approaches of Acclimatization

10.5.1.1 Non-pharmacological Acclimatization Strategies

Traditional Acclimatization Strategies

Staging Strategy

Conventionally, acclimatization to high-altitude hypoxia is accomplished through graded ascent. It is called the staging strategy of acclimatization, and staging acclimatization includes three steps of mountain ascent such as lower, middle, and high plateaus (Hillebrandt 2010). The standardized protocol for staging acclimatization was mentioned by the US Army Research Institute of Environmental Medicine in the altitude acclimatization guide (Lundby et al. 2012). Generally, the Indian army opts for three acclimatization protocols. In Stage 1, there is a 6-day halt at an altitude ranging from 2700 to 3600 m. In Stages 2 and 3, the halt of extra 4 days is recommended at the altitudes 3600 to 4500 m and > 4500 m. Soldiers must repeat the entire plan of action if they take a break of 30 days from their altitude posting. A study reported that the ascent of every 100 m beyond the altitude of 1500 m leads to a depreciation of 1% of VO_{2max}, and aerobic activity declines by 10% at 2500 m, 25% at 4000 m, and finally, 65% at 8000 m. The graded ascent or staging technique is mostly used acclimatization technique with the major drawback of continuous residency (Bärtsch and Saltin 2008).

Live High Train High (LHTH) Strategy

LHTHs are conventionally used training camps, and these camps are conducted for a month at an altitude ranging from 1800 to 2500 m. These camps are to be conducted in four phases: acclimatization phase, training phase, recovery phase, and descent phase. These camps involve the athletes mostly in outdoor activities. Usually, the acclimatization phase is conducted for 7–10 days, and the acclimatization phase is dependent on these camps' duration. High-intensity exercises are not supposed to be done in the acclimatization phase. In the training phase, athletes are supposed to be involved in cumulative training for 14–21 days. In the recovery phase, athletes are supposed to rest for 2–5 days to recover from high-altitude fatigue, and the descent phase involves the gradual descent of athletes (Bassovitch 2010). Recent LHTH camps are conducted for 3 weeks at an altitude >2000 m (Saunders et al. 2009). High-altitude training reduces the body fat percentage faster as compared to similar training at sea level (Bonne et al. 2014). High-altitude training camps at the altitude of 2000 m are reported to improve Hb_{mass} by nearly 4–7% and enhance exercise performance at baseline (Friedmann et al. 2005; Robach et al. 2006; Bonne et al. 2014).

Modern Acclimatization Strategies

Live High Train Low (LHTL) Strategy

The employment of LHTL has eradicated the traditional use of LHTH for altitude training. Primarily, Levine's research team has invented this strategy. This strategy involves high-intensity training at a plateau and low intensity at high altitudes. Till now, LHTL has been widely adopted for hypoxia training with the purpose of improving athlete performance at sea level (Brugniaux et al. 2006). Levine et al. found that altitude training at 1250 m for 27 days causes a two-fold increase in erythropoietin production and also enhances the concentration of hemoglobin in elite athletes (Levine and Stray-Gundersen 1997). Dehnert et al. (2002) reported that 2 weeks of exposure for 13 h per day of high-altitude hypoxia at 1960 m and intense training at 800 m caused an increase in the erythropoietin level and a minute decrease in Hb level in the control group, and it remained unchanged in the LHTL group. Reticulocyte count was found to be increased in the LHTL group as compared to the control group (Dehnert et al. 2002). LHTL was reported to increase oxygen transport capacity in the blood, erythropoiesis, and oxygen uptake at high-altitude/hypoxic environments at sea level (Robach et al. 2006; Wehrlin et al. 2006). Levine and Stray-Gundersen (1997) observed that acclimatization at the range of 2000 to 3000 m reduces detrimental effects like ventilatory response impairment, acute mountain sickness, and desaturation (Levine and Stray-Gundersen 2006).

Wartime and military emergencies required the rapid mobilization of soldiers to high altitudes. A high-altitude environment causes non-battle injuries to our fighters due to environmental factors and reduces their physical and operational activities. In such a situation, traditional acclimatization strategies that require continuous residency at high altitudes are not very helpful. The application of pre-acclimatization strategies that are independent of continuous residency can play an important role in reducing the acclimatization period in war-like conditions. Since years ago, various researchers have put their efforts into imitating the conditions of high-altitude environments in the simulated hypoxic chambers at baseline to prepare the individuals before their mobilization to high altitude. Recently instituted normobaric and hypobaric hypoxic tents, decompression chambers, and hypoxia sleeping units made it feasible to use simulated altitude (Levine 2002).

Simulated Hypoxic Chambers

The baseline hypoxia training centers involve the use of normobaric and hypobaric hypoxic chambers to train army personnel and athletes. Hypobaric chambers are prepared by simulating altitude exposure by reducing the barometric pressure and lowering oxygen concentration, whereas normobaric hypoxia chambers are facilitated by diluting inspired nitrogen in small spaces, such as rooms and tents. In normobaric hypoxic chambers, the oxygen concentration is maintained at nearly 20.9%, while barometric pressure is retained at 760 mmHg. These conditions reduced the partial pressure of oxygen lower than 150 mmHg (Saugy et al. 2014). The exposure of acute high-altitude hypoxia and simulated hypoxia (hypobaric and normobaric chambers) mediates some physiological changes and enhances oxygen saturation in blood to fight hypoxia (Burtscher et al. 2022). Laitinen et al. (1995) reported that exposure to the simulated chamber of 2500 m altitude to the group of 7 male trained runners and conditioning them for 6–8 h/day made an increase in EPO by 84% and RBC by 7% (Saugy et al. 2014). It is found that exposure to simulated altitude (2000 m) for 10 days about 12 h/day induced an increase of 80% in serum EPO and a 60% increase in Ret count in six athletes among the ten athletes (Laitinen et al. 1995). Comparatively, it is recorded that exercising in simulated hypobaric hypoxia causes an increase in ventilatory response than normobaric hypoxia. Sleeping in hypobaric hypoxia chambers induces more escalation in the breathing rate and oxygen desaturation as compared to the normobaric hypoxia chambers. Individuals pre-acclimatized with simulated hypobaric hypoxia, when mobilized at Pike's peak, were observed with more ventilatory acclimatization and reduced performance decrement as compared to normobaric hypoxia (Burtscher et al. 2022). In comparison to hypobaric chambers, normobaric chambers are less-costlier electricity-dependent chambers (Muza 2007).

Live Low Train High (LLTH)

The LLTH is defined as living in the low and training in the high. In the intervention of LLTH athletes are trained in discontinuous hypoxia exposure for more than 2 h in a rest or active state, which exceeds 2–5 times per week. It is independent of daily lifestyle. Modern-time application of LLTH needs newly advanced tools such as simulated hypoxia chambers of hypobaric and normobaric hypoxia (Wilber 2007). It is observed that LLTH induces favorable changes in total hemoglobin mass (Burtscher et al. 2022). LLTH upregulates the function of the HIF-1α transcriptional factor. Heavy exercises in LLTH develop more skeletal adaptations than in normoxic conditions (Vogt et al. 2001; Brocherie et al. 2017). Hypoxia stress should not exceed a non-toxic dose to prevent physiological damage. Hence, the consideration of hypoxic dose and its management is very necessary before the implementation of LLTH. Maintaining a balanced exercise-to-rest ratio and recovery time is required to avoid complications due to overtraining and undertraining (Goods et al. 2014). LLTH is associated with various sub-divisions, such as interval training, continuous repeated sprint training, resistant training, and passive exposure at rest (Millet et al. 2013; McLean et al. 2014).

Intermittent Hypoxia Exposure

Intermittent hypoxia exposure is also considered as periodic hypoxia exposure, which involves multiple times hypoxia exposure for a certain period of several days to weeks (Powell and Garcia 2000). Rodriguez et al. observed intermittent hypoxic exposure with cumulative increasing hypobaric hypoxia of 4000–5500 m for triplicate sessions of 90 min per week, which extended up to 3 weeks. He found a 7% increase in RBCs, 180% increase in Ret count, 6% increase in Hct, and 13% increase in Hb, and it enhanced the power output of the anaerobic threshold (Rodríguez et al. 2000). Julian et al. observed the progressive decrease in F_IO_2 after the effect of 70 min of exposure to intermittent hypoxia for 4 weeks where the ratio of hypoxia and normoxia was kept at 5:5 and sessions were conducted for five times per week (Julian et al. 2004). Ricart et al. (2000) observed a significant enhancement in ventilatory response and tidal volume after the effect of intermittent hypoxia at 5000 m altitude for 2 weeks (2 h/day).

Intermittent Hypoxia Training (IHT)

The intervention of the Living Low Training High (LLTH) method is known as intermittent hypoxia training (Hahn and Gore 2001). Two excessively used models of intermittent hypoxia are LLTH and LHTH (Farias et al. 2013). Since years ago, the application of intermittent hypoxia training has substituted the alternate use of hypobaric and normobaric hypoxia chambers to enhance high-altitude acclimatization. Majorly, it enhances the erythrocyte mass and aerobic capacity of the athletes. The application of normobaric chambers for the purpose of intermittent hypoxia exposure has gained unprecedented popularity because it is independent of social isolation and restricted diet plans. These exposures are known to mediate multiple physiological responses, such as mitochondrial function and ventilatory patterns (Viscor et al. 2018). IHT has been conducted in two ways: hypoxia at rest and hypoxia during exercises. Intermittent hypoxic exposure at rest (actual altitude/simulated altitude) induces some physiological changes to enhance work capacity at high altitudes, such as an increase in exercise performance at baseline, alveolar ventilation, and a decrease in the mixed venous oxygen content. Exercising in intermittent hypoxia enhances training stimulus, ventilation, heart rate, and lactate (Levine 2002). IHE associates three elements: extremity of hypoxia, duration of IHE sessions, and number of sessions. It is dependent on multiple factors, such as the risk of developing AMS and the target altitude. It is observed that IHE is capable of acclimatizing the soldiers before their mobilization to high altitudes (Muza 2007).

10.5.2 Pharmacological Approaches for Acclimatization

A few drugs aid the body in overcoming high-altitude sickness and acclimatizing to the change in environment. Dexamethasone and acetazolamide are the two commonly advised medications when dealing with hypoxia. Here, we covered the most popular acclimatization drugs.

10.5.2.1 Acetazolamide

Acetazolamide precludes the symptoms of altitude mountain sickness before sojourning the mountains. The consumption of acetazolamide mediates blood acidification and declines respiratory alkalosis by improving respiration, and arterial oxygenation and escalates acclimatization. Acetazolamide is preferred to take while ascending to avoid AMS symptoms. Acetazolamide is a renal CA inhibitor, and acetazolamide declines bicarbonate reabsorption, which causes diuresis. Acidosis caused by acetazolamide mediates the medullary respiratory center and enhances ventilation and oxygen supply to the cells. Additionally, acetazolamide prohibits CSF pressure and production and also hinders nocturnal antidiuretic hormone (ADH) secretion.

Its side effects are associated with paresthesias, myopia, impotence, drowsiness, and polyuria. It should be prescribed with vigilance to patients susceptible to allergies to sulfa drugs (Paralikar and Paralikar 2010). The possibility of arising AMS is reduced to half with acetazolamide. Recent studies recommend 125 mg of acetazolamide/12 h from the 1 day before ascent until the descent is initiated (McIntosh et al. 2019). Some researchers prescribed 250 mg of acetazolamide 2 times a day for the individual who is not responding to symptomatic treatment (Luks et al. 2014, 2017).

10.5.2.2 Dexamethasone

Dexamethasone is recommended as a substitute for those individuals who are not tolerable to acetazolamide. The most recommended dexamethasone dose is 2 mg/6 h or 4 mg/12 h. It is suggested to discontinue before initiating the de-induction. Some researchers recommend the combinatorial treatment of acetazolamide and dexamethasone for the sojourners who ascend>3500 m and to relieve moderate severe AMS patients.

10.5.2.3 Ibuprofen

When individuals are non-tolerable to acetazolamide and dexamethasone due to some complications like gastrointestinal bleeding, then it is suggested to use ibuprofen at a daily dose of 1800 mg (Aksel et al. 2019).

10.5.2.4 Nifedipine

Nifedipine both restrains and ameliorates HAPE. Based on clinical experience, two times daily consumption of 30 mg nifedipine is recommended in case of oxygen non-availability, and descent is not possible (Davis and Hackett 2017).

10.5.2.5 Other Medications

Some researchers recommend that 50 mg sildenafil or 10 mg tadalafil may be used when there is non-availability of nifedipine, portable oxygen chambers, and de-induction is not possible (Aksel et al. 2019). Previous data reported that glucocorticoids play an indispensable precautionary role in developing HAPE in susceptible people when consumed orally 1 day before ascent and proceeding up to the whole ascent. Supplemental oxygen and hyperbaric treatment are recommended to be provided immediately on hypoxia encounters at high altitudes (Paralikar 2012). Figure 10.1 represents the therapeutic doses of high-altitude manifestations.

10.6 Summary

The comprehension of this chapter expands our understanding of the various physiological alterations caused by high altitude leads and the manifestation caused by them. It covers the survival framework at a high altitude via non-pharmacological and pharmacological approaches. At high altitudes, lower atmospheric pressure and pO_2 precedence to development of AMS, HAPE, and HACE due to hypobaric

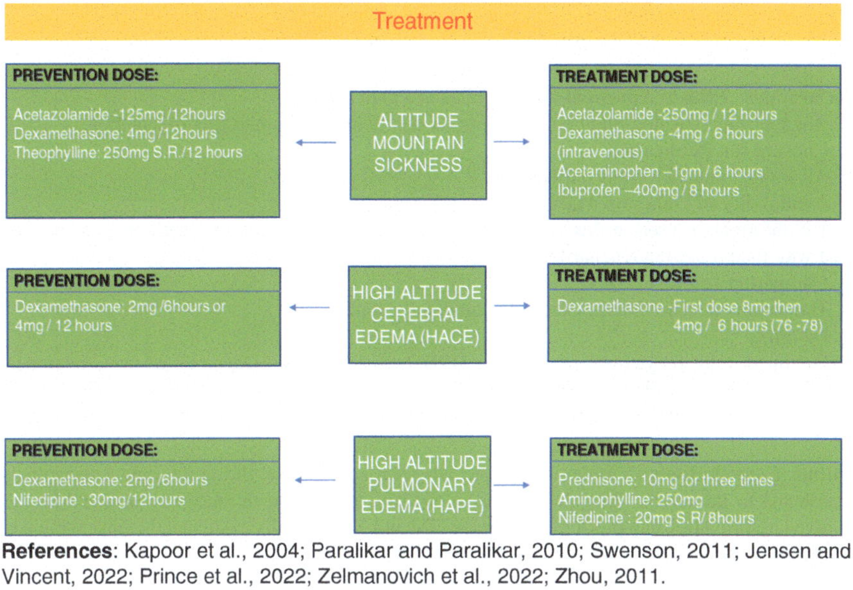

Fig. 10.1 Clinically available therapeutics for high-altitude manifestations (Kapoor et al. 2004; Paralikar and Paralikar 2010; Swenson 2011; Jensen and Vincent 2022; Prince et al. 2022; Zelmanovich et al. 2022; Zhou 2011)

hypoxia. There is a predominant role of acclimatization with the aim to evade and develop less likely altitude mountain sickness. Acclimatization is associated with those physiological changes that help the human body cope with hypobaric hypoxia at high altitudes. Non-pharmacological approaches to acclimatization involve acclimatization and pre-acclimatization strategies, such as staging strategies, intermittent hypoxia, and hypoxic chambers of hypobaric and normobaric hypoxia. Pre-acclimatization models are unprecedentedly popular among Olympic training centers in the world. To reduce the acclimatization time, which will facilitate the rapid movement of army troops and lower the probability of non-combat injuries, it is essential that we fully comprehend the idea of acclimatization. Strategies like LHTL have been found to be more cost-effective and can be conducted without hampering the daily lifestyle.

Recent protocol for ascent recommends holding the ascent after the onset of symptoms of altitude mountain sickness and severity, and people are suggested to descend at the earliest. This chapter provides a paradigm for employing both pharmacological and non-pharmacological approaches to acclimatize at high altitudes. This chapter is a framework for achieving acclimatization at high altitudes via using pharmacological and non-pharmacological approaches.

Acknowledgments We are thankful to Dr. Rajeev Varshney, the director of DIPAS-DRDO for the in-house working facility and the supporting staff Mr. Ram Niwas Meena, Mr. Alpesh Kumar

Sharma, Mr. Jai Ram Singh, Mr. Praveen Negi and Mr. Vishnu Kumar for providing the seamless digital facility to complete the article design process smoothly.

Declarations Ethical Approval and Consent to Participate

Not applicable, as this study does not involve any animal or human experimentation.

Consent for Publication

All authors agreed for this publication.

Availability of Data and Materials

The information used in this article was taken from authentic sources publicly available and cited with the proper referencing for validation.

Competing Interest

The authors declare that they have no competing interests.

Funding

Funding for this project was provided by DRDO, the Ministry of Defence, and the Government of India. SM is a recipient of the DRDO-SRF fellowship. PS is a recipient of the DRDO-JRF fellowship. The authors have not received any other financial support for publishing this article.

Author Contributions YA developed the whole conception of the article and substantially contributed to the article. PS collected the information and wrote the manuscript, and SM was involved in formatting and polishing the article. SM and YA critically evaluated the final version of the article.

Authors' Information PS is currently working as a junior research fellow, and SM is working as a senior research fellow in the areas of high-altitude physiology after completing their M.Tech in biotechnology as a major and being actively involved in biology research. YA is currently holding a position of scientist at DIPAS-DRDO and had prior research expertise in the above discipline for more than 15 years.

References

Aksel G, Çorbacıoğlu ŞK, Özen C (2019) High-altitude illness: management approach. Turk J Emerg Med 19(4):121–126. https://doi.org/10.1016/j.tjem.2019.09.002

Bagali S, Hadimani AG, Biradar MS, Das KK (2018) Introductory chapter: primary concept of hypoxia and anoxia. In: Hypoxia and anoxia. IntechOpen, London

Bärtsch P, Saltin B (2008) General introduction to altitude adaptation and mountain sickness. Scand J Med Sci Sports 18(Suppl 1):1–10. https://doi.org/10.1111/j.1600-0838.2008.00827.x

Bassovitch O (2010) "Combining hypoxic methods for peak performance": a biomedical engineering perspective. Sports Med 40(6):519–521. https://doi.org/10.2165/11535150-000000000-00000; author reply 521

Beall CM, Laskowski D, Strohl KP, Soria R, Villena M, Vargas E et al (2001) Pulmonary nitric oxide in mountain dwellers. Nature 414(6862):411–412. https://doi.org/10.1038/35106641

Bonne TC, Lundby C, Jørgensen S, Johansen L, Mrgan M, Bech SR et al (2014) "Live High–Train High" increases hemoglobin mass in Olympic swimmers. Eur J Appl Physiol 114(7): 1439–1449. https://doi.org/10.1007/s00421-014-2863-4

Brocherie F, Girard O, Faiss R, Millet GP (2017) Effects of repeated-sprint training in hypoxia on sea-level performance: a meta-analysis. Sports Med 47(8):1651–1660. https://doi.org/10.1007/s40279-017-0685-3

Brugniaux VJV, Schmitt L, Robach P, Jeanvoine H, Zimmermann H, Nicolet G et al (2006) Living high-training low: tolerance and acclimatization in elite endurance athletes. Eur J Appl Physiol 96(1):66–77. https://doi.org/10.1007/s00421-005-0065-9

Burtscher M, Millet GP, Burtscher J (2022) Hypoxia conditioning for high-altitude pre-acclimatization. J Sci Sport Exer 4(4):331–345

Calbet JAL, Lundby C (2009) Air to muscle O_2 delivery during exercise at altitude. High Alt Med Biol 10(2):123–134. https://doi.org/10.1089/ham.2008.1099

Davis C, Hackett P (2017) Advances in the prevention and treatment of high altitude illness. Emerg Med Clin North Am 35(2):241–260. https://doi.org/10.1016/j.emc.2017.01.002

Dehnert C, Hütler M, Liu Y, Menold E, Netzer C, Schick R et al (2002) Erythropoiesis and performance after two weeks of living high and training low in well trained triathletes. Int J Sports Med 23(8):561–566. https://doi.org/10.1055/s-2002-35533

Droma Y, Hanaoka M, Ota M, Katsuyama Y, Koizumi T, Fujimoto K et al (2002) Positive association of the endothelial nitric oxide synthase gene polymorphisms with high-altitude pulmonary edema. Circulation 106(7):826–830. https://doi.org/10.1161/01.cir.0000024409.30143.70

Farias JG, Jimenez D, Osorio J, Zepeda AB, Figueroa CA, Pulgar VM (2013) Acclimatization to chronic intermittent hypoxia in mine workers: a challenge to mountain medicine in Chile. Biol Res 46(1):59–67. https://doi.org/10.4067/S0716-97602013000100009

Friedmann B, Frese F, Menold E, Kauper F, Jost J, Bärtsch P (2005) Individual variation in the erythropoietic response to altitude training in elite junior swimmers. Br J Sports Med 39(3): 148–153. https://doi.org/10.1136/bjsm.2003.011387

Fukuda R, Zhang H, Kim JW, Shimoda L, Dang CV, Semenza GL (2007) HIF-1 regulates cytochrome oxidase subunits to optimize efficiency of respiration in hypoxic cells. Cell 129(1):111–122. https://doi.org/10.1016/j.cell.2007.01.047

Goods PSR, Dawson BT, Landers GJ, Gore CJ, Peeling P (2014) Effect of different simulated altitudes on repeat-sprint performance in team-sport athletes. Int J Sports Physiol Perform 9(5): 857–862. https://doi.org/10.1123/ijspp.2013-0423

Grocott M, Montgomery H, Vercueil A (2007) High-altitude physiology and pathophysiology: implications and relevance for intensive care medicine. Crit Care 11(1):203. https://doi.org/10.1186/cc5142

Hackett PH, Yarnell PR, Weiland DA, Reynard KB (2019) Acute and evolving MRI of high-altitude cerebral edema: microbleeds, edema, and pathophysiology. AJNR Am J Neuroradiol 40(3):464–469. https://doi.org/10.3174/ajnr.A5897

Hahn AG, Gore CJ (2001) The effect of altitude on cycling performance: a challenge to traditional concepts. Sports Med 31(7):533–557. https://doi.org/10.2165/00007256-200131070-00008

Harding D, Dhamrait S, Marlow N, Whitelaw A, Gupta S, Humphries S et al (2003) Angiotensin-converting enzyme DD genotype is associated with worse perinatal cardiorespiratory adaptation in preterm infants. J Pediatr 143(6):746–749. https://doi.org/10.1067/S0022-3476(03)00582-1

Hayashi T, Asano Y, Shintani Y, Aoyama H, Kioka H, Tsukamoto O et al (2015) *Higd1a* is a positive regulator of cytochrome c oxidase. Proc Natl Acad Sci U S A 112(5):1553–1558. https://doi.org/10.1073/pnas.1419767112

Hillebrandt D (2010) Wilderness medical society consensus guidelines for the prevention and treatment of acute altitude illness—a view from the other side of the Atlantic. Wilderness Environ Med 21(4):383–384. https://doi.org/10.1016/j.wem.2010.04.001; author reply 384

Hopkins SR, Bogaard HJ, Niizeki K, Yamaya Y, Ziegler MG, Wagner PD (2003) β-Adrenergic or parasympathetic inhibition, heart rate and cardiac output during normoxic and acute hypoxic exercise in humans. J Physiol 550(2):605–616. https://doi.org/10.1113/jphysiol.2003.040568

Hoppeler H, Vogt M (2001) Muscle tissue adaptations to hypoxia. J Exp Biol 204(18):3133–3139. https://doi.org/10.1242/jeb.204.18.3133

Jensen JD, Vincent AL (2022) High altitude cerebral edema. In: StatPearls. StatPearls Publishing

Julian CG, Gore CJ, Wilber RL, Daniels JT, Fredericson M, Stray-Gundersen J et al (2004) Intermittent normobaric hypoxia does not alter performance or erythropoietic markers in highly trained distance runners. J Appl Physiol (1985) 96(5):1800–1807. https://doi.org/10.1152/japplphysiol.00969.2003

Kapoor R, Narula AS, Anand AC (2004) Treatment of acute mountain sickness and high altitude pulmonary oedema. Med J Armed Forces India 60(4):384–387. https://doi.org/10.1016/S0377-1237(04)80018-2

Krzywicki HJ, Consolazio CF, Johnson HL, Nielsen WC, Barnhart RA (1971) Water metabolism in humans during acute high-altitude exposure (4,300 m). J Appl Physiol 30(6):806–809. https://doi.org/10.1152/jappl.1971.30.6.806

Laitinen H, Alopaeus K, Heikkinen R, Hietanen H, Mikkelsson L, Tikkanen H et al (1995) Acclimatization to living in normobaric hypoxia and training in normoxia at sea level in runners. Med Sci Sports Exerc 27(Suppl):S109. https://doi.org/10.1249/00005768-199505001-00617

Lee P, Chandel NS, Simon MC (2020) Cellular adaptation to hypoxia through hypoxia inducible factors and beyond. Nat Rev Mol Cell Biol 21(5):268–283. https://doi.org/10.1038/s41580-020-0227-y

Lendahl U, Lee KL, Yang H, Poellinger L (2009) Generating specificity and diversity in the transcriptional response to hypoxia. Nat Rev Genet 10(12):821–832. https://doi.org/10.1038/nrg2665

Levine BD (2002) Intermittent hypoxic training: fact and fancy. High Alt Med Biol 3(2):177–193. https://doi.org/10.1089/15270290260131911

Levine BD, Stray-Gundersen J (1997) "Living high-training low": effect of moderate-altitude acclimatization with low-altitude training on performance. J Appl Physiol (1985) 83(1): 102–112. https://doi.org/10.1152/jappl.1997.83.1.102

Levine BD, Stray-Gundersen J (2006) Dose-response of altitude training: how much altitude is enough? pp 233–247

Luks AM, McIntosh SE, Grissom CK, Auerbach PS, Rodway GW, Schoene RB et al (2014) Wilderness medical society practice guidelines for the prevention and treatment of acute altitude illness: 2014 update. Wilderness Environ Med 25(4):S4–S14. https://doi.org/10.1016/j.wem.2014.06.017

Luks AM, Swenson ER, Bärtsch P (2017) Acute high-altitude sickness. Eur Respir Rev 26(143): 160096. https://doi.org/10.1183/16000617.0096-2016

Lundby C, Millet GP, Calbet JA, Bärtsch P, Subudhi AW (2012) Does 'altitude training' increase exercise performance in elite athletes? Br J Sports Med 46(11):792–795. https://doi.org/10.1136/bjsports-2012-091231

Luo Y, Wang Y, Lu H, Gao Y (2014) "Ome" on the range: update on high-altitude acclimatization/adaptation and disease. Mol BioSyst 10(11):2748–2755. https://doi.org/10.1039/C4MB00119B

Malacrida S, Giannella A, Ceolotto G, Reggiani C, Vezzoli A, Mrakic-Sposta S et al (2019) Transcription factors regulation in human peripheral white blood cells during hypobaric hypoxia exposure: an in-vivo experimental study. Sci Rep 9(1):9901. https://doi.org/10.1038/s41598-019-46391-6

McIntosh SE, Hemphill M, McDevitt MC, Gurung TY, Ghale M, Knott JR et al (2019) Reduced acetazolamide dosing in countering altitude illness: a comparison of 62.5 vs 125 mg (the radical trial). Wilderness Environ Med 30(1):12–21. https://doi.org/10.1016/j.wem.2018.09.002

McLean BD, Gore CJ, Kemp J (2014) Application of 'live low-train high' for enhancing normoxic exercise performance in team sport athletes. Sports Med 44(9):1275–1287. https://doi.org/10.1007/s40279-014-0204-8

Michiels C (2004) Physiological and pathological responses to hypoxia. Am J Pathol 164(6): 1875–1882. https://doi.org/10.1016/S0002-9440(10)63747-9

Millet GP, Faiss R, Brocherie F, Girard O (2013) Hypoxic training and team sports: a challenge to traditional methods? Br J Sports Med 47(Suppl 1):i6–i7. https://doi.org/10.1136/bjsports-2013-092793

Muza SR (2007) Military applications of hypoxic training for high-altitude operations. Med Sci Sports Exer 39(9):1625–1631. https://doi.org/10.1249/mss.0b013e3180de49fe

Netzer N, Strohl K, Faulhaber M, Gatterer H, Burtscher M (2013) Hypoxia-related altitude illnesses. J Travel Med 20(4):247–255. https://doi.org/10.1111/jtm.12017

Palmer BF (2010) Physiology and pathophysiology with ascent to altitude. Am J Med Sci 340(1): 69–77. https://doi.org/10.1097/MAJ.0b013e3181d3cdbe
Paralikar SJ (2012) High altitude pulmonary edema-clinical features, pathophysiology, prevention and treatment. Indian J Occup Environ Med 16(2):59–62. https://doi.org/10.4103/0019-5278.107066
Paralikar SJ, Paralikar JH (2010) High-altitude medicine. Indian J Occup Environ Med 14(1):6–12. https://doi.org/10.4103/0019-5278.64608
Paula de P, Niebauer J (2012) Effects of high altitude training on exercise capacity: fact or myth. Sleep Breath 16(1):233–239. https://doi.org/10.1007/s11325-010-0445-1
Pham K, Parikh K, Heinrich EC (2021) Hypoxia and inflammation: insights from high-altitude physiology. Front Physiol 12:676782. https://doi.org/10.3389/fphys.2021.676782
Powell FL, Garcia N (2000) Physiological effects of intermittent hypoxia. High Alt Med Biol 1(2): 125–136. https://doi.org/10.1089/15270290050074279
Prince TS, Thurman J, Huebner K (2022) Acute mountain sickness. In: StatPearls. StatPearls Publishing
Ricart A, Casas H, Casas M, Pagés T, Palacios L, Rama R et al (2000) Acclimatization near home? Early respiratory changes after short-term intermittent exposure to simulated altitude. Wilderness Environ Med 11(2):84–88. https://doi.org/10.1580/1080-6032(2000)011[0084:anherc]2.3.co;2
Robach P, Schmitt L, Brugniaux VJV, Roels B, Millet G, Hellard P et al (2006) Living high–training low: effect on erythropoiesis and aerobic performance in highly-trained swimmers. Eur J Appl Physiol 96(4):423–433. https://doi.org/10.1007/s00421-005-0089-1
Rodríguez FA, Ventura JL, Casas M, Casas H, Pagés T, Rama R et al (2000) Erythropoietin acute reaction and haematological adaptations to short, intermittent hypobaric hypoxia. Eur J Appl Physiol 82(3):170–177. https://doi.org/10.1007/s004210050669
San T, Polat S, Cingi C, Eskiizmir G, Oghan F, Cakir B (2013) Effects of high altitude on sleep and respiratory system and their adaptations. Sci World J 2013:1–7
Saugy JJ, Schmitt L, Cejuela R, Faiss R, Hauser A, Wehrlin JP et al (2014) Comparison of 'live high-train low' in normobaric versus hypobaric hypoxia. PLoS One 9(12):e114418. https://doi.org/10.1371/journal.pone.0114418
Saunders PU, Pyne DB, Gore CJ (2009) Endurance training at altitude. High Alt Med Biol 10(2): 135–148. https://doi.org/10.1089/ham.2008.1092
Smedley T, Grocott MP (2013) Acute high-altitude illness: a clinically orientated review. Br J Pain 7(2):85–94. https://doi.org/10.1177/2049463713489539
Smith EMD, Baillie JK, Thompson AAR, Irving JB, Porteous D, Webb DJ (2006) Endothelial nitric oxide synthase polymorphisms do not influence pulmonary artery systolic pressure at altitude. High Alt Med Biol 7(3):221–227. https://doi.org/10.1089/ham.2006.7.221
Swenson ER (2011) High-altitude pulmonary edema. In: Textbook of pulmonary vascular disease. Springer, Boston, MA, pp 871–888
Tannheimer M, Fusch C, Böning D, Thomas A, Engelhardt M, Schmidt R (2010) Changes of hematocrit and hemoglobin concentration in the cold Himalayan environment in dependence on total body fluid. Sleep Breath 14(3):193–199. https://doi.org/10.1007/s11325-009-0284-0
Tello D, Balsa E, Acosta-Iborra B, Fuertes-Yebra E, Elorza A, Ordóñez Á et al (2011) Induction of the mitochondrial NDUFA4L2 protein by HIF-1α decreases oxygen consumption by inhibiting complex I activity. Cell Metab 14(6):768–779. https://doi.org/10.1016/j.cmet.2011.10.008
Turner REF, Gatterer H, Falla M, Lawley JS (2021) High-altitude cerebral edema: its own entity or end-stage acute mountain sickness? J Appl Physiol (1985) 131(1):313–325. https://doi.org/10.1152/japplphysiol.00861.2019
Viscor G, Torrella JR, Corral L, Ricart A, Javierre C, Pages T et al (2018) Physiological and biological responses to short-term intermittent hypobaric hypoxia exposure: from sports and mountain medicine to new biomedical applications. Front Physiol 9:814. https://doi.org/10.3389/fphys.2018.00814

Vogt M, Puntschart A, Geiser J, Zuleger C, Billeter R, Hoppeler H (2001) Molecular adaptations in human skeletal muscle to endurance training under simulated hypoxic conditions. J Appl Physiol (1985) 91(1):173–182. https://doi.org/10.1152/jappl.2001.91.1.173

Wehrlin JP, Zuest P, Hallén J, Marti B (2006) Live high-train low for 24 days increases hemoglobin mass and red cell volume in elite endurance athletes. J Appl Physiol (1985) 100(6):1938–1945. https://doi.org/10.1152/japplphysiol.01284.2005

Weiss J, Haefeli WE, Gasse C, Hoffmann MM, Weyman J, Gibbs S et al (2003) Lack of evidence for association of high altitude pulmonary edema and polymorphisms of the NO pathway. High Alt Med Biol 4(3):355–366. https://doi.org/10.1089/152702903769192313

West JB (2006) Human responses to extreme altitudes. Integr Comp Biol 46(1):25–34. https://doi.org/10.1093/icb/icj005

Wilber RL (2007) Application of altitude/hypoxic training by elite athletes. Med Sci Sports Exer 39(9):1610–1624. https://doi.org/10.1249/mss.0b013e3180de49e6

Zelmanovich R, Pierre K, Felisma P, Cole D, Goldman M, Lucke-Wold B (2022) High altitude cerebral edema: improving treatment options. Biologics (Basel) 2(1):81–91. https://doi.org/10.3390/biologics2010007

Zheng Y, Huang J (2020) Angiotensin-converting enzyme gene insertion/deletion polymorphism and high-altitude pulmonary edema: an updated meta-analysis. J Renin-Angiotensin-Aldosterone Syst 21(1):1470320319900039. https://doi.org/10.1177/1470320319900039

Zhou Q (2011) Standardization of methods for early diagnosis and on-site treatment of high-altitude pulmonary edema. Pulm Med 2011:190648. https://doi.org/10.1155/2011/190648

Performance Improvement In Extremely Stressful Environment

11

Rajkumar Tulsawani

Abstract

Ascending to high altitude poses a hypobaric hypoxia environment, and at extreme altitude, this may lead to an imbalanced redox homeostasis and ultimately affect performance in a sojourn. This study was carried out to understand the role of the aqueous extract of *Cordyceps sinensis* in the enhancement of physical performance under an extremely simulated high-altitude environment. Sprague–Dawley rats were exposed to hypobaric hypoxia at 7620 m for 7 days. A decrease in exhaustion time was observed in rats exposed to hypobaric hypoxia compared to control rats that were prevented after supplementation of extract. Skeletal muscle extracted from exposed rats to hypoxia has higher levels of malondialdehyde, compromised antioxidant status, and higher lactate levels indicating stressed and exhausted muscles. The ratio of NAD/NADH and ADP/ATP was altered in rat muscle exposed to hypobaric hypoxia in comparison to respective control suggesting disturbed redox and bio-energetic homeostasis. Treatment of rats with the extract, having high antioxidant activity, enhanced exhaustion time significantly and prevented performance loss and all changes in studied variables observed following hypobaric hypoxia. To conclude, the adaptive response to performance loss under hypobaric hypoxia by *Cordyceps sinensis* is associated with balancing redox homeostasis.

Keywords

Hypobaric hypoxia · *Cordyceps sinensis* · Performance enhancement

R. Tulsawani (✉)
Defence Institute of Physiology and Allied Sciences (DIPAS), Defence Research and Development Organization (DRDO), Delhi, India
e-mail: rktulsawani@yahoo.com

R. Tulsawani, D. Vohora (eds.), *Adaptation under Stressful Environments through Biological Adjustments and Interventions*,
https://doi.org/10.1007/978-981-99-7652-2_11

11.1 Introduction

Environmental conditions at high altitude include cold and hypoxia due to decreased partial pressure (Pugh 1963; Peacock 1998; Massey et al. 2015; Santocildes et al. 2021), causing altitude sickness and finally leading to decreased physical and mental impairments in soldiers and mountaineers who are un-acclimatized individuals (Schoene et al. 1984; Cerretelli and di Prampero PE 1985; Hoppeler et al. 1990; Cerretelli 1992; Peacock 1998; Howald and Hoppeler 2003; Grocott et al. 2007; Sharma and Tulsawani 2022; Szymczak et al. 2021). The capacity to perform is reduced at extremely high altitudes in un-acclimatized persons and soldiers posted at high altitudes (Pugh et al. 1964; West et al. 1983; Sutton et al. 1988; Cymerman et al. 1989; Shukitt-Hale et. 1998; Singh et al. 2013; Sharma and Tulsawani 2020; Jatwani and Tulsawani 2021). This loss of performance is restored after acclimatization at moderate altitude (Buskirk et al. 1967); however, the performance loss is restored partially at higher and extreme altitudes irrespective of the duration of acclimatization (Pugh 1963; West et al. 1983; Sutton et al. 1988; Cymerman et al. 1989). At extremely stressful altitudes, physical and cognitive performance loss occurs (Hornbein et al. 1989; Sharma and Tulsawani 2020). These problems may escalate rapidly at higher and extreme altitudes and sometimes may be life-threatening.

To adapt to such harsh conditions of hypoxic and cold conditions, the body must sense oxygen change at particular altitudes with altitude and cold stress. The body needs to initiate responses to maintain biochemical homeostasis and induce transcription factors and gene expression appropriately to adverse changes in the body due to these stressors. The vital redox-sensitive transcriptional factors involved in oxygen sensing machinery are hypoxia-inducible factor 1 (HIF1), nuclear factor kappa-B (NFκB), and nuclear factor (erythroid-derived 2)-like 2 (Nrf2). Expression of antioxidant-responsive-element (ARE)-driven genes and enzymes is directed by Nrf2 cap'n'collar bZIP transcription factors including glutathione peroxidase (GPx), heme oxygenase (HO), and superoxide dismutase (SOD). HIF1 is selectively stabilized in hypoxia, and this, in turn, leads to the activation of several genes, such as erythropoietin (EPO) promotes erythropoiesis, vascular endothelial growth factor (VEGF) promotes angiogenesis, and glucose transporter-1 (GLUT1) regulate energy metabolism (Singh et al. 2013).

Current treatments, such as acetazolamide and dexamethasone, are only effective at moderate altitudes and at higher and extreme altitudes, and these therapeutic regimens prove ineffective, fail to restore work capacity, and are contraindicated. Moreover, there are no specific drugs to negate the impact of high altitude stressors, and hence, nutraceuticals having adaptogenic potential might be safer and more effective alternatives. Many herbs are reported to enhance the adaptation process against various stressors, including *Gingko biloba*, *Hippophae rhamnoides*, *Rhodiola imbricata*, etc. In our earlier study, we reported that *Cordyceps sinensis* induces tolerance to hypoxia in vitro (Singh et al. 2013), and therefore, the present study was designed to evaluate whether *Cordyceps sinensis* could initiate an adaptive response in vivo and delineate its mechanism of action.

11.2 Material and Methods

11.2.1 Collection of Plant Material

Cordyceps sinensis mycelia were supplied by the Defence Institute of Bio-Energy Research, Haldwani. *Cordyceps sinensis* was collected during the rainy season from wood logs and tree stumps at an altitude of over 4000 m from the North-West Himalayas, India, where the plant grows widely under natural conditions.

11.2.2 Extraction Procedure

Aqueous extract of *Cordyceps sinensis* was prepared by environmentally friendly technique Accelerated Solvent System ASE 350 equipped with a solvent controller unit (Dionex Corporation, CA, USA). Finally, the supernatant solution was lyophilized (Allied Frost, India), and the dried extract was stored at 4 °C.

11.2.3 Experimental Animals

SD rats (180–200 g) bred in the animal facility of the Defence Institute of Physiology and Allied Sciences (DIPAS) were used for all experiments. The animals were maintained on a bedding of rice husk in polypropylene cages under a controlled environment in the institute's animal house at 22–25 °C, 55 $\pm$ 10% humidity, and a 12-h light-dark cycle. The study has the approval of the Institutional Ethical Committee on Animal Experimentations and conformed to national guidelines on the care and use of laboratory animals, in India.

11.2.4 Rodent Treadmill Test

The experimental animals were habituated on the rodent treadmill at varying speeds (10 m/min to 30 m/min) for 1 week, and trained rats were divided into the following groups: untreated group and extract treated groups (25, 50, and 100 mg/kg body weight) were exposed to hypobaric hypoxia at 7620 m for 7 days. The extract was administered daily for 7 days. After the end of the treatment period, all rats were allowed to run on a treadmill until exhaustion, and this duration on the treadmill was recorded as exhaustion time (min).

11.2.5 Biochemical Analysis

After this, rats were sacrificed, and muscle was collected for analysis of MDA, TAC, Lactate, ADP/ATP ratio, and NAD/NADH ratio. MDA estimated as previously described elsewhere. The other parameters studied were conducted using commercial kits (Randox, UK; Biovision, USA) as per the kit manual.

11.2.6 Data Analysis

The data are presented as mean ± SE of ten animals in each group. One-way ANOVA was utilized to analyze statistical significance set at a 5% level (SigmaStat, USA), and the treatment group was compared with the control group using Dunnett's test.

11.3 Results

11.3.1 Effect of *Cordyceps sinensis* on the Performance of Rats on Rodent Treadmill Test

Figure 11.1 depicts the performance-enhancing potential of *Cordyceps sinensis* extracts in rats exposed to hypobaric hypoxia. Experimental animals exposed to hypobaric hypoxia for 7 days @ 7620 m revealed a loss of physical performance than the unexposed rats (Fig. 11.1, bar 2 vs. bar 1). Hypoxic rats were exhausted earlier than the normoxic rats. The rats treated with the extract revealed improved physical performance (Fig. 11.1, bars 3–5) evidencing the performance-enhancing potential of the aqueous extract of *Cordyceps sinensis*.

11.3.2 Effect of *Cordyceps sinensis* on Stress Marker and Anti-Oxidative Status

Figures 11.2 and 11.3 show the effects of hypobaric hypoxia on stress marker, malondialdehyde, and total antioxidant status in the absence or presence of *Cordyceps sinensis* extract. Skeletal muscles of rats exposed to hypobaric hypoxia resulted in elevated levels of stress marker, malondialdehyde (Fig. 11.2, bar 2). In contrast, the level of total antioxidant content was found compromised in rat skeletal

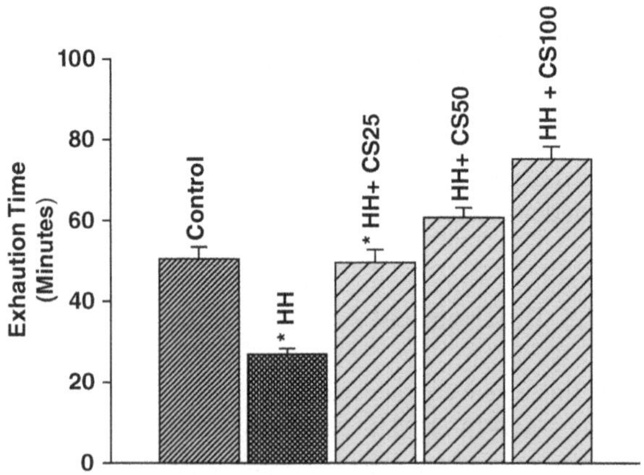

Fig. 11.1 Efficacy of *Cordyceps sinensis* on exhaustion time following hypobaric hypoxia. *HH* hypobaric hypoxia. *p > 0.05

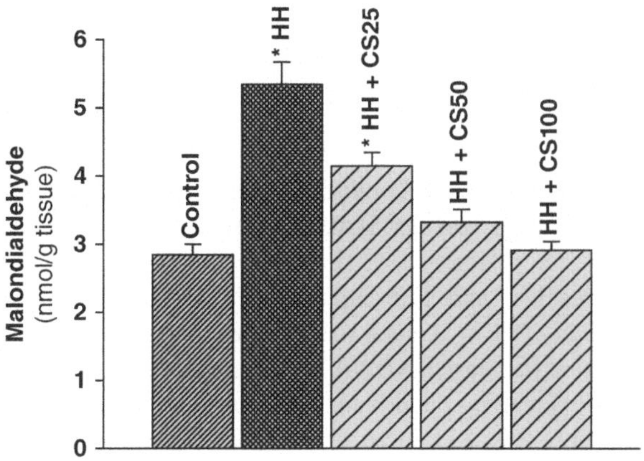

Fig. 11.2 Effects of *Cordyceps sinensis* on oxidative stress marker, malondialdehyde in rat skeletal muscles exposed to hypobaric hypoxia. *HH* hypobaric hypoxia. *p > 0.05

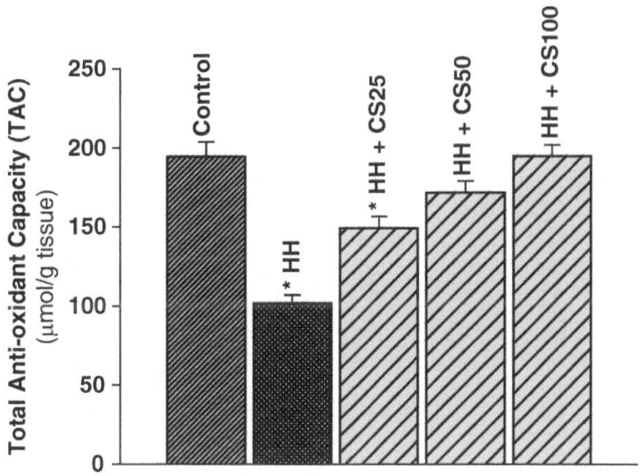

Fig. 11.3 Effects of *Cordyceps sinensis* on total antioxidant status in rat skeletal muscles exposed to hypobaric hypoxia stress. *HH* hypobaric hypoxia. *p > 0.05

muscle exposed to hypobaric hypoxia (Fig. 11.3, bar 2). Skeletal muscles of experimental animals treated with extract prevented the rise of malondialdehyde and decrease in total antioxidant status observed after hypobaric hypoxia exposure indicating the strong anti-oxidative potential of aqueous extract of *Cordyceps sinensis* (Fig. 11.2 and 11.3, bars 3–5).

11.3.3 Effect of *Cordyceps sinensis* on Muscle Lactic Acid Exposed to Hypobaric Hypoxia

Figure 11.4 depicts the efficacy of *Cordyceps sinensis* on lactate, fatigue marker, of skeletal muscles of rats exposed to hypobaric hypoxia in the absence or presence of extract. Lactate levels were recorded to be significantly higher in skeletal muscle exposed to hypobaric hypoxia (Fig. 11.2, bar 2), and lactate levels were restored with the treatment of extract (Fig. 11.4, bars 3–5) in a dose-dependent manner. This finding indicates the anti-fatigue potential of the aqueous extract of *Cordyceps sinensis*.

11.3.4 Effect of *Cordyceps sinensis* on Muscle Bioenergetics and Redox Status Exposed to Hypobaric Hypoxia

Figures 11.5 and 11.6 depict the effects of *Cordyceps sinensis* on ADP/ATP and NAD/NADH in skeletal muscle after hypobaric hypoxia. Hypoxic skeletal muscle revealed decreased ADP/ATP which was concomitant with a higher NAD/NADH ratio evidencing disturbances in bioenergetics and redox homeostasis after hypoxia exposure (Figs. 11.5 and 11.6, bar 2). Skeletal muscles treated with extract restored changes induced by hypobaric hypoxia (Figs. 11.5 and 11.6, bars 3–5).

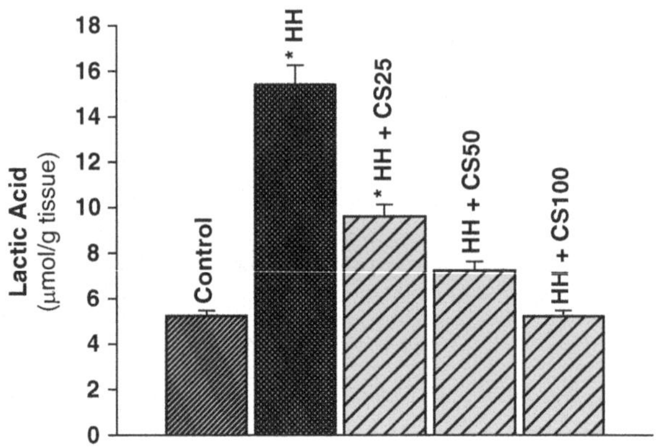

Fig. 11.4 Effects of *Cordyceps sinensis* on lactate level of rat skeletal muscles exposed to hypobaric hypoxia. *HH* hypobaric hypoxia. *p > 0.05

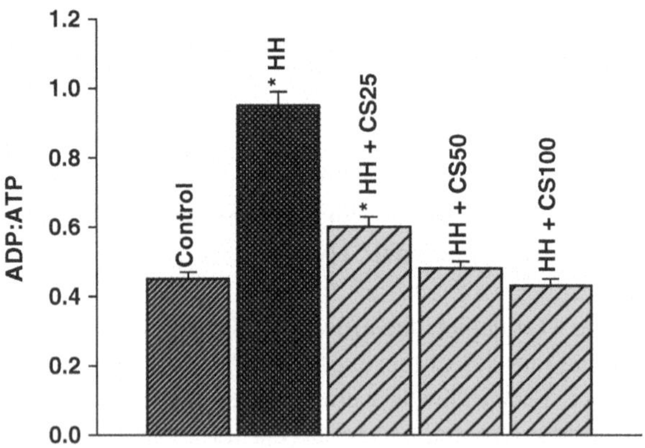

Fig. 11.5 Effects of *Cordyceps sinensis* on bioenergetics in rat skeletal muscles exposed to hypobaric hypoxia. *HH* hypobaric hypoxia. *p > 0.05

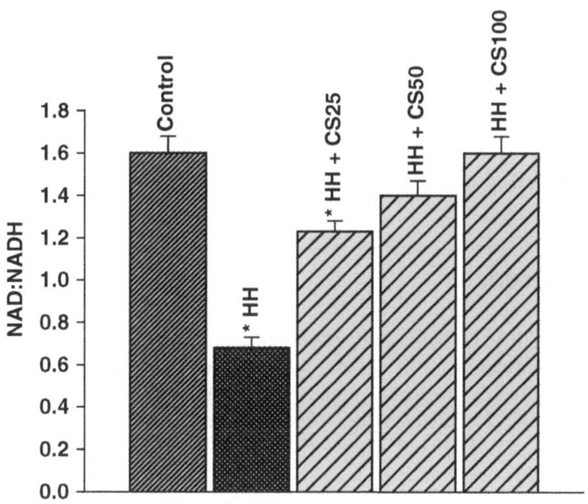

Fig. 11.6 Effects of *Cordyceps sinensis* on redox homeostasis in rat skeletal muscles exposed to hypobaric hypoxia. *HH* hypobaric hypoxia. *$p > 0.05$

11.4 Discussion

Extreme stress, particularly hypobaric hypoxia, disturbs body homeostasis and leads to loss of physical performance. It is also reported that extreme altitude exposure results in loss of skeletal muscle mass. Exposure to hypobaric hypoxia is the cause of concern at high altitude adaptation and affects work capacity resulting in fatigue. The extract of *Cordyceps sinensis* was found to be rich in phenolics and flavonoids and had good antioxidant activities in the chemical system in our earlier work (Singh et al. 2013; Sharma and Tulsawani 2022). Therefore, we hypothesized that supplementation of extract may fasten adaptation to high altitude stressors and enhance performance under high altitude environments. In the present study, hypobaric hypoxia extreme stress was used to prove the hypothesis. Furthermore, *Cordyceps sinensis* is reported to contain bioactive compounds including polysaccharides, nucleotides, cordycepin, cordymin, cerevisterol, d-mannitol, glycosides etc. (Hobbs 1995; Zhang et al. 2005; Kim et al. 2006; Chen et al. 2010; Jeong et al. 2010; Kim 2010; Wang et al. 2014; Chang et al. 2015; Hu et al. 2015; Huang et al. 2018; Ashraf et al. 2020), which might be responsible for the observed better-maintained antioxidant status and redox homeostasis, anti-fatigue, and performance enhancing potential in this study.

Treatment of experimental rats with extract increased exhaustion time in comparison to the rats exposed to hypobaric hypoxia without extract, indicating the performance-enhancing potential of *Cordyceps sinensis* extract. Furthermore, the accumulation of fatigue marker, lactic acid, in the skeletal muscles of rats exposed to hypobaric hypoxia was prevented with the extract treatment. Furthermore, extract treatment improved the antioxidant status of rat muscle highlighting the anti-oxidative potential of *Cordyceps sinensis* extract. The skeletal muscle of hypobaric hypoxia-exposed rats showed disturbed redox and bio-energetic homeostasis, which was maintained with the supplementation of the *Cordyceps sinensis* extract.

Overall, treatment of hypobaric hypoxia exposed rats with an aqueous extract of *Cordyceps sinensis* exhibited anti-stress, anti-fatigue, and physical performance-enhancing activity.

References

Ashraf SA, Elkhalifa AEO, Siddiqui AJ, Patel M, Awadelkareem AM, Snoussi M et al (2020) Cordycepin for health and wellbeing: a potent bioactive metabolite of an Entomopathogenic cordyceps medicinal fungus and its nutraceutical and therapeutic potential. Molecules 25 (12):2735. https://doi.org/10.3390/molecules25122735

Buskirk ER, Kollias J, Akers RF, Prokop EK, Reategui EP (1967) Maximal performance at altitude and return from altitude in conditioned runners. J Appl Physiol 23(2):259–266. https://doi.org/10.1152/jappl.1967.23.2.259, PMID 6033527

Cerretelli and di Prampero PE (1985) Aerobic and anaerobic metabolism during exercise at altitude in High Altitude Deterioration. In: Rivolier J, Cerretelli P, Foray J, Segantini P (eds) International Congress for Mountain Medicine, Chamonix, March 1984. Basel, Karger, pp 1–19

Chang Y, Hsu WH, Lu WJ, Jayakumar T, Liao JC, Lin MJ et al (2015) Inhibitory mechanisms of CME-1, a novel polysaccharide from the mycelia of Cordyceps sinensis, in platelet activation. Curr Pharm Biotechnol 16(5):451–461. https://doi.org/10.2174/1389201016666150303152237

Chen W, Zhang W, Shen W, Wang K (2010) Effects of the acid polysaccharide fraction isolated from a cultivated Cordyceps sinensis on macrophages in vitro. Cell Immunol 262(1):69–74. https://doi.org/10.1016/j.cellimm.2010.01.001

Cymerman A, Reeves JT, Sutton JR, Rock PB, Groves BM, Malconian MK et al (1989) Operation EVEREST II: maximal oxygen uptake at extreme altitudes. J Appl Physiol (1985) 66(5): 2446–2453. https://doi.org/10.1152/jappl.1989.66.5.2446, PMID 2745305

Grocott M, Montgomery H, Vercueil A (2007) High-altitude physiology and pathophysiology: implications and relevance for intensive care medicine. Crit Care 11(1):203. https://doi.org/10.1186/cc5142. PMID: 17291330; PMCID: PMC2151873

Hobbs C (1995) Medicinal mushrooms: an exploration of tradition, healing, and culture, 2nd edn. Botanica Press, Inc., Santa Cruz, CA

Hornbein TF, Townes BD, Schoene RB, Sutton JR, Houston CS (1989) The cost on the central nervous system of climbing to extremely high altitude. N Engl J Med 321(25):1714–1719. https://doi.org/10.1056/NEJM198912213212505, PMID 2512483

Hoppeler H, Howald H, Cerretelli P (1990) Human muscle structure after exposure to extreme altitude. Experientia 46(11–12):1185–1187. https://doi.org/10.1007/BF01936933. PMID: 2253720

Howald H, Hoppeler H (2003) Performing at extreme altitude: muscle cellular and subcellular adaptations. Eur J Appl Physiol 90:360–364. https://doi.org/10.1007/s00421-003-0872-9

Hu H, Xiao L, Zheng B, Wei X, Ellis A, Liu YM (2015) Identification of chemical markers in Cordyceps sinensis by HPLC-MS/MS. Anal Bioanal Chem 407(26):8059–8066. https://doi.org/10.1007/s00216-015-8978-6. Epub 2015 Aug 25. PMID: 26302964; PMCID: PMC4596796

Huang D, Meran S, Nie SP, Midgley A, Junqiao W, Cui SW et al (2018) Cordyceps sinensis: anti-fibrotic and inflammatory effects of a cultured polysaccharide extract. Bioact Carbohydr Diet Fibre 14:2–8. https://doi.org/10.1016/j.bcdf.2017.07.012. ISSN 2212-6198

Jatwani A, Tulsawani R (2021) Ganoderma lucidum induces myogenesis markers to avert damage to skeletal muscles in rats exposed to hypobaric hypoxia. High Alt Med Biol. https://doi.org/10.1089/ham.2020.0172

Jeong JW, Jin CY, Kim GY, Lee JD, Park C, Kim GD, Kim WJ, Jung WK, Seo SK, Choi IW, Choi YH (2010) Anti-inflammatory effects of cordycepin via suppression of inflammatory mediators in BV2 microglial cells. Int Immunopharmacol 10(12):1580–1586. https://doi.org/10.1016/j.intimp.2010.09.011. Epub 2010 Oct 14. PMID: 20937401

Kim SD (2010) Isolation, structure and cholesterol esterase inhibitory activity of a polysaccharide, PS-A, from Cordyceps sinensis. J Korean Soc Appl Biol Chem 53(6):784–789. https://doi.org/10.3839/jksabc.2010.118

Kim HG, Shrestha B, Lim SY, Yoon DH, Chang WC, Shin DJ et al (2006) Cordycepin inhibits lipopolysaccharide-induced inflammation by the suppression of NF-kappaB through Akt and p38 inhibition in RAW 264.7 macrophage cells. Eur J Pharmacol 545(2–3):192–199. https://doi.org/10.1016/j.ejphar.2006.06.047

Massey HC, House JR, Tipton MJ (2015) Vasomotor responses to hypoxia and cold air. Extrem Physiol Med 4(Suppl 1):A53. https://doi.org/10.1186/2046-7648-4-S1-A53. PMCID: PMC4580823

Peacock AJ (1998) ABC of oxygen: oxygen at high altitude. BMJ 317(7165):1063–1066. https://doi.org/10.1136/bmj.317.7165.1063. PMID: 9774298; PMCID: PMC1114067

Pugh LG (1963) Tolerance to extreme cold at altitude in a Nepalese pilgrim. J Appl Physiol 8:1234–1238. https://doi.org/10.1152/jappl.1963.18.6.1234. PMID: 14080750

Pugh LG (1964) Cardiac output in muscular exercise at 5,800 meters (19,000 ft). J Appl Physiol 19:441–447

Pugh LG, Gill MB, Lahiri S, Milledge JS, Ward MP, West JB (1964) Muscular exercise at great altitudes. Appl Phys 19:431–440. https://doi.org/10.1152/jappl.1964.19.3.431. PMID: 14173539

Santocildes G, Viscor G, Pagès T, Ramos-Romero S, Torres JL, Torrella JR (2021) Physiological effects of intermittent passive exposure to hypobaric hypoxia and cold in rats. Front Physiol 12:673095. https://doi.org/10.3389/fphys.2021.673095. PMID: 34135770; PMCID: PMC8201611

Schoene RB, Lahiri S, Hackett PH, Peters RM Jr, Milledge JS, Pizzo CJ, Sarnquist FH, Boyer SJ, Graber DJ, Maret KH, West JB (1984) Relationship of hypoxic ventilatory response to exercise performance on Mount Everest. J Appl Physiol Respir Environ Exerc Physiol 56(6):1478–1483. https://doi.org/10.1152/jappl.1984.56.6.1478. PMID: 6735806

Sharma P, Tulsawani R (2020) Ganoderma lucidum aqueous extract prevents hypobaric hypoxia induced memory deficit by modulating neurotransmission, neuroplasticity and maintaining redox homeostasis. Sci Rep 10(1):8944. https://doi.org/10.1038/s41598-020-65812-5, PMID: 32488040; PMCID: PMC7265456

Sharma P, Tulsawani R (2022) Efficacy of aqueous extract of Chinese caterpillar mushroom Ophiocordyceps sinensis (Ascomycota) against simulated altitude stress and subacute toxicity studies. Int J Med Mushrooms 24(11):21–34. https://doi.org/10.1615/IntJMedMushrooms.2022045194

Shukitt-Hale B et al (1998) Elevation-dependent symptom mood and performance changes produced by exposure to hypobaric hypoxia. Int J Aviat Psychol 8(4):319–334. https://doi.org/10.1207/s15327108ijap0804_1

Singh M, Tulsawani R, Koganti P, Chauhan A, Manickam M, Misra K (2013) Cordyceps sinensis increases hypoxia tolerance by inducing heme oxygenase-1 and metallothionein via Nrf2 activation in human lung epithelial cells. Biomed Res Int 2013:569206. https://doi.org/10.1155/2013/569206, PMID: 24063008; PMCID: PMC3770031, 1

Sutton JR, Reeves JT, Wagner PD, Groves BM, Cymerman A, Malconian MK et al (1988) Operation EVEREST II: oxygen transport during exercise at extreme simulated altitude. J Appl Physiol (1985) 64(4):1309–1321. https://doi.org/10.1152/jappl.1988.64.4.1309, PMID 3132445

Szymczak RK, Grzywacz T, Ziemann E, Sawicka M, Laskowski R (2021) Prolonged Sojourn at Very High Altitude Decreases Sea-Level Anaerobic Performance, Anaerobic Threshold, and Fat Mass. Front Physiol 12:743535. https://doi.org/10.3389/fphys.2021.743535. PMID: 34675820; PMCID: PMC8523780

Wang Y, Liu D, Zhao H, Jiang H, Luo C, Wang M et al (2014) Cordyceps sinensis polysaccharide CPS-2 protects human mesangial cells from PDGF-BB-induced proliferation through the PDGF/ERK and TGF-β1/Smad pathways. Mol Cell Endocrinol 382(2):979–988. https://doi.org/10.1016/j.mce.2013.11.018

West JB, Boyer SJ, Graber DJ, Hackett PH, Maret KH, Milledge JS et al (1983) Maximal exercise at extreme altitudes on Mount Everest. J Appl Physiol Respir Environ Exerc Physiol 55(3): 688–698. https://doi.org/10.1152/jappl.1983.55.3.688, PMID 6415008

Zhang W, Yang J, Chen J, Hou Y, Han X (2005) Immunomodulatory and antitumour effects of an exopolysaccharide fraction from cultivated Cordyceps sinensis (Chinese caterpillar fungus) on tumour-bearing mice. Biotechnol Appl Biochem 42(1):9–15. https://doi.org/10.1042/BA20040183

Military Footwear and Extreme Environment Operations: An Ergonomics Perspective

12

Debojyoti Bhattacharyya and Tirthankar Chatterjee

Abstract

Footwear holds immense importance in various terrain-specific military operations. Considering Indian defence scenarios, the soldiers are operating across all the challenging terrains, i.e. high altitudes, deserts, jungles and coastal areas. Application of ergonomic principles in military footwear design and development can contribute to better protection, reduced risk of musculoskeletal injuries, optimal physiological and biomechanical responses, and improved mobility and agility. For over six decades, substantial ergonomics/human factors research has been conducted for the design and development of operation-specific military footwear in countries like the USA, UK and Australia. However, the Indian context of human-centric/ergonomics research on military footwear is relatively a nascent concept. The literature available to date is merely limited. There is enormous scope to establish the ergonomic database and generate improved designs of operation-specific military footwear. The present article has critically reviewed the important global literature on the design and development of military footwear, as well as the recent research endeavor conducted in the Indian Military context from the perspective of ergonomics and human factors.

Keywords

Snow boot · Combat boot · Physiology · Biomechanics · Balance · Stability · Plantar pressure · Thermal comfort · High altitude

D. Bhattacharyya (✉) · T. Chatterjee
Defence Institute of Physiology and Allied Sciences (DIPAS), Defence Research and Development Organization (DRDO), Delhi, India
e-mail: rajdeb8206@gmail.com

R. Tulsawani, D. Vohora (eds.), *Adaptation under Stressful Environments through Biological Adjustments and Interventions*,
https://doi.org/10.1007/978-981-99-7652-2_12

12.1 Introduction

Environmental austerity, i.e. extremely hot, cold, humidity and hypoxia as well as the surface condition of various terrains, i.e. undulating hard rocky and snow-covered terrains of high altitudes/extreme altitudes, shifting sand dunes of deserts, slimy and bushy terrains of jungles and submersed wetlands of coastal areas, make the soldiers' every step challenging towards successful accomplishment of the task. Similarly, military training conditions are equally challenging where exercises are mostly conducted by simulation of real-time scenarios. At the same time, the ease of performance with sustained mobility and desired comfort of the wearer is of critical concern for securing the occupational health of the soldiers.

It is noteworthy to observe that because of the enormous diversity in defence operations, military populations have frequently been engaged in the epidemiological studies of footwear (Cavanagh 1980a). Previous studies have shown that footwear is one of the contributing factors for lower extremity-related injuries (Bensel 1976; Bensel and Kish 1983; Jones 1983; Tomlinson et al. 1987) reported during various military operations including training sessions, real-time combat scenarios and war scenarios. Investigations in the early 70s (dominantly in the American and British Army) have indicated that combat boot is the main causal factor towards various training-related malaises like Achilles tendinitis, stress fractures and blisters (Akers and Sulzberger 1972; Comaish 1975; Bensel 1976; Worthen and Yanklowitz 1978; Reinker and Ozburne 1979). Therefore, it can be depicted that the combat boot holds a pivotal position in the ancestry of footwear research. Over time, it was felt by the investigators that running shoes could be the better alternative to combat boots in reducing the rate of lower extremity injuries. Consequently, enhanced aerobic fitness and morale could also be expected (Jones 1983). However, switching the type of footwear, based on the preliminary observations, was not found to be the best anticipatory strategy for footwear-related maladies. By that time, there were no scientific studies available conducted under controlled laboratory conditions with a larger population across various shoes, demonstrating the reasonable relationship between injury prevention and shoe design (Cavanagh 1980b).

However, simultaneously, it was also recognized that running shoes are exclusively meant for running/training purposes and are the prominent choice of civilians. This was in contrast with the various combat activities of the military population, where every activity is not essentially linked with a running event. Therefore, it was obvious that completely switching to running shoes was not possible from the practical perspective of combat activities (Jones 1983). Activity-wise combat boots are designed for providing protection to the feet and ankles over a variety of surface conditions, whereas, running shoes should have characteristics like shock absorption (impact properties), flexibility, stable heel counter and supportive insole. Controlled laboratory studies on various biomechanical properties involving five running shoes and a combat boot revealed that a combat boot was inferior to running shoes specifically in terms of impact absorption and flexibility (deMoya 1982). At the same time, these two properties are considered to be essential for the combat boot when used during running conditions under combat scenarios. Therefore, the integration of these two features in the design of the combat boot is considered to be indispensable.

Furthermore, challenges are multifaceted when we specifically talk about military footwear. Aside from the type of footwear worn, there are other essential contributors that play significant roles and should, therefore, be considered throughout design.

12.2 Users' Need

Evaluation of the users' need for a specific activity is regarded as the most important requirement for the successful completion of the prototype design and development (Rosenblad-Wallin 1988; Kaulio 1998; Alcántara et al. 2005; Luh et al. 2013; Lin and Chen 2015; Keukenkamp et al. 2021). The *population characteristics* like gender, age group, specific occupational group (e.g. in the present context, the military population is being considered), differently abled, athletes, etc. In order to address the users' need for footwear design, it is essential to collect the *users' feedback* through a survey by applying a standardized structured questionnaire. Gao et al. (2008) conducted the questionnaire based survey including samples from newspaper delivery, mining, military and construction personnel and reported the following: (1) Ice covered with snow was rated as the most (4.5) slippery surface (1 = not slippery, 5 = extremely slippery); (2) military personnel provided a 2.7 rating for the slip resistance property of the protective shoes (safety shoe with steel toecap and winter boot) on a Likert scale of 1 = strongly disagree, 5 = strongly agree on the statement of "The footwear provided by the employer is slip resistant on ice and snow in winter"; (3) On a scale of 1 = most important, 12 = least important, designing a new shoe based on thermal insulation received the highest ranking (2.68) and inexpensiveness received the lowest ranking (10.18); and (4) the spread of anti-slip materials (6.5) and the use of slip-resistant materials (6.3) received the highest ranking on the preventive measure on a scale of 1 = least preferred, 9 = most preferred.

12.3 Terrains

Terrain characteristics are of paramount importance in footwear design, specifically for the military population. *The coefficient of friction (COF)* of surfaces of various terrains poses a tough challenge to the outsole of the footwear as well as to the physiological and biomechanical load of the wearer (Ramaswamy et al. 1966; Pandolf et al. 1976; Lejeune et al. 1998). Snow-covered/icy terrain (Gao and Abeysekera 2004) and sandy terrain (Yoo 2014), are considered to be the toughest terrains for military operations. According to Gao and Abeysekera (2004), there are a number of factors that contribute to slip and fall accidents on snowy surfaces.

1. Footwear (sole) properties include sole material, hardness, roughness, worn/unworn, tread (geometry) design, the centre of gravity, anti-slip devices, wearability (weight, height, flexibility, ease of walking and comfort), etc.
2. Underfoot surface characteristics, covered with ice, snow, contaminants, anti-slip materials, uneven, ascending/descending slope, etc.
3. Footwear (sole)/surface interface—tribological aspect, i.e. coefficient of friction (static, transitional and dynamic)
4. Human gait biomechanics: muscle strength, postural control, musculoskeletal function, postural reflex and sway, balance capability, acceleration, deceleration, stride length, step length, heel contact velocity, the transition of the whole-body centre of mass, vertical and horizontal forces, required COF, etc.
5. Human physiological and psychological aspects, i.e. the so-called intrinsic factors, including a decline in visual, vestibular and proprioceptive functions, ageing, perception of slipperiness, information processing, experience, training, diabetes, drug and alcohol usage, unsafe behaviour (rushing, reading while walking), etc.
6. Environment (extrinsic factors): temperature, snowfall, Gulf Stream, lighting conditions, etc.

Furthermore, orthodox in-depth studies from the US and Canadian military vividly depicted the biomechanical requirements of military footwear based on the interactions among the material characteristics, footwear and ground (MacDonald and Swider 1974; Dyck 1992; Hamill and Bensel 1992, 1996; Pisani 2002; DeBusk 2016).

12.4 Extreme Environments

A common strategy for increasing soldiers' resilience for real-time combat operations is to expose them to environmental extremes repeatedly through structured training schedules. The extreme factors may impose a temperature range from +50 to −50 °C, relative humidity up to 95%, hypobaric hypoxia of naturally high altitudes, windchill due to strong wind velocity and dry-hot strong wind. All of these factors are known to be integral characteristics of high altitudes, deserts, jungles and coastal environments. The interaction of environmental extremes, unstructured hostile terrain conditions and a number of human factors issues places critical demands on soldier operators, pushing physiological, mental, biochemical and biomechanical systems to their breaking point (Arya et al. 2022; Chatterjee et al. 2022a, b; Yadav et al. 2022).

Likewise, footwear meant for these extreme interactions should provide outstanding shock absorption, adequate thermal comfort, adequate balance and stability, proper fitting, etc. The use of poorly designed footwear may negatively impact the soldiers' performance, which can lead to various complicated health hazards and life-threatening injuries (Fig. 12.1). *Heavyweight footwears* increase the physiological cost through the increased cardiorespiratory demands and metabolic cost (Martin 1984; Jones et al. 1984; Legg and Mahanty 1986; Krings et al. 2018). Additional musculoskeletal work to support each foot lifting with heavy-weight shoes against gravity raises muscle metabolism, which requires more oxygen via blood

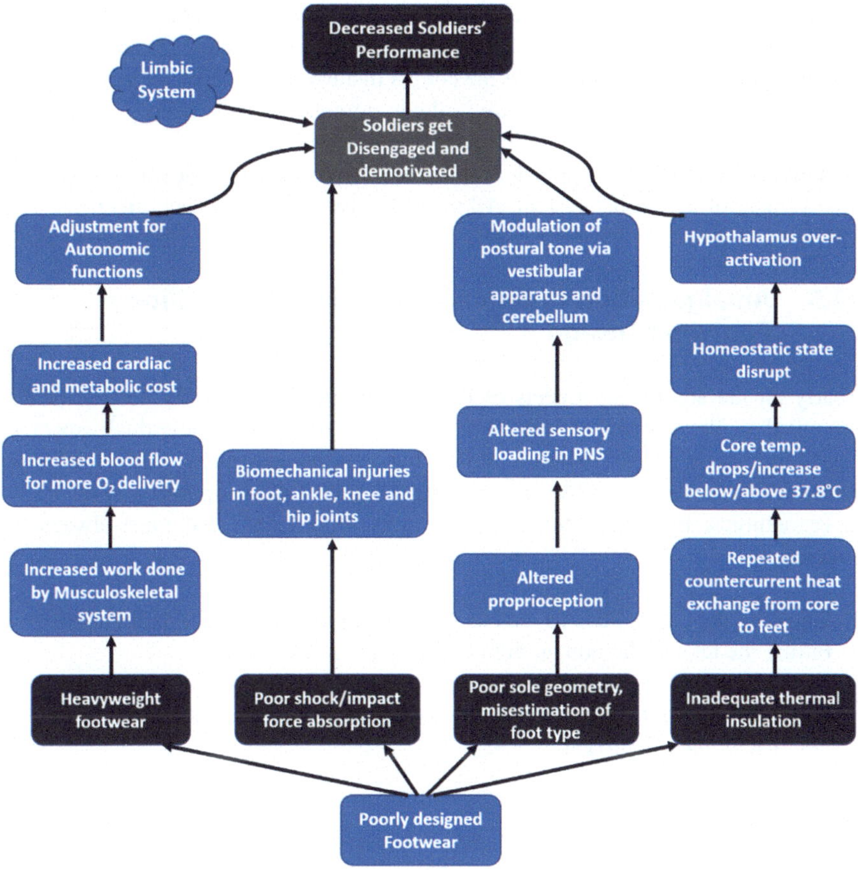

Fig. 12.1 The multifaceted consequences of wearing poorly designed footwear, which may ultimately result in performance deterioration

redistribution to the working muscle. Poor shock absorption at the sole system of the footwear may lead to biomechanical injuries at the ankle, knee and hip joints due to cumulative impact force absorption over a period of sustained usage (Windle et al. 1999; Dixon 2007; Muniz and Bini 2017; Lullini et al. 2020). Poor outsole geometry and misestimation of the users' foot type may lead to perturbed balance and stability (Hill et al. 2017; DeBusk et al. 2018). A series of alterations/adjustments happen at the level of the peripheral nervous system (PNS) through altered proprioception. Depending on the severity of the case, modulation of postural tone via the vestibular apparatus and cerebellum may also be levied. Lack of adequate/desired thermal insulation to the footwear either meant for extremely cold or extremely hot climates may induce cold and heat-related physiological consequences, respectively (Endrusick et al. 2005). Blood circulating from the core towards the feet comes in contact with the coldest part of the lower extremities, lowering its temperature and again going back to the core. With every cycle of circulation, there is a drop in body core temperature by means of a counter-current heat exchange mechanism among the blood vessels. The same fact is true for extremely hot climates, which leads to an increase in core temperature. Once the core is disturbed (i.e. above or below 37.8 °C), the temperature regulation centre in the hypothalamus is activated. Under such circumstances, overall homeostasis gets disrupted. Soldiers may feel disengaged and demotivated to their task, which can cause an emotional spillover through the involvement of the limbic system. All the executive functioning get affected.

12.5 Amalgamation of Ergonomics Principle in Military Footwear Design

Nearly all the significant aspects of footwear design have already been critically portrayed in the earlier sections of this article. Now every individual aspect is considered as a single building block that can be cemented with ergonomic principles to create a tangible product with a human-centric approach (Fig. 12.2). In ergonomics, there are four major "schools of thought" that must always be considered-

1. Knowing the capabilities as well as the limitations of the users' population
2. Fitting the task to the human but not fitting the human to task
3. Taking researchers, scientists, academicians, doctors, engineers, designers, industrial partners, and, if possible, all stakeholder groups on board to ensure an interdisciplinary approach

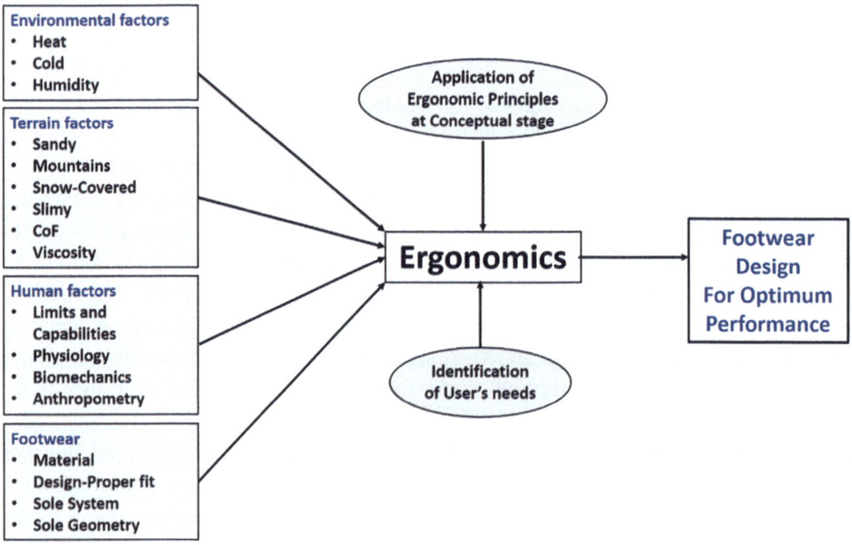

Fig. 12.2 Application of ergonomic principles is the bridging element among multiple contributing factors, users' needs and footwear design for optimum performance

4. A proactive ergonomics approach ensures consideration of ergonomic principles from the conceptual phase

12.6 An International Perspective on Ergonomics Aspects of Military Footwear

In this section, we are citing some important studies (exclusively based on the authors' understanding and discretion) that critically discussed various ergonomics aspects of military footwear. The critical insights of these studies have been tabulated in Table 12.1, highlighting the *type of footwear, new design/evaluation of existing boots, military of which country, brief study protocol, major parameters and major outcomes.*

12.7 The Indian Perspective

The geopolitical conditions of India are extremely challenging, especially at the border regions, which mainly constitute high altitude, desert and jungle areas. Soldiers need to have proper footwear that can provide adequate thermal support, desired mobility and improved performance under intimidating conditions. After evaluating a number of disadvantages in the existing military footwear, like, the inadequate inner lining of thermal insulation, poor grip at the outsole, heavyweight, etc., two footwear, i.e. snow boots and combat boots were developed indigenously based on ergonomic principles addressing elucidation to most of the existing problems. The following steps were taken considerations to address the problems

1. Investigations on the users' needs for working under particular environments and terrain conditions
2. Detailed investigation of the design and material specification of the existing boots
3. Scientific evaluation with the human participants
4. Proposing the new design concept and material specification in the search for a light weight good fit boot
5. Use of indigenous anthropometric database
6. Development of the three-dimensional model of the proposed design and two-dimensional drafting with dimensions for manufacturing
7. Laboratory and field studies with the newly developed boot in comparison with the existing boots and feedback from the users.
8. Any necessary modification that will further improve the performance of the boot as per the feedback
9. Low-cost production (cost-effective)

Table 12.1 Significant global literature on the ergonomics aspects of military footwear design

Reference	Footwear type	New design/ evaluation of existing	Military of which country	Study protocol (in brief)	Major parameters	Outcome
Harman et al. (1999)	Military boots and commercially available boots	Evaluation of two existing Army boots, five prototype Army boots and five commercial hiking boots	United States of America	Comparative studies on the physiological, biomechanical and subjective responses based on comfort and functionality analysis while wearing each boot type while carrying a 60-lb backpack load, without load, walking on both straight and zigzag 400 m grass courses and walking 6 miles at 3 mph over pavement and wooded trail in each boot-type	Oxygen consumption, kinematic changes, ground reaction force (GRF), posture, gait, lower extremities joints forces and torques, maximum speed runs and post march trauma analysis	Based on the overall performance analysis, the Salomon Adventure 9 Ultralight was found to be the best. The poorest performance with the Army combat and jungle boot depicted the need for new design and development. The best features of each boot were selected by the designers and developers and incorporated into the future military boot Five prototype boots were designed and produced to improve the wearer's locomotor efficiency and likelihood of lower extremity injury
Kodithuwakku Arachchige et al. (2020)	Military footwear	Evaluation of existing design	United States of America	The participants in standard (STD) and minimalist (MIN)	Muscle activity from knee flexors, extensors, ankle	An increased requirement of muscle activity was observed

				footwear were measured using electromyography, before (PRE) and after (POST) a simulated workload during sensory organization and motor control tests	dorsiflexors and plantar flexors	in STD footwear, post condition. MIN showed lesser balance decrement post workload, which could be attributed to its design characteristics. Therefore, critical care should be given to minimize muscular exertion while anticipating better postural control and thus preventing postural instability
Kopitar et al. (2019)	Military footwear (Desert boot)	Evaluation of existing design focusing on the temperature difference between day and night	Croatia	Thermal resistance of the footwear was determined using Sweating Guarded Hotplate equipment, according to the standard ISO 11092 in desert conditions of 15 and 30 °C with a relative humidity of 40%	Thermal resistance, thermal resistance difference and leather thickness	The leather thickness influences the change between the temperature range of 15–30 °C. Increased temperature leads to 5.9% decrease in the thermal resistance at the two-layered membranes of the tongue. The thermal resistance difference for temperature ranging from 15 to 30 °C between min and max values for quarter and lining membranes

(continued)

Table 12.1 (continued)

Reference	Footwear type	New design/ evaluation of existing	Military of which country	Study protocol (in brief)	Major parameters	Outcome
						are 17% and 13% lining
Torrens et al. (2012)	Military footwear	Evaluation of existing design	United Kingdom	First, the nature and level of protection of the military footwear in different circumstances is assessed The fit, physiological maintenance, task performance and other factors influencing the design are considered Finally, future trends in design and production are introduced	Identification of foot anthropometric measurements for a good fit, identification of safe pressure zones in the foot and 3D modeling	The best practice from Human Factors and the application of available technologies are emphasized to provide the optimum compromise within a final design solution
Sinclair et al. (2015)	Cross trainer and running shoe	Comparison of existing traditional military boots with two newly introduced shoes	United Kingdom (recreational runners)	The participants ran at 4.0 ms^{-1} in cross trainer shoes, running trainer shoes and military boot Knee joint kinetics was recorded	Peak knee extensor moment, Patellofemoral contact force, Patellofemoral loading rate, Patellofemoral pressure and Peak knee abduction moment	The study recommended that the recruits susceptible to injuries mediated through excessive knee loads should select the cross-trainer's shoe for their running activities

Chander et al. (2020)	Standard and minimalist military boots	Evaluation of existing design	United States of America	The participants were tested with two military boots, standard (STD) and minimalist (MIN). The before (PRE) and after (POST) load carrying task during normal gait (NG), unexpected (US) and expected (ES) slips were recorded. EMG of four muscles [vastus medialis, medial hamstrings, tibialis anterior and medial gastrocnemius] of the left lower extremity (slipping leg) was recorded	Joint kinematics, ground reaction forces and muscle activity	Gait modifications due to footwear and gait were observed STD exhibited greater ankle dorsiflexion and knee flexion, lowered ground reaction forces and increased muscle activity compared to MIN US and ES elicited greater plantar flexion and knee flexion angles while requiring greater muscle activation compared to NG
Chander et al. (2019)	Standard boot and minimalist boot	Evaluation of existing design	United States of America	The postural stability of the participants was tested before (pre) and after (post) load carriage task wearing standard boot (STB) and minimalist boot (MTB), using sensory organization and motor control tests on Neurocom Equitest™	Postural sway, equilibrium scores and postural latencies	MTB exhibited greater balance due to its ergonomic design when compared to STB in balance conditions Decrements in postural stability could be attributed to boot design characteristics and muscular exertion

(continued)

Table 12.1 (continued)

Reference	Footwear type	New design/ evaluation of existing	Military of which country	Study protocol (in brief)	Major parameters	Outcome
						due to the load carriage task
Dixon (2007)	Combat assault boot	Evaluation of existing design	British	Running trials were performed by military trainees under two conditions: boot with the standard-issue insole and boot with the test orthotic	Peak Impact Force, Time of Peak Impact Force, Average Loading Rate, Peak Loading Rate and frontal plane rearfoot angle	The commercially available orthotic provides benefits typically associated with moulded prescription orthoses
Hill et al. (2020)	Minimalist and standard military footwear	Evaluation of existing design	United States of America	Recreationally active male participants wore minimalist and standard military footwear and performed military style workloads. Parameters of balance and right lower extremity were recorded before and after the workload	Unilateral static balance, electromyography of right lower extremity's medial gastrocnemius, tibialis anterior, peroneus longus and tibialis posterior	Minimalist footwear increased muscle activation prior to the workload, whereas standard footwear increased muscle activity after the workload thus footwear design characteristics affect lower extremity muscle activity differently depending on the workload condition
Rosenblad-Wallin (1988)	Military boots	New design	Sweden	The footwear requirement was listed based on literature	Last design, material selection, physical characterization of the	Ergonomically designed new boot concept that was

				reviewed on in vivo biomechanical and physiological studies. Information was also collected through users' interview sessions. Prototypes were developed. They were tested and evaluated in wear trials	materials, model design and soling techniques of the footwear	lighter, softer and more flexible that best fulfilled the requirements, minimalized the conflicts and basis of cost-effective production was established
Oakley (1984)	Infantry footwear	Evaluation of existing design and recommendation on new design	England	The review examines the footwear design aspects in relation to function and elucidates the many conflicting requirements of ideal boot design	Mobility, protection, insulation, waterproofing, vapour permeability, durability, weight, fit and supply of the footwear	A suggestion for an ergonomic modular infantry 'footwear package' consisting of an inner and a number of different middle and outer boots

12.8 Ergonomically Designed Snow Boot (EDSB)

12.8.1 Problem in the Existing Scenario

The existing indigenous snow boot (RI Boot) is heavy weight (3.5 kg, size 8), lacks adequate thermal insulation in the foot bed, vamp and ankle areas, lacks a breathable inner lining and demonstrates a poor grip on the snow-covered terrains due to a soft outer sole. An imported boot like SCARPA, LOWA and other brands address many of these problems; however, these are heavy weight (>3 kg size 8) at exorbitant cost (Rs. >50,000/approx.)

12.8.2 Innovative Approach to Solve the Problem

The newly developed snow boot (Fig. 12.3) is based on Ergonomic principles, which will facilitate/enhance soldiers' performance, comfort and combat capabilities. The following are the beneficial features (Table 12.2):

12.8.3 Scientific Exploration with EDSB

Scientific studies conducted on the natural military operational terrains using EDSB at Khardung La (5602 m), Siachen (3810 m) and Chang La (5384 m), in comparison with the existing imported snow boot (Legacy boot) (Bhattacharyya et al. 2021). In this study, the thermal comfort of the boots was assessed over a period of 24 h by monitoring the toe temperature inside the boot in comparison with the environmental temperature. The methodology of this study allowed the soldiers to conduct their

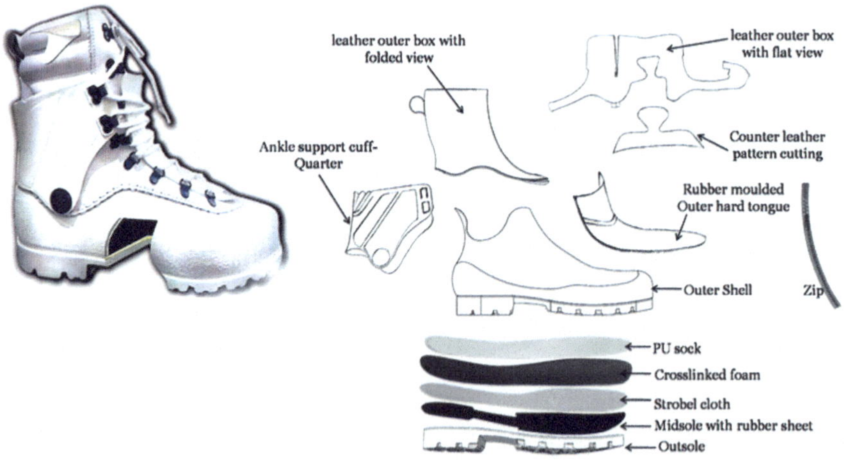

Fig. 12.3 Solid view and exploded view with various components of EDSB

Table 12.2 Important features of EDSB

Important features	Potential benefits
Ergonomically designed	Better fit, comfort, safety and usability
Lightweight	Decreased energy expenditure, improved mobility and low fatigue
The ankle support Cuff quarter is attached to the shell with a flexible hinge	Providing support and strength to the ankle during the power step
Footbed with midsole PU/EVA/PP sheet, strobel insole and PU insocks Of rubber cloth, PE	Adequate thermal protection and advanced impact shock absorption
Inner lining of breathable fabric	Keeps the foot dry and prevents microbial infection
Outsole with Zic pattern and sand texture high cleats	Improved slip resistance over snow covered terrain
Zip at the medial aspect of the boot	Quick wear and removal of the boot during emergency
Grooving at outsole	Crampon attachment for better mobility
Cost	Cost-effective
Design Registration: 315321-001 (Granted), 312511 (Under examination)	
Indian Patent No.: 398166	

regular activities, i.e. vigilance and other logistics duties in outdoor and intermittent short indoor exposure etc., in order to achieve more realistic data collection. The result showed that toe temperature varied over a range of 35–15 and 35–14 °C wearing the EDSB and 'Legacy boot' respectively, while the environmental temperature varied from −5 to −15 °C. Therefore, an effective temperature delta of 30 °C was maintained under sub-zero conditions (considering the lowest environmental temperature −15 °C and lowest toe temperature inside the boot +15 °C). Furthermore, the observations from the IR thermographic study are also in corroboration with the above data. Subjective feedback received also showed adequate thermal comfort with the EDSB.

Hofer et al. (2014) conducted the microclimate assessment of ski boots inside the climate chamber measuring the toe temperature under +10, 0, −10 and −20 °C ambient temperature over a period of 90 min of exposure. It was observed that toe temperature dropped from ≈26 to ≈24, 25 to ≈20, 25 to ≈15 and ≈24 to ≈14 °C over a period of 90 min of exposure under +10, 0, −10 and −20 °C temperatures, respectively. Therefore, an effective temperature delta of 25 and 34 °C was maintained under −10 and −20 °C ambient temperatures, respectively.

Kuklane et al. (1998) also evaluated the thermal performance of the insulated boot inside the climate chamber and found that the temperature declined from ≈32 to ≈12 °C under the −25 °C ambient temperature over a period of 120 min. A temperature delta of 37 °C was observed.

An illustration of the comparison of thermal efficacy across the boots mentioned in the three studies is presented in Fig. 12.4. A significant difference observed among these three studies is that Bhattacharyya et al. (2021) conducted the study in the real field condition over a period of 24 h compared to Hofer et al. (2014) and Kuklane

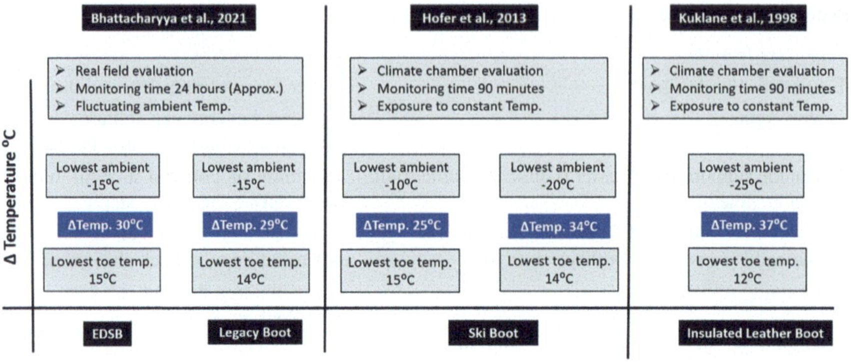

Fig. 12.4 Comparison of study methodologies, boot types and delta temperature changes across three studies

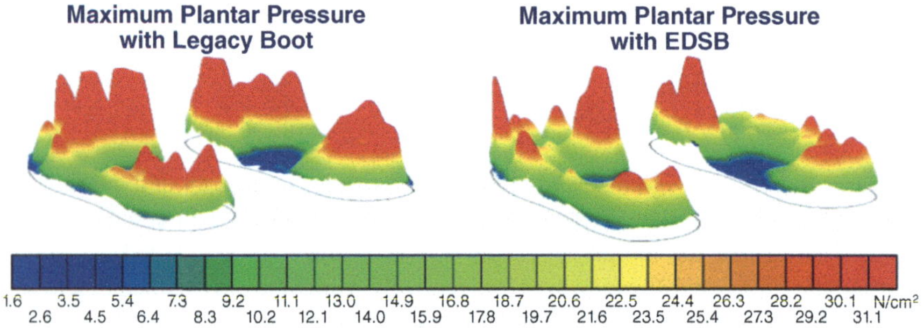

Fig. 12.5 A pedobarographical presentation of the plantar pressure while wearing both types of boots. The red color represents the maximum values, which span a wider area beneath the toes and heel with the legacy boot compared to the EDSB

et al. (1998), where studies were conducted inside the climate chamber over a shorter period of exposure. Moreover, Bhattacharyya et al. (2021) also found that toe temperature was maintained between 15 and 20 °C inside the EDSB over a period of approximately 5 h with a constant ambient temperature of−15 °C. Therefore, it can be depicted that the thermal efficacy of the EDSB was found to be adequate even under the natural extreme cold weather conditions.

Changes in plantar pressure (PP) were also recorded with both boots using foot scan insole during short-duration walking trials on the natural terrains (Fig. 12.5). This was the first scientific attempt with the snow boot where the objective evaluation was conducted in the real field condition in predicting the injury risks (may be related to different design construct of the snow boot). The in-depth data analysis is in progress.

12.9 Ergonomically Designed Combat Boot (EDCB)

12.9.1 Problem in the Existing Scenario

The existing Direct Moulded Sole (DMS) boot lacks multi-layered sole systems and the advanced impact force absorbing insock has increased energy expenditure, not ergonomically designed, which results in injuries to lower extremities (specifically at ankle joints, knee joints and hip joints) and reduced performance.

12.9.2 Innovative Approach to Solve the Problem

The newly developed combat boot (Fig. 12.6) is based on ergonomic principles that take care of the user's requirements for flexibility, durability, proper fit, comfort, an improved sole system, breathability, etc. The ultimate target is to improve performance and safety and reduce the risk of injury.

Feature-wise potential benefits of the Ergonomically designed combat boot are mentioned in Table 12.3.

12.9.3 Scientific Exploration with EDSB

The indigenously developed EDCB was subjected to biomechanical evaluation with a structured methodology in comparison with the existing combat boot. A compact

Fig. 12.6 Solid view and exploded view with a multi-layered sole system of EDCB

Table 12.3 Important features of EDCB

Important features	Potential benefits
Ergonomically designed	Better fit, comfort, safety and usability
Lightweight—weighs <1.5 kg a pair (size 8)	Decreases energy expenditure, improves mobility and low fatigue
Advanced sole system with the outsole, midsole, anti-perforation insole and visco-elastic insock	Efficient impact shock absorbing
Chrome tanned leather and nylon textile	Water repellent Improves flexibility at the ankle and toes
Anti-penetration insole textile (by replacing traditional steel plate)	Lighter in weight Protects feet from sharp objects like nails
Inner lining of breathable fabric	Keeps the foot dry and prevents microbial infection
Outsole impregnated with five different varieties of cleats	Improves slip resistance over a variety of terrains and reduces the risk of fall and injury.
Antistatic outsole	Protects feet from electrical shock
Oil/acid resistant	Inactive towards chemical reaction
Zip at the medial aspect of the boot	Quick wear and removal of the boot during emergency
Design Registration: 315321-001 (Granted)	
Indian Patent Application No.: 201811047736	

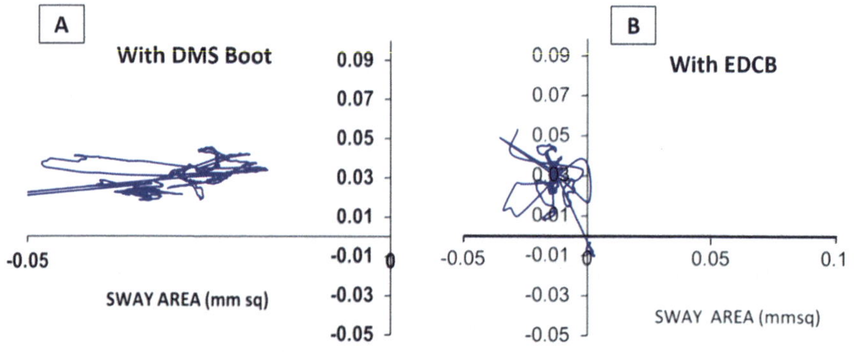

Fig. 12.7 With the existing DMS boot, the sway area is seen to be dispersed and away from the body midline (**a**), which depicts poor stability; whereas with EDCB, the sway area is seen to be more concentrated and towards the body midline, which depict a better stability (**b**)

and concentrated sway area and shifting of the centre of pressure (COP) towards the body midline with the EDCB depicted improved balance and stability in comparison with the existing boot (Sen et al. 2019) (Fig. 12.7). This was mainly due to the fact that EDCB has an advanced sole system (outsole, midsole, insole and insertable insock), which is completely absent in the existing boot (only the outsole and a thinner insole). More precisely, it can be reported that the presence of an advanced shock-absorbing insock resulted in the further reduction of vertical ground reaction

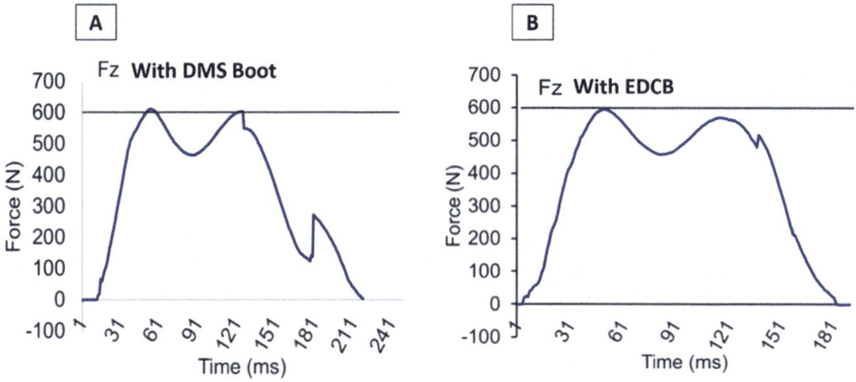

Fig. 12.8 A higher and peaked vertical ground reaction force value observed with the DMS (**a**) boot as compared to EDCB (**b**)

force while wearing the EDCB in comparison with the existing DMS boot (Fig. 12.8). However, the beneficial effects of EDCB have been shown over the DMS boot only with laboratory test trials under an unloaded walking condition. However, this beneficial effect of the EDCB has been shown over the DMS boot only in laboratory test trials under an unloaded walking condition. There are other military situations, like Army foot drills, loaded and unloaded marches, vertical jumps during training sessions, etc., where reduction of vertical force impact is considered to be the biggest challenge. For instance, Carden et al. (2015), after observing the vertical impact forces across various conditions, interpreted that Army foot drill generates greater forces, loading rates and accelerations than running and load carriage, whereas marching is comparable to moderate jogging (10.8 km/h). The high occurrence of overuse injuries during preliminary military training may be caused by the significant biomechanical loading of foot drills; hence, measures to control or lessen this loading should be investigated. Wang et al. (2012) experimented with a 32 kg load in the backpack and found that loaded walking is a significant source of muscle fatigue, which resulted in an increase in ground reaction forces (vertical and braking) as well as in loading rate, leading to higher chances of lower extremity injuries. Likewise, we can also explore the beneficial effects of the combat boot, through the experimental approach adopted by Polcyn et al. (2002), across various load carriage configurations (fighting, approach and sustainment) and load carriage ensembles (LW I, LW II, MOLLE and ALICE) and record the detailed biomechanical responses involving a larger sample size.

In another study, Sen et al. (2020) assessed the changes in plantar pressure at the forefoot, mid foot, lateral, medial and heel regions during a walking trial at 4 km/h speed across 0%, 5% and 10% gradients and observed that pressure increased gradually with the increasing gradient of both the feet. This study's findings may be of particular importance at the conceptual stage of shoe design, where wearers anticipate frequent uphill walking.

12.10 Conclusion and Way Forward for Indian Perspective

In summarizing the entire article, it can be said very conspicuously that many other developed nations have progressed significantly and judiciously in terms of advanced ergonomics and human factors research, database generations, developing standards and joint venturing with all the associated stakeholders through various integrated policies. As a consequence, the design and development of various types of military footwear are becoming very smooth and hassle free. Moreover, occupational health and safety have been improved in terms of reduced risk of lower extremities injuries, less fatigue and improved and sustained performance. This is in contrast with the Indian scenario, where the design and development of military footwear based on ergonomics and human factors research is found to be at its nascent stage. At the same time, advertently, it is felt that there is enormous scope from the Indian perspective to conduct integral ergonomics research for the design and development of military footwear. A relatively larger military population and variety of task performances under various terrain and climatic conditions can be explored to an exclusive level for the creation of a stronger path of research and development.

It is also observed that the research activities conducted and data published from the Indian perspective mostly venture into the real operational terrains, specifically for the EDSB. The literature available is very scanty regarding the real terrain evaluation of various army boots. Real terrain objective evaluation conducted with the EDSB under the extreme environment could be claimed as the rarest one or first of its kind. Findings of this kind of study can pave the stronger path to acquire merely practical/real-time feedback on the newly designed and developed product. Any need for further improvement/modifications can easily be felt due to direct interaction with the users' group and conducting experiments on the operational field. Furthermore, it is also observed that under most of the Indian defence scenarios, the environmental conditions as well as the soldiers' performance (physical and cognitive) both reach up to the extreme limit. Designing footwear for such extreme conditions is always very challenging. Keeping this status in mind, the following points are considered/recommended for the more stable way forward for Indian defence scenarios-

1. An in-depth survey on military footwear related injuries and reporting of data. The database must include all the relevant information during various training sessions, simulated and real-time combat exercises.
2. Being the fourth largest Army in the world with an approximate strength of 5,137,500, there is an utmost need to generate a database on foot anthropometry. Accordingly, we can generate our own size standards and lasts for footwear design and development. Most importantly, revisiting the data set at an interval of one decade is highly required in order to find out any secular trends in foot anthropometry, as observed by Greiner and Gordon (1990). If changes are noticed, all the standards must be modified accordingly.

3. Development of State-of-the-Art simulation facilities enabling the simulation of climatic conditions as well as terrain factors, as reported by Bagheri et al. (2019) and Hsu et al. (2016).
4. Certainly, there is a need to increase the sample size of scientific studies, specifically under simulated conditions. Experiments under austere real operational terrains should be an exception.
5. Bidirectional exchange of information on the various aspects of scientific evaluations and user trials for various products among the users, scientists, designers and production partners Accordingly, customized scientific methodologies and trial directives should be finalized.
6. In order to consider the ergonomics factors right from the conceptual stage of product design, it is imperative to formulate the specification, e.g. in General, Staff Qualitative Requirements (GSQR), accordingly.

Finally, it can be concluded that challenges are multifaceted when we are talking about military footwear. Three major strategies should always be considered. The preliminary strategy should be targeted towards footwear that can enhance the locomotor capabilities of the wearer, minimize the occurrence of lower extremity injury and pain and provide comfort. Secondarily, the weight of the boot, ankle height, design of the closures, water resistance and durability of the material should be considered. The cost and rate of production should be considered tertiary.

Acknowledgments The authors extend their wholehearted thanks to all the participants for their unconditional support. The authors also express their gratitude to all the officers of concerned field locations for understanding the need for the study and providing all possible logistics support. We are also thankful to the Defence Research and Development Organization (DRDO) for providing funding for these studies.

References

Akers WA, Sulzberger MB (1972) The friction blister. Mil Med 137(1):1–7. https://doi.org/10.1093/milmed/137.1.1, PMID 4624239

Alcántara E, Artacho MA, González JC, García AC (2005) Application of product semantics to footwear design. Part I—identification of footwear semantic space applying differential semantics. Int J Ind Ergon 35(8):713–725. https://doi.org/10.1016/j.ergon.2005.02.005

Arya K, Yadav A, Bhattacharyya D, Chatterjee S, Pal M (2022) Effect of long-distance load marching on physiological responses at desert environment. BLDE Univ J Health Sci 7(2): 204–210. https://doi.org/10.4103/bjhs.bjhs_73_22

Bagheri ZS, Patel N, Li Y, Rizzi K, Lui KYG, Holyoke P et al (2019) Selecting slip resistant winter footwear for personal support workers. Work 64(1):135–151. https://doi.org/10.3233/WOR-192947, PMID 31450526

Bensel CK (1976) The effects of tropical and leather combat boots on lower extremity disorders among US Marine Corps recruits. Army Natick Research and Development Labs Ma Clothing Equipment and Materials Engineering Laboratory

Bensel CK, Kish RN (1983) Lower extremity disorders among men and women in Army basic training and effects of two types of boots. Army Natick Research and Development Labs Ma Individual Protection Laboratory

Bhattacharyya D, Pal M, Chatterjee T, Varshney R (2021) Extreme interaction among human–environment–equipment: a pilot study on the ergonomic design of military snow boots. Ergon Des:106480462110208. https://doi.org/10.1177/10648046211020851

Carden PPJ, Izard RM, Greeves JP, Lake JP, Myers SD (2015) Force and acceleration characteristics of military foot drill: implications for injury risk in recruits. BMJ Open Sport Exerc Med 1(1):bmjsem-2015-000025. https://doi.org/10.1136/bmjsem-2015-000025, PMID 2 7900110

Cavanagh PR (1980a) The running shoe book. Anderson Publishing World, Mountain View

Cavanagh PR (1980b) Focus on shoes, new eng. running, fall, pp 22–24

Chander H, Knight AC, Garner JC, Wade C, Carruth D, Wilson SJ et al (2019) Impact of military type footwear and load carrying workload on postural stability. Ergonomics 62(1):103–114. https://doi.org/10.1080/00140139.2018.1521528, PMID 30196761

Chander H, Arachchige SN, Wilson SJ, Knight AC, Carruth DW, Wade C et al (2020) Impact of military footwear type and a load carriage workload on slip initiation biomechanics. Int J Hum Factors Ergon 7(2):125–143. https://doi.org/10.1504/ijhfe.2020.10031690

Chatterjee T, Bhattacharyya D, Yadav A, Pal M (2022a) Heart rate variability, task load and perceived exertion associated with a long-distance military ski exercise: a pilot study. Indian J Physiol Pharmacol 66:196–202. https://doi.org/10.25259/IJPP_101_2022

Chatterjee T, Bhattacharyya D, Yadav A, Pal M (2022b) Quantification of physiological and mental workloads of faster and slower finishers of a long-distance military training activity. BMJ Mil Health e002154. https://doi.org/10.1136/military-2022-002154. , PMID 36283744

Comaish JA (1975) The study of friction blisters in man. Ministry of Defense Publication, Royal Victoria Infirmary, Newcastle Upon Tyne

DeBusk JH (2016) Impact of military boots and military workload on balance. Mississippi State University, Starkville

DeBusk H, Hill CM, Chander H, Knight AC, Babski-Reeves K (2018) Influence of military workload and footwear on static and dynamic balance performance. Int J Ind Ergon 64:51–58. https://doi.org/10.1016/j.ergon.2017.11.003

deMoya RG (1982) A biomechanical comparison of the running shoe and the combat boot. Mil Med 147(5):380–383. https://doi.org/10.1093/milmed/147.5.380, PMID 6810212

Dixon SJ (2007) Influence of a commercially available orthotic device on rearfoot eversion and vertical ground reaction force when running in military footwear. Mil Med 172(4):446–450. https://doi.org/10.7205/milmed.172.4.446, PMID 17484322

Dyck W (1992) A review of footwear for cold/wet scenarios. Part 1. The boot

Endrusick TL, Cole ID, Matonich PM (2005) Effects of simulated sustained operations on the thermal insulation of military footwear. In: Environmental ergonomics—the ergonomics of human comfort, health and performance in the thermal environment. Elsevier, Amsterdam, pp 389–393. https://doi.org/10.1016/S1572-347X(05)80062-X

Gao C, Abeysekera J (2004) A systems perspective of slip and fall accidents on icy and snowy surfaces. Ergonomics 47(5):573–598. https://doi.org/10.1080/00140130410081658718, PMID 15204304

Gao C, Holmér I, Abeysekera J (2008) Slips and falls in a cold climate: underfoot surface, footwear design and worker preferences for preventive measures. Appl Ergon 39(3):385–391. https://doi.org/10.1016/j.apergo.2007.08.001, PMID 17880907

Greiner TM, Gordon CC (1990) An assessment of long-term changes in anthropometric dimensions: secular trends of US Army males. Army Natick Research Development and Engineering Center

Hamill J, Bensel CK (1992) Biomechanical analysis of military boots. Phase 1. Materials testing of military and commercial footwear. Massachusetts Univ Amherst Biomechanics Laboratory

Hamill J, Bensel CK (1996) Biomechanical analysis of military boots: phase III. United States Army technical report Natick, TR-96.013, vol 11, p 42

Harman E, Frykman P, Pandorf C, LaFiandra M, Smith TY (1999) A comparison of 2 current-issue army boots, 5 prototype military boots, and 5 commercial hiking boots: performance, efficiency, biomechanics, comfort and injury. Army Research Institute of Environmental Medicine, Natick

Hill CM, DeBusk H, Knight AC, Chander H (2017) Influence of military-type workload and footwear on muscle exertion during static standing. Footwear Sci 9(3):169–180. https://doi.org/10.1080/19424280.2017.1403968

Hill C, DeBusk H, Knight A, Chander H (2020) Military-type workload and footwear alter lower extremity muscle activity during unilateral static balance: implications for tactical athletic footwear design. Sports (Basel) 8(5):58. https://doi.org/10.3390/sports8050058, PMID 32354166

Hofer P, Hasler M, Fauland G, Bechtold T, Nachbauer W (2014) Microclimate in ski boots—temperature, relative humidity, and water absorption. Appl Ergon 45(3):515–520. https://doi.org/10.1016/j.apergo.2013.07.007, PMID 23932378

Hsu J, Shaw R, Novak A, Li Y, Ormerod M, Newton R et al (2016) Slip resistance of winter footwear on snow and ice measured using maximum achievable incline. Ergonomics 59(5): 717–728. https://doi.org/10.1080/00140139.2015.1084051, PMID 26555738

Jones BH (1983) Overuse injuries of the lower extremities associated with marching, jogging, and running: a review. Mil Med 148(10):783–787. https://doi.org/10.1093/milmed/148.10.783, PMID 6417562

Jones BH, Toner MM, Daniels WL, Knapik JJ (1984) The energy cost and heart-rate response of trained and untrained subjects walking and running in shoes and boots. Ergonomics 27(8): 895–902. https://doi.org/10.1080/00140138408963563, PMID 6489316

Kaulio MA (1998) Customer, consumer and user involvement in product development: a framework and a review of selected methods. Total Qual Manag 9(1):141–149. https://doi.org/10.1080/0954412989333

Keukenkamp R, van Netten JJ, Busch-Westbroek TE, Nollet F, Bus SA (2021) Users' needs and expectations and the design of a new custom-made indoor footwear solution for people with diabetes at risk of foot ulceration. Disabil Rehabil 44:8493. https://doi.org/10.1080/09638288.2021.2003878

Kodithuwakku Arachchige SNK, Chander H, Turner AJ, Wilson SJ, Simpson JD, Knight AC et al (2020) Muscle activity during postural stability tasks: role of military footwear and load carriage. Safety 6(3):35. https://doi.org/10.3390/safety6030035

Kopitar D, Akalovic J, Skenderi Z (2019) Thermal resistance of leather and membranes for summer desert military footwear under different climate conditions. In: Functional textiles and clothing. Springer Singapore, Singapore, pp 3–9

Krings BM, Miller BL, Chander H, Waldman HS, Knight AC, McAllister MJ et al (2018) Impact of occupational footwear during simulated workloads on energy expenditure. Footwear Sci 10(3): 157–165. https://doi.org/10.1080/19424280.2018.1460623

Kuklane K, Geng Q, Holmér I (1998) Effect of footwear insulation on thermal responses in the cold. Int J Occup Saf Ergon 4(2):137–152. https://doi.org/10.1080/10803548.1998.11076386, PMID 10602614

Legg SJ, Mahanty A (1986) Energy cost of backpacking in heavy boots. Ergonomics 29(3): 433–438. https://doi.org/10.1080/00140138608968276, PMID 3698970

Lejeune TM, Willems PA, Heglund NC (1998) Mechanics and energetics of human locomotion on sand. J Exp Biol 201(13):2071–2080. https://doi.org/10.1242/jeb.201.13.2071, PMID 9622579

Lin LY, Chen CH (2015) Innovation and ergonomics consideration for female footwear design. Procedia Manuf 3:5867–5873. https://doi.org/10.1016/j.promfg.2015.07.891

Luh YP, Wang JB, Chang JW, Chang SY, Chu CH (2013) Augmented reality-based design customization of footwear for children. J Intell Manuf 24(5):905–917. https://doi.org/10.1007/s10845-012-0642-9

Lullini G, Giangrande A, Caravaggi P, Leardini A, Berti L (2020) Functional evaluation of a shock absorbing insole during military training in a group of soldiers: a pilot study. Mil Med 185(5–6): e643–e648. https://doi.org/10.1093/milmed/usaa032, PMID 32175582

MacDonald JL, Swider E (1974) Conceptual combat footwear study. IIT Research Institute, Chicago
Martin PE (1984) The effect of lower extremity loading on mechanical and physiological measures of running performance. Med Sci Sports Exerc 16(2):184. https://doi.org/10.1249/00005768-198404000-00365
Muniz AMS, Bini RR (2017) Shock attenuation characteristics of three different military boots during gait. Gait Posture 58:59–65. https://doi.org/10.1016/j.gaitpost.2017.07.037, PMID 2873 8226
Oakley EH (1984) The design and function of military footwear: a review following experiences in the South Atlantic. Ergonomics 27(6):631–637. https://doi.org/10.1080/00140138408963535, PMID 6148238
Pandolf KB, Haisman MF, Goldman RF (1976) Metabolic energy expenditure and terrain coefficients for walking on snow. Ergonomics 19(6):683–690. https://doi.org/10.1080/00140137608931583, PMID 1009916
Pisani R (2002) Footwear and soft ground interaction. University of Salford, Salford
Polcyn AF, Bensel CK, Harman EA, Obusek JP, Pandorf C (2002) Effects of weight carried by soldiers: combined analysis of four studies on maximal performance, physiology, and biomechanics. Army Natick Soldier Center Ma Supporting Science and Technology Directorate
Ramaswamy SS, Dua GL, Raizada VK, Dimri GP, Viswanathan KR, Madhaviah J et al (1966) Effect of looseness of snow on energy expenditure in marching on snow-covered ground. J Appl Physiol 21(6):1747–1749. https://doi.org/10.1152/jappl.1966.21.6.1747, PMID 5929299
Reinker KA, Ozburne S (1979) A comparison of male and female orthopaedic pathology in basic training. Mil Med 144(8):532–536. https://doi.org/10.1093/milmed/144.8.532, PMID 116171
Rosenblad-Wallin EFS (1988) The design and evaluation of military footwear based upon the concept of healthy feet and user requirement studies. Ergonomics 31(9):1245–1263. https://doi.org/10.1080/00140138808966765
Sen S, Chatterjee S, Sahrawat TR, Singh SB, Pal M (2019) Effect of military footwear on balance and stability: a pilot study. IJRAR 6(2):464–469
Sen S, Bhattacharyya D, Chaterjee S, Kumar B, Pal M (2020) Effect of shod walking on plantar pressure with varying uphill gradients. Asian J Med Sci 11(6):38–45. https://doi.org/10.3126/ajms.v11i6.29902
Sinclair J, Taylor PJ, Atkins S (2015) Effects of new military footwear on knee loading during running. Footwear Sci 7(3):165–171. https://doi.org/10.1080/19424280.2015.1066879
Tomlinson JP, Lednar WM, Jackson JD (1987) Risk of injury in soldiers. Mil Med 152(2):60–64. https://doi.org/10.1093/milmed/152.2.60, PMID 3103018
Torrens G, Campbell I, Tutton W (2012) Design issues in military footwear and handwear. In: Advances in military textiles and personal equipment. Elsevier, Amsterdam, pp 139–164
Wang H, Frame J, Ozimek E, Leib D, Dugan EL (2012) Influence of fatigue and load carriage on mechanical loading during walking. Mil Med 177(2):152–156. https://doi.org/10.7205/milmed-d-11-00210, PMID 22360059
Windle CM, Gregory SM, Dixon SJ (1999) The shock attenuation characteristics of four different insoles when worn in a military boot during running and marching. Gait Posture 9(1):31–37. https://doi.org/10.1016/s0966-6362(99)00002-8, PMID 10575068
Worthen BM, Yanklowitz BA (1978) The patho-physiology and treatment of stress fractures in military personnel. J Am Podiatry Assoc 68(5):317–325. https://doi.org/10.7547/87507315-68-5-317, PMID 25911
Yadav A, Arya K, Malhari A, Meena R, Chatterjee T, Bhattacharyya D et al (2022) Effect of intense military exercise on physical proficiency and hormonal responses of soldiers: a pilot study. BLDE Univ J Health Sci 7(1):35–40. https://doi.org/10.4103/bjhs.bjhs_14_21
Yoo B (2014) The effect of carrying a military backpack on a transverse slope and sand surface on lower limb during gait. The University of Utah, Salt Lake City. [Original source: https://studycrumb.com/alphabetizer]

Therapeutic Significance of Tanshinone 13

Smriti Sudha and Mrinalini Singh

Abstract

Danshen, the dried root of *Salvia miltiorrhiza*, has been used for thousands of years as a traditional Chinese remedy, mainly for the treatment of cardiovascular and cerebrovascular diseases. Tanshinone is the major bioactive component of Danshen. It is a lipophilic compound, synthesized via two different pathways: the methylerythritol phosphate (MEP) pathway in plasmids and the mevalonate (MVA) pathway in the cytosol. Many in vitro approaches can be applied to boost its production such as metabolic engineering using elicitors and recombinant DNA technology. Tanshinone and its derivatives can be administered orally, intravenously, and intraperitoneally but have poor absorption due to low solubility and membrane permeability. This can be enhanced through structural modifications of the tanshinones. Pharmaceutical properties of tanshinones include antioxidant activity, anti-inflammatory and anti-tumor activity, immune modulation, phytoestrogen, anti-fibrosis, anti-microbial activity, neuroprotection, and bone health. These properties help tanshinone play a therapeutic role in various cardiovascular diseases including cardiac fibrosis, hypertension, myocardial infarction, and atherosclerosis. Under hypoxic conditions, tanshinone protects cardiomyocytes and microvessels from damage, helps in the recovery of ventricle dysfunction, controls hypertension, and improves heart function.

Keywords

Tanshinone · Hypoxia · Cardiovascular diseases · Hypertension

S. Sudha · M. Singh (✉)
Defence Institute of Physiology and Allied Sciences (DIPAS), Defence Research and Development Organisation (DRDO), Delhi, India
e-mail: mrinalini.dipas@gov.in

R. Tulsawani, D. Vohora (eds.), *Adaptation under Stressful Environments through Biological Adjustments and Interventions*,
https://doi.org/10.1007/978-981-99-7652-2_13

13.1 Introduction

The dried root of *Salvia miltiorrhiza*, commonly known as Danshen in China, has been widely used as traditional herbal medicine since the Eastern Han dynasty, 2000 years ago. It grows in northeastern countries like China, Japan, Korea, and the Magnolia Mountain areas and is one of more than 1000 species of the *Salvia* genus across the globe, belonging to the family Labiatae. More than 200 compounds have been identified from *Salvia miltiorrhiza* (Danshen), according to the Chinese Academy of Sciences Chemical Database (www.organchem.csdb.cn) and the Chinese herbal medicine database (Wang et al. 2015).

There are around 40 lipophilic compounds and 50 hydrophilic compounds present in Danshen (Table 13.1). Among these, a few lipophilic bioactive compounds are majorly used in pharmaceutics such as tanshinone I, tanshinone IIa, tanshinone IIb, dihydrotanshinone, tanshinlactone, and cryptotanshinone (Dong et al. 2011; Hao et al. 2016; Huang et al. 2022). Caffeic acid-derived water-soluble compound of *Salvia*, sodium tanshinone IIa sulfonate, sulfonated at C-16, is also used in pharmacology (Chen et al. 2009). Among all these, tanshinone IIa has comparatively higher therapeutic potential. Tanshinone IIA (TanIIA) abundantly occurs in the rhizomes and roots of *Salvia miltiorrhiza* (Danshen). It is widely used for various pathological conditions, such as congenital heart defects, hypertension, arrhythmia, lung disease, left ventricular remodeling, dilated coronary artery, angina pectoris, and polycystic ovary syndrome (Zhou et al. 2005; Guo et al. 2020).

Structure of Tanshinone Tanshinone, as a bioactive compound, was first isolated in 1934 by a Japanese research scholar, *Nakao*, from the root of *Salvia* (Kakisawa et al. 1968). Tanshinone is a fat-soluble compound, belonging to abietane diterpenoids. The molecular weight is 276.3 kDa. It consists of four fused rings (A, B, C, D); two of these (A and B) are tetrahydronaphthalene, ring C is paraquinone or lactone, and ring D is dihydrofuran. The overall structure of all the tanshinones is common but depending on ring saturation and different substituents present, they metabolize differently. For example, the saturated A and D rings of cryptotanshinone (CYT) are metabolized via dehydrogenation, the D ring of dihydrotanshinone I (DHTI) is metabolized through hydrolysis, and the saturated A ring of TanIIA is metabolized by hydroxylation. Alteration in the carbon substituents alters the metabolic reaction (Jiang et al. 2019; Huang et al. 2022).

13.2 Biosynthesis of Tanshinone

Tanshinone is mainly synthesized by the two universal precursors, isopentenyl pyrophosphate (IPP) (Rohmer et al. 1993), and its isomer dimethylallyl diphosphate (DMAPP), via two different pathways, as shown in Fig. 13.1 (Schwarz 1994) The methylerythritol phosphate (MEP) pathway occurs in the plastids. Since tanshinone is a diterpene, it is especially synthesized from geranylgeranyl diphosphate (Rodríguez-Concepción and Boronat 2002). Other isoterpenoids are synthesized

Table 13.1 Chemical compounds of *Salvia miltiorrhiza* (Wang et al. 2007)

S. no	Lipophilic compounds	Hydrophilic compounds
1	Tanshinone I	Denshensu (Salvianic acid A)
2	Tanshinone IIA	Protocatechuic aldehyde
3	Tanshinone IIB	Salvianic acid B
4	Cryptotanshinone	Isoferulic acid
5	Dihydrotanshinone	Methyl Rosmarinate
6	Sodium tanshinone IIA sulfonate (STS)	Rosmarinic acid
7	Methyl tanshinonate	Salvianolic acid D
8	1,2-Dihydro-tanshinquinone	Salvianolic acid F
9	Methylene-tanshinquinone	Salvianolic acid G
10	Nortanshinone	Salvianolic acid H
11	Dihydro-nortanshinone	Salvianolic acid I
12	Methyl dihydro-nortanshinonate	Salvianolic acid J
13	Hydroxytanshinone IIA	Salvianolic acid K
14	Tanshinol A and B	Salvianolic acid E
15	Przewaquinone A, B and C	Prolithospermic acid
16	3-Hydroxy tanshinone IIA	Salvinal
17	3-Hydroxymethylene tanshinquinone	Salviaflaside
18	Tanshinaldehyde	Lithospermic acid
19	Tanshindiol A, B and C	Ethyl lithospermate
20	Dihydroisotanshinone I	Dimethyl lithospermate
21	1-Ketoiso-cryptotanshinone	Isosalvianolic acid C
22	Tanshinlactone	Lithospermic acid B
23	Neo-tanshinlactone	Ethyl lithospermate B
24	Miltirone	
25	Dehydromiltirone	
26	Miltirone I	
27	Tanshinone VI	
28	Neocryptotanshinone	
29	Danshenxinkun A and Danshenxinkun B	
30	Sugiol	
31	Ferruginol	

by the mevalonate pathway (MVA pathway) in the cytosol, from 3-hydroxy-3-methylglutaryl CoA (HMG-CoA) (Miziorko 2011).

IPP isomerase (IPI) may also trigger the interconversion of IPP and DMAPP. They are then converted to geranylgeranyl diphosphate (GGPP) in a stepwise manner by geranyl diphosphate synthase (GPPS), farnesyl diphosphate synthase (FPPS), and geranylgeranyl diphosphate synthase (GGPPS). GGPP acts as the common precursor of all diterpenoids. Biosynthesis of the tanshinone skeletons is carried out by tanshinone synthases. Production of ferruginol from miltiradene was shown to be catalyzed by *CYP76AH1* through expression in yeast whole cells in vitro as well as in vivo (Cai et al. 2013). However, overexpression of key

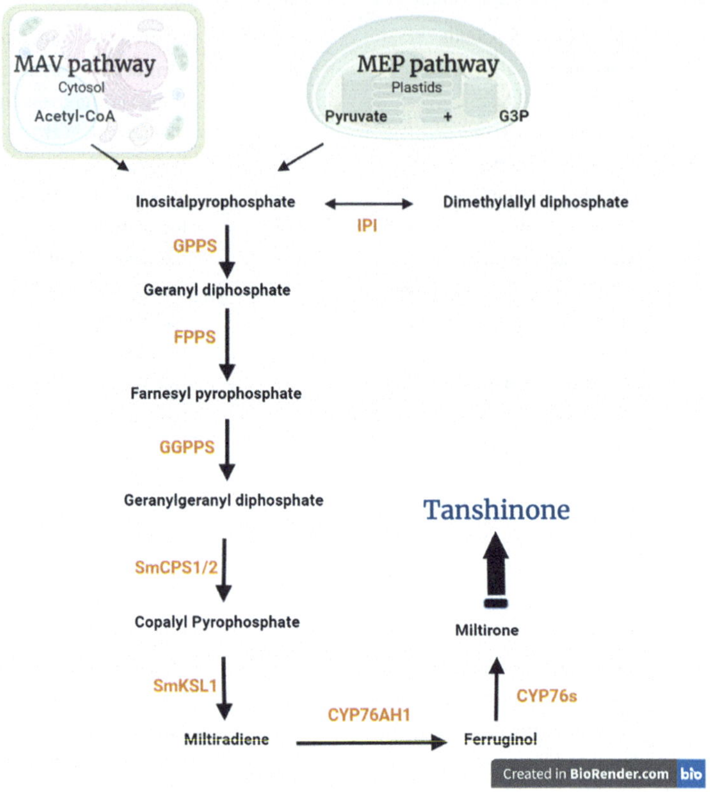

Fig. 13.1 Tanshinone biosynthesis pathway

regulatory enzymes such as 1-deoxy-D-xylulose-5-phosphate synthase (DXS) in the MEP pathway and 3-hydroxy-3-methylglutaryl-CoA reductase (HMGR) in the MAV pathway can help to enhance the tanshinone production.

13.3 Approaches to Increase Tanshinone Production

Many in vitro approaches can be used to enhance the production of tanshinone. One such approach is metabolic engineering using elicitors, which can increase tanshinone yield by approximately sixfold (Kai et al. 2012). Two types of elicitors, biotic elicitors (methylejasmonate, yeast elicitor, jasmonic acid), (Hao et al. 2015), and abiotic elicitors (Ag^+, Co^{2+}, and Cd^{2+}) are mainly used to elevate the genes responsible for tanshinone synthesis (Wang et al. 2016a).

Methyl jasmonate treatment can be applied in conjugation with plant hormones such as polyethyl glycol (PEG) and abscisic acid (ABA). PEG and ABA stimulate methyl jasmonate accumulation, which in turn enhances tanshinone production

(Yang et al. 2012). PEG and ABA also trigger nitric oxide (NO) formation. NO signaling plays an important part in the water stress-induced production of tanshinone. An NO donor, such as sodium nitroprusside, can also stimulate production (Du et al. 2015).

As opposed to applying a single elicitor, combining two different elicitors proves to be more successful in enhancing tanshinone production because of a synergistic effect. For example, yeast extract and hyperosmotic stress (through sorbitol addition) together increased tanshinone production remarkably more than applied separately in *Salvia miltiorrhiza* hairy root cultures (Wang and Wu 2010).

Another approach to increase the yield of tanshinone is recombinant DNA technology using *Agrobacterium rhizogenes* in hairy root cultures. There are more than 200 genes related to 32 enzymes involved in the biosynthesis of terpenoids. Overexpressing the genes identified for tanshinone can help increase the yield in vitro. The overexpression of a single *MDS* gene, which is a key regulator for tanshinone production, can result in a huge increase in tanshinone concentration. Likewise, other tanshinone-specific genes, SmGGPPS, SmDXR, SmDXS, and SmHMGR, can increase tanshinone yield by ~2.5-fold than the control.

13.4 Pharmacokinetics of Tanshinone

Tanshinone and its derivatives can be administered orally, intravenously, and intraperitoneally. It has been reported that the half-life and bio-availability of tanshinone are low, approximately 0.475–0.613 mg/g. Tan IIA has very low solubility and membrane permeability, resulting in poor absorption. The intravenous injection of tanshinone at a significant concentration (2 mg/mL) in a nanosuspension, in order to make it suitable as a parenteral solution, suggested that the Tan IIA distribution was higher in tissue (liver and lung) than in the plasma. Lipoproteins in plasma are crucial for transporting Tan IIA and other lipophilic molecules. The distribution of Tan IIA in other organs, such as the heart and brain, is comparatively lower than in the plasma (Lam et al. 2003).

However, under the pathological state of the brain, it is reported that Tan IIA can transport to the cerebral tissue due to a broken blood–brain barrier. There is a poor accumulation of Tan IIA in kidneys which may account for the meager amounts of Tan IIA and its metabolites excreted through urine. Despite being extremely lipophilic, the distribution of Tan IIA could not be seen in fat repositories. This could be due to the close association of Tan IIA with the reticuloendothelial system and the restricted blood supply to fat tissues (Hao et al. 2006).

In order to boost intestinal adsorption and solve the issue of poor solubility, structural modifications can be made to tanshinones. One such formulation is the water-soluble sodium tanshinone IIA sulfonate (STS), which shows antioxidant activity and has a wide range of pharmacological uses. Another avenue for improving the bioavailability of poorly water-soluble compounds such as tanshinones is using solid lipid nanoparticles as carriers. It has been shown to improve the circulation and residence time of Tan IIA in plasma (Table 13.2).

Table 13.2 Pharmaceutical properties of tanshinone

S. no	Therapeutic activity	Tanshinone	Mode of action	References
1.	Antioxidant activity	Tanshinone IA, Tanshinone IIA, Tanshinone IIB, Cryptotanshinone and Dihydrotanshinone I	Prevents reactive oxygen species (ROS) accumulation by inhibiting lipid peroxidation and Ca^{2+} influx in myocardial cells, and by increasing levels of superoxide dismutase (SOD) and growth stimulating hormone (GSH) Inhibits mitochondrial autophagy, and decreases malondialdehyde (MDA) and inducible nitric oxide synthase level. Increases level of anti-oxidant enzymes	Guo et al. (2018), Kubli et al. (2013), Li et al. (2006), Wang et al. (2020)
2.	Anti-inflammatory	Tanshinone IIA and Cryptotanshinone	Inhibits expression of pro-inflammatory cytokine gene, TLR2/Nf-κB signaling. Reduces expression of adhesion molecules and infiltration of neutrophils	Chen et al. (2001), Yang et al. (2016)
3.	Anti-tumor	Tanshinone IIA, Tanshinone I and Cryptotanshinone	Inhibits cancer cell growth and proliferation, promoting apoptosis by regulating p53 gene and cell cycle arrest. Inhibits E6 protein, in case of the Human Papilloma Virus (HPV) Controls acute leukemia through the tumor suppressor gene and AKT/mTOR pathways	Lu et al. (2013), Xu et al. (2018), Yintao (2010)
4.	Immune modulation	Tanshinone IIA	Induces dendritic cell-mediated innate immunity response, upregulates regulatory	Gong et al. (2020), Qin et al. (2010)

(continued)

Table 13.2 (continued)

S. no	Therapeutic activity	Tanshinone	Mode of action	References
			cytokines such as TGF-β1, promotes anti-inflammatory cytokines production and reduces LPS-induced TLR4/ NF-κB signaling cascade. Induces T helper cell polarization into T reg cells	
5.	Phyto-estrogen	Tanshinone I, Tanshinone IIA, Cryptotanshinone and Dihydrotanshinone	Activates and modulates G-protein-coupled estrogen receptor to regulate transcription of collagen and elastin genes. Can also activate PXR, PPARγ, and AR	Mao et al. (2014)
6.	Anti-fibrosis	Tanshinone IIA	Decreases myofibroblast proliferation by activating Nrf2/ glutathione signaling. Blocks ROS-mediated PKCδ/smad3 pathway and miRNA-mediated TGF-β1/smad3 pathway	Feng et al. (2019), Meng et al. (2016), Shu et al. (2016)
7.	Neuroprotectant	Tanshinone IIA, Tanshinone I and Cryptotanshinone	Improves learning and memory deficits by shielding neuronal loss through inhibition of ERK and GSK-3β pathway. Downregulates p53 protein and maintains membrane potential to inhibit apoptosis of neuronal cells. Protects neurons from beta-amyloid toxicity.	Bi et al. (2008), Chen et al. (2018)
8.	Anti-microbial	Dihydrotanshinone and Cryptotanshinone	Anti-bacterial activity against a wide range of Gram-positive bacteria	Honda et al. (1988)

(continued)

Table 13.2 (continued)

S. no	Therapeutic activity	Tanshinone	Mode of action	References
9.	Bone health	Tanshinone IIa	Regulates RANKL-MCSF signaling and inhibits excessive bone resorption. Controls osteoporosis	Guo et al. (2014), Lee et al. (2005), Wang et al. (2019)

13.5 Therapeutic Role of Tanshinone in Cardiovascular Diseases

13.5.1 Cardiac Fibrosis (Hinderer and Schenke-Layland 2019)

Cardiac fibrosis causes various heart diseases such as arrhythmias, hypertensive heart diseases, hypertrophic cardiomyopathy, and coronary heart diseases. It involves fibrotic scar development due to excessive and abnormal extracellular matrix (ECM) formation, especially collagen type I, which results in endomyocardial fibrosis and impaired normal tissue function (Murtha et al. 2017). Mast cell accumulation and degranulation also promote excess collagen and fibroblast synthesis (Kong et al. 2014). TanIIA inhibits NF-κB mediated signaling cascade to reduce inflammatory cytokine production (Wu et al. 2018) and upregulates matrix metalloproteinases to promote collagen metabolism (Mao et al. 2014). As protective activity against cardiac fibrosis, TanIIA further inhibits β-tubulin expression (Zhang et al. 2018b) and reduces oxidative stress causing fibroblast formation by activating Nrf2, which inhibits TGF-β generation (Bi et al. 2021; Zhou et al. 2012). Cardiac fibrosis also results in cell death due to inflammatory response, by increased TGF-β (Desmoulière et al. 1993). TanIIA has shown a significant anti-apoptotic response, by regulating caspase protein and mitochondrial membrane potential, which results in regulated cytochrome *c* secretion (Zhang et al. 2014).

13.5.2 Hypertension

Globally, one in every four adults suffers from systemic arterial hypertension, and the percentage of individuals with non-optimum blood pressure (BP) is increasing every year (Forouzanfar et al. 2017). The average range of BP is around 115/75 mmHg, which is a ratio of systolic and diastolic pressures. However, there could be an increase in normal BP due to several influences such as environment, lifestyle, excessive sodium consumption, as well as genetics (Poulter et al. 2015). Many internal factors are involved in the human body that maintain physiological blood pressure, such as the renin–angiotensin–aldosterone system, the sympathetic nervous system, as well as the immune system. Interruption of these parameters can lead to cardiovascular disease. It is reported in many studies that the onset condition of high blood pressure can be controlled by tanshinone IIA. It is observed that tanshinone inhibits angiotensin-converting enzymes to lower the blood pressure.

Sodium plays a very crucial role in regulating blood pressure by maintaining blood volume. Imbalance in sodium concentrations will lead to the production of nitric oxide to vasodilate blood vessels. However, some individuals show salt-sensitivity with endothelial dysfunction, due to which there is an overproduction of TGF-β through the voltage- and calcium-activated potassium channels (Feng et al. 2017) and MAPK pathway, leading to fibrosis and oxidative stress (Hovater and Sanders 2012). Vasoconstriction is very prominent under hypoxia since the voltage-gated potassium channel is closed, causing membrane depolarization, and thus increasing Ca^{2+} ion concentration. Tan IIA induces the PI3K/Akt-eNOS signaling pathway for vasodilation and aortic relaxation. It is also found that tanshinone opens ATP-sensitive K^+ channels to decrease calcium ion levels, leading to vasodilation (Sonkusare et al. 2006; Chan et al. 2011). Moreover, disruption in several genes and their allelic variants can also be involved in hypertension and cardiovascular disease. To improve ventricular remodeling, Tan IIA can downregulate TGF-β1 through miR-205-3p and thus maintain physiological blood pressure (Qiao et al. 2021).

13.5.3 Myocardial Infarction

Myocardial infarction has affected more than 15 million people. It occurs due to excessive reactive oxygen species (ROS) production and inflammation; leading to cardiomyocyte death. The antioxidant and anti-inflammatory properties of TanIIA help to control myocardial injury (Lu et al. 2022). TanIIA inhibits inflammation generated during ischemia by reducing the expression of inflammatory cytokines such as MCP-1 and TGF-β1, as well as infiltration of monocytes, and macrophages to the infarcted area, to improve heart function. TanIIA has been used for years to treat myocardial diseases, as it can increase coronary flow by dilating coronary arteries, which prevents ischemia. TanIIA also shows an anti-inflammatory effect by preventing NF-κB-mediated TNF-α production (Ren et al. 2010). It has been reported that TanIIA helps recovery from MI-induced heart injury by decreasing infarct size and collagen deposition.

13.5.4 Atherosclerosis

Atherosclerosis is a multi-arterial inflammatory disease, resulting mainly from hypertension and hypercholesterolemia. It is caused by the hardening of arteries due to excessive deposition of plaque. The increased plasma lipids promote the accumulation of low-density lipids C (LDL-C) on the walls of the arteries and escalate the expression of adhesion molecules. This leads to adherence of monocytes to the sub-endothelial spaces and conversion into foamy macrophages. They express

CD68, which works as a chemoattractant for the accumulation of intracellular cholesterol (Sakakura et al. 2013).

Anti-oxidant activity of TanIIA prevents ROS-mediated LDL oxidation (Hartley et al. 2019) and the migration of monocytes into the arterial endothelium (Tabas and Bornfeldt 2016). It also increases GPx and SOD levels to avoid foamy macrophage accumulation (Tang et al. 2007) and prevents the accumulation of platelets and cholesterols (Ahmadsei et al. 2015) through anti-inflammatory activity.

Overexpression of Nf-κB causes an inflammatory response; thus, blocking the activation of Nf-κB could help cure inflammation-mediated atherosclerosis. Research findings confirmed that TanIIA has the potential to regulate Nf-κb activation under diseased conditions by attenuating the expression of adhesion molecules such as VCAM and ICAM (Chang et al. 2014).

TanIIA also possesses immunomodulatory effects. It controls the expression of pro-inflammatory cytokines, DC maturation, and T-cell maturation (Li et al. 2014).

Calcium influx results in the mineralization of plaque, leading to the narrowing of the arteries. Under stress or hypoxic conditions, calcium influx is inhibited by TanIIA, which helps to improve energy production in cardiomyocytes and maintain cardiac functioning (Satoshi et al. 1990).

13.6 Effect of Tanshinone on Cardiovascular Diseases Under Hypoxia

High altitude is characterized by a decrease in the partial pressure of oxygen, resulting in several pathological conditions (Avellanas Chavala 2018), such as acute mountain sickness (AMS), high-altitude pulmonary edema (HAPE), high-altitude cerebral edema (HACE), and high-altitude pulmonary hypertension (HAPH) (Simonneau et al. 2009; Bärtsch and Swenson 2013). HAPH is caused by pulmonary vasoconstriction under chronic exposure to hypobaric hypoxia (Hakim et al. 1983). The mean pulmonary arterial pressure is less than 25 mmHg; however, under HAPH, it exceeds 30 mmHg (León-Velarde et al. 2005).

Vascular resistance is increased due to long-term vasoconstriction under hypobaric hypoxia (Siques et al. 2021). Vascular remodeling is triggered, which results in right ventricle hypertrophy. Vasoconstrictors, such as reactive oxygen species (ROS) and endothelin-1, also contribute to vascular remodeling. The endogenous antioxidant system becomes inefficient at removing excess ROS due to atmospheric stress at high altitudes. The enhanced ROS results in oxidative stress on the cell, causing mitochondrial dysfunction and cell death (Farías et al. 2016). In cardiomyocytes and other cells, NADPH is responsible for the production of ROS. Under hypoxia, the expression of the NADPH complex (Nox4) is increased, which induces TGF-β level causing pulmonary hypertension (Sturrock et al. 2006; Zhang et al. 2018a, b; Siques et al. 2021). Vascular remodeling is regulated by MAPK signaling and the activator protein-1 (AP-1) transcription factor which controls proliferation, inflammation, and apoptosis. Thus, AP-1 subunits could be a good therapeutic target (Yadav et al. 2017; Singh et al. 2018). Tanshinone IIA has been shown to modulate and promote cell survival in human lung epithelial cells under hypoxia via the AP-1-Nrf2 transcription factor (Yadav et al. 2020).

During hypoxia, cardiomyocytes undergo apoptosis and narcosis due to the activation of specific pathways. Excessive apoptosis results in cardiovascular disorders such as heart failure and myocardial infarction (Chen et al. 2014). TanIIA protects cardiomyocytes, and microvessels from damage from excessive superoxide production, and intracellular calcium influx, under hypoxia. Left ventricle dysfunction, resulting due to chronic intermittent hypoxia, is also recovered by TanIIA (Wang et al. 2017).

Activation of hypoxia-induced factor (HIF) is required to maintain normal physiology under hypoxia (Forsythe et al. 1996). In the case of cardiovascular diseases, TanIIA improves heart function by upregulating HIF and further enhancing angiogenesis (Xu et al. 2009). However, in cancer conditions, TanIIA exhibits anti-angiogenesis by decreasing the HIF-1α level and inhibiting growth factors like VEGH and β-FGF (Sui et al. 2017).

Classical transient receptor potential canonical (TRPC) channels are closely associated with cardiac function and play a role in cardiac hypertrophy, myocardial infarction, arrhythmias, and fibrosis. Under chronic hypoxia, the TanIIA derivative STS reduces Ca^{2+} influx and prevents the expression of TRPC1 and TRPC6 (Wang et al. 2013). STS also inhibits the BMPR-induced smad pathway to control unnecessary cell apoptosis. TanIIA inhibits TGF-β induced smad3 pathway to control hypertension under hypoxia. In vivo studies have reported that TanIIA can induce apoptosis to reverse the pulmonary artery remodeling by suppressing the JAK2/STAT3 pathway and enhancing the connexin 43 level (Chen et al. 2014) under hypoxia.

Various classes of non-coding RNA, such as small RNAs and long non-coding RNAs, could be potential targets for cardiovascular therapeutics as many non-coding RNAs are involved in cardiovascular diseases. miR-1 is an important non-coding RNA (miRNA) that regulates a number of genes in myocytes, including reduced expression of the apoptotic genes and Akt phosphorylation. Misregulation of miR-1 can cause myoblast apoptosis, arrhythmogenesis, cardiac hypertrophy, etc. A study found that TanIIA can suppress miR-1 expression by inhibiting p38 MAPK, which improves heart function during ischemia and hypoxic injury (Zhang et al. 2010). miR-133 regulates apoptosis-related proteins such as caspase9 and reduced apoptosis rate can be observed in cardiac cells by upregulating miR-133 mediated MAPK ERK1/2 expression. This can be activated by both hypoxia and TanIIA.

Also, hypoxia followed by re-oxygenation results in the upregulation of pro-apoptotic pathways. However, pre-treatment with TanIIA increases cell viability, decreases the expression of pro-apoptotic genes, and protects against hypoxia-induced damage (Cui et al. 2016).

References

Ahmadsei M, Lievens D, Weber C, von Hundelshausen P, Gerdes N (2015) Immune-mediated and lipid-mediated platelet function in atherosclerosis. Curr Opin Lipidol 26(5):438–448. https://doi.org/10.1097/MOL.0000000000000212, PMID 26270811

Avellanas Chavala ML (2018) Un viaje entre la hypoxia de la gran altitud y la hypoxia del enfermo crítico: ¿quépuede enseñarnos en la compresión y manejo de las enfermedades críticas? Med Intensiva 42(6):380–390. https://doi.org/10.1016/j.medin.2017.08.006, PMID 28919307

Bärtsch P, Swenson ER (2013) Clinical practice: acute high-altitude illnesses. N Engl J Med 368(24):2294–2302. https://doi.org/10.1056/NEJMcp1214870, PMID 23758234

Bi HC, Zuo Z, Chen X, Xu CS, Wen YY, Sun HY et al (2008) Preclinical factors affecting the pharmacokinetic behaviour of tanshinone IIA, an investigational new drug isolated from Salvia miltiorrhiza for the treatment of ischaemic heart diseases. Xenobiotica 38(2):185–222. https://doi.org/10.1080/00498250701767675, PMID 18197559

Bi Z, Wang Y, Zhang W (2021) A comprehensive review of tanshinone IIA and its derivatives in fibrosis treatment. Biomed Pharmacother 137:111404. https://doi.org/10.1016/j.biopha.2021.111404, PMID 33761617

Cai Y, Guo J, Zhou YJ, Zhu ZW, Wu WY, Huang LQ et al (2013) [Construction of Saccharomyces cerevisiae whole-cell biocatalyst system for conversion miltiradiene]. Yao Xue Xue Bao 48(10): 1618–1623, PMID 24417091

Chan P, Liu IM, Li YX, Yu WJ, Cheng JT (2011) Antihypertension induced by tanshinone IIA isolated from the roots of Salvia miltiorrhiza. Evid Based Complement Alternat Med 2011: 392627. https://doi.org/10.1093/ecam/nep056, PMID 19542183

Chang CC, Chu CF, Wang CN, Wu HT, Bi KW, Pang JH et al (2014) The anti-atherosclerotic effect of tanshinone IIA is associated with the inhibition of TNF-α-induced VCAM-1, ICAM-1 and CX3CL1 expression. Phytomedicine 21(3):207–216. https://doi.org/10.1016/j.phymed.2013.09.012, PMID 24157079

Chen YH, Lin SJ, Ku HH, Shiao MS, Lin FY, Chen JW et al (2001) Salvianolic acid B attenuates VCAM-1 and ICAM-1 expression in TNF-alpha-treated human aortic endothelial cells. J Cell Biochem 82(3):512–521. https://doi.org/10.1002/jcb.1176, PMID 11500927

Chen Y, Tu JH, He YJ, Zhang W, Wang G, Tan ZR et al (2009) Effect of sodium tanshinone II A sulfonate on the activity of CYP1A2 in healthy volunteers. Xenobiotica 39(7):508–513. https://doi.org/10.1080/00498250902951763, PMID 19534587

Chen M, Liu Y, Yi D, Wei L, Li Y, Zhang L (2014) Tanshinone IIA promotes pulmonary artery smooth muscle cell apoptosis in Vitroby inhibiting the JAK2/STAT3 signaling pathway. Cell Physiol Biochem 33(4):1130–1138. https://doi.org/10.1159/000358682, PMID 24733040

Chen J, Bi Y, Chen L, Zhang Q, Xu L (2018) Tanshinone IIA exerts neuroprotective effects on hippocampus-dependent cognitive impairments in diabetic rats by attenuating ER stress-induced apoptosis. Biomed Pharmacother 104:530–536. https://doi.org/10.1016/j.biopha.2018.05.040, PMID 29800917

Cui ZT, Liu JP, Wei WL (2016) The effects of tanshinone IIA on hypoxia/reoxygenation-induced myocardial microvascular endothelial cell apoptosis in rats via the JAK2/STAT3 signaling pathway. Biomed Pharmacother 83:1116–1126. https://doi.org/10.1016/j.biopha.2016.07.054, PMID 27551758

Desmoulière A, Geinoz A, Gabbiani F, Gabbiani G (1993) Transforming growth factor-beta 1 induces alpha-smooth muscle actin expression in granulation tissue myofibroblasts and in quiescent and growing cultured fibroblasts. J Cell Biol 122(1):103–111. https://doi.org/10.1083/jcb.122.1.103, PMID 8314838

Dong Y, Morris-Natschke SL, Lee KH (2011) Biosynthesis, total syntheses, and antitumor activity of tanshinones and their analogs as potential therapeutic agents. Nat Prod Rep 28(3):529–542. https://doi.org/10.1039/c0np00035c, PMID 21225077

Du X, Zhang C, Guo W, Jin W, Liang Z, Yan X et al (2015) Nitric oxide plays a central role in water stress-induced tanshinone production in Salvia miltiorrhiza hairy roots. Molecules 20(5): 7574–7585. https://doi.org/10.3390/molecules20057574, PMID 25919278

Farías JG, Herrera EA, Carrasco-Pozo C, Sotomayor-Zárate R, Cruz G, Morales P et al (2016) Pharmacological models and approaches for pathophysiological conditions associated with hypoxia and oxidative stress. Pharmacol Ther 158:1–23. https://doi.org/10.1016/j.pharmthera.2015.11.006, PMID 26617218

Feng W, Dell'Italia LJ, Sanders PW (2017) Novel paradigms of salt and hypertension. J Am Soc Nephrol 28(5):1362–1369. https://doi.org/10.1681/ASN.2016080927, PMID 28220030

Feng F, Cheng P, Zhang H, Li N, Qi Y, Wang H, Wang Y, Wang W (2019) The protective role of tanshinone IIA in silicosis rat model via TGF-β1/Smad signaling suppression, NOX4 inhibition and Nrf2/ARE signaling activation. Drug Des Devel Ther 13:4275–4290. https://doi.org/10.2147/DDDT.S230572

Forouzanfar MH, Liu P, Roth GA, Ng M, Biryukov S, Marczak L et al (2017) Global burden of hypertension and systolic blood pressure of at least 110 to 115 mmHg, 1990-2015. JAMA 317(2):165–182. https://doi.org/10.1001/jama.2016.19043, PMID 28097354

Forsythe JA, Jiang BH, Iyer NV, Agani F, Leung SW, Koos RD et al (1996) Activation of vascular endothelial growth factor gene transcription by hypoxia-inducible factor 1. Mol Cell Biol 16(9): 4604–4613. https://doi.org/10.1128/MCB.16.9.4604, PMID 8756616

Gong Y, Liu YC, Ding XL, Fu Y, Cui LJ, Yan YP (2020) Tanshinone IIA ameliorates CNS autoimmunity by promoting the differentiation of regulatory T cells. Neurotherapeutics 17(2): 690–703. https://doi.org/10.1007/s13311-019-00789-2, PMID 31845175

Guo Y, Li Y, Xue L, Severino RP, Gao S, Niu J et al (2014) Salvia miltiorrhiza: an ancient Chinese herbal medicine as a source for anti-osteoporotic drugs. J Ethnopharmacol 155(3):1401–1416. https://doi.org/10.1016/j.jep.2014.07.058, PMID 25109459

Guo Z, Yan M, Chen L, Fang P, Li Z, Wan Z et al (2018) Nrf2dependent antioxidant response mediated the protective effect of tanshinone IIA on doxorubicin induced cardiotoxicity. Exp Ther Med 16(4):3333–3344. https://doi.org/10.3892/etm.2018.6614, PMID 30233680

Guo R, Li L, Su J, Li S, Duncan SE, Liu Z et al (2020) Pharmacological activity and mechanism of Tanshinone IIA in related diseases. Drug Des Dev Ther 14:4735

Hakim TS, Michel RP, Minami H, Chang HK (1983) Site of pulmonary hypoxic vasoconstriction studied with arterial and venous occlusion. J Appl Physiol Respir Environ Exerc Physiol 54(5): 1298–1302. https://doi.org/10.1152/jappl.1983.54.5.1298, PMID 6863090

Hao H, Wang G, Cui N, Li J, Xie L, Ding Z (2006) Pharmacokinetics, absorption and tissue distribution of tanshinone IIA solid dispersion. Planta Med 72(14):1311–1317. https://doi.org/10.1055/s-2006-951698, PMID 17024606

Hao X, Shi M, Cui L, Xu C, Zhang Y, Kai G (2015) Effects of methyl jasmonate and salicylic acid on tanshinone production and biosynthetic gene expression in transgenic Salvia miltiorrhiza hairy roots. Biotechnol Appl Biochem 62(1):24–31. https://doi.org/10.1002/bab.1236, PMID 24779358

Hao W, Chen L, Wu LF, Yang F, Niu JX, Kaye AD et al (2016) Tanshinone IIA exerts an antinociceptive effect in rats with cancer-induced bone pain. Pain Phys 19(7):456–476, PMID 27676663

Hartley A, Haskard D, Khamis R (2019) Oxidized LDL and anti-oxidized LDL antibodies in atherosclerosis—novel insights and future directions in diagnosis and therapy. Trends Cardiovasc Med 29(1):22–26. https://doi.org/10.1016/j.tcm.2018.05.010, PMID 29934015

Hinderer S, Schenke-Layland K (2019) Cardiac fibrosis—a short review of causes and therapeutic strategies. Adv Drug Deliv Rev 146:77–82. https://doi.org/10.1016/j.addr.2019.05.011, PMID 31158407

Honda G, Koezuka Y, Tabata M (1988) Isolation of an antidermatophytic substance from the root of Salvia miltiorrhiza. Chem Pharm Bull (Tokyo) 36(1):408–411. https://doi.org/10.1248/cpb.36.408, PMID 3378301

Hovater MB, Sanders PW (2012) Effect of dietary salt on regulation of TGF-β in the kidney. Semin Nephrol 32(3):269–276. https://doi.org/10.1016/j.semnephrol.2012.04.006, PMID 22835458
Huang X, Jin L, Deng H, Wu D, Shen QK, Quan ZS et al (2022) Research and development of natural product tanshinone I: pharmacology, total synthesis, and structure modifications. Front Pharmacol 13:920411. https://doi.org/10.3389/fphar.2022.920411, PMID 35903340
Jiang Z, Gao W, Huang L (2019) Tanshinones, critical pharmacological components in Salvia miltiorrhiza. Front Pharmacol 10:202. https://doi.org/10.3389/fphar.2019.00202, PMID 30923 500
Kai G, Liao P, Xu H, Wang J, Zhou C, Zhou W et al (2012) Molecular mechanism of elicitor-induced tanshinone accumulation in Salvia miltiorrhiza hairy root cultures. Acta Physiol Plant 34(4):1421–1433. https://doi.org/10.1007/s11738-012-0940-z
Kakisawa H, Hayashi T, Okazaki I, Ohashi M (1968) Isolation and structures of new tanshinones. Tetrahedron Lett 9(28):3231–3234. https://doi.org/10.1016/S0040-4039(00)89532-5
Kong P, Christia P, Frangogiannis NG (2014) The pathogenesis of cardiac fibrosis. Cell Mol Life Sci 71(4):549–574. https://doi.org/10.1007/s00018-013-1349-6, PMID 23649149
Kubli DA, Zhang X, Lee Y, Hanna RA, Quinsay MN, Nguyen CK et al (2013) Parkin protein deficiency exacerbates cardiac injury and reduces survival following myocardial infarction. J Biol Chem 288(2):915–926. https://doi.org/10.1074/jbc.M112.411363, PMID 23152496
Lam BYH, Lo ACY, Sun X, Luo HW, Chung SK, Sucher NJ (2003) Neuroprotective effects of tanshinones in transient focal cerebral ischemia in mice. Phytomedicine 10(4):286–291. https://doi.org/10.1078/094471103322004776, PMID 12809358
Lee SY, Choi DY, Woo ER (2005) Inhibition of osteoclast differentiation by tanshinones from the root of Salvia miltiorrhiza Bunge. Arch Pharm Res 28(8):909–913. https://doi.org/10.1007/BF02973876, PMID 16178416
León-Velarde F, Maggiorini M, Reeves JT, Aldashev A, Asmus I, Bernardi L et al (2005) Consensus statement on chronic and subacute high altitude diseases. High Alt Med Biol 6(2): 147–157. https://doi.org/10.1089/ham.2005.6.147, PMID 16060849
Li P, Wang GJ, Li J, Hao HP, Zheng CN (2006) Identification of tanshinone IIA metabolites in rat liver microsomes by liquid chromatography–tandem mass spectrometry. J Chromatogr A 1104(1–2):366–369. https://doi.org/10.1016/j.chroma.2005.12.047, PMID 16405978
Li HZ, Lu YH, Huang GS, Chen Q, Fu Q, Li ZL (2014) Tanshinone II A inhibits dendritic cell-mediated adaptive immunity: potential role in anti-atherosclerotic activity. Chin J Integr Med 20(10):764–769. https://doi.org/10.1007/s11655-012-1213-9, PMID 23001464
Lu L, Li C, Li D, Wang Y, Zhou C, Shao W et al (2013) Cryptotanshinone inhibits human glioma cell proliferation by suppressing STAT3 signaling. Mol Cell Biochem 381(1–2):273–282. https://doi.org/10.1007/s11010-013-1711-x, PMID 23740516
Lu TC, Wu YH, Chen WY, Hung YC (2022) Targeting oxidative stress and endothelial dysfunction using tanshinone IIA for the treatment of tissue inflammation and fibrosis. Oxidative Med Cell Longev 2022:2811789. https://doi.org/10.1155/2022/2811789, PMID 35432718
Mao S, Wang Y, Zhang M, Hinek A (2014) Phytoestrogen, tanshinone IIA diminishes collagen deposition and stimulates new elastogenesis in cultures of human cardiac fibroblasts. Exp Cell Res 323(1):189–197. https://doi.org/10.1016/j.yexcr.2014.02.001, PMID 24525372
Meng XM, Nikolic-Paterson DJ, Lan HY (2016) TGF-β: the master regulator of fibrosis. Nat Rev Nephrol 12(6):325–338. https://doi.org/10.1038/nrneph.2016.48, PMID 27108839
Miziorko HM (2011) Enzymes of the mevalonate pathway of isoprenoid biosynthesis. Arch Biochem Biophys 505(2):131–143. https://doi.org/10.1016/j.abb.2010.09.028, PMID 20932 952
Murtha LA, Schuliga MJ, Mabotuwana NS, Hardy SA, Waters DW, Burgess JK et al (2017) The processes and mechanisms of cardiac and pulmonary fibrosis. Front Physiol 8:777. https://doi.org/10.3389/fphys.2017.00777, PMID 29075197
Poulter NR, Prabhakaran D, Caulfield M (2015) Hypertension. Lancet 386(9995):801–812. https://doi.org/10.1016/S0140-6736(14)61468-9

Qiao P, Xu J, Liu X, Li X (2021) Tanshinone IIA improves ventricular remodeling following cardiac infarction by regulating miR-205-3p. Dis Markers 2021:8740831. https://doi.org/10.1155/2021/8740831, PMID 34880957

Qin XY, Li T, Yan L, Liu QS, Tian Y (2010) Tanshinone IIA protects against immune-mediated liver injury through activation of T-cell subsets and regulation of cytokines. Immunopharmacol Immunotoxicol 32(1):51–55. https://doi.org/10.1080/08923970903120997, PMID 19653860

Ren ZH, Tong YH, Xu W, Ma J, Chen Y (2010) Tanshinone II A attenuates inflammatory responses of rats with myocardial infarction by reducing MCP-1 expression. Phytomedicine 17(3–4): 212–218. https://doi.org/10.1016/j.phymed.2009.08.010, PMID 19800776

Rodríguez-Concepción M, Boronat A (2002) Elucidation of the methylerythritol phosphate pathway for isoprenoid biosynthesis in bacteria and plastids. A metabolic milestone achieved through genomics. Plant Physiol 130(3):1079–1089. https://doi.org/10.1104/pp.007138, PMID 12427975

Rohmer M, Knani M, Simonin P, Sutter B, Sahm H (1993) Isoprenoid biosynthesis in bacteria: a novel pathway for the early steps leading to isopentenyl diphosphate. Biochem J 295(2): 517–524. https://doi.org/10.1042/bj2950517, PMID 8240251

Sakakura K, Nakano M, Otsuka F, Ladich E, Kolodgie FD, Virmani R (2013) Pathophysiology of atherosclerosis plaque progression. Heart Lung Circ 22(6):399–411. https://doi.org/10.1016/j.hlc.2013.03.001, PMID 23541627

Satoshi T, Kouichi T, Katsumi H, Kazuko K, Mika O, Akira Y et al (1990) Beneficial effect of Tan-Shen, an extract from the root of salvia, on post-hypoxic recovery of cardiac contractile force. Biochem Pharmacol 40(5):1137–1143. https://doi.org/10.1016/0006-2952(90)90504-E

Schwarz MKV (1994) Terpen-Biosynthese in Ginkgo biloba eine überraschende Geschichte

Shu M, Hu XR, Hung ZA, Huang DD, Zhang S (2016) Effects of tanshinone IIA on fibrosis in a rat model of cirrhosis through heme oxygenase-1, inflammation, oxidative stress and apoptosis. Mol Med Rep 13(4):3036–3042. https://doi.org/10.3892/mmr.2016.4886, PMID 26936326

Simonneau G, Robbins IM, Beghetti M, Channick RN, Delcroix M, Denton CP et al (2009) Updated clinical classification of pulmonary hypertension. J Am Coll Cardiol 54(1 Suppl): S43–S54. https://doi.org/10.1016/j.jacc.2009.04.012, PMID 19555858

Singh M, Yadav S, Kumar M, Saxena S, Saraswat D, Bansal A et al (2018) The MAPK-activator protein-1 signaling regulates changes in lung tissue of rat exposed to hypobaric hypoxia. J Cell Physiol 233(10):6851–6865. https://doi.org/10.1002/jcp.26556, PMID 29665093

Siques P, Pena E, Brito J, el Alam S (2021) Oxidative stress, kinase activation, and inflammatory pathways involved in effects on smooth muscle cells during pulmonary artery hypertension under hypobaric hypoxia exposure. Front Physiol 12:690341. https://doi.org/10.3389/fphys.2021.690341, PMID 34434114

Sonkusare S, Palade PT, Marsh JD, Telemaque S, Pesic A, Rusch NJ (2006) Vascular calcium channels and high blood pressure: pathophysiology and therapeutic implications. Vasc Pharmacol 44(3):131–142. https://doi.org/10.1016/j.vph.2005.10.005, PMID 16427812

Sturrock A, Cahill B, Norman K, Huecksteadt TP, Hill K, Sanders K et al (2006) Transforming growth factor-β1 induces Nox4 NAD(P)H oxidase and reactive oxygen species-dependent proliferation in human pulmonary artery smooth muscle cells. Am J Physiol Lung Cell Mol Physiol 290(4):L661–L673. https://doi.org/10.1152/ajplung.00269.2005, PMID 16227320

Sui H, Zhao J, Zhou L, Wen H, Deng W, Li C et al (2017) Tanshinone IIA inhibits β-catenin/VEGF-mediated angiogenesis by targeting TGF-β1 in normoxic and HIF-1α in hypoxic microenvironments in human colorectal cancer. Cancer Lett 403:86–97. https://doi.org/10.1016/j.canlet.2017.05.013, PMID 28602978

Tabas I, Bornfeldt KE (2016) Macrophage phenotype and function in different stages of atherosclerosis. Circ Res 118(4):653–667. https://doi.org/10.1161/CIRCRESAHA.115.306256, PMID 26892964

Tang F, Wu X, Wang T, Wang P, Li R, Zhang H et al (2007) Tanshinone II A attenuates atherosclerotic calcification in rat model by inhibition of oxidative stress. Vasc Pharmacol 46(6):427–438. https://doi.org/10.1016/j.vph.2007.01.001, PMID 17337361

Wang JW, Wu JY (2010) Tanshinone biosynthesis in Salvia miltiorrhiza and production in plant tissue cultures. Appl Microbiol Biotechnol 88(2):437–449. https://doi.org/10.1007/s00253-010-2797-7, PMID 20694462

Wang X, Morris-Natschke SL, Lee KH (2007) New developments in the chemistry and biology of the bioactive constituents of Tanshen. Med Res Rev 27(1):133–148. https://doi.org/10.1002/med.20077, PMID 16888751

Wang J, Jiang Q, Wan L, Yang K, Zhang Y, Chen Y et al (2013) Sodium tanshinone IIA sulfonate inhibits canonical transient receptor potential expression in pulmonary arterial smooth muscle from pulmonary hypertensive rats. Am J Respir Cell Mol Biol 48(1):125–134. https://doi.org/10.1165/rcmb.2012-0071OC, PMID 23065131

Wang et al (2015) Herba epimedii: an ancient chinese herbal medicine in the prevention and treatment of osteoporosis. Curr Pharm Des 22(3):328–349. https://doi.org/10.2174/1381612822666151112145907

Wang CH, Zheng LP, Tian H, Wang JW (2016a) Synergistic effects of ultraviolet-B and methyl jasmonate on tanshinone biosynthesis in Salvia miltiorrhiza hairy roots. J Photochem Photobiol B 159:93–100. https://doi.org/10.1016/J.JPHOTOBIOL.2016.01.012, PMID 27043259

Wang L, Li Y, Guo Y, Ma R, Fu M, Niu J et al (2016b) Herba Epimedii: an ancient Chinese herbal medicine in the prevention and treatment of osteoporosis. Curr Pharm Des 22(3):328–349. https://doi.org/10.2174/1381612822666151112145907, PMID 26561074

Wang N, Chang Y, Chen L, Guo YJ, Zhao YS, Guo QH et al (2017) Tanshinone IIA protects against chronic intermittent hypoxia-induced myocardial injury via activating the endothelin 1 pathway. Biomed Pharmacother 95:1013–1020. https://doi.org/10.1016/j.biopha.2017.08.036, PMID 28922718

Wang L, Cheng L, Zhang B, Wang N, Wang F (2019) Tanshinone prevents alveolar bone loss in ovariectomized osteoporosis rats by up-regulating phosphoglycerate dehydrogenase. Toxicol Appl Pharmacol 376:9–16. https://doi.org/10.1016/j.taap.2019.05.014, PMID 31108107

Wang X, Yang Y, Liu X, Gao X (2020) Pharmacological properties of tanshinones, the natural products from Salvia miltiorrhiza. Adv Pharmacol 87:43–70. https://doi.org/10.1016/bs.apha.2019.10.001

Wu DM, Wang YJ, Han XR, Wen X, Li L, Xu L et al (2018) Retracted. J Cell Mol Med 22(6): 3058–3072. https://doi.org/10.1111/jcmm.13557, PMID 29524303

Xu W, Yang J, Wu LM (2009) Cardioprotective effects of tanshinone IIA on myocardial ischemia injury in rats. Pharmazie 64(5):332–336, PMID 19530445

Xu Z, Chen L, Xiao Z, Zhu Y, Jiang H, Jin Y et al (2018) Potentiation of the anticancer effect of doxorubicinin drug-resistant gastric cancer cells by tanshinone IIA. Phytomedicine 51:58–67. https://doi.org/10.1016/j.phymed.2018.05.012, PMID 30466628

Yadav S, Kalra N, Ganju L, Singh M (2017) Activator protein-1 (AP-1): a bridge between life and death in lung epithelial (A549) cells under hypoxia. Mol Cell Biochem 436(1–2):99–110. https://doi.org/10.1007/s11010-017-3082-1, PMID 28589371

Yadav S, Singh M, Singh SN, Kumar B (2020) Tanshinone IIA pretreatment promotes cell survival in human lung epithelial cells under hypoxia via AP-1-Nrf2 transcription factor. Cell Stress Chaperones 25(3):427–440. https://doi.org/10.1007/s12192-020-01083-3, PMID 32144684

Yang D, Ma P, Liang X, Wei Z, Liang Z, Liu Y et al (2012) PEG and ABA trigger methyl jasmonate accumulation to induce the MEP pathway and increase tanshinone production in Salvia miltiorrhiza hairy roots. Physiol Plant 146(2):173–183. https://doi.org/10.1111/j.1399-3054.2012.01603.x, PMID 22356467

Yang JX, Pan YY, Ge JH, Chen B, Mao W, Qiu YG et al (2016) Tanshinone II A attenuates TNF-α-induced expression of VCAM-1 and ICAM-1 in endothelial progenitor cells by blocking activation of NF-κB. Cell Physiol Biochem 40(1–2):195–206. https://doi.org/10.1159/000452537, PMID 27855363

Yintao Y (2010) Effects of cryptotanshinone on proliferation and apoptosis of Hela cell line of cervical cancer. Chin J Chin Matera Med 35:35. https://doi.org/10.4268/cjcmm20100125

Zhang Y, Zhang L, Chu W, Wang B, Zhang J, Zhao M et al (2010) Tanshinone IIA inhibits miR-1 expression through p38 MAPK signal pathway in post-infarction rat cardiomyocytes. Cell Physiol Biochem 26(6):991–998. https://doi.org/10.1159/000324012, PMID 21220930

Zhang J, Wang J, Jiang JY, Liu SD, Fu K, Liu HY (2014) Tanshinone IIA induces cytochrome c-mediated caspase cascade apoptosis in A549 human lung cancer cells via the JNK pathway. Int J Oncol 45(2):683–690. https://doi.org/10.3892/ijo.2014.2471, PMID 24888720

Zhang N, Dong M, Luo Y, Zhao F, Li Y (2018a) Danshensu prevents hypoxic pulmonary hypertension in rats by inhibiting the proliferation of pulmonary artery smooth muscle cells via TGF-β-smad3-associated pathway. Eur J Pharmacol 820:1–7. https://doi.org/10.1016/j.ejphar.2017.12.010, PMID 29221952

Zhang Y, Zhang S, Chen X (2018b) Tanshinone IIA protects against cardiac fibrosis through inhibition of β-tubulin expression. J Biol Regul Homeost Agents 32(6):1451–1455, PMID 30 574749

Zhou L, Zuo Z, Chow MSS (2005) Danshen: an overview of its chemistry, pharmacology, pharmacokinetics, and clinical use. J Clin Pharmacol 45(12):1345–1359. https://doi.org/10.1177/0091270005282630, PMID 16291709

Zhou D, Li Z, Zhang L, Zhan C (2012) Inhibitory effect of tanshinone II A on TGF II-β1-induced cardiac fibrosis. J Huazhong Univ Sci Technolog Med Sci 32(6):829–833. https://doi.org/10.1007/s11596-012-1042-2, PMID 23271281

14 Acclimatization: A Way to Cope Up with a Stressful Environment

Abhishek Bharadwaj, Pooja Chaudhary, Rajinder Kumar Gupta, and Sachin Kumar

Abstract

Climate is the most important ecological factor determining growth and development. The strategy by which mammals and birds maintain body temperature irrespective of the surrounding ambient temperature is by three mechanisms requiring thermal gradient (conduction, convection, and radiation); collectively called "sensible heat loss.". When the ambient environment exceeds body temperature these routes of exchange are lost, and only heat loss is through evaporation (sweating and panting). Animals have developed coping mechanisms for these environmental stressors in their biological systems. These responses are known as acclimation, acclimatization, and adaptation. Acclimatization is induced by the environment and is considered phenotypic, not genotypic. The responses decay if the stress is removed. Acclimatization tends to improve animal fitness to the environment and is not a process that involves evolutionary adaptations or natural selection. Acclimatization is a process taking several days to weeks and this process occurs via homeorhetic and not homeostatic mechanisms.

Keywords

Acclimatization · Heat stress · Cold stress

A. Bharadwaj (✉) · P. Chaudhary · R. K. Gupta · S. Kumar
Defence Institute of Physiology and Allied Sciences (DIPAS), Defence Research and Development Organization (DRDO), Delhi, India
e-mail: abhishek.dipas@gov.in

R. Tulsawani, D. Vohora (eds.), *Adaptation under Stressful Environments through Biological Adjustments and Interventions*,
https://doi.org/10.1007/978-981-99-7652-2_14

14.1 Introduction

Habitability in extreme terrestrial environments is a challenging task for humans. The concern is both survival and optimal performance under such environments in everyday life, occupation, and sports. This requires coping with specific environmental stressors through adaptive biological processes, such as adaptation, acclimatization, acclimation, and habituation (Folk 1966; IUPS 1987; Gunga 2015).

For meeting challenges of extreme climate and geographical constraints, humans have the principal task of maintaining homeostasis and operating equilibrium for attaining metabolic balance. This equilibrium is obtained by adjustments of physiological alterations, which are reflected in morphology. Adjustments include physiological, morphological, and behavioral responses (Moen 1968).

14.2 Morphological Adjustments

14.2.1 Body Size and Proportions

A principle based on Bergmann's size rule states that "stating that similar bodies but larger has smaller skin surface in proportion to mass" (Allee and Schmidt 1951). A bigger body increases body-mass/body-surface ratio and is of some advantage where retention of metabolically generated heat is desirable. The larger size of chest and lung is found in high-altitude natives, eventually enabling higher ventilatory capacity.

14.2.2 Subcutaneous Fat

Baker (1959) reported if blood flows to the body surface greater, the body is insulated according to the thickness and distribution of subcutaneous fat. Such insulation appears to be advantageous where retention of internally generated heat is prime, whereas it seems disadvantageous for heat dissipation.

14.2.3 Skin Pigmentation

Gloger's pigmentation rule is followed in humans, which states that melanin increases among those living in warmer, humid climates, whereas dwellers of arid regions tend to be yellow or red-brown in pigmentation. In areas of extreme heat and light, heavily pigmented skin is an advantage.

14.2.4 Sweat Glands

The only means of dissipating body heat in an ambient temperature warmer than the body is by evaporation of sweat. Thus in areas of extreme heat, efficient sweat physiology is necessary for survival. Whereas, it is very important for cold-climate habitants to deal with the problem of insensible respiration, after a rate of energy production based on ambient temperature is reached, with actual thermal sweating (Mitciiei and Edman 1951).

14.3 Physiological Adjustments

In heat stress, cooling mechanisms include increased skin blood flow and sweating, simultaneously, circulatory measures get initiated to maintain central blood pressure. In cold stress, heat loss is prevented by peripheral vasoconstriction and heat production by shivering. In hypoxia stress, which is the mark of high altitude, hyperventilation, hemo-concentration, and stimulated erythropoiesis are physiological responses to compensate for oxygen delivery to tissues (West 2006, 2017).

14.4 Behavioral Adjustments

Behavioral adjustments are important for humans in extreme environmental conditions. There are advances in technology pertaining to appropriate clothing, shelter, air conditioning, and oxygen enrichment (Makinen 2010; West 2015, 2017). Various adaptive processes take place to cope with stressors of extreme terrestrial environments like cold, heat, and hypoxia (high altitude). The following decade was labeled the "Decade of Extremes" by the World Meteorological Organization (Coumou and Rahmstorf 2012). The subsequent decade continued this alarming trend with 2014 being the hottest (globally averaged) year.

14.5 Physiological Adjustments to Heat

Lower sweating rates were observed among Nigerians as compared with European troops, having lower core temperatures under moderate stress conditions (Thomson 1954). Sweating starts at lower core body temperature with higher production rate but, after prolonged exposure there is fall in sweating due to sweat-gland fatigue (Ladell 1951). Nigerians susceptibility to heat cramps and prostration is less compared to Europeans despite low salt intake (Ladell 1951). Whereas, Bushmen hunters of Kalahari and acclimatized Europeans were at par in heat tolerance under moderate stressing. Under severe stress, it is believed that Bushmen would have thermo-regulatory advantage (Wyndham and Morrison et al. 1956).

An ideal "Desert Man" has been described by Baker (1958a) who stated that the person would be linear built, having low subcutaneous fat with brunet skin color and high tanning potential. This description fits the "Solubbies" of the Arabian desert exactly. Similarly, although no physiological data were collected, Roberts (1960) considered "Nilotes of Savannas of Sudan" to be best adapted in the world to hot-dry heat. So from the observations, Baker inferred that Afro-American of predominantly West African origin may have superior tolerance to hot-humid climates but not to hot-dry climates.

14.6 Physiological Adjustments to Cold

It is known for some time that the lowest air temperature at which naked men maintain core temperature without metabolic increase is 25–27 °C, and this is considered as "Thermo-neutral temperature." Relative to many other mammalian species, man has limited physiological means of defense against cold. Physiological adjustment to cold is brought about by two basic mechanisms:

1. Reduction of heat loss (heat conserving mechanisms) and
2. Increase in heat production when ambient temperature continues to fall below a "critical" value.

The ability to maintain constant core temperature even during changes in metabolic heat production and ambient temperature is termed as "thermal homeostasis." A sudden rise of ambient temperature results in impaired ability to perform work. If such conditions prevail for several days there is a gradual return of the ability to perform work with minimal discomfort, thus becoming "acclimatized to heat.".

Thermo-regulation and acclimatization are core components of humans' mechanism to withstand extreme environmental temperatures.

A measured response to changes in the environment of an organism can be termed acclimatization. Such responses are to an extent habitual and reversible if conditions return to the initial stage. Individuals regulate internal processes and thus sustain within the range of environmental changes. This rapid regulation is limited to a small range of environmental changes. Acclimatization does not affect the genetic mechanisms of the organism. In dealing with acclimatization, climatic influence can be treated under adjustments to (a) temperature, (b) humidity, (c) salinity, (d) light, (e) pressure, and (f) certain chemical substances.

Organisms do not have unlimited combinations for adaptations. For example, resulting in acclimatization to hypoxia, the capacity of blood to transport oxygen gets improved by increasing red blood cells; similarly under pathological conditions of emphysema, inadequate oxygen supply is compensated by similar polycythemia.

Thus, acclimatization is characterized as a cycle where an organism changes its behavior or physiological responses because of changes in the current circumstance. It is usually reversible across most cases.

In other words:

- Change in physiological, anatomical, or morphological changes in an organism due to environmental pressure is known as acclimatization.
- These changes are reversible when a living organism is placed back in a favorable environment.
- Changes occur within a short period of time (days to a week).

14.7 Acclimatization to Heat

Protection against hyperthermia is effectively accomplished by two mechanisms—(a) peripheral vasodilation and (b) sweating.

Despite these capacities, it is common that sudden hot weather results in an impaired ability to work. If adverse conditions prevail for several days, there is a gradual ability to return to work with little or no discomfort. This improvement is not merely psychological, but well-defined physiological adaptations. Simultaneously, other parallel supporting adaptations also occur thus improving the ability of the individual to counter the decrement in work performance. Cardiovascular response improves with continued heat stress, it is generally felt that improvement is the result of increased blood flow (Robinson et al. 1941).

The role of sweat rate has been extensively studied for acclimatization. McArdle (1945) reported that with successive heat exposure, the initiation of sweating is at a lower core body temperature.

Certain physiological indices have been identified for acclimatization. When acclimatized, the individual performs a given workload under heat with (a) lower pulse rate, (b) lower core temperature, (c) lower skin temperature, and (d) higher sweat rate compared to non-acclimatized. Based on these, the process of acclimatization is considered a physiological entity. Studies by Bean and Eichna (1943) indicated that heat acclimatization:

1. begins from the first day and is well developed by the fourth day,
2. is enhanced by the good physical condition,
3. is enhanced by exercise in the heat. Just exposure with no exercise results in mild acclimatization.
4. Short daily exposure along with activity gives appreciable acclimatization.
5. Acclimatization is the same for short, medium, or long periods with a lesser workload.
6. Acclimatization is retained for 1–2 weeks after which there is variable degradation.

Works by Robinson et al. (1943) confirmed these findings and found that 80% acclimatization is attained within the first 2–3 days. They also reported that after acclimatization, working was more efficient as the energy cost reduced.

From all studies by different researchers, it can be conferred that the process of acclimatization behaves in many ways as a single physiological response and can readily be characterized in terms of rate, retention, and degree.

14.8 Acclimatization to Cold

Relative to many other mammalian species, humans have limited physiological means of defense against the cold. Although the response to cold stress is essentially behavioral, there are situations where these cannot be used to protect against the cold. In the life of an individual, there are two particular periods where susceptibility to cold is increased, during a few months after birth and at old age.

Physiological response varies with the age and physical fitness of an individual (Buskirk 1978), along with the intensity of cold and the duration of exposure, whether in air or water (Carlson and Hsieh 1974; Folk 1974; Webster 1974; Alexander 1979; Stanier et al. 1984).

The physiological effects of cold are also largely modified in cold-acclimated or cold-acclimatized subjects.

Heat-conservation mechanisms, termed physical thermoregulatory mechanisms, are the first to act on cold exposure. They include alterations in peripheral blood flow and countercurrent heat exchange between arterial and venous blood in limbs.

When the heat-conservation mechanism is unable to maintain "normal" body-core temperature, metabolic processes get activated, to generate more heat under thermoneutral conditions in order to compensate for increased heat loss from the body.

Maintaining homeothermia in a cold environment is an expensive process energetically. It requires additional energy expenditure and substrate, and cold thermogenesis may decline significantly during periods of food deprivation, particularly in infants (Brooke and Cocks 1973; Folk 1981) or slender individuals, leading to extreme cases of hypothermia (Maclean and Emslie-Smith 1977).

14.9 Physiological Responses to Chronic Cold Exposure: Adaptive Changes

The definition of acclimatization to cold has been evolving since Hart's (1957) introduction of the term in 1957. As mentioned earlier, effective maintenance of homeothermy during cold exposure is brought about by two basic mechanisms:

1. Increased heat production through the enhancement of metabolic processes (metabolic adaptation) and
2. Decreased heat loss by means of increased insulation (insulative adaptation).

Besides, with prolonged or repeated cold exposures, an increased tolerance to the lowered heat content of the body develops, causing a shift of threshold for shivering to lower temperatures (hypothermic adaptation).

The temperature gradient between the core and skin surface can be modified by alterations in the insulation of skin and superficial tissues, which helps regulate heat loss by convection and conduction. With the body being exposed to cold, cutaneous blood flow promptly reduces by intense constriction of skin resistance vessels, arteriovenous anastomoses, and veins in the skin. By this mechanism, heat transfer from the core to the body surface is attenuated. Skin temperature decreases and less heat is lost to the environment. Cutaneous vasoconstriction also decreases heat loss by directing venous blood through deep veins that are close to the arteries. In this way arterial blood, entering the limb at a relatively high temperature comes in contact with cooled venous blood.

When cold is severe, the skin temperature dips below 10 °C, and cutaneous blood vessels in areas with arterio-venous anastomoses (e.g., in the hands and feet) dilate periodically. Vasodilations are particularly pronounced when the core is warm and only peripheral parts are exposed to cold. Under such conditions, brief episodes of vasodilation in the skin help to maintain the normal function of the extremities and prevent severe damage to their tissues.

The pattern of cold-induced vasodilation (CIVD) was first described in 1930 by Lewis who called it the hunting phenomenon (Lewis 1930). Compared with other mammalian species, human's capacity for CIVD is limited, and as such frostbite or prolonged ischemia in extremities have been frequently reported when unacclimatized men were exposed to severe cold. Increase in muscle tone and involuntary muscle contractions: Shivering-Involuntary contractions of the skeletal muscle groups throughout the body is an important source for regulation of heat production and can be initiated rapidly and continued for a long period of cold exposure. Since very little external work is performed during shivering, almost all energy released appears promptly as heat. An increase in muscle tone, termed the "preshivering" or "thermal" muscle tone, is primary the muscle response to cold exposure.

Primary "center" involved in control of shivering is located in the dorsomedial posterior hypothalamus, near the wall of third ventricle. Besides the primary center, a few secondary centers, involved mainly in the suppression of shivering, were demonstrated in the brain. There are at least four suppression regions for shivering:

1. in the ventro-lateral posterior hypothalamus
2. in the ventro-medial forebrain septum
3. in the ventro-medial preoptic region of the anterior hypothalamus (4) at some points of the cerebral cortex

14.10 Acclimatization to Local Cold Exposure

It has been proved that people who habitually expose their hands to cold respond to local cooling with less pronounced cutaneous vasoconstriction, and with more rapid onset of vasodilation than that occurs in unacclimatized men. Thus, they tolerate cold-stress better than men unadapted to cold.

Increase in heat production by shivering or by voluntary muscular activity makes little contribution to the maintenance of thermal balance in cold water. During muscular contractions, muscle blood flow is markedly increased resulting in higher heat conductance to the skin. This, in turn, elevates heat loss by an amount proportional to the difference between the body core and water temperature.

14.11 Limits to Acclimatization

Acclimatization offers protection against heat exposure, however, that protection is bounded by caveats. As depicted in the Thermoregulatory Model, core temperature is the combined result of BMR, physical intensity and duration, subsequent internal heat generation, and external climatic conditions contributing to heat exchange, level of acclimatization, and capacity for dissemination of excess heat to the environment. These all have an effect on resultant heat gain and determine whether core temperature surpasses thresholds. Thus, progression to morbidity and death, or not, depends on the interrelation between heat generation, ambient climatic factors, and the body's capacity to regulate core temperature within a safe range. Physiological cooling limitations rest with upper limits of heat loss mechanisms as maximal sweat rates, and maximal cardiac and perfusion responses.

There is temperature–humidity combinations beyond which body cooling will not be possible regardless of the state of acclimatization, even with minimal exertion. Hence, there exists maximum heat exposure level and duration, beyond which even full acclimatization will not prevent the lethal rise in core body temperature.

Breaching those upper thermo-tolerance limits by one of the following: (a) exposure to particularly intense heat, (b) prolonged heat exposure, or (c) exercise of high intensity in the heat will induce marked hyperthermia.

Acclimatization is not a permanent state. Acclimatization capacity decays in the absence of ongoing heat exposure (Weller et al. 2007). Furthermore, individual variations are very much prevalent for time required to acclimatize, duration required for the acclimatization to decay and again to re-acclimatize.

In the long term, specific morphological, physiological, and behavioral adjustments enable people to live an almost normal life in such extreme areas. In contrast to the strategies adopted by indigenes, people acutely exposed to extreme conditions may experience impaired performance, at least initially, and different mechanisms of adjustment will later on enable them to cope with cold, hot, and/or hypoxic environments.

References

Alexander G (1979) Cold thermogenesis. Int Rev Physiol 111(D):43–155

Allee WC, Schmidt KP (1951) Ecological animalgeography, 2nd edn. Wiley, New York, p 154

Baker PT (1958a) Racial differences in heat tolerance. Am J Phys Anthropol 16(3):287–306. https://doi.org/10.1002/ajpa.1330160303, PMID 13649898

Baker PT (1959) American Negro-white differences in the thermal insulative aspects of body fat. Hum Biol 31(4):316–324, PMID 13796025

Bean WB, Eichna LW (1943) Performance in relation to environmental temperature. Fed Proc 2: 144

Brooke G, Cocks T (1973) The metabolic rate of malnourished children. J Physiol (Lond) 231:18

Buskirk ER (1978) Cold stress: a selective review. Environmental stress: individual human adaptation. Academic, New York, pp 249–266

Carlson LD, Hsieh ACL (1974) Temperature and humidity, part A. Cold. In: Balfour N, Slonim AR (eds) Environmental physiology. CV Mosby Company, Saint Louis, pp 61–84

Coumou D, Rahmstorf S (2012) A decade of weather extremes. Nat Clim Change 2(7):491–496. https://doi.org/10.1038/nclimate1452
Folk G Jr (1966) Introduction to environmental physiology. Lea & Febiger, Philadelphia
Folk GE (1974) Textbook of environmental physiology. H. Kimpton Publishing, London
Folk GE Jr (1981) Climatic change and acclimatization. In: Cena K, Clark JA (eds) Bioengineering, thermal physiology and comfort: studies in environmental science, vol 10. Elsevier Scientific Publishing Company, Amsterdam, pp 157–168
Gunga H (2015) Human physiology in extreme environments. Elsevier, Amsterdam
Hart JS (1957) Climatic and temperature-induced changes in the energetics of homeotherms. Rev Can Biol 16(2):133–174, PMID 13454073
IUPS T (1987) Glossary of terms for thermal physiology. Pflugers Arch. 2nd ed. Revised by the commission for thermal physiology of the International Union of Physiological Sciences (IUPS thermal commission) 410(4–5):567–587. https://doi.org/10.1007/BF00586542
Ladell WS (1951) Assessment of group acclimatization to heat and humidity. J Physiol 115(3): 296–312. https://doi.org/10.1113/jphysiol.1951.sp004672, PMID 14898513
Lewis T (1930) Observations upon the reactions of the vessels of the human skin to cold. Heart 15: 177–208
Maclean D, Emslie-Smith D (1977) Accidental hypothermia. Blackwell Scientific Publications, Oxford
Makinen TM (2010) Different types of cold adaptation in humans. Front Biosci (Schol Ed) 2(3): 1047–1067. https://doi.org/10.2741/s117, PMID 20515840
Mitciiei HH, Edman E (1951) Nutrition and climatic stress. Thomas, Springfield, p 12
Moen AN (1968) The critical thermal environment: a new look at an old concept. BioScience 18(11):1041–1043. https://doi.org/10.2307/1294554
Robinson S, Dill DB, Wilson JW, Nielson M (1941) Adaptations of white men and Negroes to prolonged work in humid heat. Am J Trop Med 21(2):261–288
Robinson S, Turrell ES, Belding HS, Horvath SM (1943) Rapid acclimatization to hot climates. Am J Physiol 140:168–176
Stanier MW, Mount LE, Bligh J (1984) Energy balance and temperature regulation. Cambridge University Press, Cambridge
Thomson ML (1954) A comparison between the number and distribution of functioning eccrine sweat glands in Europeans and Africans. J Physiol. 123(2):225–233. https://doi.org/10.1113/jphysiol.1954.sp005045, PMID 13143505
Webster AJF (1974) Physiological effects of cold exposure. Part 2. Environmental physiology/ physiology. In: Butterworths Robertshaw D (ed) Series one, vol 7. University Park Press, Baltimore, pp 33–69
Weller AS, Linnane DM, Jonkman AG, Daanen HAM (2007) Quantification of the decay and re-induction of heat acclimation in dry-heat following 12 and 26 days without exposure to heat stress. Eur J Appl Physiol 102(1):57–66. https://doi.org/10.1007/s00421-007-0563-z, PMID 1 7891541
West JB (2006) Human responses to extreme altitudes. Integr Comp Biol 46(1):25–34. https://doi.org/10.1093/icb/icj005, PMID 21672720
West JB (2015) A strategy for oxygen conditioning at high altitude: comparison with air conditioning. J Appl physiol 119:719–723. https://doi.org/10.1152/japplphysiol.00421.2015
West JB (2017) Physiological effects of chronic hypoxia. N Engl J Med 376(20):1965–1971. https://doi.org/10.1056/NEJMra1612008, PMID 28514605
Wyndham CH, Morrison JF (1956) Heat regulation of MaSarwa (Bushmen). Nature 178:869–870

Part II

Oxidative and Genotoxic Stress: Health Implications and/or Approaches to Management

Genotoxin Stress and the Role of Alkaloids 15

Vaishali Yadav and Divya Vohora

Abstract

Humans are subjected to an array of external and internal stimuli during their lives, which injure genetic material and generate somatic mutations that lead to accumulating genomic stress. Several human pathologies like chronic degenerative diseases, including hepatic, neurodegenerative, and cardiovascular disorders; diabetes; arthritis; cancer; chronic inflammation; and aging, are associated with genotoxic stress. In the past few decades, a great deal of research has been done to find substances that can shield humans from DNA damage and its impact. Natural substances continue to show a beneficial impact on genome integrity. Scientists have identified and reviewed unique bioactive plant-derived compounds, such as flavonoids, phenolic compounds, and carotenoids, that can mitigate the impact of genotoxic stress; however, the genoprotective potential of alkaloids is not being reviewed so far. Herbs high in alkaloidal content have a plethora of pharmacological traits, which are utilized as drugs, and are significant to the pharmaceutical sector. This chapter summarizes the literature on the genoprotective/antigenotoxic action of alkaloids acting against endogenously and exogenously produced genetic damage. The current knowledge regarding the function of alkaloids to guard the genome can be exploited to create dietary plans and natural health products that are beneficial in the sequel of chemotherapeutic-associated toxicities.

V. Yadav
Neurobehavioral Pharmacology Laboratory, Department of Pharmacology, School of Pharmaceutical Education and Research (SPER), Jamia Hamdard, New Delhi, India

D. Vohora (✉)
Neurobehavioral Pharmacology Laboratory, Department of Pharmacology, School of Pharmaceutical Education and Research (SPER), Jamia Hamdard, New Delhi, India
e-mail: dvohra@jamiahamdard.ac.in

R. Tulsawani, D. Vohora (eds.), *Adaptation under Stressful Environments through Biological Adjustments and Interventions*,
https://doi.org/10.1007/978-981-99-7652-2_15

Keywords

Genotoxicity · Genoprotective · Anti-genotoxicity · Alkaloids · Piperine

15.1 Introduction

Deoxyribonucleic acid (DNA) in living organisms is constantly exposed to harmful substances. It is assessed that every cellular component in the human body experiences millions of genomic lesions per day (Luca et al. 2016). Genotoxic stress can be defined as a situation that initiates DNA damage compromising the cellular genomic integrity. The capacity of various substances to cause harm to genetic material is known as genotoxicity, while genome instability refers to a series of events associated with an elevated rate of DNA damage and associated mutations, arising from defects in the DNA damage response, DNA double-strand breaks, DNA replication, cell cycle control, or chromosome segregation (Coates et al. 2005; Kovalchuk 2021). Genotoxic stress can be induced by intrinsic processes (such as reactive oxygen species generated during physiological and pathogenic progressions) and extrinsic compounds (such as UV light, ionizing radiation, anti-cancer drugs, and even air, water, and food pollutants). Human exposure to numerous genotoxic stressors has escalated considerably in recent years. Specifically, cancer patients undertaking radiotherapy and chemotherapy are overexposed to cyto(geno)toxic agents. Furthermore, several pathological conditions exhibit genome instability events and DNA damage, such as aging, carcinogenesis, immune deficiencies, infertility, cardiovascular diseases, metabolic syndrome, and neuromuscular and neurodegenerative diseases like Alzheimer's disease (Georgoulis et al. 2017). Therefore, it is of significant interest to identify compounds, particularly from traditional medicinal plants, that can guard against genotoxic harm and its consequences.

Several reviews related to the anti-genotoxic activity of phytochemicals like flavonoids, polysaccharides, phenolic compounds, and carotenoids are already available in the public domain (Hosseinimehr 2010; George et al. 2017; Izquierdo-Vega et al. 2017; López-Romero et al. 2018); however, we did not find any such information regarding genoprotective action of alkaloids. In view of the above, this chapter summarizes research studies on the genoprotective/antigenotoxic effects of alkaloids against both endogenous and exogenously induced genetic damage, specifically, DNA damage induced by various carcinogenic drugs. Further, this chapter also sheds light on sources of genotoxic stress, physiological response to genotoxic stress, general methods to detect DNA damage, the purpose of pharmacological interventions, and the role of specific alkaloids in the same.

15.2 Source and Types of Genotoxic Stress

A cell experiences thousands to lakhs of molecular lesions per day as a result of external influences and regular metabolic activities. Although this only accounts for 0.000165% of the roughly 6 billion (3 billion base pair) bases that make up the human genome, if left unrepaired, it could lead to mutations in important genes (such as tumor suppressor genes), which would significantly increase the risk of tumor development and disease states like cancer (Luca et al. 2016).

There are various types of DNA damage that can happen as a result of either regular cellular functions or cells being exposed to external factors that cause DNA damage. DNA bases can be damaged by: (1) Oxidative processes, (2) Alkylation of bases, (3) Base loss caused by the hydrolysis of bases, (4) Bulky adduct formation, (5) DNA crosslinking, and (6) DNA strand breaks, including single- and double-stranded breaks. Table 15.1 represents DNA damage inducers from various sources. Below is a description of genetic lesions induced by several factors:

15.2.1 Oxidative Damage

Reactive oxygen species (ROS) have the ability to produce substantial physiological stress and harm cells, comprising DNA oxidative injury. Among the most susceptible and electrophilic reactive oxygen species are the hydroxyl radicals ($^{\bullet}OH$), which can be created by UV and ionizing radiations as well as other radicals generated during enzymatic activities. Hydroxyl radicals subsequently react with nucleo-bases, as guanine has the lowest ionization potential, it is quickly oxidized to form 8-oxo-7,8-dihydroguanine (8-oxoG); a major prevalent genomic lesion and a potential hallmark of oxidative stress (Cooke et al. 2003).

15.2.2 Alkylation of Bases

Alkylating substances are abundant in ecosystems and are generated internally by cells as waste products of metabolism. They cause cytotoxic lesions, and mutagenic or neutral lesions to be incorporated into genomic bases. Cytotoxic lesions inhibit the genomic replication process, interfere with the transcription process, or activate the cascade for apoptosis, while mutagenic lesions lead to miscoding and mutations in a newly synthesized genome. The most common type of alkylation is methylation with the major products including N7-methylguanine (7meG), N3-methyladenine (3meA), and O6-methylguanine (O6meG) (Kondo et al. 2010).

15.2.3 Base Loss

When the Nitrogen-glycosyl bond connecting the nitrogenous base and the sugar-phosphate backbone is disrupted by water molecules spontaneously or is broken by a

Table 15.1 Compounds inducing genotoxic stress and their DNA-damaging effects

S. no.	Compounds	DNA damaging action	Type of DNA damage	References
Environmental toxicants				
1	Exposure of Ultraviolet B radiation	Formation of dimeric DNA photoproducts, specifically cyclobutane pyrimidine dimers and pyrimidine-6,4-pyrimidone photoproducts	DSSB and DDSB	Chatterjee and Walker (2017)
2	Gamma rays	ROS formation	DSSB and DDSB	
3	Hydrogen peroxide	Formation of hydroxyl radicals via a Fenton-type reaction	DSSB, DDSB, and CA	
4	Polycyclic aromatic hydrocarbons	Activation of the aryl hydrocarbon receptor	DNA adduct formation	
5	Styrene	Generation of electrophilic compound: styrene-7,8-oxide	DSSB and DDSB	
6	*N*-nitrosamines	Induction of CYP450	DNA adduct formation	
7	Heterocyclic aromatic amines	Inducer of CYP450 1A superfamily, *N*-acetyltransferases, and sulfotransferases	DSSB, DDSB, CA, and SCE	
8	Heterocyclic aromatic nitro derivatives	Inducer of cytochrome-P450	DSSB, alkali-labile sites formation, and CA	
9	Asbestos	Generation of ROS	DSSB, DDSB	
Chemotherapeutic agents				
10	Alkylating agents (nitrogen mustards)	Base alkylation-monofunctional DNA adducts	Inhibits DNA replication process	Swift and Golsteyn (2014)
11	Platinum drugs (cisplatin, carboplatin, and oxaliplatin)	Monofunctional DNA adducts	Inhibits DNA replication process	
12	Topoisomerase poisons (Camptothecin and etoposide)	Blocks topoisomerase enzymes in DNA	DSSB and DDSB	
13	Antimetabolites (5-fluorouracil)	Misincorporates into DNA	Inhibits DNA replication process	

DSSB DNA single-strand breaks, *DDSB* DNA double-strand breaks, *ROS* reactive oxygen species, *CA* chromosomal aberrations, *SCE* sister chromatid exchanges genomic glycosylase to produce an intermediate in the base excision repair pathway, abasic or AP (apurinic/apyrimidic) sites are continually generated in the DNA. These are fundamentally unstable and easily undergo a beta-elimination reaction that destroys the phosphodiester link of the remaining deoxyribose, resulting in single-strand breaks (Chatterjee and Walker 2017).

15.2.4 Bulky Adduct Formation

Certain substances are physiologically responsive and generate covalent bonds with biological macromolecules like DNA and proteins, which will result in the development of big, bulky appendages, or adducts, that split off from the original molecule. For example, benzo[a]pyrene is exposed to humans via cigarette smoking, activated into (+) benzo[a]pyrene-7,8-dihydrodiol-9,10-epoxide, which generates a DNA adduct with guanine residues and disrupts the genomic backbone (Poirier 2012).

15.2.5 DNA Crosslinking

When cross-bridging chemicals covalently join two nucleotide residues from the same genomic strand (intra-strand) or from the opposing strands (inter-strand), they cause DNA crosslinking damage. The nucleotide excision repair mechanism is a simple and effective method, undergoing in a physiological state, for removing intra-strand crosslinks. However, DNA strand detachment is completely prevented by inter-strand crossing, which obstructs crucial DNA metabolic functions including transcription and replication. Further, this can be exceedingly harmful if not corrected, especially in dividing cells, as it can halt DNA replication and result in cell death. Chemotherapeutic agents like melphalan and cisplatin are examples of inter-strand cross-forming compounds (Lehmann and Wennerberg 2021).

15.2.6 DNA Strand Breaks

DNA double-strand breaks (DSBs) can occur as a consequence of endogenous mechanisms like DNA replication and repair along with external factors like radiation and certain chemicals. DSBs are especially problematic as, when left unrepaired, they can result in cell death. Additionally, incorrectly repaired DSBs can result in DNA deletions, translocations, and fusions. These outcomes are referred to as genomic rearrangements collectively, and they are frequently observed in malignant cells (Chatterjee and Walker 2017).

15.3 Physiological Response to Genotoxic Stress: DNA Damage Response

The process by which cells react to genotoxic stress is extremely complicated. However, it is explained as a series of activities, sequentially engaged to cause the DNA damage response (DDR) after recognition of genotoxic stress (Mohanan et al. 2021). It involves detecting genomic damage via a variety of sensor molecules, engaging on checkpoint proteins, and then activating repair mechanisms. The elements of DDR encompass "sensors," "transducers," and "effectors"; these groups mediate final outcomes such as repair and/or apoptosis. Genomic injury/lesions serve as an initial signal, which is perceived by "sensors," and escalated via "transducers." The "effectors" ultimately catch the signal and carry out various cellular processes (Yang et al. 2004).

ATM and ATR-Dependent DNA Damage Response The members of the phosphatidylinositol 3-kinase (PI-3) superfamily are stimulated at the initial phase of the DDR, which might function as sensors, as well as initiators of the physiological genotoxic stress response, including ataxia-telangiectasia mutated (ATM) and ATM and RAD3 (ATR) activation. ATM and ATR can be activated by interaction with genomic lesions or by interaction with DNA repair proteins (Yang et al. 2004). The involvement of ATM and ATR is well studied in DNA double-strand break formation. Mre11–Rad50–Nbs1 complex recognizes this injury, following the recruitment of ATM. The active ATM subsequently initiates two processes, which are crucial for checkpoint signaling and DSB repair, that organize the chromatin and handle the DNA close to the DSB lesion. The very first processing of ATM is phosphorylation of histone H2AX, followed by activation of Mdc1 and formation of protein complexes encompassing Mre11–Rad50–Nbs1 complex, p53-binding protein, BRCA1, and ATM. This is followed by the targeting of ATM toward substrate-like kinase checkpoint kinase-2, which results in the dispersal of injury signaling across the nucleus (Shiloh 2006; Li and Zou 2005). The chromatin alteration signaling is widespread in the cell cycle, independent of cyclin-dependent kinase activity, while ATR stimulation is cyclin-dependent kinase-dependent, which is constrained to S and G2 phases. Double-strand breaks expand the single strand of DNA, which is covered by replication protein A complex, an intermediate for DNA repair and ATR-dependent signaling. Further, ATR is activated following independent recruitment of the ATR-ATR-interacting protein complex and DNA topoisomerase 2-binding protein 1. The effector signaling kinase, checkpoint kinase-1, whose activation also needs Claspin and ATR-dependent phosphorylation of Rad17, is one of the substrates that ATR then targets. The ATM-Chk2 and ATR-Chk1 transmission molecules then cooperatively orchestrate the checkpoint responses and DNA repair in the damaged cell (Bartek and Lukas 2007; Zhou and Elledge 2000; Sirbu and Cortez 2013). A schematic presentation of the genotoxic stress response is presented in Fig. 15.1, as a simplified version of the "p53 signaling pathway" obtained and developed from Kanehisa Laboratories in the Institute for Chemical Research, Kyoto University. They are dedicated to establishing KEGG (Kyoto

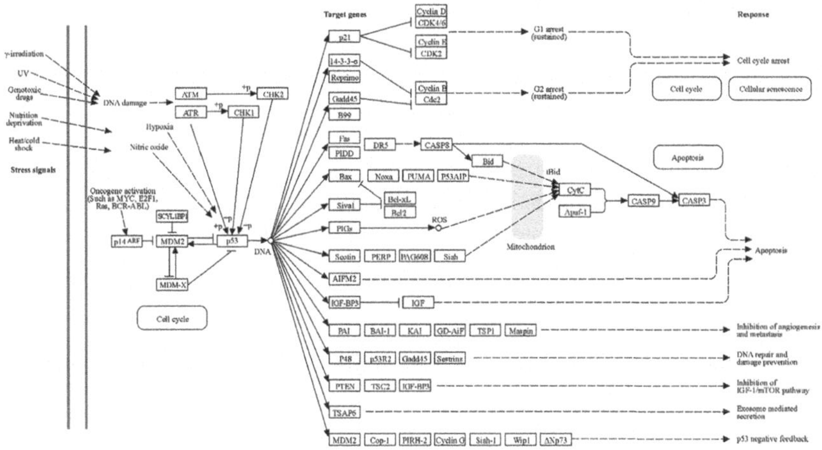

Fig. 15.1 Physiological response to genotoxic stress: The figure represents sources of genomic injury (such as radiations and genotoxic drugs) which generates a range of genomic lesions (like oxidative DNA damage, DNA adduct, alkylation of the genome, and DNA single- and double-strand break formation), these lesions when left unrepaired cause a severe malfunction in DNA machinery consequently leading to cell death. These genomic lesions are sensed by the members of the phosphatidylinositol 3-kinase (PI-3) superfamily, mainly including ataxia-telangiectasia mutated (ATM) and ATM and RAD3 (ATR) activation. Stimulation of ATM and ATR further leads to the recruitment of Chk1, Chk2, and p53. These proteins via the involvement of various genes carry out the final cellular responses like control of the cell cycle, apoptosis, DNA repair, and damage prevention. The figure is developed by the Kanehisa Laboratory (now part of the Bioinformatics Center) in the Institute for Chemical Research, Kyoto University

Encyclopedia of Genes and Genomes), a tool to derive cellular response from its genomic data (Kanehisa et al. 2004).

15.4 Methods to Detect DNA Damage

The Organization for European Cooperation and Development (OECD), US Environmental Protection Agency (EPA), and International Council for Harmonization (ICH) have given various guidelines to predict genotoxicity, which is an essential step to be followed during the drug development process of any substance. In the pharmaceutical industry, genotoxicity testing, toxicology, and the detection of genomic abnormalities in clinical specimens depend on the precision/accuracy of the method to recognize DNA damage.

In context to this chapter, the preventive or damaging effects of alkaloids can be evaluated in the following system:

1. Cell-free medium like plasmid relaxation assay.
2. Prokaryotes-based test system, which consists of SOS chromotest, Umu assay, and Vitotox assay.

3. Yeast-based test system.
4. Higher plant-based genetic method consisting of *Allium cepa* assay.
5. Cytogenetic testing strategies such as micronucleus assay (cytokinesis block micronucleus assay-CBMN and in vivo rodent micronucleus assay), chromosome aberration test (in vitro and rodent bone marrow chromosome aberration test), and sister chromatid exchange assay.
6. Unicellular-based methods including comet assay.
7. DNA repair-centered strategies such as unscheduled DNA synthesis (UDS) assay.
8. DNA fragmentation tests like TUNEL assay, DNA ladder assay, annexin V Staining, and diphenylamine (DPA) assay.
9. Biomarker-based strategies, which evaluate 8-Hydroxy-2-deoxy guanosine (8-OHdG), γH2AX, and DNA adduct detections.
10. Cell cycle alterations, detected via flow cytometry.
11. Genetic/molecular strategies like DNA microarray, polymerase chain reaction (PCR), and random amplification of polymorphic DNA (RAPD), and
12. High throughput/advanced strategies like 3D cell culture test system and fluorescence in situ hybridization (FISH) test system (Figueroa-González and Pérez-Plasencia 2017; Mohamed et al. 2017; Turkez et al. 2017; Nelson et al. 2017; Reddig et al. 2018; Firsanov et al. 2021).

15.5 Purpose of Pharmacological Intervention

An agent that combats genotoxic stress has the ability to stop, block, or avoid the spread of genetic damage. The development of efficient anticancer drugs is drawing increasing attention as cancer is the main cause of human mortality, accounting for around 9.6 million deaths worldwide in 2018. While chemotherapy and radiotherapy are the pillars of cancer treatment, their significant adverse reactions frequently limit their effectiveness and utilization. Most chemotherapy drugs are genotoxic, which are likely to cause epigenetic and genetic damage. Chemotherapeutic agents like cyclophosphamide, cisplatin, 5-fluorouracil, bleomycin, doxorubicin, methotrexate, etc., are the cornerstones in cancer treatment, which are designed to directly target DNA and are associated with high levels of DNA damage. With the significant rise in the number of cancer patients each year, it is predicted that by 2040, there will be 15.3 million cancer patients to require first-line chemotherapy treatment (Sak 2012; George et al. 2017). Thus, it is essential to search for compounds that can counter the genotoxic damage of such drugs. The floral world is composed of plants with a wide range of chemical properties, most of which have a favorable toxicity profile. The most commonly utilized sources of phytochemicals for cancer therapy are herbal formulations, plant crude extracts, and even food. These have demonstrated unmatched benefits in minimizing the side effects of anticancer medicines (Aprotosoaie et al. 2019; Liu et al. 2021). These phytochemicals may function as adaptogens, shielding the body from the harmful consequences of intervention, or as biologic response modifiers, adding to or even enhancing the effects of pharmacological medicines. Further, numerous genome-protecting substances are emerging as putative genoprotectors and have been linked to the management of other human pathologies like neurodegeneration and neuromuscular disease. As a result, researchers have generated a renewed interest in this area and broadened the horizon of such compounds (Proshkina et al. 2020).

Alkaloids and other phytochemicals might show protection against genotoxic stress by following mechanisms: (a) suppression of genotoxin uptake; (b) complexation or deactivation of toxin; (c) blocking and competition mechanisms such as scavenging of reactive oxygen species and protection of DNA nucleophilic sites; (d) alteration in transmembrane transport; (e) alteration in DNA repair mechanism; and (f) alterations in xenobiotic metabolizing enzymes (López-Romero et al. 2018; Morales-González et al. 2019; de Kok et al. 2010).

15.6 Alkaloids Against Genotoxic Stress

Several bioactive chemicals found in nature can be transformed into the medications we use today. Most chemicals, which help plants thrive and propagate, are recognized to be synthesized by plants. Alkaloids, phenols, steroids, glycosides, tannins, terpenoids, and phytoalexins are examples of such bioactive metabolites. These metabolites display biodynamic behaviors that are beneficial to both human and animal well-being. Alkaloids are amid the immense class of natural compounds, which constitutes a plethora of chemical structures, including over 12,000 entities. The main prerequisite for being classified as an alkaloid is the presence of a basic nitrogen atom in any location throughout the molecule, which excludes nitrogen found in peptides (Debnath et al. 2018). Alkaloids are universally scattered in the plant kingdom and belong to families like Ranunculaceae, Apocynaceae, Leguminosae, Erythroxylaceae, Papaveraceae, Amaryllidaceae, Menispermaceae, Loganiaceae, etc. Alkaloids are typically categorized into (a) true alkaloid, nitrogen-containing heterocyclic ring, for instance, cocaine, morphine, quinine, etc. (b) Protoalkaloids, where nitrogen is not a part of heterocyclic ring, for instance, yohimbine, mescaline, and hordenine; and (c) pseudoalkaloids, where they are not derived from amino acids, for instance capsaicin, caffeine, and ephedrine. Further, alkaloids can also be categorized depending on their chemical structure such as tropane alkaloids(example: cocaine), pyrrolizidine alkaloids (example: senecionine), piperidine alkaloids (example: lobeline), quinolines alkaloid (example: quinine), isoquinoline alkaloid (example: berbamine), indole alkaloids (example: ergotamine), steroidal alkaloids (example: jervine), purine alkaloids (example: caffeine), and pyrrolidine alkaloids (example: hygrine) (Dey et al. 2020). Herbs high in alkaloidal content have a plethora range of pharmacological traits. Alkaloids have been used as drugs throughout human history, which highlights their significance to the pharmaceutical sector. The below mentioned are the alkaloids that have shown evidence to be utilized against various genotoxic stress. The chemical structure of these alkaloids is presented in Fig. 15.2, while Table 15.2 represents the specific details of the experiments conducted on them.

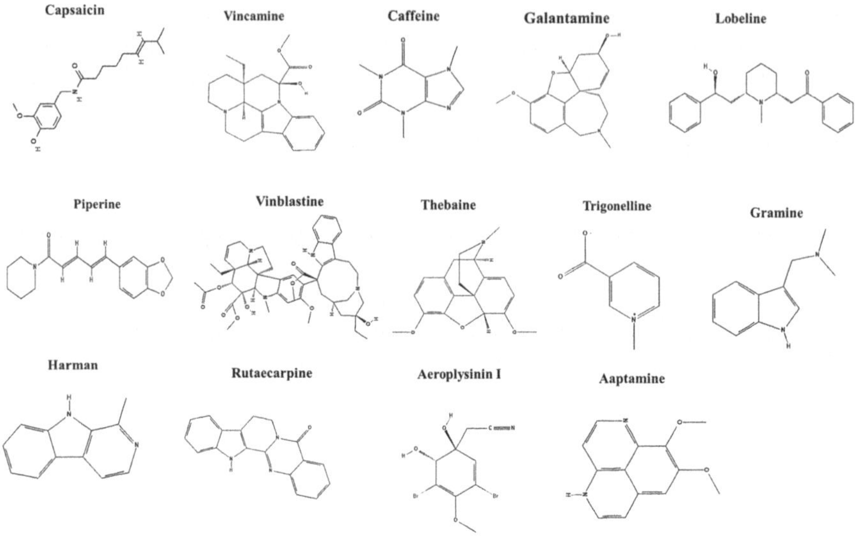

Fig. 15.2 Chemical structure of various alkaloids acting against genotoxic stress

15.6.1 Capsaicin

The main spicy alkaloid present in chili peppers is referred by capsaicin, a homovanillic acid derivate, which is chemically 8-methyl-*N*-vanillyl-*trans*-6-nonenamide. The chili pepper is the fruit of the herbaceous, Solanacea family, endemic to the Americas, Capsicum genus plants. Since Mesoamerican civilizations cultivated chilies, they have been used in food and medicine all throughout the world. A multitude of human disorders, including chronic pain, obesity, diabetes, cardiovascular conditions, respiratory diseases, and cancer have been identified as possible therapeutic targets for capsaicin. Capsaicin is commercialized as topical analgesic cream, used for the treatment of medically diagnosed painful diabetic neuropathy or postherpetic neuralgia. Conflicting results have been found on the genotoxicity/genoprotective potential of capsaicin. Díaz Barriga Arceo et al. (1995) have demonstrated that a low dose of capsaicin (1.46 mg/kg) resulted in genomic injury after 32 days of consecutive intraperitoneal administration to mice, while at a higher dose (1.94 mg/kg), the injury occurred from day 16 onward, as observed by calculating the frequency of micronucleated normochromic erythrocytes. Lewinska and coworkers (2015) have also demonstrated the induction of DNA double-strand breaks in human prostate carcinoma cells and human lung carcinoma cells by capsaicin at a concentration of 100 μM. Contrary to these, it was reported that capsaicin did not show any genomic lesions up to the concentration of 250 μM in human endothelial cells, as evaluated by comet assay, while stimulation of inflammatory cascade was observed (Richeux et al. 2000). Caetano et al. (2018) further explored the role of capsaicin in the 1,2-dimethyl-hydrazine-associated colon carcinogenesis model of rats, where it resulted in the protection of genomic injury.

Table 15.2 Protective effect of alkaloids against various genotoxins

S. no	Alkaloid	Inducer of genotoxic stress	Analysis system	Evaluating parameters	Result	References
1	Aaptamine	Cisplatin	Rat kidney proximal tubular cells and human bladder cancer cells	γH2AX, p53, and checkpoint kinase-1	Geno-protection observed in tubular and glomerular rat kidney cells	Funk et al. (2014)
2	Aaptamine and Aeroplysinin-1	Daunorubicin	Acute myeloid leukemia	γH2AX, Kap-1, and Chk-1	Reduction in DNA damage response	Stuhldreier et al. (2015)
3	β-carboline alkaloids (Harmane, Harmine, Harmol, Harmaline, and Harmalol)	Hydrogen peroxide; 100 μM	Chinese hamster lung fibroblasts (V79) cells	Comet assay	Hydroxyl radical-scavenging action is responsible for genome protection	Moura et al. (2007)
4	Caffeine	Vinblastine (1.5 mg/kg)	Swiss male albino mice	Mitotic metaphase chromosome aberration assay and micronucleus test from bone marrow	A reduction in cytogenetic markers was observed	Geriyol et al. (2015)
5	Capsaicin	1,2-Dimethylhydrazine	Male Wistar rats	Leukocyte and fecal water genotoxicity	Reduced cyto- and geno-toxicity, promotes antiproliferative and proapoptotic responses	Caetano et al. (2018)
		Mitomycin C	Albino rats	Micronucleus test and chromosomal aberrations	Absences of genomic lesions	Raymond et al. (2004)
6	Gramine	2-Aminofluorene (20 μg/plate), sodium azide (1 μg/plate), and 2-nitrofluorene (1 μg/plate)	Bacteria strain of *Salmonella typhimurium* and Swiss albino mice	% mutagenic inhibition, bone marrow micronucleus assay and chromosomal aberrations	Genome protection activity due to antioxidant potential	Pathak et al. (2018)

(continued)

Table 15.2 (continued)

S. no	Alkaloid	Inducer of genotoxic stress	Analysis system	Evaluating parameters	Result	References
7	Galantamine	Amyloid beta $A\beta_{(1\text{-}42)}$	SH-SY5Y cell line	Clonogenic assay, cytokinesis block micronucleus cytome and comet assay	Genome protection was observed in neuronal cells	Castillo et al. (2016)
8	Lobeline	Ethanol (2.5 g/kg), hydrogen peroxide, cyclophosphamide	CF-1 male mice	Comet assay, micronucleus test, *Salmonella*/microsome mutagenicity assay	Genome protection was observed	da Costa e Silva et al. (2014)
		Pilocarpine	CF-1 male mice	Comet assay, DPPH activity, superoxide dismutase and catalase	Genome protection was observed in hippocampus and cortex	da Costa e Silva et al. (2018)
9	Thebaine	Mitomycin C	TK6 cells	Cytokinesis-block micronucleus test	Genome protection was observed only after co-treatment of toxicant and drug	Reimann et al. (2021)
10	Trigonelline	UV-B irradiation	Primary Human Dermal Fibroblasts and BALB/C mice	P53, ATM, ATR, γH2AX, Chk1, and Chk2	Provides geno-protection through PI3K-AKT-Nrf2 pathway	Tanveer et al. (2023)
11	Piperine	7,12-dimethylbenz[*a*] anthracene	Male golden Syrian hamsters	Formation of micronucleus polychromatic erythrocytes and chromosomal aberrations	Antioxidant action was responsible for genome protection	Balakrishnan et al. (2008)
		Benzo(*a*)pyrene	Male Swiss albino mice	8-oxo-20-deoxyguanosine, Comet assay	Prevented DNA damage in lungs and liver	Sehgal et al. (2011)

		Cyclophosphamide	Wistar albino rats	Mammalian erythrocyte micronucleus test, diphenylamine assay, quantification of 8-OHdG and γH2AX	Prevented DNA damage in bone marrow cells, liver, and hippocampus. Reversed the markers of oxidative DNA damage and DNA double-strand breaks	Yadav et al. (2023a, b)
12	Rutaecarpine	Hydrogen peroxide; 500 μM	Hepa-1c1c7 cells	Comet assay	Reduced DNA-damaged strands	Lee et al. (2012)
		Benzo[*a*]pyrene; 5 μM	Mucoepidermoid pulmonary carcinoma cells and hepatocellular carcinoma (Hep3B) cells	Commercial DNA damage kit	Prevention of genomic damage	Li et al. (2019)
13	Vincamine	Cisplatin, 10 mg/kg, i. p.	Albino rats	Urea, creatinine, kidney injury molecule-1, malondialdehyde, myeloperoxidase, nuclear factor erythroid 2-related factor 2, Hemeoxygenase-1, Toll like receptor 4, interferon gamma-CD44, transforming growth factor beta and DNA fragmentation	Anti-inflammatory and antioxidant activity are attributed to prevention in DNA damage	El-Sayed et al. (2021)

Capsaicin at the dose of 50 mg/kg, for 4 weeks (3 times a week) reduced the 1,2-dimethyl-hydrazine-associated genomic damage, as evaluated in rat leukocytes by comet assay. The findings were analogous to fecal water genotoxicity in CaCO-2 cells at 50 mg/kg. The report further established a reduction in the genetic expression of proliferation and proapoptotic markers by capsaicin.

15.6.2 Caffeine

Chemically, caffeine is 1,3,7-trimethyl xanthine, a purine alkaloid, which is obtained from *Camellia sinensis* and *Coffea arabica* commonly referred to as tea and coffee. Studies suggested that caffeine has resulted in genomic damage, when tested in various in vitro (CHO cells, rat kidney cell line, and cultured mouse preimplantation embryo cells) and in vivo systems (albino mice); however, the ability of caffeine to cause genomic damage is observed when the concentration is raised from ten to hundred times more than consumed in humans (Nehlig and Debry 1994). As a result, caffeine is not believed to be genotoxic to animals and does not appear to be mutagenic to people when consumed. Stoilov and coworkers (1994) studied the temperature-dependent effect of caffeine on radiation-associated genomic damage. At a lower temperature, it exhibited genome protective effects, while at a higher temperature, an increased rate of damage was observed in lymphocytes. Caffeine's capability to either exacerbate or decrease radiation-associated genomic damage appears to be correlated with its capacity to bind to DNA and neutralize radicals. Further, the experiment of Geriyol et al. (2015) demonstrated that caffeine administered to mice at the dose of 25, 50, and 100 mg/kg, counteracted vinblastine-associated cytogenotoxicity, as indicated by a decline in chromosomal aberrations in erythrocytes derived from bone marrow. The effect was postulated to be the free radical scavenging property of caffeine. Kumar et al. (2001), have also suggested similar action of caffeine, taking advantage of its antioxidant property, where it efficiently shielded DNA against ultraviolet radiation (at a dose rate of 1.2 Gy/min), when exposed to plasmid pBR322 DNA. Further, to comprehend the antioxidant mechanisms, the reaction of caffeine with various radicals (such as superoxide, hydrogen peroxide, and deoxyribose peroxyl radical) was studied, where caffeine scavenged the primary as well as secondary radicals generated by radiations.

15.6.3 Galantamine

Galantamine can be isolated from bulbs and flowers of *Galanthusnivalis* (Common snowdrop), *Galanthuscaucasicus* (Caucasian snowdrop), *Galanthusworonowii* (Voronov's snowdrop), and some other members of the family Amaryllidaceae, such as *Narcissus* (daffodil), *Leucojumaestivum* (snowflake), and Lycoris including *Lycoris radiata* (red spider lily). It is an acetylcholinesterase inhibitor used for improving cognitive functions in Alzheimer's disease patients at mild, moderate, and advanced moderate stages (Villarroya et al. 2007). The research team of Castillo has recently demonstrated the efficacy of galantamine against genotoxic stress exacerbated by amyloid beta, in traditional assays of cytogenotoxicity such as clonogenic assay, cytokinesis block micronucleus assay, and comet assay in SH-SY5Y human neuroblastoma cell line. The authors speculated that the positive outcome of galantamine is associated with its antioxidant property. Galantamine could possibly protect against genomic injuries by stimulating DNA repair enzymes and the replication process (Castillo and Aristizabal-Pachon 2017; Castillo et al. 2016, 2018).

15.6.4 Piperine

Piperine is chemically an alkaloid, which can be isolated from the *Piper* genus, family Piperaceae. Piperine gained the attention of health professionals because of its wide range of biological actions. Also, humans are regularly exposed to piperine, majorly via diet, in the form of black pepper. An average intake of 333 mg per person of pepper results in an intake of 16.5–29.7 mg of piperine per person per day. Reen et al. (1997) have provided in vitro validation of piperine to counter genomic damage. The authors used genetically engineered V79 Chinese hamster cell constructs of rat liver CYP4502B1, where piperine reversed the genomic damage against aflatoxin B1. The study established that piperine is able to suppress CYP4502B1 action while combating pro-carcinogens (such as aflatoxins) triggered by cytochrome enzymes.

In 2005, Selvendiran et al. conducted a preliminary experiment to assess the genopreventive action of piperine in mice. Piperine when administered to mice at the dose of 75 mg/kg for 5 days, reversed the frequency of micronucleus in erythrocytes of bone marrow after single-dose administration of benzo(alpha)pyrene (25 mg/kg, intraperitoneal administration) and cyclophosphamide (20 mg/kg, intraperitoneal administration). The action was attributed to the antioxidant effects and alteration of phase II enzymes (Selvendiran et al. 2005). Sehgal et al. (2011) assessed the effect of piperine against benzo(a)pyrene-induced injury in the lungs and liver of mice. Exposure of piperine before the administration of toxicant, for 7 days, substantially counteracted the levels of 8-oxo-2′-deoxyguanosine and the percentage of genomic damage in the comet assay. In a separate report, they also showed the mitigating effect of piperine on the antioxidant defense system by assessing biochemical parameters such as lipid peroxides, superoxide dismutase, catalase, and reduced glutathione (Sehgal et al. 2012). In 2013, it was further demonstrated that piperine causes a reduction in the activities of ethoxyresorufin-*o*-deethylase (specific marker for liver CYP1A1), pentoxyresorufin *o*-depentylase (specific marker for liver CYP1B1), and levels of benzo(alpha)pyrene-diol epoxide DNA adducts (Sehgal et al. 2013). These studies suggested that piperine could be a good genoprotectant against carcinogenic compounds via modulating xenobiotic-metabolizing enzymes and free radical scavenging properties.

Recently, in our laboratory, we examined the activity of piperine against cyclophosphamide in a rodent model. We observed that cyclophosphamide, 5 mg/kg, resulted in significant genomic damage and tissue-specific toxicity. Genomic damage was observed in high-proliferating tissues like bone marrow cells as well as in vital tissues like the liver and hippocampus. Interestingly, piperine treatment at the dose of 50 mg/kg for 28 days could efficiently reverse genomic damage and tissue-specific toxicities induced by cyclophosphamide. Genomic damage was quantified as the frequency of micronucleus polychromatic erythrocytes in bone marrow erythrocytes, while in hepatocytes and hippocampus, the percentage of DNA fragmentation was measured. The mechanistic approach was deciphered by evaluating the biomarkers of genotoxicity. 8-OHdG and γH2AX are the hallmarks of genotoxicity, where 8-hydroxy-2′-deoxyguanosine (8-OHdG) is a biomarker used to detect oxidative DNA base damage provoked by reactive oxygen species while γH2AX is a sensitive and early indicator of DNA double-strand breaks, which plays a critical role in DNA damage response. Piperine treatment reduced the levels of 8-OHdG as well as suppressed the expression of γH2AX in hepatocytes and hippocampus, as compared to cyclophosphamide-treated rats. Further, piperine treatment also reversed the oxidative status of tissues toward normal, as presented by a reduction in reactive oxygen species formation and malondialdehyde levels, while enhancing the levels of glutathione, thus restoring the redox homeostasis disturbed as a result of cyclophosphamide treatment. On histopathological lesions in the liver and hippocampus, we observed that cyclophosphamide treatment resulted in dreadful lesions, such as a large number of apoptotic cells in both tissues, while several other specific histological lesions in the liver such as steatosis, blood filled in sinusoidal space, inflammation of cells, and dilatation of central vein with hypermia. In subregions of the hippocampus (CA1 and DG), cyclophosphamide induced high pyknotic cells along with loss of neuronal layer in both the subregions. Such lesions are also observed in clinical scenarios with patients receiving cyclophosphamide in their chemotherapeutic regimen suggesting that our model could mimic the clinical condition. The effects were efficiently counteracted by piperine, with better reversal observed in liver cells (Yadav et al. 2023a, b).We also investigated *Piper longum* extract under similar conditions, where the extract has provided protection against cyclophosphamide-associated genomic injury and the phytochemical investigation of the extract demonstrated the presence of piperine, as one of its bioactive compounds (Yadav et al. 2023a, b).

The studies conducted in our laboratory clearly indicated that piperine treatment can reverse the DNA damage in high proliferating tissues like bone marrow, reduce tissue-specific genomic injury, diminish the markers of oxidative damage and DNA double-strand breaks, and restore the redox balance while providing protection against cytoarchitectural damage induced by cyclophosphamide, a routinely used chemotherapeutic. These findings are providing evidence to translate and harness the utilization of phytochemicals like piperine from the preclinical to the clinical testing phase.

The preventive effect of piperine is also beneficial as phytobiotics to chicken feed. da Silva Cardoso et al. (2016) performed an experiment where piperine at the dose of 60 mg/kg was fed to 9 days old male chicks, in combination with aflatoxin B1 at the dose of 0.5 mg/kg. They analyzed the peripheral blood at various time intervals, such as 24, 48, 72, and 96 h on the first day of the experiment as well as on the last day of the experiment (on the 26th day), which was evaluated by comet assay and micronuclei formation. Animals exposed to aflatoxins for an extended period of time developed lesions resembling aflatoxicosis, featuring genomic injuries and problems with bone marrow blood production. The findings suggested that the addition of piperine to chicks after aflatoxin intoxication diminished the genomic damage and cytotoxic effects. However, future studies must replicate the effect on a larger sample size.

15.6.5 Lobeline

It is a lipophilic, pyridine alkaloid obtained from *Lobelia inflata*, which is also referred to as *Rapuntiuminflatum* or Indian weed. The pharmaceutical formulation of lobeline is commercialized as tablet preparation for use as a smoking cessation aid, despite the lack of scientific data to back up this claim. Pharmacological evidence suggests the direct interaction between lobeline and nicotinic acetylcholine receptor. It suppresses psychostimulant-associated action via modulation at the nicotinic acetylcholine receptor, along with the capability to modify presynaptic dopamine storage and release, via the interaction with vesicular monoamine transporter-2 (VMAT2) (Dwoskin and Crooks 2002). The team of Jaqueline Nascimento Picada has studied the effect of lobeline in ethanol-associated genomic instability and pilocarpine-induced seizure model. Lobeline administration to mice, at the doses of 5/10/20 mg/kg, did not change the ethanol-associated genotoxicity in peripheral blood, while preventing pilocarpine-associated DNA damage in the hippocampus, as revealed in the comet assay. The result was hypothesized to be caused by a decrease in oxidative injury, alternatively, the inhibitory action of lobeline on glutamatergic excitotoxicity may have suppressed the production of free radicals, hence lowering genomic damage (da Costa e Silva et al. 2014, 2018).

15.6.6 Vinblastine

Vincristine and vinblastine are the natural vinca alkaloid, extracted from *Vinca rosea*, and are frequently utilized as single or in conjugation with other chemotherapeutics or radiations for several malignancies. These classes of alkaloids are aneugens (producing anomalous chromosome number) and mitotic spindle poison, they associate with tubulin and suppress the gathering of microtubules. Aneugens are predicted to possess a lower degree of clastogenicity because they prevent the development of the mitotic spindle. However, vincristine is reported to augment genomic damage; vinblastine, on the other hand, demonstrated defense to normal cell gamma-radiation-associated genome injury. Rajagopalan et al. (2003) demonstrated that vinblastine when administered to mice, before gamma-radiation contact, suppressed genome injury in fibrosarcoma, leukocytes, and bone marrow cells, as revealed through comet assay. The action was postulated due to the differential feature of cancer and non-cancer cells, where vinblastine condensed mitosis in cancer cells, which has its accessibility to interact with the genome, thereby preventing radiation-induced damage.

15.6.7 Vincamine

Vincamine is a monoterpenoid indole alkaloid extracted from the leaves of *Vinca minor*. Clinically, it is widely utilized in Europe, as a prescription medicine for the treatment of primary degenerative and vascular dementia. Vincamine is also demonstrated to possess multiple biological activities like antioxidant, anti-inflammatory, and antiapoptotic characteristics. In a very recent study, prior oral administration of vincamine, at the dose of 40 mg/kg for 7 consecutive days, with a single exposure of cisplatin to rats, prevented genomic injury in renal tissue. Vincamine antioxidant and anti-inflammatory properties are suggested to be the cause of this effect, which was achieved by engaging the Nrf2/HO-1 pathway and preventing the TLR4-IFN-γ-CD44 cell's inflammatory state (El-Sayed et al. 2021).

15.6.8 Trigonelline

Chemically, trigonelline is *N*-methyl nicotinic acid ($C_7H_7NO_2$), a heterocyclic alkaloid, which is extracted from *Trigonellafoenum-graecum* (fenugreek); from the family, Fabaceae. It offers a wide range of therapeutic benefits and is widely known to treat conditions like diabetes mellitus and neurological disorders (Zhou et al. 2012). A detailed analysis of trigonelline was conducted by Tanveer et al. (2023), where its genome protective action was established against ultraviolet B radiation using human dermal fibroblasts and mice models. The effect was demonstrated to be regulated by main hallmarks of DNA damage consisting of ATM (ataxia-telangiectasia mutated), ATR (ATM- and Rad3-Related), p53, Gamma-H2AX (H2A histone family member X), and the Checkpoint kinases (Chk1 and Chk2). Further, trigonelline treatment also modulated the PI3K-AKT-Nrf2 signaling pathway and offered genomic stability. The study suggests that this property can be utilized in the cosmeceutical industry to guard dermis cells against exposure to ultraviolet damage.

15.6.9 Thebaine

Thebaine, also known as paramorphine, is an isoquinoline derivative and an alkaloid that is isolated from the dried latex of immature *Papaver somniferum*, often known as the opium poppy or bread seed poppy and is a member of the Papaveraceae family. In 2018, the European Food Safety Authority arrived at the decision that there was insufficient evidence to carry out a quantitative risk categorization of thebaine (Eisenreich et al. 2020). Reimann et al. (2021) showed that a high dose of thebaine induced micronucleus formation and no effects were observed in the comet assay; however, when co-exposed with mitomycin-C, a reduction in micronucleus frequency was observed in TK6 cells. Thus, further analysis is required to conclude the findings.

15.6.10 Gramine

Gramine is an indole nucleus containing alkaloid, also referred to as donaxine, and can be isolated from a variety of plants like *Avenasativa* (commonly known as oat, from family Gramineae), *Hordeum vulgare*, *Triticumaestivum*, etc. The pharmacological properties of gramine are comparable to those of ephedrine and include the relaxation of bronchial smooth muscle, vasorelaxation, and the treatment of bronchitis, nephritis, and bronchial asthma. Pathak et al. (2018) conducted various assays to evaluate the genotoxicity of gramine under in vivo conditions. Bacterial mutagenicity was tested in *S. typhimurium* strain, where gramine (at the concentration range of 1250–10,000 μg/plate) did not induce any mutagenic response, while in the presence of a metabolizing enzyme, gramine co-exposed with sodium azide, attenuated the colonies in TA1535 strains. Also, when administered to mice at the dose of 150 mg/kg, the results of genotoxicity were negative as determined by micronucleus polychromatic erythrocyte and chromosomal aberrations. However, this evidence is not comprehensive and needs to be assessed in various cell culture/preclinical models for detailed mechanistic exploration.

15.6.11 Beta-Carbolines Alkaloid

Beta-carboline alkaloids are indole alkaloid derivatives, with a major effect on cerebrum functioning, chemically containing tricyclicpyrido[3,4-b] indole cycle. Moura et al. (2007) investigated the action of aromatic (harmane, harmine, and harmol) and dihydro-beta-carboline (harmaline and harmalol) alkaloids for genoprotection in Chinese hamster lung fibroblasts. They observed that harmane, harmine, and harmol were able to produce genomic damage only at the concentration of 40 μg/mL, no damage was observed at 10 and 20 μg/mL, as calculated in the comet assay, while harmaline and harmalol did not induce any damage at the same concentration. However, both aromatic and dihydro-beta-carbolines showed a reduction in DNA damage index and damage frequency when treated against 100 μM concentration of hydrogen peroxide. These reports point toward the genoprotective potential of beta-carboline, where the action is attributed to the hydroxyl radical-scavenging property; however, in vivo investigations are required to validate these findings.

15.6.12 Rutaecarpine

Rutecarpine or rutaecarpine, known for potent cyclooxygenase-2 inhibition, is derived from *Tetradiumruticarpum* and categorized as a nonbasic alkaloid. Li et al. (2019) investigated the action of rutaecarpine against benzo[a] pyrene-associated genotoxic stress. When rutaecarpine, at the concentration of 10 μM, was exposed in combination with benzo[a]pyrene (5 μM), to mucoepidermoid pulmonary carcinoma cells and hepatocellular carcinoma cells, a significant reduction in genomic injury was observed. The effect was examined by ethoxyresorufin-*O*-deethylase assay and Muse™Multi-Color DNA Damage Kit. The findings suggested that rutaecarpine plays a significant role in controlling the cytochrome activity of hepatocytes and lung cancer cells.

15.6.13 Aeroplysinin-1 and Aaptamine

Marine organisms are an incredibly rich source of small molecules with pharmacological properties that may be good for human health. Marine sponges, belonging to phylum Poriphera, especially of Aplysina genus, serve as a reservoir for brominated isoxazoline alkaloids, amid which aeroplysinin-1, isolated from *Aplysiniacavernicola*, is a very well-known anti-inflammatory, anti-tumor, and anti-angiogenic compound (García-Vilas et al. 2015). Similarly, Genus *Aaptos*, is also a reservoir for a bioactive benzo[de][1,6]-naphthyridine alkaloids, known for anticancer and antiviral properties, is referred to as aaptamine (Nadar et al. 2022). Funk et al. (2014),established the genoprotective property of aaptamine and aeroplysinin-1 in rat renal tubular cells and rat glomerular endothelial cells against cisplatin. They confirmed the alteration in DNA adduct formation and nuclear gamma H2AX expression in pretreated cells with aaptamine and aeroplysinin-1, at the concentration of 2 μmol/L, prior to cisplatin exposure (30 μmol/L). Similar observations were presented by Stuhldreier et al. (2015), against daunorubicin (0.12 μM) in acute myeloid leukemia, where aaptamine and aeroplysinin-1 diminished gamma H2AX expression, phosphorylation of KRAB domain protein (KAP1), and phosphorylation of checkpoint kinase 1 (Chk1). Thus, both studies provide conclusive results that aaptamine and aeroplysinin-1 protect against genotoxic stress and should be explored in further experiments. Also, such activities might be explored to broaden the therapeutic window of chemotherapeutic agents like cisplatin and daunorubicin, consequently rescuing the cells from associated side effects.

15.7 Potential Problems in Using Alkaloids

The above literature is suggesting that alkaloids have the potential to mitigate the genotoxic sequelae specifically caused by chemotherapeutic drugs. Despite the large number of studies that have been done and published about the advantages of alkaloids or phytochemicals in chemotherapy, there remain numerous critics and skeptics among medical professionals and patients alike. This may be the result of ignorance on the one hand, but the spread of myths and erroneous beliefs is also a significant factor. However, the toxicity of many alkaloids remains another critical issue when translating their use to clinical scenarios. The primary function of alkaloids in flora is generally acknowledged to be toxicity toward predators and diseases. Alkaloids like caffeine, which is consumed to increase mental clarity; nicotine, to improve physical performance; morphine, a common analgesic; codeine, which is used to treat coughs; and strychnine, a potent tetanic toxin, are known to have serious side effects or toxic effects. Alkaloids are further known to demonstrate organ-specific toxicity (such as neuro-, cardio-, hepato-, and nephrotoxicity), while the mechanistic pathways for such toxicity appear to be associated with enzymatic modifications affecting physiological progressions, respiratory suppression, interaction/alkylating with genome and proteins, prevention of DNA synthesis and repair pathways, suppression of cell division, etc. However, they are also known to possess pleiotropic nature where they show toxic as well as preventive potential as demonstrated in cell line and in vivo models (Matsuura and Fett-Neto 2015). In a similar context, their capacity to interact with the genome has been thoroughly investigated. The majority of these chemicals exhibit DNA-damaging or cytotoxic effects and are consistent with genome interaction. They are shown to interact with DNA via highly specific non-covalent, reversible, and formation of molecular complexes with the genome (Basu and Kumar 2018; Sjakste et al. 2020). Likewise, chemotherapeutic medicines are designed to target DNA and this could result in possible interactions with alkaloids; hence, careful examination and study design is required to draw any conclusions.

15.8 Conclusion

The present chapter has provided a glimpse of major sources and signaling pathways involved in mediating genotoxic stress. However, genotoxic stress can also be exaggerated by several other pathophysiological processes and secondary consequences of genotoxins (such as oxidative stress and inflammatory response) (George et al. 2017; Cristóbal-Luna et al. 2018; Yubo et al. 2021). Therefore, an understanding of the molecular mechanisms by which genotoxic stress is progressed in physiology is required to produce effective therapies with fewer adverse effects. In the case of alkaloids, studies clearly indicate that alkaloids are efficient in reversing genotoxic stress as demonstrated in cell culture and rodent models. Experimental data from in vitro and in vivo studies suggest that alkaloids have been able to combat various biomarkers and cellular responses against DNA damage. However, there are also a few studies that contradict the same. To avoid undesired consequences such as increasing genomic injury via oxidative stress, alterations in autophagy to remove toxic biomolecules, and/or inhibition of critical pro-survival trails to cell stress, it is of great significance to cautiously characterize the hormetic impact of these alkaloids. Further, most of the experiments are often limited in providing mechanistic approaches to these compounds, where the studies did not reveal the target signaling pathway of DNA damage or provide very little information on DNA repair machinery. Additionally, there is negligible to limited evidence from clinical trials with controlled human interventions, where alkaloids are showing effectiveness against genotoxic stress. While the comparatively short period of these interventions may have contributed to the lack of impact of alkaloids on DNA stability in human investigations, it is also possible that these chemicals, even at dietary-appropriate amounts, are ineffective genoprotectants in vivo. Thus, although findings across the globe are showing encouragement, more studies must be conducted emphasizing mechanistic approaches and clinical translation of these compounds.

References

Aprotosoaie AC, Luca VS, Trifan A, Miron A (2019) Antigenotoxic potential of some dietary non-phenolic phytochemicals. Stud Nat Prod Chem 60:223–297. https://doi.org/10.1016/B978-0-444-64181-6.00007-3

Balakrishnan S, Vellaichamy L, Menon VP, Manoharan S (2008) Antigenotoxic effects of curcumin and piperine alone or in combination against 7,12-dimethylbenz(a)anthracene induced genotoxicity in bone marrow of Golden Syrian hamsters. Toxicol Mech Methods 18(9): 691–696. https://doi.org/10.1080/15376510701781520, PMID 20020926

Bartek J, Lukas J (2007) DNA damage checkpoints: from initiation to recovery or adaptation. Curr Opin Cell Biol 19(2):238–245. https://doi.org/10.1016/j.ceb.2007.02.009, PMID 17303408

Basu A, Kumar GS (2018) Nucleic acids binding strategies of small molecules: lessons from alkaloids. Biochim Biophys Acta Gen Subj 1862(9):1995–2016. https://doi.org/10.1016/j.bbagen.2018.06.010, PMID 29908208

Caetano BFR, Tablas MB, Pereira NEF, de Moura NA, Carvalho RF, Rodrigues MAM et al (2018) Capsaicin reduces genotoxicity, colonic cell proliferation and preneoplastic lesions induced by 1,2-dimethylhydrazine in rats. Toxicol Appl Pharmacol 338:93–102. https://doi.org/10.1016/j.taap.2017.11.008, PMID 29155087

Castillo WO, Aristizabal-Pachon AF (2017) Galantamine protects against beta amyloid peptide-induced DNA damage in a model for Alzheimer's disease. Neural Regen Res 12(6):916–917. https://doi.org/10.4103/1673-5374.208572, PMID 28761423

Castillo WO, Aristizabal-Pachon AF, de Lima Montaldi AP, Sakamoto-Hojo ET, Takahashi CS (2016) Galanthamine decreases genotoxicity and cell death induced by β-amyloid peptide in SH-SY5Y cell line. Neurotoxicology 57:291–297. https://doi.org/10.1016/j.neuro.2016.10.013, PMID 27793617

Castillo WO, Aristizabal-Pachon AF, Sakamoto-Hojo E, Gasca CA, Cabezas-Fajardo FA, Takahashi C (2018) Caliphruria subedentata (Amaryllidaceae) decreases genotoxicity and cell death induced by β-amyloid peptide in SH-SY5Y cell line. Mutat Res Genet Toxicol Environ Mutagen 836(Pt B):54–61. https://doi.org/10.1016/j.mrgentox.2018.06.010, PMID 30442346

Chatterjee N, Walker GC (2017) Mechanisms of DNA damage, repair, and mutagenesis. Environ Mol Mutagen 58(5):235–263. https://doi.org/10.1002/em.22087, PMID 28485537

Coates PJ, Lorimore SA, Wright EG (2005) Cell and tissue responses to genotoxic stress. J Pathol 205(2):221–235. https://doi.org/10.1002/path.1701, PMID 15643669

Cooke MS, Evans MD, Dizdaroglu M, Lunec J (2003) Oxidative DNA damage: mechanisms, mutation, and disease. FASEB J 17(10):1195–1214. https://doi.org/10.1096/fj.02-0752rev, PMID 12832285

Cristóbal-Luna JM, Álvarez-González I, Madrigal-Bujaidar E, Chamorro-Cevallos G (2018) Grapefruit and its biomedical, antigenotoxic and chemopreventive properties. Food Chem Toxicol 112:224–234. https://doi.org/10.1016/j.fct.2017.12.038, PMID 29284137 da Costa E Silva LD, Rodrigues LC, Dos Santos VR, da Costa AM, de Barros Falcão Ferraz A, Rolla HC et al (2014) Evaluation of mutagenic and genotoxic activities of lobeline and its modulation on genomic instability induced by ethanol. Life Sci 103(2):73–78. https://doi.org/10.1016/j.lfs.2014.03.034, PMID 24727238

da Costa E Silva LD, Pereira P, Regner GG, Boaretto FBM, Hoffmann C, Pflüger P et al (2018) DNA damage and oxidative stress induced by seizures are decreased by anticonvulsant and neuroprotective effects of lobeline, a candidate to treat alcoholism. Metab Brain Dis 33(1): 53–61. https://doi.org/10.1007/s11011-017-0130-1, PMID 29032429

da Silva Cardoso V, Vermelho AB, Ribeiro de Lima CA, Mendes de Oliveira J, Freire de Lima ME, Pinto da Silva LH et al (2016) Antigenotoxic effect of piperine in broiler chickens intoxicated with aflatoxin b1. Toxins 8(11):316. https://doi.org/10.3390/toxins8110316, PMID 27809242

de Kok TM, de Waard P, Wilms LC, van Breda SG (2010) Antioxidative and antigenotoxic properties of vegetables and dietary phytochemicals: the value of genomics biomarkers in molecular epidemiology. Mol Nutr Food Res 54(2):208–217. https://doi.org/10.1002/mnfr.200900288, PMID 20024933

Debnath B, Singh WS, Das M, Goswami S, Singh MK, Maiti D et al (2018) Role of plant alkaloids on human health: a review of biological activities. Mater Today Chem 9:56–72. https://doi.org/10.1016/j.mtchem.2018.05.001

Dey P, Kundu A, Kumar A, Gupta M, Lee BM, Bhakta T et al (2020) Analysis of alkaloids (indole alkaloids, isoquinoline alkaloids, tropane alkaloids). In: Recent advances in natural products analysis. Elsevier, Amsterdam, pp 505–567

Díaz Barriga Arceo S, Madrigal-Bujaidar E, Calderón Montellano E, Ramírez Herrera L, Díaz García BD (1995) Genotoxic effects produced by capsaicin in mouse during subchronic treatment. Mutat Res Genet Toxicol 345(3–4):105–109. https://doi.org/10.1016/0165-1218(95)90046-2

Dwoskin LP, Crooks PA (2002) A novel mechanism of action and potential use for lobeline as a treatment for psychostimulant abuse. Biochem Pharmacol 63(2):89–98. https://doi.org/10.1016/s0006-2952(01)00899-1, PMID 11841781

Eisenreich A, Sachse B, Gürtler R, Dusemund B, Lindtner O, Schäfer B (2020) What do we know about health risks related to thebaine in food? Food Chem 309:125564. https://doi.org/10.1016/j.foodchem.2019.125564, PMID 31670127

El-Sayed RM, Abo El Gheit RE, Badawi GA (2021) Vincamine protects against cisplatin induced nephrotoxicity via activation of Nrf2/HO-1 and hindering TLR4/IFN-γ/CD44 cells inflammatory cascade. Life Sci 272:119224. https://doi.org/10.1016/j.lfs.2021.119224, PMID 33610575

Figueroa-González G, Pérez-Plasencia C (2017) Strategies for the evaluation of DNA damage and repair mechanisms in cancer. Oncol Lett 13(6):3982–3988. https://doi.org/10.3892/ol.2017.6002, PMID 28588692

Firsanov DV, Kulikova VA, Solovjeva LV, Mikhailov VM, Nikiforov AA, Svetlova MP (2021) Methods for the detection of DNA damage. In: Genome stability. Academic, New York, pp 679–697

Funk F, Krüger K, Henninger C, Wätjen W, Proksch P, Thomale J et al (2014) Spongean alkaloids protect rat kidney cells against cisplatin-induced cytotoxicity. Anti-Cancer Drugs 25(8): 917–929. https://doi.org/10.1097/CAD.0000000000000119, PMID 24820908

García-Vilas JA, Martínez-Poveda B, Quesada AR, Medina MÁ (2015) Aeroplysinin-1, a sponge-derived multi-targeted bioactive marine drug. Mar Drugs 14(1):1. https://doi.org/10.3390/md14010001, PMID 26703630

George VC, Dellaire G, Rupasinghe HPV (2017) Plant flavonoids in cancer chemoprevention: role in genome stability. J Nutr Biochem 45:1–14. https://doi.org/10.1016/j.jnutbio.2016.11.007, PMID 27951449

Georgoulis A, Vorgias CE, Chrousos GP, Rogakou EP (2017) Genome instability and γH2AX. Int J Mol Sci 18(9):1979. https://doi.org/10.3390/ijms18091979, PMID 28914798

Geriyol P, Basavanneppa HB, Dhananjaya BL (2015) Protecting effect of caffeine against vinblastine (an anticancer drug) induced genotoxicity in mice. Drug Chem Toxicol 38(2):188–195. https://doi.org/10.3109/01480545.2014.928720, PMID 24956935

Hosseinimehr SJ (2010) Flavonoids and genomic instability induced by ionizing radiation. Drug Discov Today 15(21–22):907–918. https://doi.org/10.1016/j.drudis.2010.09.005, PMID 20 933097

Izquierdo-Vega JA, Morales-González JA, SánchezGutiérrez M, Betanzos-Cabrera G, Sosa-Delgado SM, Sumaya-Martínez MT et al (2017) Evidence of some natural products with antigenotoxic effects. Part 1: fruits and polysaccharides. Nutrients 9(2):102. https://doi.org/10.3390/nu9020102, PMID 28157162

Kanehisa M, Goto S, Kawashima S, Okuno Y, Hattori M (2004) The KEGG resource for deciphering the genome. Nucleic Acids Res 32(Database issue):D277–D280. https://doi.org/10.1093/nar/gkh063, PMID 14681412

Kondo N, Takahashi A, Ono K, Ohnishi T (2010) DNA damage induced by alkylating agents and repair pathways. J Nucleic Acids 2010:543531. https://doi.org/10.4061/2010/543531, PMID 21113301

Kovalchuk I (2021) Cancer and genomic instability. In: Genome stability. Academic, New York, pp 495–519

Kumar SS, Devasagayam TP, Jayashree B, Kesavan PC (2001) Mechanism of protection against radiation-induced DNA damage in plasmid pBR322 by caffeine. Int J Radiat Biol 77(5): 617–623. https://doi.org/10.1080/09553000110034649, PMID 11382340

Lee SJ, Ahn H, Nam KW, Kim KH, Mar W (2012) Effects of rutaecarpine on hydrogen peroxide-induced apoptosis in murine hepa-1c1c7 cells. Biomol Ther (Seoul) 20(5):487–491. https://doi.org/10.4062/biomolther.2012.20.5.487, PMID 24009839

Lehmann F, Wennerberg J (2021) Evolution of nitrogen-based alkylating anticancer agents. Processes 9(2):377. https://doi.org/10.3390/pr9020377

Lewinska A, Jarosz P, Czech J, Rzeszutek I, Bielak-Zmijewska A, Grabowska W et al (2015) Capsaicin-induced genotoxic stress does not promote apoptosis in A549 human lung and DU145 prostate cancer cells. Mutat Res Genet Toxicol Environ Mutagen 779:23–34. https://doi.org/10.1016/j.mrgentox.2015.02.003, PMID 25813723

Li L, Zou L (2005) Sensing, signaling, and responding to DNA damage: organization of the checkpoint pathways in mammalian cells. J Cell Biochem 94(2):298–306. https://doi.org/10.1002/jcb.20355, PMID 15578575

Li Y, Park MS, Chan WK, Livesey J, Zhou Z (2019) Effect of rutaecarpine on the benzo(a)pyrene-induced DNA damage in vitro. FASEB J 33(Suppl 1):672–679. https://doi.org/10.1096/fasebj.2019.33.1_supplement.672.9

Liu YQ, Wang XL, He DH, Cheng YX (2021) Protection against chemotherapy- and radiotherapy-induced side effects: a review based on the mechanisms and therapeutic opportunities of phytochemicals. Phytomedicine 80:153402. https://doi.org/10.1016/j.phymed.2020.153402, PMID 33203590

López-Romero D, Izquierdo-Vega JA, Morales-González JA, Madrigal-Bujaidar E, Chamorro-Cevallos G, Sánchez-Gutiérrez M et al (2018) Evidence of some natural products with antigenotoxic effects. Part 2: plants, vegetables, and natural resin. Nutrients 10(12):1954. https://doi.org/10.3390/nu10121954, PMID 30544726

Luca VS, Miron A, Aprotosoaie AC (2016) The antigenotoxic potential of dietary flavonoids. Phytochem Rev 15(4):591–625. https://doi.org/10.1007/s11101-016-9457-1

Matsuura HN, Fett-Neto AG (2015) Plant alkaloids: main features, toxicity, and mechanisms of action. Plant Toxins 2(7):1–5

Mohamed SA, Upreti S, Rajendra SV, Dang R (2017) Genotoxicity: mechanisms, testing guidelines and methods. Glob J Pharm Pharm Sci 1(5):133–138

Mohanan G, Das A, Rajyaguru PI (2021) Genotoxic stress response: what is the role of cytoplasmic mRNA fate? BioEssays 43(8):e2000311. https://doi.org/10.1002/bies.202000311, PMID 340 96096

Morales-González JA, Madrigal-Bujaidar E, Sánchez-Gutiérrez M, Izquierdo-Vega JA, Valadez-Vega MDC, Álvarez-González I et al (2019) Garlic (Allium sativum L.): a brief review of its antigenotoxic effects. Foods 8(8):343. https://doi.org/10.3390/foods8080343, PMID 31412555

Moura DJ, Richter MF, Boeira JM, Pêgas Henriques JA, Saffi J (2007) Antioxidant properties of β-carboline alkaloids are related to their antimutagenic and antigenotoxic activities. Mutagenesis 22(4):293–302. https://doi.org/10.1093/mutage/gem016, PMID 17545209

Nadar VM, Manivannan S, Chinnaiyan R, Govarthanan M, Ponnuchamy K (2022) Review on marine sponge alkaloid, aaptamine: a potential antibacterial and anticancer drug. Chem Biol Drug Des 99(1):103–110. https://doi.org/10.1111/cbdd.13932, PMID 34331335

Nehlig A, Debry G (1994) Potential genotoxic, mutagenic and antimutagenic effects of coffee: a review. Mutat Res 317(2):145–162. https://doi.org/10.1016/0165-1110(94)90022-1, PMID 7511793

Nelson BC, Wright CW, Ibuki Y, Moreno-Villanueva M, Karlsson HL, Hendriks G et al (2017) Emerging metrology for high-throughput nanomaterial genotoxicology. Mutagenesis 32(1): 215–232. https://doi.org/10.1093/mutage/gew037, PMID 27565834

Pathak MP, Policegoudra RS, Goyary D, Das A, Mandal S, Chakraborti S et al (2018) Safety evaluation of an oat grain alkaloid gramine by genotoxicity assays. Drug Chem Toxicol 41(2): 147–154. https://doi.org/10.1080/01480545.2017.1322605, PMID 28523965

Poirier MC (2012) Chemical-induced DNA damage and human cancer risk. Discov Med 14(77): 283–288. https://doi.org/10.1038/nrc1410, PMID 23114584

Proshkina E, Shaposhnikov M, Moskalev A (2020) Genome-protecting compounds as potential geroprotectors. Int J Mol Sci 21(12):4484. https://doi.org/10.3390/ijms21124484, PMID 32 599754

Rajagopalan R, Ranjan SK, Nair CK (2003) Effect of vinblastine sulfate on γ-radiation-induced DNA single-strand breaks in murine tissues. Mutat Res 536(1–2):15–25. https://doi.org/10.1016/s1383-5718(03)00015-9, PMID 12694742

Raymond P, Thompson C, Longstaff E (2004) Examination of the potential genotoxicity of pure capsaicin in bacterial mutation chromosome aberration and rodent micronucleus tests. Environ Mol Mutagen 44(5):441–447. 10.1002/(ISSN)1098-2280. 10.1002/em.v44:5 https://doi.org/10.1002/em.20072

Reddig A, Rübe CE, Rödiger S, Schierack P, Reinhold D, Roggenbuck D (2018) DNA damage assessment and potential applications in laboratory diagnostics and precision medicine. J Lab Precis Med 3:31. https://doi.org/10.21037/jlpm.2018.03.06

Reen RK, Wiebel FJ, Singh J (1997) Piperine inhibits aflatoxin b1-induced cytotoxicity and genotoxicity in V79 Chinese hamster cells genetically engineered to express rat cytochrome P4502B1. J Ethnopharmacol 58(3):165–173. https://doi.org/10.1016/s0378-8741(97)00104-9, PMID 9421252

Reimann H, Bankoglu EE, Stopper H, Hintzsche H (2021) In vitro evaluation of chromosomal damage and DNA strand breaks after treatment with the poppy seed alkaloid thebaine. Mutat Res Genet Toxicol Environ Mutagen 870–871:503393. https://doi.org/10.1016/j.mrgentox.2021.503393, PMID 34583825

Richeux F, Cascante M, Ennamany R, Sanchez D, Sanni A, Saboureau D et al (2000) Implications of oxidative stress and inflammatory process in the cytotoxicity of capsaicin in human endothelial cells: lack of DNA strand breakage. Toxicology 147(1):41–49. https://doi.org/10.1016/s0300-483x(00)00184-0, PMID 10837931

Sak K (2012) Chemotherapy and dietary phytochemical agents. Chemother Res Pract 2012:282570. https://doi.org/10.1155/2012/282570, PMID 23320169

Sehgal A, Kumar M, Jain M, Dhawan DK (2011) Combined effects of curcumin and piperine in ameliorating benzo(a)pyrene induced DNA damage. Food Chem Toxicol 49(11):3002–3006. https://doi.org/10.1016/j.fct.2011.07.058, PMID 21827816

Sehgal A, Kumar M, Jain M, Dhawan DK (2012) Synergistic effects of piperine and curcumin in modulating benzo(a)pyrene induced redox imbalance in mice lungs. Toxicol Mech Methods 22(1):74–80. https://doi.org/10.3109/15376516.2011.603392, PMID 21859361

Sehgal A, Kumar M, Jain M, Dhawan DK (2013) Modulatory effects of curcumin in conjunction with piperine on benzo(a)pyrene-mediated DNA adducts and biotransformation enzymes. Nutr Cancer 65(6):885–890. https://doi.org/10.1080/01635581.2013.805421, PMID 23909733

Selvendiran K, Padmavathi R, Magesh V, Sakthisekaran D (2005) Preliminary study on inhibition of genotoxicity by piperine in mice. Fitoterapia 76(3–4):296–300. https://doi.org/10.1016/j.fitote.2005.03.016, PMID 15890459

Shiloh Y (2006) The ATM-mediated DNA-damage response: taking shape. Trends Biochem Sci 31(7):402–410. https://doi.org/10.1016/j.tibs.2006.05.004, PMID 16774833

Sirbu BM, Cortez D (2013) DNA damage response: three levels of DNA repair regulation. Cold Spring Harb Perspect Biol 5(8):a012724. https://doi.org/10.1101/cshperspect.a012724, PMID 23813586

Sjakste N, Djelić N, Dzintare M, Živković L (2020) LDNA-BINDING and DNA-protecting activities of small natural organic molecules and food extracts. Chem Biol Interact 323: 109030. https://doi.org/10.1016/j.cbi.2020.109030, PMID 32205154

Stoilov LM, Mullenders LH, Natarajan AT (1994) Caffeine potentiates or protects against radiation-induced DNA and chromosomal damage in human lymphocytes depending on temperature and concentration. Mutat Res 311(2):169–174. https://doi.org/10.1016/0027-5107(94)90174-0, PMID 7526181

Stuhldreier F, Kassel S, Schumacher L, Wesselborg S, Proksch P, Fritz G (2015) Pleiotropic effects of spongean alkaloids on mechanisms of cell death, cell cycle progression and DNA damage response (DDR) of acute myeloid leukemia (AML) cells. Cancer Lett 361(1):39–48. https://doi.org/10.1016/j.canlet.2015.02.030, PMID 25697484

Swift LH, Golsteyn RM (2014) Genotoxic anti-cancer agents and their relationship to DNA damage, mitosis, and checkpoint adaptation in proliferating cancer cells. Int J Mol Sci 15(3): 3403–3431. https://doi.org/10.3390/ijms15033403, PMID 24573252

Tanveer MA, Rashid H, Nazir LA, Archoo S, Shahid NH, Ragni G et al (2023) Trigonelline, a plant derived alkaloid prevents ultraviolet-B-induced oxidative DNA damage in primary human dermal fibroblasts and BALB/c mice via modulation of phosphoinositide 3-kinase-Akt-Nrf2 signalling axis. Exp Gerontol 171:112028. https://doi.org/10.1016/j.exger.2022.112028, PMID 36384201

Turkez H, Arslan ME, Ozdemir O (2017) Genotoxicity testing: progress and prospects for the next decade. Expert Opin Drug Metab Toxicol 13(10):1089–1098. https://doi.org/10.1080/17425255.2017.1375097, PMID 28889778

Villarroya M, García AG, Marco-Contelles J, López MG (2007) An update on the pharmacology of galantamine. Expert Opin Investig Drugs 16(12):1987–1998. https://doi.org/10.1517/13543784.16.12.1987, PMID 18042006

Yadav V, Krishnan A, Zahiruddin S, Ahmad S, Vohora D (2023a) Amelioration of cyclophosphamide-induced DNA damage, oxidative stress, and hepato-and neurotoxicity by Piper longum extract in rats: the role of γH2AX and 8-OHdG. Front Pharmacol 14:1147823. https://doi.org/10.3389/fphar.2023.1147823, PMID 36969834

Yadav V, Krishnan A, Vohora D (2023b) Altered hallmarks of DNA double-strand breaks, oxidative DNA damage and cytogenotoxicity by piperlongumine in hippocampus and hepatocytes of rats intoxicated with cyclophosphamide. Life Sci 316:121391. https://doi.org/10.1016/j.lfs.2023.121391, PMID 36657641

Yang J, Xu ZP, Huang Y, Hamrick HE, Duerksen-Hughes PJ, Yu YN (2004) ATM and ATR: sensing DNA damage. World J Gastroenterol 10(2):155–160. https://doi.org/10.3748/wjg.v10.i2.155, PMID 14716813

Yubo L, Chen Q, Zhang N, Zhang K, Dou T, Cao Y, Liu Y, Li K, Hao X, Xie X, Li W, Ren Y, Zhang J (2021) Author correction: proteomic profiling and genome-wide mapping of O-GlcNAc chromatinassociated proteins reveal an O-GlcNAc-regulated genotoxic stress response. Nat Commun 12(1). https://doi.org/10.1038/s41467-021-26886-5

Zhou BB, Elledge SJ (2000) The DNA damage response: putting checkpoints in perspective. Nature 408(6811):433–439. https://doi.org/10.1038/35044005, PMID 11100718

Zhou J, Chan L, Zhou S (2012) Trigonelline: a plant alkaloid with therapeutic potential for diabetes and central nervous system disease. Curr Med Chem 19(21):3523–3531. https://doi.org/10.2174/092986712801323171, PMID 22680628

Cellular Stress, Survival Mechanism and Adaptation

16

Sarika Singh and Mukesh Nandave

Abstract

At the cellular level, stress can be defined as an array of processes that get sparked as a result of an external stimulus, thereby leading to acute or chronic deviation from the normal cellular milieu and homeostasis with the intent of negating the impairment together with fixing the injury, thereby ultimately safeguarding the cell or the organism against any insults. Therefore, the induction of stress response signifies the presence of the stressor and the induced damage; however, stress itself is not the insult. When exposed to a particular stress, the cells try to counter it in multiple ways, like the oxidative stress response system, heat shock response system, genotoxic response system, metabolic/energetic response system, hypoxia response system, and so on, which ensure the continuity or may lead to programmed cell death by way of which the injured cells are ultimately removed. So, the various cellular death mechanisms such as necrosis, apoptosis, etc., bank on the cell's capacity to combat cellular stress.

Keywords

Cellular stress · Survival mechanisms · Adaptation · Oxidative stress response system · Heat shock response system · Genotoxic response system · Hypoxia response system · Metabolic/energetic response system

S. Singh
School of Allied Health Sciences and Management, Delhi Pharmaceutical Sciences and Research University (DPSRU), New Delhi, India

M. Nandave (✉)
Department of Pharmacology, Delhi Pharmaceutical Sciences and Research University (DPSRU), New Delhi, India
e-mail: mukesh.nandave@dpsru.edu.in

R. Tulsawani, D. Vohora (eds.), *Adaptation under Stressful Environments through Biological Adjustments and Interventions*,
https://doi.org/10.1007/978-981-99-7652-2_16

16.1 Introduction

Health, well-being and performance are affected by the interaction with different stresses at living and working environment. The degree of interaction determines the severity of the hazard. Exposure to adverse climatic conditions such as heat, cold, humidity, wind velocity, solar radiation, barometric pressure, toxicants, isolation, noise, vibrations, electromagnetic fields, etc. may lead to alterations in metabolic, behavioral and cognitive processes due to stress-induced demands posed on the body's resources. Research is discovering an increasing number of stressors capable of adversely affecting the physical and mental performance of populations and leading to stress-induced disorders.

Health has been appropriately defined by the World Health Organization as 'a state of complete physical, mental and social well-being and not merely the absence of disease or infirmity'. This definition entails the ability of the body and mind to quickly adapt itself to constantly changing environmental conditions and perform optimally. Since human beings remain in a state of dynamic equilibrium, exposure to varied stress factors may give rise to destabilization with changed behavioral, psychological, neurological, physiological and biochemical profiles. Though a minimal amount of stress is necessary for evolution, development and higher achievements, prolonged and continuous stress beyond a certain limit can temporarily or permanently impair the functioning of vital body functions. To control these effects, the threshold of stress, that should not be crossed, needs to be recognized in conjunction with risk assessment. Failure to recognize such a threshold of stress on the part of a person through appropriate risk assessment techniques/markers is associated with stress-induced maladjustments and ill health/disease.

Hans Selye (1936) aptly identified a stressor as any stimulus, which brings about measurable changes in the biochemistry, physiology and histopathology of the neuroendocrine system along with effects on body organs and its functions as a result of the acclimatization process. However, when this stress is extended for longer durations, the homeostatic equilibrium gets unbalanced thereby impairing functions of several organ systems (Takeda et al. 2004). The consequences could be the manifestation of various physiopathologies. Hence, along with the subjective evaluation, the physiological or hormonal or cellular stress responses merit attention to discern the stress response.

16.2 Cellular Stress

Cellular stress can be defined as an array of processes which get sparked as a result of the external stimulus, thereby leading to acute or chronic deviation from the normal cellular milieu and homeostasis with the intent of negating the impairment together with fixing the injury, thereby ultimately safeguarding the cell or the organism against any insults. Therefore, the induction of stress response signifies the presence of the stressor and the induced damage; however, itself it is not the insult. The cells may be subjected to various kinds of stress. When exposed to a particular stress, the cells try to counter it in multiple ways like invigoration of cellular pathways which ensure the continuity or may lead to programmed cell death by way of which the injured cells are ultimately removed. Initially, the cell tries to cope up with the stress and restore the normal functioning and injury. Nonetheless, in case the stress keeps continuing, then the cellular death signaling pathways get sparked. Therefore, the cell's survival is a function of its potential to soar through the stress with correct responses and this response is extremely conserved in evolution as evident from the presence of the heat shock proteins and antioxidant machinery in higher as well as lower organisms.

To add, the cellular reaction to encounter the stress banks on the type as well as the degree of injury. For instance, safeguarding cellular retorts like the expression of heat shock or unfolded protein reaction arbitrate a rise in chaperone protein activity that boosts the protein folding ability of the cell, thereby negating the stress and enhancing cellular recovery. Hence, the cellular acclimatizing potential basically depicts its fate. So, based on the amount and stress conditions, the cell takes up varied defense mechanisms together with prosurvival approaches. Furthermore, in case these defense mechanisms do not work, then the cellular death programs get sparked up so as the injured cells can be taken care of. Therefore, the various cellular death mechanisms such as necrosis, apoptosis, etc., bank on the cell's capacity to combat cellular stress.

16.3 Survival Mechanism and Adaptation

As a result of the stressors, reflected by way of the cellular stress, alterations in the extracellular conditions may percolate into structural and functional modifications at the molecular level, thereby affecting the vital biomolecules like lipids, proteins, DNA and various other biomolecules. To cope up, organisms have also developed defense mechanisms to cellular stress, thereby safeguarding the existence of stress-acclimatized cells in arduous situations together with elimination of injured cells. The cellular defense mechanisms back up the various functions of the cell, thereby preserving homeostasis. However, they can also lead to disturbances after a while, consequently accelerating senescence and/or diseases like cardiovascular abnormalities, diabetes, neurodegeneration, cancer to name a few (Kourtis and Tavernarakis 2011). The body adapts against any stress by way of activating the cellular retort processes which in turn are guided by the stress-sparked signaling pathways ultimately deciding the destiny of the cell. These cellular retort processes comprise

1. The oxidative stress response system—sparked by surplus reactive oxygen species (ROS) and disturbances in the oxidant/antioxidant profile within cells (Kehrer and Lund 1994; Slimen et al. 2014);

2. The heat shock response system-induced as a result of heat and exposure to other factors negatively affecting protein folding and RNA metabolism (Richter et al. 2010);
3. The genotoxic response system—sparked by DNA damage (Ciccia and Elledge 2010);
4. The metabolic/energetic response system—caused due alterations in metabolites (Kroemer et al. 2010); and
5. The hypoxia response system—gets activated in reduced oxygen levels (Majmundar et al. 2010).

Therefore, the cell by way of the cellular stress response tries to weigh and neutralize the injury caused as a result of the ongoing stress, thereby enhancing the strength to combat the damage for some time and eliminating out the injured cells by apoptosis.

16.3.1 Oxidative Stress Response System

Oxidative stress can be defined as the shift in the cellular environment from the native reducing state toward an oxidizing state because of increased reactive oxygen species (ROS) and reactive nitrogen species (RNS) and decreased antioxidant levels. The capacity of biological systems to combat the excess production of ROS/RNS via the antioxidant defense mechanism is altered resulting in damage to the biomolecules like proteins, lipids and nucleic acids (Bartsch and Nair 2000). Hence, a condition of oxidative stress occurs when the critical cellular redox balance is disrupted due to depletion of antioxidants or excess accumulation of ROS or both beyond the ability of the defenses to cope, and thus causes its detrimental consequences. Such stress occurs when severely adverse environmental or physiologic conditions overwhelm biological systems. Therefore, one rapid and clear indicator of oxidative stress is the induction of antioxidant defenses and/or increases in endogenous ROS/RNS levels. Besides, ROS production and accumulation is a common denominator in many diseases and environmental insults and can lead to severe cellular damage leading to physiological dysfunctions together with the manifestation of subjective symptoms and signs. Therefore, evaluation of cellular stress retort is a much more sensitive tool to outline the biological effects of stressors as it is one of the well-recognized biological means toward self-defense and protection from varied harmful environmental stimuli.

16.3.1.1 Free Radicals and Reactive Species

A free radical can be identified as a molecule possessing an unpaired electron and having the potential of independent existence (Halliwell and Gutteridge 1989). The unpaired electron in the atomic orbital makes these species unstable, and to gain stability, it can give or pull out an electron from other molecules and hence acts like an oxidant or reductant. For the most part, oxygen-derived free radicals like superoxide and hydroxyl radical are involved in the pathophysiology of majority of ailments (Young and Woodside 2001). Together the endogenous (mitochondrial leakage, respiratory burst, enzyme and auto-oxidation reactions) and environmental factors (smoking, alcohol, pollutants, UV light, ionizing radiation, iron-overload, pesticides, pharmacological agents) control the generation of these free radicals by several mechanisms in the body. For instance, these radicals and reactive species are derivatives of oxygen metabolism responsible for the injury and obliteration caused to biomolecules, thus depleting the efficiency of biological systems.

The persisting damage is a threat to the life of cell organelles and the cells ultimately culminating into apoptosis or necrosis. The superoxide anion generated as a result of normal cellular respiration or stress further leads to the generation of a range of more reactive and harmful varieties. Superoxide radicals pair with nitric oxide (NO) to give peroxynitrite anion ($ONOO^-$) having the capability to cause far-reaching molecular obliteration (Radi et al. 1991). Additionally, superoxide radicals can undergo enzymatic dismutation by superoxide dismutases (SOD) to yield lesser toxic hydrogen peroxide (H_2O_2). H_2O_2 is not a free radical, but owing to its comparatively longer half-life along with the capacity to cross the biomembranes, it can disperse the injury produced by free radical over larger region. In addition, the formation of hydroxyl ($^{\bullet}OH$) radical from H_2O_2 via the Fenton reaction spreads the damage to distant sites. Fenton reaction is a process in which hydroxyl free radicals are generated from hydrogen peroxide produced during mitochondrial oxidative respiration in the presence of transition metals like iron (Phillips et al. 2009; Jajte et al. 2001).

$$Fe^{2+} + H_2O_2 \rightarrow Fe^{3+} + {}^{\bullet}OH + OH^-$$

The activity of enzymes like cytochrome p450 oxidase in liver and enzymes engaged in the production of adrenal hormones also cause the escape of electrons leading to superoxide generation.

16.3.1.2 Antioxidant Defense System

The cellular machinery is well equipped with the multiple antioxidant defenses to prevent free radical and reactive species-induced injury. The various antioxidants behave as radical scavengers, electron and hydrogen donors, peroxide decomposer, singlet oxygen quencher as well as enzyme inhibitor in order to provide the requisite protection. The enzymatic and the nonenzymatic defense mechanisms work in cooperation in the intra and extracellular compartments of the cell to combat free radical induced obliteration as illustrated in Fig. 16.1.

The antioxidant defense system of the biological organism functions on various lines and on that basis are categorized into four categories (Valko et al. 2007). The first line of defense comprises the preventive antioxidants like SOD, glutathione peroxidase (GSH-Px), catalase (CAT), etc., which restrict the production of free radical itself. The second line constitutes chain initiation or propagation reaction inhibiting antioxidants (vitamin C, E, GSH, albumin) in case of radical generation. The protein-bound thiol units in plasma also serve as important chain-breaking

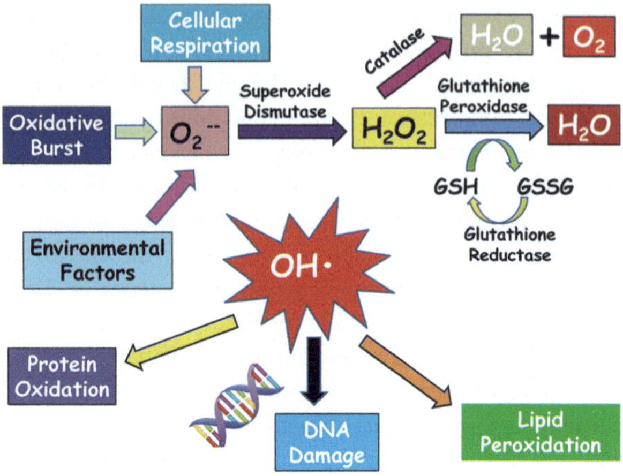

Fig. 16.1 Schematic diagram showing free radicals generation and damage along with major endogenous antioxidants in cells

antioxidants as the sulphydryl groups defuse the free radical by giving an electron with the consequent generation of protein thiyl radical (Young and Woodside 2001).

Albumin being the predominant plasma protein provides significant part of the sulphydryl groups along with numerous additional antioxidant functions (Halliwell 1988). The subsequent class includes the DNA repair enzymes often termed as repair and de novo antioxidants. The remaining one is the adaptation line that comprises the manufacturing of needed antioxidants together with their shipping to the place of action in order to obliterate the radicals. Nevertheless, in certain conditions the antioxidant levels may get perturbed because of surplus ROS and free radical production, inactivation of detoxification defenses and disparity in their demand and supply subsequently leading to oxidative stress situation in the cell.

The enzymatic antioxidants like GSH-Px, CAT and GST are well established to forage the ROS manufactured in the body. GSH-Px being an important endogenous enzymatic antioxidant occupies a significant part in H_2O_2 and lipid hydroperoxide removal along with the reduction of peroxides by GSH utilization as a hydrogen donor (Irmak et al. 2002; Ilhan et al. 2004). Hence, GSH-Px activity depression may be regarded as an oxidative upshot due to severe oxidant concentration of H_2O_2. Besides GSH-Px, CAT also converts the toxic H_2O_2 into water, thereby protecting the cells from oxidative stress insults. CAT mainly occupies the peroxiosomes of the cell, which also houses the majority of H_2O_2-producing enzymes. GSH-Px oxidizes GSH to the detriment of hydroperoxide, which may be hydrogen peroxide or other lipid hydroperoxide molecules as well. During the course of action of GSH-Px, GSH is consumed and is oxidized to yield oxidized glutathione (GSSG). In order to maintain the redox status of the cell and GSH-Px activity, the GSSG molecule is converted back into GSH by another enzyme Glutathione reductase (GR). GR is a flavin nucleotide-dependent enzyme and chiefly responsible for maintaining the very high ratio of GSH to GSSG (Young and Woodside 2001). GSH therefore acts as an antioxidant by playing a crucial part in xenobiotic detoxification and foraging of free radicals or ROS, thereby preserving the redox status of cells (Reiter et al. 2000). The detoxifying enzyme glutathione-*S*-transferase (GST) principally catalyzes the conjugation of p450 metabolites with endogenous ligand (reduced glutathione) (Hartman and Shankel 1990).

On the contrary, the chain-breaking antioxidants are small biomolecules, which combine with the radicals by either donation or acceptance of an electron to form stable by-products (Halliwell 1995). These nonenzymatic antioxidants mainly constitute the lipid or aqueous phase chain-breaking antioxidants. The lipid-phase antioxidants like vitamin E, carotenoids, flavonoids, etc., defuse radicals in biomembranes and lipoprotein molecules and hence protect from lipid peroxidation insult. On the other hand, the aqueous phase chain-breaking antioxidants such as vitamin C, uric acid, GSH, plasma proteins like albumin, etc., directly forage the radicals in the aqueous compartments.

16.3.1.3 Biomarkers of Oxidative Stress

The specification and oxidative stress effects in conjunction to knowledge concerning ROS/RNS nature can be accumulated by distinct evaluation of oxidative/nitrosative stress biomarkers in the biological samples; in addition, these biomarkers also signify the normal/pathological state of the biological system (Palmieri and Sblendorio 2007). Because of the short half-life of these ROS/RNS species owing to greater reactivity, generally these species are traced by their comparatively stable by-products formed as a result of their interaction with other biomolecules like the lipid peroxidation and protein oxidation end products or target oxidant levels. The increase in malondialdehyde (MDA) levels and reduction of reduced GSH molecules and GSH/GSSG ratio are some well-documented biomarkers of oxidative stress.

High MDA levels are a consequence of lipid peroxidation, which is the domino effect of oxidative stress injury. MDA is a well-known oxidative stress biomarker reflecting lipid peroxidation generated as a result of the oxidation of polyunsaturated fatty acids in a chain reaction. MDA is the end-product of lipid peroxidation and its reaction with thiobarbituric acid (TBA) is used as a marker for MDA quantification. Polyunsaturated fatty acids contained in phospholipids are the chief constituent of biological cell membranes. Cell membrane fluidity is influenced because of lipid peroxidation, which further increases membrane permeability and depolarization. Free radicals are assumed to alter the function of membrane channels viz. sodium, potassium and calcium (Bartosz 2008).

Total antioxidant status (TAS) is also a key biomarker, which can give fruitful insights about the health condition or any disease or stress in the body. The biological system employs these antioxidant reserves to combat the stress-inducing free radicals and ROS, thereby protecting the body from the oxidative insult. The antioxidant levels if checked periodically, may aid in the early recognition of any disorder. Hence, a thorough antioxidant profile can yield the true picture of all the antioxidant levels and their variation with respect to any exposure or disease. Any reduction in the body's antioxidant defense mechanism (TAS etc.) and rise in oxidative stress markers reflects the generalized drop in antioxidant status. The consequent fall in antioxidant status can be due to the consumption of the various antioxidants to counteract the excess free radicals generated. The extent of oxidative/nitrosative stress injury can also be assessed by the measurement of antioxidant enzyme activities in addition to the concentration of mid- and end-products of molecular oxidation or lipid peroxidation.

16.3.2 Heat Shock Response System (HSRS)

HSRS is one of the important prosurvival or cytoprotective mechanism of the cells. It was assumed to be activated in response to mild heat stress, having temperature rises ranging between 3 and 5 °C above normal with the rapid expression of a type of protein called as heat shock proteins (Craig 1985; Lindquist 1986). Now, it is well known that various nonthermal stressors such as heavy metals, oxidative stress, infections, pharmacological agents, etc., can also induce this cellular reaction. This cytoprotective retort is an evolutionarily preserved transcriptional program mediated by the Hsf1 (heat shock factor 1) protein (Morimoto 1998). Trimers are produced by the combination of three copies of Hsf1 protein which then bind with the promoters of Hsf1-dependent genes. As a result, there is overexpression of heat shock proteins in the cell including the chaperones. Protein injury being the prime cellular outcome of these stresses give rise to agglomeration of unfolded proteins. The cell offsets this by raising the chaperone protein expression. The chaperone proteins address the cellular stress by folding and refolding the misfolded proteins together with the diminution of protein agglomeration, as well as inducing autophagy, thereby offering short-term protection (Lindquist 1986). This gives rise to a state termed as thermotolerance characterized by increased resistance in the cells against varied stresses like oxidative stress, rise in temperatures, different anticancer medicines, etc. (Samali and Orrenius 1998; Samali et al. 1999). The manufacturing of heat shock protein dwindles to the baseline values as soon as the cell cancels out the ill effects of the stress.

To add, as the HSRS gets triggered, the normal processes of protein transcription and translation gets interrupted so as to diminish the burden of misfolded proteins. But, the particular group of safeguarding genes, i.e. the heat shock factors (HSFs), are selectively stimulated and their expression is raised by the respective transcription factors (Morimoto et al. 1996). There are three varied types of HSFs in cells of vertebrates: HSF1 is vital for the heat shock response together with its role in developmental processes, HSF2 and HSF4 are crucial for differentiation as well as development, whereas HSF3 is restricted to avian cells (Pirkkala et al. 2001; Shabtay and Arad 2006). It has been found that the cells deficient in HSF1 are susceptible to stress and are incapable of establishing thermotolerance (McMillan et al. 1998; Zhang et al. 2002), corroborating the significance of HSF1 specifically for heat shock response. Heat shock proteins (Hsps) are evolutionary conserved proteins. They are categorized into subfamilies having molecular mass of roughly 110, 90, 70, 60, 40, and 15–30 kDa (Samali and Orrenius 1998; Jäättelä 1999). Among these, for instance, Hsp90 are mainly expressed, playing the role of molecular chaperones in the cell, thereby limiting the untimely folding of nascent proteins (Lindquist and

Craig 1988). Remaining Hsps, specifically Hsp27 and Hsp70, are normally produced at low basal levels; however, stressors may lead to their surge and therefore they are called as inducible Hsps and are also important element of the heat shock response (Samali and Orrenius 1998).

Hsp27 is part of a stress protein subfamily, the small Hsps, found essentially in all organisms and can be also managed by phosphorylation as well as energetic association/dissociation into multimers starting from dimers to large oligomers (Garrido 2002). At the same time, Hsp70 is the inducible part of the 70 kDa Hsps family. Hsp27 and Hsp70 both have been documented to safeguard the cells against the initiation of cell death by varied stressors and by diverse mechanisms of cell death, comprising apoptosis (Samali and Cotter 1996; Jäättelä et al. 1998) and necrosis (Kabakov and Gabai 1995; Mehlen et al. 1995; Champagne et al. 1999). These Hsps work directly by way of suppressing the cell death pathways, or indirectly by way of the normal prosurvival proceedings. For instance, these Hsps as molecular chaperones adhere to and promote the refolding of unfolded proteins as a result checking the assemblage of protein (Hartl and Hayer-Hartl 2002). As Hsp27 can collaborate with actin and hence is significant for preserving the cytoskeleton integrity which may have a function in fostering survival (Concannon et al. 2003). Besides these indirect processes, Hsp27 and Hsp70 can precisely check apoptosis through the regulation of both intrinsic and extrinsic apoptotic pathways as well as meddling with caspase stimulation at various levels (Pandey et al. 2000; Concannon et al. 2001). To add, both of these Hsps have been documented to directly obstruct the liberation of proapoptotic elements, comprising cytochrome *c* from the mitochondria (Samali et al. 2001; Chauhan et al. 2003; Steel et al. 2004). Further, because of their capacity to attach with cytochrome *c* and procaspase-3 (in the case of Hsp27) (Pandey et al. 2000; Concannon et al. 2001) and procaspases-3, -7 and Apaf-1 (in the case of Hsp70) (Jäättelä et al. 1998; Saleh et al. 2000; Beere et al. 2000; Komarova et al. 2004), these Hsps can obstruct apoptosome development as well as the stimulation of downstream caspases in the cytoplasm. Additionally, Hsp70 checks the apoptotic nuclear alterations by way of collaborating with as well as hindering apoptosis-inducing factor (AIF) (Creagh et al. 2000; Ravagnan et al. 2001). Together, Hsps possess the capacity to regulate the death receptor pathway. DAXX, an adaptor protein joining the Fas death receptor and the ER stress sensor IRE1 to ASK-1 and downstream JNK proapoptotic signaling is documented to get hindered by Hsp27 (Charette and Landry 2000). It has been also found that Hsp70 checks JNK activity (Gabai et al. 2000; Park and Liu 2001). Furthermore, cell survivability is also governed by the collaboration of Hsp27 and 70 with additional proteins.

For instance, Hsp27 corroborates with prosurvival Ser/Thr kinase Akt, known for its significance for continued Akt activity (Konishi et al. 1997; Rane et al. 2003). To add, Hsp70 can form complex with co-chaperones, comprising DnaJ/Hsp40 and BAG-1, thereby regulating apoptosis (Takayama et al. 1997; Qiu et al. 2006). On the whole, Hsps are established prosurvival as well as antiapoptotic biomolecules, which get triggered by diverse variety of stressors, and in turn preserve the cellular milieu by affecting varied cellular processes, thereby depicting the destiny of the cell.

16.3.3 Genotoxic Response System (GRS)

The integrity of cellular genome is incredibly pivotal for maintaining appropriate cellular growth together with prosperous transfer of genetic data to the future generation (Anand et al. 2020). At the same time, various genotoxic stressors challenge the genome integrity (Christmann et al. 2003; Roos and Kaina 2006), thereby leading to the emergence as well as development of suitable compensatory mechanisms to counter DNA insult and hence, preserve genomic integrity (Ferreira and Dutreix 2019; Shukla et al. 2020). Therefore, the Genotoxic Response System (GRS) comes into action as soon as the cell's blue print, i.e. the DNA gets injured thereby endangering the cell's genomic integrity. GRS primarily comprises the various processes such as signaling, fixing DNA insults and injury tolerance in order to restore or repair the harm caused due to the genotoxic stress (Sirbu and Cortez 2013; Basu 2018; Turgeon et al. 2018).

DNA damage is a ubiquitous phenomenon in the cellular milieu arising due to varied endogenous stressors like faults in DNA replication, insertion of wrong nucleotide in the DNA sequence, mutation in repair enzymes as well as exogenous stressors such as environmental genotoxic agents [polycyclic hydrocarbons or ultraviolet (UV) light], radiation exposure, chemotherapeutic agents to name a few (Christmann et al. 2003; Roos and Kaina 2006). DNA injury can originate in a number of ways such as due to alterations in base (generation of 8-oxo-7,8-dihydroguanine from 8-oxoguanine), genesis of pyrimidine dimer, etc. Nevertheless, regardless of the myriad kinds of DNA adducts and base alterations happening in the genetic material, the DNA insults can be classified into some main categories like double-strand breaks (DSBs), single-strand breaks (SSBs), nucleotide adduct generation together with base alteration and base pair mismatch. Therefore, correspondingly the important categories of DNA repair comprise DSB restoration by homologous recombination or non-homologous end-joining (NHEJ), nucleotide excision repair (NER) and nucleotide mismatch repair (MMR) (Kültz 2005).

A good deal of the cellular apparatus responsible for the detection of DNA insult is considerably conserved in eukaryotes and prokaryotes, however, with substantial variation. Nonetheless, some elements of this intricate nexus like MutL/MLH, MutS/MSH, RecA/RadA/Rad51, Mre11/Rad32, Top I/III, Rad50 and MutT/MTH are extremely maintained in all the three super kingdoms (Kültz 2005). Although, the last-mentioned two proteins exhibit a diminished level of homology amid prokaryotes and eukaryotes. Genomic integrity gets compromised as a result of the mutations in all these proteins, thereby affecting cellular stress response signaling networks. Bacterial MutS and eukaryotic MSH proteins possess the ability to distinguish and attach to the lesions formed as a result of ill-matched DNA base pairing (Sixma 2001). In mammalian cells, BRCA1-associated genome surveillance supercomplex (BASC) plays a crucial role in the identification as well as the restoration of DNA base mismatches along with the detection of remaining kinds of injuries (Wang et al. 2000). The MMR proteins such as MSH2, MSH6 and MLH1 belong to the huge BASC in the eukaryotic cell. Once MutS/MSH proteins attach to the locations of DNA insults, they employ MutL (in bacterial cells) or MLH (in eukaryotic cells), thereby starting the congregation of MMR complex. Therefore, in all life forms, MutS/MSH and MutL/MLH proteins play a crucial role in discerning DNA ill-matches and their detection potential is enhanced by overexpression of MutS/MSH proteins at the time of stress.

DNA damage response (DDR) gets instigated primarily because of DNA double- and single-strand breaks (Roos and Kaina 2006). The DNA is more susceptible to chemical insults or nucleases when it gets unwind into two single-stranded DNA strands, for instance, during DNA duplication and transcription. Under these situations, mainly, the single-strand breaks (SSBs) originate (Richard et al. 2009). To add, SSBs may also originate at the time of DNA repair, for instance, during the process of NER (Fulda et al. 2010). Once the DNA insult is identified, dual incision 5′ to the DNA injury by ERCC1-XPF as well as 3′ to the insult by XPG eliminates the same (Staresincic et al. 2009). On the other hand, double-strand breaks (DSBs) generate as a result of several anticancer medicines involving topoisomerase blockers, DNA alkylating, intercalating or crosslinking agents, together with nucleotide analogs (Roos and Kaina 2006). When DSBs originate, MRE-11-Rad50-NBS1 (MRN) complex engages ataxia telangiectasia mutated (ATM) at the points of DNA incision followed by downstream substrate phosphorylation, for instance, checkpoint kinase 2 and p53 (Christmann et al. 2003; Harper and Elledge 2007). Furthermore, as p53 prompts transcriptional actuation of various functional schemes, such as cell cycle regulatory proteins like p21 together with proapoptotic factors, for instance, CD95, PUMA and BAX (Vousden and Lane 2007). Additionally, researchers have documented the non-transcriptional proapoptotic role of p53 which controls the mitochondria intervened apoptotic pathway (Vaseva and Moll 2009). Deleterious injuries to the DNA biomolecule in the cell may ultimately put the cell's subsistence at stake; however, this is averted by activating the DNA restoration procedures (Roos and Kaina 2006). However, in case the injury is too acute to be mended, the cellular GRS induces the effector systems to arbitrate cell death (Roos and Kaina 2006).

16.3.4 Hypoxia Response System (HRS)

Oxygen is vital for the cellular survival and cellular oxygen sensing mechanism has been well elucidated (Kaelin and Ratcliffe 2008; Ratcliffe 2013; Lee et al. 2020; Semenza 2020; Wicks and Semenza 2022). Hypoxia, insufficient oxygen supply, leads to oxidative burst (Smith et al. 2017) resulting in changes in metabolic and function state at cellular level (Taylor and Scholz 2022). Hypoxia response system (HRS) is another stress responsive system that is triggered in mammalian cells subjected to low oxygen environment. The low oxygen environment creates hypoxia within cellular setup and under hypoxic conditions, cells sense and respond through

HRS. The HRS response under hypoxic microenvironment is mediated via activation of redox-sensitive transcriptional factor, hypoxia-inducible transcription factor that was purified from its binding to hypoxia-responsive element (HRE) within the promoter of the erythropoietin gene (Wang and Semenza 1995; Wang et al. 1995). The lack of oxygen at cellular level results in redox imbalance, energetic stress and metabolic changes within the cells. This is accompanied by reprogramming of transcription factors, particularly, HIF1 and gene expression. HIF1 is master regulator of cellular adaptation to low oxygen environment (Iommarini et al. 2017; Li et al. 2019; Yfantis et al. 2023). Under normal conditions, prolyl hydroxylases (PHDs) hydroxylate hypoxia-inducible factor 1, HIF-1, on two proline residues, triggering pVHL-mediated ubiquitination and proteasomal degradation of hydroxylated HIF. However, in hypoxia, hydroxylation of HIF1a by PHDs is inhibited that dimerizes with HIF-1β (HIF-1 complex) selectively stabilized HIF1 and activating several genes for survival and maintaining redox and energy homeostasis including erythropoietin (EPO) for controlling erythropoiesis, glucose transporter-1 (GLUT1) and lactate dehydrogenase A (LDH-A) to balance energy homeostasis and cellular adaptation, nitric oxide synthase (NOS) to modulate vascular tone, vascular endothelial growth factor (VEGF) for regulating angiogenesis (Salceda and Caro 1997; Wenger 2002; Giaccia et al. 2004; Cavadas et al. 2013; Iommarini et al. 2017; Li et al. 2019; Chen et al. 2020; Liao and Zhang 2020; Semenza 2022; Taylor and Scholz 2022; Yfantis et al. 2023). Besides, in chronic hypoxic stress, HIF-1 promote cellular apoptosis under conditions of chronic hypoxia (Carmeliet et al. 1998) via induction of Nip3, pro-death factor (Bruick 2000). Overall, HIF-dependent hypoxic stress response is key feature and target in many ischemic diseases, pulmonary hypertension and cellular adaptation to hypoxic stress (Wenger 2002; Iommarini et al. 2017; Li et al. 2019; Strowitzki et al. 2019; Yfantis et al. 2023) and hypoxia-induced oxidative stress (Li et al. 2019).

16.3.5 Metabolic/Energetic Response System (M/ERS)

Energy is essential for sustaining cellular metabolic processes and also for stress adaptation (Picard et al. 2018). M/ERS stress response system involves energy mobilization from energy stores to maintain energy homeostasis to maintain metabolism processes (van der Kooij 2020). The metabolic response to stimuli is complex phenomenon that requires involvement of endocrine, immune and nervous system. However, persistent stress causes imbalance in energy homeostasis resulting in energetic stress which disarranges stress response and results in metabolic disorder. The metabolic changes depend on the stress stimuli and duration and frequency of the exposure. The MRS behaves differently in acute and chronic stress conditions. In acute stress case, stress stimuli causes hypothalamus-pituitary-adrenal (HPA) axis to respond and lead to the release of corticotropin-releasing hormone (CRH) that stimulates the sympathetic nervous system (SNS) and catecholamine release resulting in stress fasting as an adaptive mechanism to maintain energy homeostasis (Ketchesin et al. 2017; Mifsud and Reul 2018; van der Valk et al. 2018). CRH released following acute stress exposure can suppress appetite through inhibition of neuropeptide Y (NPY) that facilitates food intake (Tsigos et al. 2000; Spencer and Tilbrook 2011; Fardet and Fève 2014). Chronic stress leads to disturbed energy homeostasis and uncontrolled release of glucocorticoids (GCs) such as cortisol and neuropeptide Y resulting in stress eating and finally causing stress-linked metabolic dysfunction and disorders such as obesity and diabetes. In chronic stress, GCs stimulate NPY and it is also reported that secretion of GCs switch food preferences toward fat and sucrose-rich foods (Tataranni et al. 1996; Tsigos et al. 2000; George et al. 2010; Spencer and Tilbrook 2011; van der Valk et al. 2018). Glucocorticoid receptor (GR) is an essential component of negative feedback loop in stressful environment (Herman et al. 2016).

It is documented that energy availability and stress response are interdependent reciprocally and any changes in energy availability and demand may disturb energy homeostasis at cellular level and results in energetic stress. Harrell and authors have summarized that M/ERS stress is based on the energy homeostasis state at particular situation and involves multiple mechanism. M/ERS is a responsive system to dietary imbalance, GCs signaling within or outside hypothalamic-pituitary-adrenal (HPA) axis, sympathetic nervous system (SNS) activity, gut-brain, endocrine system and inflammation that control metabolic activities (Emery Thompson et al. 2010; Bland and Birnbaum 2011; Miller and Hamilton 2012; Picard et al. 2018).

16.4 Conclusion

The cellular stress retort is a highly intricate process meant for the subsistence of robust cells as well as the elimination of cells injured at the time of stress. This bodily retort gets induced as a consequence of external stressors of varied types, thereby leading to acute or chronic deviation from the normal cellular milieu and homeostasis with the aim of minimizing the damage together with restoring the insults, thereby ultimately ensuring the cell survival. Hence, inauguration of stress response denotes the existence of the stressor as well as the induced damage; however, itself it is not the insult. In a stressful situation, depending on the amount, cell type and stress conditions, the cell copes with the ongoing stress through multiple ways such as invigoration of the various cellular pathways—oxidative stress response system, heat shock response system, genotoxic response system, hypoxia response system, metabolic/energetic response system to name some, which ensure the continuity or cell death. Therefore, the various cellular death mechanisms such as necrosis, apoptosis, etc., bank on the cell's capacity to combat cellular stress. The cellular defense mechanisms support the varied cellular roles, thereby preserving homeostasis, though, they may later cause problems, thereby speeding senescence and/or diseases. Therefore, the body acclimatizes against a particular stress through stimulating the specific cellular retort processes which in turn are steered by the stress-sparked signaling pathways eventually determining the fate of the cell.

References

Anand SK, Sharma A, Singh N, Kakkar P (2020) Entrenching role of cell cycle checkpoints and autophagy for maintenance of genomic integrity [DNA rep]. DNA Repair (Amst) 86:102748. https://doi.org/10.1016/j.dnarep.2019.102748, PMID 31790874

Bartosz G (2008) The other face of oxygen. PWN, Warszawa, pp 99–107. [in Polish]

Bartsch H, Nair J (2000) Ultrasensitive and specific detection methods for exocytic DNA adducts; markers for lipid peroxidation and oxidative stress. Toxicology 153(1–3):105–114. https://doi.org/10.1016/s0300-483x(00)00307-3, PMID 11090950

Basu AK (2018) DNA damage, mutagenesis and cancer. Int J Mol Sci 19(4):970. https://doi.org/10.3390/ijms19040970, PMID 29570697

Beere HM, Wolf BB, Cain K, Mosser DD, Mahboubi A, Kuwana T et al (2000) Heat-shock protein 70 inhibits apoptosis by preventing recruitment of procaspase-9 to the Apaf-1 apoptosome. Nat Cell Biol 2(8):469–475. https://doi.org/10.1038/35019501, PMID 10934466

Bland ML, Birnbaum MJ (2011) Cell biology. ADaPting to energetic stress. Science 332(6036): 1387–1388. https://doi.org/10.1126/science.1208444, PMID 21680830

Bruick RK (2000) Expression of the gene encoding the proapoptotic Nip3 protein is induced by hypoxia. Proc Natl Acad Sci U S A 97(16):9082–9087. https://doi.org/10.1073/pnas.97.16.9082, PMID 10922063, PMCID PMC16825

Carmeliet P, Dor Y, Herbert JM, Fukumura D, Brusselmans K, Dewerchin M et al (1998) Role of HIF-1alpha in hypoxia-mediated apoptosis, cell proliferation and tumour angiogenesis. Nature 394(6692):485–490. https://doi.org/10.1038/28867. Erratum in: Nature Keshet E [corrected to Keshert E]. 1998;395(6701):525, PMID 9697772

Cavadas MA, Nguyen LK, Cheong A (2013) Hypoxia-inducible factor (HIF) network: insights from mathematical models. Cell Commun Signal 11(1):42. https://doi.org/10.1186/1478-811X-11-42, PMID 23758895, PMCID PMC3686674

Champagne MJ, Dumas P, Orlov SN, Bennett MR, Hamet P, Tremblay J (1999) Protection against necrosis but not apoptosis by heat-stress proteins in vascular smooth muscle cells: evidence for distinct modes of cell death. Hypertension 33(3):906–913. https://doi.org/10.1161/01.hyp.33.3.906, PMID 10082507

Charette SJ, Landry J (2000) The interaction of HSP27 with Daxx identififies a potential regulatory role of HSP27 in Fas. Ann N Y Acad Sci 926:126–131. https://doi.org/10.1111/j.1749-6632.2000.tb05606.x, PMID 11193028

Chauhan D, Li G, Hideshima T, Podar K, Mitsiades C, Mitsiades N et al (2003) Hsp27 inhibits release of mitochondrial protein Smac in multiple myeloma cells and confers dexamethasone resistance. Blood 102(9):3379–3386. https://doi.org/10.1182/blood-2003-05-1417, PMID 12 855565

Chen PS, Chiu WT, Hsu PL, Lin SC, Peng IC, Wang CY et al (2020) Pathophysiological implications of hypoxia in human diseases. J Biomed Sci 27(1):63. https://doi.org/10.1186/s12929-020-00658-7, PMID 32389123, PMCID PMC7212687

Christmann M, Tomicic MT, Roos WP, Kaina B (2003) Mechanisms of human DNA repair: an update. Toxicology 193(1–2):3–34. https://doi.org/10.1016/s0300-483x(03)00287-7, PMID 14 599765

Ciccia A, Elledge SJ (2010) The DNA damage response: making it safe to play with knives. Mol Cell 40(2):179–204. https://doi.org/10.1016/j.molcel.2010.09.019, PMID 20965415

Concannon CG, Orrenius S, Samali A (2001) Hsp27 inhibits cytochrome c-mediated caspase activation by sequestering both pro-caspase-3 and cytochrome c. Gene Expr 9(4–5):195–201. https://doi.org/10.3727/000000001783992605, PMID 11444529

Concannon CG, Gorman AM, Samali A (2003) On the role of Hsp27 in regulating apoptosis. Apoptosis 8(1):61–70. https://doi.org/10.1023/a:1021601103096, PMID 12510153

Craig EA (1985) The heat shock response. CRC Crit Rev Biochem 18(3):239–280. https://doi.org/10.3109/10409238509085135, PMID 2412760

Creagh EM, Carmody RJ, Cotter TG (2000) Heat shock protein 70 inhibits caspase-dependent and -independent apoptosis in Jurkat T cells. Exp Cell Res 257(1):58–66. https://doi.org/10.1006/excr.2000.4856, PMID 10854054

Emery Thompson ME, Muller MN, Kahlenberg SM, Wrangham RW (2010) Dynamics of social and energetic stress in wild female chimpanzees. Horm Behav 58(3):440–449. https://doi.org/10.1016/j.yhbeh.2010.05.009, PMID 20546741

Fardet L, Fève B (2014) Systemic glucocorticoid therapy: a review of its metabolic and cardiovascular adverse events. Drugs 74(15):1731–1745. https://doi.org/10.1007/s40265-014-0282-9, PMID 25204470

Ferreira S, Dutreix M (2019) DNA repair inhibitors to enhance radiotherapy: progresses and limitations. Cancer Radiother 23(8):883–890. https://doi.org/10.1016/j.canrad.2019.08.008, PMID 31615730

Fulda S, Gorman AM, Hori O, Samali A (2010) Cellular stress responses: cell survival and cell death. Int. J Cell Biol 2010:214074. https://doi.org/10.1155/2010/214074, 23 pages, PMID 201 82529

Gabai VL, Yaglom JA, Volloch V, Meriin AB, Force T, Koutroumanis M et al (2000) Hsp72-mediated suppression of c-Jun N-terminal kinase is implicated in development of tolerance to caspase-independent cell death. Mol Cell Biol 20(18):6826–6836. https://doi.org/10.1128/MCB.20.18.6826-6836.2000, PMID 10958679

Garrido C (2002) Size matters: of the small HSP27 and its large oligomers. Cell Death Differ 9(5): 483–485. https://doi.org/10.1038/sj.cdd.4401005, PMID 11973606

George SA, Khan S, Briggs H, Abelson JL (2010) CRH-stimulated cortisol release and food intake in healthy, non-obese adults. Psychoneuroendocrinology 35(4):607–612. https://doi.org/10.1016/j.psyneuen.2009.09.017, PMID 19828258, Google Scholar

Giaccia AJ, Simon MC, Johnson R (2004) The biology of hypoxia: the role of oxygen sensing in development, normal function, and disease. Genes Dev 18(18):2183–2194. https://doi.org/10.1101/gad.1243304, PMID 15371333, PMCID PMC517513

Halliwell B (1988) Albumin an important extracellular antioxidant? Biochem Pharmacol 37(4): 569–571. https://doi.org/10.1016/0006-2952(88)90126-8, PMID 3277637

Halliwell B (1995) How to characterize an antioxidant: an update. Biochem Soc Symp 61:73–101. https://doi.org/10.1042/bss0610073, PMID 8660405

Halliwell B, Gutteridge JM (1989) Free radicals in biology and medicine, 2nd edn. Clarendon Press, Oxford

Harper JW, Elledge SJ (2007) The DNA damage response: ten years after. Mol Cell 28(5):739–745. https://doi.org/10.1016/j.molcel.2007.11.015, PMID 18082599

Hartl FU, Hayer-Hartl M (2002) Molecular chaperones in the cytosol: from nascent chain to folded protein. Science 295(5561):1852–1858. https://doi.org/10.1126/science.1068408, PMID 11884 745

Hartman PE, Shankel DW (1990) Antimutagens and anticarcinogens a survey of putative interceptor molecules. Environ Mol Mutat Res 202:285–306

Herman JP, McKlveen JM, Ghosal S, Kopp B, Wulsin A, Makinson R et al (2016) Regulation of the hypothalamic-pituitary-adrenocortical stress response. Compr Physiol 6(2):603–621. https://doi.org/10.1002/cphy.c150015, PMID 27065163

Ilhan A, Gurel A, Armutcu F, Kamisli S, Iraz M, Akyol O et al (2004) Ginkgo biloba prevents mobile phone induced oxidative stress in rat brain. Clin Chim Acta 340(1–2):153–162. https://doi.org/10.1016/j.cccn.2003.10.012, PMID 14734207

Iommarini L, Porcelli AM, Gasparre G, Kurelac I (2017) Non-canonical mechanisms regulating hypoxia-inducible factor 1 alpha in cancer. Front Oncol 27(7):286. https://doi.org/10.3389/fonc.2017.00286, PMID 29230384, PMCID PMC5711814

Irmak MK, Fadillioğlu E, Güleç M, Erdoğan H, Yağmurca M, Akyol O (2002) Effects of electromagnetic radiation from a cellular telephone on the oxidant and antioxidant levels in rabbits. Cell Biochem Funct 20(4):279–283. https://doi.org/10.1002/cbf.976, PMID 12415560

Jäättelä M (1999) Heat shock proteins as cellular lifeguards. Ann Med 31(4):261–271. https://doi.org/10.3109/07853899908995889, PMID 10480757

Jäättelä M, Wissing D, Kokholm K, Kallunki T, Egeblad M (1998) Hsp70 exerts its anti-apoptotic function downstream of caspase-3-like proteases. EMBO J 17(21):6124–6134. https://doi.org/10.1093/emboj/17.21.6124, PMID 9799222

Jajte J, Zmyślony M, Palus J, Dziubałtowska E, Rajkowska E (2001) Protective effect of melatonin against in vitro iron ions and 7 mT 50-Hz magnetic field-induced DNA damage in rat lymphocytes. Mutat Res 483(1–2):57–64. https://doi.org/10.1016/s0027-5107(01)00230-5, PMID 11600133

Kabakov AE, Gabai VL (1995) Heat-shock-induced accumulation of 70-kDa stress protein (HSP70) can protect ATP-depleted tumor cells from necrosis. Exp Cell Res 217(1):15–21. https://doi.org/10.1006/excr.1995.1058, PMID 7867713

Kaelin WG Jr, Ratcliffe PJ (2008) Oxygen sensing by metazoans: the central role of the HIF hydroxylase pathway. Mol Cell 30(4):393–402. https://doi.org/10.1016/j.molcel.2008.04.009, PMID 18498744

Kehrer JP, Lund LG (1994) Cellular reducing equivalents and oxidative stress. Free Radic Biol Med 17(1):65–75. https://doi.org/10.1016/0891-5849(94)90008-6, PMID 7959167

Ketchesin KD, Stinnett GS, Seasholtz AF (2017) Corticotropin-releasing hormone-binding protein and stress: from invertebrates to humans. Stress 20(5):449–464. https://doi.org/10.1080/10253890.2017.1322575, PMID 28436309

Komarova EY, Afanasyeva EA, Bulatova MM, Cheetham ME, Margulis BA, Guzhova IV (2004) Downstream caspases are novel targets for the antiapoptotic activity of the molecular chaperone Hsp70. Cell Stress Chaperones 9(3):265–275. https://doi.org/10.1379/csc-27r1.1, PMID 1 5544164

Konishi H, Matsuzaki H, Tanaka M, Takemura Y, Kuroda S, Ono Y et al (1997) Activation of protein kinase B (Akt/RAC-protein kinase) by cellular stress and its association with heat shock protein Hsp27. FEBS Lett 410(2–3):493–498. https://doi.org/10.1016/s0014-5793(97)00541-3, PMID 9237690

Kourtis N, Tavernarakis N (2011) Cellular stress response pathways and ageing: intricate molecular relationships. EMBO J 30(13):2520–2531. https://doi.org/10.1038/emboj.2011.162, PMID 21 587205

Kroemer G, Mariño G, Levine B (2010) Autophagy and the integrated stress response. Mol Cell 40(2):280–293. https://doi.org/10.1016/j.molcel.2010.09.023, PMID 20965422

Kültz D (2005) Molecular and evolutionary basis of the cellular stress response. Annu Rev Physiol 67:225–257. https://doi.org/10.1146/annurev.physiol.67.040403.103635, PMID 15709958

Lee CC, Wu CY, Yang HY (2020) Discoveries of how cells sense oxygen win the 2019 Nobel Prize in Physiology or Medicine. Biom J 43(5):434–437. https://doi.org/10.1016/j.bj.2020.05.019, PMID 33012698

Li HS, Zhou YN, Li L, Li SF, Long D, Chen XL et al (2019) HIF-1α protects against oxidative stress by directly targeting mitochondria. Redox Biol 25:101109. https://doi.org/10.1016/j.redox.2019.101109, PMID 30686776, PMCID PMC6859547

Liao C, Zhang Q (2020) Understanding the oxygen-sensing pathway and its therapeutic implications in diseases. Am J Pathol 190(8):1584–1595. https://doi.org/10.1016/j.ajpath.2020.04.003, PMID 32339495, PMCID PMC7416076

Lindquist S (1986) The heat-shock response. Annu Rev Biochem 55:1151–1191. https://doi.org/10.1146/annurev.bi.55.070186.005443, PMID 2427013

Lindquist S, Craig EA (1988) The heat-shock proteins. Annu Rev Genet 22:631–677. https://doi.org/10.1146/annurev.ge.22.120188.003215, PMID 2853609

Majmundar AJ, Wong WJ, Simon MC (2010) Hypoxia-inducible factors and the response to hypoxic stress. Mol Cell 40(2):294–309. https://doi.org/10.1016/j.molcel.2010.09.022, PMID 20965423

McMillan DR, Xiao X, Shao L, Graves K, Benjamin IJ (1998) Targeted disruption of heat shock transcription factor 1 abolishes thermotolerance and protection against heat-inducible apoptosis. J Biol Chem 273(13):7523–7528. https://doi.org/10.1074/jbc.273.13.7523, PMID 9516453

Mehlen P, Preville X, Chareyron P, Briolay J, Klemenz R, Arrigo AP (1995) Constitutive expression of human hsp27, Drosophila hsp27, or human αB-crystallin confers resistance to TNF- and oxidative stress-induced cytotoxicity in stably transfected murine L929 fifibroblasts. J Immunol 154(1):363–374. https://doi.org/10.4049/jimmunol.154.1.363, PMID 7995955

Mifsud KR, Reul JMHM (2018) Mineralocorticoid and glucocorticoid receptor-mediated control of genomic responses to stress in the brain. Stress 21(5):389–402. https://doi.org/10.1080/10253890.2018.1456526, PMID 29614900

Miller BF, Hamilton KL (2012) A perspective on the determination of mitochondrial biogenesis. Am J Physiol Endocrinol Metab 302(5):E496–E499. https://doi.org/10.1152/ajpendo.00578.2011, PMID 22205627

Morimoto RI (1998) Regulation of the heat shock transcriptional response: cross talk between a family of heat shock factors, molecular chaperones, and negative regulators. Genes Dev 12(24):3788–3796. https://doi.org/10.1101/gad.12.24.3788, PMID 9869631

Morimoto RI, Kroeger PE, Cotto JJ (1996) The transcriptional regulation of heat shock genes: a plethora of heat shock factors and regulatory conditions. EXS 77:139–163. https://doi.org/10.1007/978-3-0348-9088-5_10, PMID 8856973

Palmieri B, Sblendorio V (2007) Oxidative stress tests: overview on reliability and use. Part II. Eur Rev Med Pharmacol Sci 11(6):383–399, PMID 18306907

Pandey P, Farber R, Nakazawa A, Kumar S, Bharti A, Nalin C et al (2000) Hsp27 functions as a negative regulator of cytochrome c-dependent activation of procaspase-3. Oncogene 19(16):1975–1981. https://doi.org/10.1038/sj.onc.1203531, PMID 10803458

Park J, Liu AYC (2001) JNK phosphorylates the HSF1 transcriptional activation domain: role of JNK in the regulation of the heat shock response. J Cell Biochem 82(2):326–338. https://doi.org/10.1002/jcb.1163, PMID 11527157

Phillips JL, Singh NP, Lai H (2009) Electromagnetic fields and DNA damage. Pathophysiology 16(2–3):79–88. https://doi.org/10.1016/j.pathophys.2008.11.005, PMID 19264461

Picard M, McEwen BS, Epel ES, Sandi C (2018) An energetic view of stress: focus on mitochondria. Front Neuroendocrinol 49:72–85. https://doi.org/10.1016/j.yfrne.2018.01.001, PMID 29339091, PMCID PMC5964020

Pirkkala L, Nykänen P, Sistonen L (2001) Roles of the heat shock transcription factors in regulation of the heat shock response and beyond. FASEB J 15(7):1118–1131. https://doi.org/10.1096/fj00-0294rev, PMID 11344080

Qiu XB, Shao YM, Miao S, Wang L (2006) The diversity of the DnaJ/Hsp40 family, the crucial partners for Hsp70 chaperones. Cell Mol Life Sci 63(22):2560–2570. https://doi.org/10.1007/s00018-006-6192-6, PMID 16952052

Radi R, Beckman JS, Bush KM, Freeman BA (1991) Peroxynitrite oxidation of sulfhydryls. The cytotoxic potential of superoxide and nitric oxide. J Biol Chem 266(7):4244–4250. https://doi.org/10.1016/S0021-9258(20)64313-7, PMID 1847917

Rane MJ, Pan Y, Singh S, Powell DW, Wu R, Cummins T et al (2003) Heat shock protein 27 controls apoptosis by regulating Akt activation. J Biol Chem 278(30):27828–27835. https://doi.org/10.1074/jbc.M303417200, PMID 12740362

Ratcliffe PJ (2013) Oxygen sensing and hypoxia signalling pathways in animals: the implications of physiology for cancer. J Physiol 591(8):2027–2042. https://doi.org/10.1113/jphysiol.2013.251470, PMID 23401619

Ravagnan L, Gurbuxani S, Susin SA, Maisse C, Daugas E, Zamzami N et al (2001) Heat-shock protein 70 antagonizes apoptosis-inducing factor. Nat Cell Biol 3(9):839–843. https://doi.org/10.1038/ncb0901-839, PMID 11533664

Reiter RJ, Tan DX, Osuna C, Gitto E (2000) Actions of melatonin in the reduction of oxidative stress. A review. J Biomed Sci 7(6):444–458. https://doi.org/10.1007/BF02253360, PMID 110 60493

Richard DJ, Bolderson E, Khanna KK (2009) Multiple human single-stranded DNA binding proteins function in genome maintenance: structural, biochemical and functional analysis. Crit Rev Biochem Mol Biol 44(2–3):98–116. https://doi.org/10.1080/10409230902849180, PMID 1 9367476

Richter K, Haslbeck M, Buchner J (2010) The heat shock response: life on the verge of death. Mol Cell 40(2):253–266. https://doi.org/10.1016/j.molcel.2010.10.006, PMID 20965420

Roos WP, Kaina B (2006) DNA damage-induced cell death by apoptosis. Trends Mol Med 12(9): 440–450. https://doi.org/10.1016/j.molmed.2006.07.007, PMID 16899408

Salceda S, Caro J (1997) Hypoxia-inducible factor 1α (HIF-1α) protein is rapidly degraded by the ubiquitin-proteasome system under normoxic conditions. Its stabilization by hypoxia depends on redox-induced changes. J Biol Chem 272(36):22642–22647. https://doi.org/10.1074/jbc.272.36.22642, PMID 9278421

Saleh A, Srinivasula SM, Balkir L, Robbins PD, Alnemri ES (2000) Negative regulation of the Apaf-1 apoptosome by Hsp70. Nat Cell Biol 2(8):476–483. https://doi.org/10.1038/35019510, PMID 10934467

Samali A, Cotter TG (1996) Heat shock proteins increase resistance to apoptosis. Exp Cell Res 223(1):163–170. https://doi.org/10.1006/excr.1996.0070, PMID 8635489

Samali A, Orrenius S (1998) Heat shock proteins: regulators of stress response and apoptosis. Cell Stress Chaperones 3(4):228–236. https://doi.org/10.1379/1466-1268(1998)003<0228:hspros>2.3.co;2, PMID 9880235

Samali A, Holmberg CI, Sistonen L, Orrenius S (1999) Thermotolerance and cell death are distinct cellular responses to stress: dependence on heat shock proteins. FEBS Lett 461(3):306–310. https://doi.org/10.1016/s0014-5793(99)01486-6, PMID 10567716

Samali A, Robertson JD, Peterson E, Manero F, van Zeijl L, Paul C et al (2001) Hsp27 protects mitochondria of thermotolerant cells against apoptotic stimuli. Cell Stress Chaperones 6(1): 49–58. https://doi.org/10.1379/1466-1268(2001)006<0049:hpmotc>2.0.co;2, PMID 1152 5243

Selye H (1936) A syndrome produced by diverse noxious agents. Nature 138(3479):32–33. https://doi.org/10.1038/138032a0

Semenza GL (2020) The genomics and genetics of oxygen homeostasis. Annu Rev Genomics Hum Genet 21:183–204. https://doi.org/10.1146/annurev-genom-111119-073356, PMID 32255719

Semenza GL (2022) Breakthrough science: hypoxia-inducible factors, oxygen sensing, and disorders of hematopoiesis. Blood 139(16):2441–2449. https://doi.org/10.1182/blood.2021011043, PMID 34411243

Shabtay A, Arad Z (2006) Reciprocal activation of HSF1 and HSF3 in brain and blood tissues: is redundancy developmentally related? Am J Physiol Regul Integr Comp Physiol 291(3):R566–R572. https://doi.org/10.1152/ajpregu.00685.2005, PMID 16497816

Shukla V, Halabelian L, Balagere S, Samaniego-Castruita D, Feldman DE, Arrowsmith CH et al (2020) HMCES functions in the alternative end-joining pathway of the DNA DSB repair during class switch recombination in B cells. Mol Cell 77(2):384–394.e4. https://doi.org/10.1016/j.molcel.2019.10.031

Sirbu BM, Cortez D (2013) DNA damage response: three levels of DNA repair regulation. Cold Spring Harb Perspect Biol 5(8):a012724. https://doi.org/10.1101/cshperspect.a012724, PMID 23813586

Sixma TK (2001) DNA mismatch repair: MutS structures bound to mismatches. Curr Opin Struct Biol 11(1):47–52. https://doi.org/10.1016/s0959-440x(00)00169-x, PMID 11179891

Slimen IB, Najar T, Ghram A, Dabbebi H, Ben Mrad M, Abdrabbah M (2014) Reactive oxygen species, heat stress and oxidative-induced mitochondrial damage. A review. Int J Hyperthermia 30(7):513–523. https://doi.org/10.3109/02656736.2014.971446, PMID 25354680

Smith KA, Waypa GB, Schumacker PT (2017) Redox signaling during hypoxia in mammalian cells. Redox Biol 13:228–234. https://doi.org/10.1016/j.redox.2017.05.020, PMID 28595160

Spencer SJ, Tilbrook A (2011) The glucocorticoid contribution to obesity. Stress 14(3):233–246. https://doi.org/10.3109/10253890.2010.534831, PMID 21294656, Google Scholar

Staresincic L, Fagbemi AF, Enzlin JH, Gourdin AM, Wijgers N, Dunand-Sauthier I et al (2009) Coordination of dual incision and repair synthesis in human nucleotide excision repair. EMBO J 28(8):1111–1120. https://doi.org/10.1038/emboj.2009.49, PMID 19279666

Steel R, Doherty JP, Buzzard K, Clemons N, Hawkins CJ, Anderson RL (2004) Hsp72 inhibits apoptosis upstream of the mitochondria and not through interactions with Apaf-1. J Biol Chem 279(49):51490–51499. https://doi.org/10.1074/jbc.M401314200, PMID 15371421

Strowitzki MJ, Cummins EP, Taylor CT (2019) Protein hydroxylation by hypoxia-inducible factor (HIF) hydroxylases: unique or ubiquitous? Cell 8(5):384. https://doi.org/10.3390/cells8050384, PMID 31035491, PMCID PMC6562979

Takayama S, Bimston DN, Matsuzawa SI, Freeman BC, Aime-Sempe C, Xie Z et al (1997) BAG-1 modulates the chaperone activity of Hsp70/Hsc70. EMBO J 16(16):4887–4896. https://doi.org/10.1093/emboj/16.16.4887, PMID 9305631

Takeda E, Terao J, Nakaya Y, Miyamoto K, Baba Y, Chuman H et al (2004) Stress control and human nutrition. J Med Investig 51(3–4):139–145. https://doi.org/10.2152/jmi.51.139, PMID 1 5460899

Tataranni PA, Larson DE, Snitker S, Young JB, Flatt JP, Ravussin E (1996) Effects of glucocorticoids on energy metabolism and food intake in humans. Am J Phys 271(2 Pt 1): E317–E325. https://doi.org/10.1152/ajpendo.1996.271.2.E317, PMID 8770026, Google Scholar

Taylor CT, Scholz CC (2022) The effect of HIF on metabolism and immunity. Nat Rev Nephrol 18(9):573–587. https://doi.org/10.1038/s41581-022-00587-8, PMID 35726016, PMCID PMC9208707

Tsigos C, Kyrou I, Kassi E, Chrousos GP (2000) Stress, endocrine physiology and pathophysiology. MDText.com, Inc., South Dartmouth

Turgeon MO, Perry NJS, Poulogiannis G (2018) DNA damage, repair, and cancer metabolism. Front Oncol 8:15. https://doi.org/10.3389/fonc.2018.00015, PMID 29459886

Valko M, Leibfritz D, Moncol J, Cronin MT, Mazur M, Telser J (2007) Free radicals and antioxidants in normal physiological functions and human disease [review]. Int J Biochem Cell Biol 39(1):44–84. https://doi.org/10.1016/j.biocel.2006.07.001, PMID 16978905

van der Kooij MA (2020) The impact of chronic stress on energy metabolism. Mol Cell Neurosci 107:103525. https://doi.org/10.1016/j.mcn.2020.103525, PMID 32629109

van der Valk ES, Savas M, van Rossum EFC (2018) Stress and obesity: are there more susceptible individuals? Curr Obes Rep 7(2):193–203. https://doi.org/10.1007/s13679-018-0306-y, PMID 29663153, PMCID PMC5958156

Vaseva AV, Moll UM (2009) The mitochondrial p53 pathway. Biochim Biophys Acta 1787(5): 414–420. https://doi.org/10.1016/j.bbabio.2008.10.005, PMID 19007744

Vousden KH, Lane DP (2007) p53 in health and disease. Nat Rev Mol Cell Biol 8(4):275–283. https://doi.org/10.1038/nrm2147, PMID 17380161

Wang GL, Semenza GL (1995) Purification and characterization of hypoxia-inducible factor 1. J Biol Chem 270(3):1230–1237. https://doi.org/10.1074/jbc.270.3.1230, PMID 7836384

Wang GL, Jiang BH, Rue EA, Semenza GL (1995) Hypoxia-inducible factor 1 is a basic helix-loop-helix-PAS heterodimer regulated by cellular O2 tension. Proc Natl Acad Sci U S A 92(12): 5510–5514. https://doi.org/10.1073/pnas.92.12.5510, PMID 7539918

Wang Y, Cortez D, Yazdi P, Neff N, Elledge SJ, Qin J (2000) BASC, a super complex of BRCA1-associated proteins involved in the recognition and repair of aberrant DNA structures. Genes Dev 14(8):927–939. https://doi.org/10.1101/gad.14.8.927, PMID 10783165

Wenger RH (2002) Cellular adaptation to hypoxia: O2-sensing protein hydroxylases, hypoxia-inducible transcription factors, and O2-regulated gene expression. FASEB J 16(10):1151–1162. https://doi.org/10.1096/fj.01-0944rev, PMID 12153983

Wicks EE, Semenza GL (2022) Hypoxia-inducible factors: cancer progression and clinical translation. J Clin Invest 132(11):e159839. https://doi.org/10.1172/JCI159839, PMID 35642 641

Yfantis A, Mylonis I, Chachami G, Nikolaidis M, Amoutzias GD, Paraskeva E et al (2023) Transcriptional response to hypoxia: the role of HIF-1-associated co-regulators. Cell 12(5): 798. https://doi.org/10.3390/cells12050798, PMID 36899934

Young IS, Woodside JV (2001) Antioxidants in health and disease. J Clin Pathol 54(3):176–186. https://doi.org/10.1136/jcp.54.3.176, PMID 11253127

Zhang Y, Huang L, Zhang J, Moskophidis D, Mivechi NF (2002) Targeted disruption of hsf1 leads to lack of thermotolerance and defines tissue-specific regulation for stress-inducible hsp molecular chaperones. J Cell Biochem 86(2):376–393. https://doi.org/10.1002/jcb.10232, PMID 12112007

Malnutrition-Induced Oxidative Stress in Nervous System and Its Health Implications

17

Shobi Veleri

Abstract

Neural development, especially in the brain, is a rapid metabolic process that demands a huge supply of building blocks and energy. Nutrients are loaded with essential ingredients for this need. Early brain development requires an ample supply of proteins and calories. Fatty acids are necessary to build the elaborate membrane structures of neurons. The neurons in the brain consume huge amounts of energy produced in the body and are readily generated from glucose. Metabolic pathways rely on many enzymes to generate energy or building blocks for the cells. Many enzymes' functions depend on vitamins and minerals that are obtained from micronutrient sources. Malnutrition is a serious environmental stress, and it creates oxidative stress in cells. Foods rich in antioxidants such as leaves and vegetables are necessary for early development and for functional maintenance of the nervous system. The lack of balanced maternal nutrition during in utero development and the first 1000 days of life can compromise the structural and functional integrity of neurons. Severe malnutrition induces oxidative stress in the developing neurons that eventually damages cellular functioning. This may lead to neural dysfunctions, psychological disorders, and cognitive impairment in childhood and adulthood.

Keywords

Malnutrition · Nutrients · Neuronal diseases · Neural development

S. Veleri (✉)
Drug Safety Division, ICMR-National Institute of Nutrition, Department of Health Research, Ministry of Health & Family Welfare, Government of India, Hyderabad, Telangana, India
e-mail: shobi.veleri@icmr.gov.in

R. Tulsawani, D. Vohora (eds.), *Adaptation under Stressful Environments through Biological Adjustments and Interventions*,
https://doi.org/10.1007/978-981-99-7652-2_17

17.1 Introduction

The brain develops as a small streak in the embryo at an early stage in the human fetus. The streak is called neural crescent, and it starts at 3–4 weeks after fertilization. It further develops and folds into a neural fold and later into a neural tube. By 20 weeks of development, it becomes a proto-brain. Most importantly, during the in utero development the neurons rapidly divide and produce millions of cells. It demands a huge supply of energy and building blocks. This is entirely met by maternal nutrition via the placenta. Therefore, during embryonic development ample supply of protein, carbohydrates, lipids, vitamins, and minerals is necessary for proper development.

Fetal development is early on marked by the critical event of the development of the brain. The following 2 years into infant growth are also critical for brain development. Post-conception, the first 1000 days of development in humans are a crucial period. The initial 280 days of fetal life and 720 days of infancy show the fastest growth of the nervous system. Therefore, the pregnant mother and the infant up to 2 years of age require a tremendous supply of nutrition to meet the energy needs of the growing baby.

In the developing fetus, a neural tube streak develops around 16 days post-conception (Hill 2020). The brain completes its full shape in 210 days, i.e., 7th month (McCartney and Phillips 2006). The events in the embryonic developmental period are fixed and irreversible and are termed developmental milestones of brain development. Brain development is impaired if these milestones are missed. The brain development in the fetus witnesses a rapid increase in neuronal cell number from 10,000 cells at 4th week of pregnancy to 10 billion by the 24th week (McCall et al. 2000). This profound increase in the number and growth of neurons in a fetus demands an enormous amount of nutrients. This is solely obtained from maternal nutrition passed via the placenta. Therefore, a malnourished mother cannot adequately supply nutrients for the fetus and it could severely impair fetal development, especially of the nervous system.

17.2 Nutrition in Neural Development

The pregnant and lactating mothers have large requirement of calorie. Therefore, they have enhanced requirement of macronutrients such as carbohydrate, protein, and fatty acids during this period. In addition, micronutrients are also necessary for proper development of fetal and infant nervous system because micronutrients are essential for neural development and function. Fatty acids are inevitable for nervous system development. Lipids of the class polyunsaturated fatty acids (PUFA) are very important for neural development. Not surprisingly, breast milk is a rich source of these essential nutrients. In addition, egg yolk and fish oils also contain such essential PUFA. These nutrients are necessary for proper functioning of the brain. In this regard, breastfeeding provides a natural superfood required for rapid growth of the developing brain. Breast milk has growth factors and vitamins in addition to other nutrients. It is estimated that in the early development of infant brain one million neural connections are created in a second (Child 2007). In fact, the toddler brain has double the number of synapses than in adult brain. In the second year of infancy, the language center of brain develops. During this period, more synaptic connections are established. The increased synaptic connections enable the infant to quickly grasp language skills. While the network of neural circuits is rapidly made in toddler brain, huge supply of proteins and lipids is required. Besides macronutrients, micronutrients such as iron, zinc, and iodine are also essential. The inadequate availability of iron in maternal diet and its deficiency until 2 years of infancy contribute to irrecoverable neural damage in the developing brain (Cerami 2017). The impact of iron deficiency could impair learning capacity and cause abnormal social and emotional behaviors. This is because the milestones in early fetal and infant brain development are critical and irreversible. For example, neural tube closure is such a critical event. Once it is missing, it cannot be repaired. Thus, nutrients are the most critical abiotic factor to accomplish the milestones in early development. The inevitability of balanced nutrition in pregnancy and first 1000 days of life post-conception is reflected in the abnormally developed children who suffered severe malnutrition.

17.2.1 Protein

An adequate supply of protein is required for growth of the body. The maternal body needs enough store of protein even to conceive and start embryogenesis. Protein is required for the development of neurons, oligodendrocytes, and astrocytes. It is also required for neurotransmitter homeostasis in neurons. The physiological and metabolic functions in neurons and oligodendrocytes also largely depend on the protein (Cusick and Georgieff 2016). Early-life protein or protein-energy restriction limits the DNA replication and RNA synthesis, which in turn reduces the cell division. This reduces the total number of neurons and synapses formed in the brain, and ultimately, brain growth is affected (Wiggins et al. 1984). The altered synaptic neurochemistry can be often traced to the basic cause of neuropsychiatric disorders.

The restricted protein intake during embryogenesis is known to alter the neural development because it affects the epigenetic signature in the neural progenitor cells (Frapin et al. 2020). The restricted protein diet affected expression of genes involved in the mitochondrial functions, such as genes required in electron transport chain. It also affected mitochondrial metabolic activities in the hypothalamus. This furthermore altered leptin function (Delahaye et al. 2008) and insulin signaling and leads to defects in food intake. The protein restriction increased the committed neuronal progenitor cells in embryos (Frapin et al. 2020). This means that under protein restriction the differentiation is accelerated in the embryo, and it is generally accomplished by increasing the epigenetic marks in the genome (Veleri 2021). Why protein restriction increases differentiation or fast-forwarding development is not clear.

Protein restriction in utero reduced the proliferation of neural stem cells in the embryo, and it enhanced fate determination of progenitor cells in cortex. As a consequence, the cortex thickness has increased (Gould et al. 2018). The maternal levels of amino acid can be sensed by the mitochondria, and it reshapes its structure, and as a consequence, the metabolic activity alters. In nutrient deficiency, mitochondrial fusion happens to increase oxidative phosphorylation (Gomes et al. 2011).

17.2.2 Carbohydrate

The brain consumes 20% of energy produced in the body though it accounts for only 2% of the total body mass. The calorie requirement of the developing brain is much more than in adult. In this regard, brain mainly uses glucose for its energy needs, especially the developing brain, where the undifferentiated cells mainly rely on glycolysis (Khacho et al. 2019). Therefore, ample supply of carbohydrates should be available for the developing brain of the fetus or infant. It is known that the developing brain has high demand for aerobic glycolysis, and it increases in case of malnutrition (Goyal and Raichle 2018). However, the differentiating neurons switch to oxidative phosphorylation for its increasing energy requirement (Khacho et al. 2016, 2019).

17.2.3 Lipid

Docosahexaenoic acid (DHA) is a long-chain polyunsaturated fatty acid essential for developing neuronal membranes, myelination, and synaptogenesis during embryogenesis. Eicosapentaenoic acid (EPA) and choline are required for the development of the membranes in neurons. Maternal supplement of DHA from 18 weeks to 3 months postpartum had significant effect on cognitive functions (Helland et al. 2003), but supplementation of arachidonic acid in the same period did not show any added advantage. The breast milk has higher concentration of long-chain PUFA. Therefore, breastfeed children show higher IQ than non-breastfeed controls (Belkind-Gerson et al. 2008). Lipids such as lysolipids are required in axon guidance during neural circuitry development (Guy and Kamiguchi 2021). The brain has the highest level of lipids such as cholesterol, glycosphingolipids, and phosphatidyl glucoside. The functional integrity of brain largely depends on lipid homeostasis (Aureli et al. 2015). Lipids serve multi-purpose roles such as a store for energy and a signaling molecule for tissue physiology (Roy and Tedeschi 2021). Its uncontrolled accumulation in tissue leads to metabolic disorders (Gunay et al. 2021). Lipid rafts serve as the scaffolds for signal transduction mediated by many neurotrophic factors (Tsui-Pierchala et al. 2002).

17.2.4 Vitamins, Minerals, and Antioxidants

Vitamin E It is a fat-soluble vitamin. The α-tocopherol is the principal form of vitamin E used in metabolic functions in human body, and its insufficiency caused ataxia in humans (Jiang 2014). However, α-tocopherol's role as antioxidant is debatable. On the other hand, γ-tocopherol, δ-tocopherol, and γ-tocotrienol are involved in antioxidant and anti-inflammatory roles (Jiang 2014). Some forms of vitamin E have scavenging activity on reactive nitrogen species (RNS) and block enzymatic activity of cyclooxygenase and 5-lipoxygenase that produce eicosanoids from PUFA. Thus, they could stop inflammatory signaling through NF-κB and STAT3/6 (Jiang 2014). The γ-tocopherol and tocotrienol may work as an anti-inflammatory agent because they suppressed inflammation in animal and human studies (Jiang et al. 2022). The γ-tocopherol supplementation could reduce inflammation markers in kidney diseases and asthma (Jiang et al. 2022). It can protect hydroxycholesterol-induced neuronal death (Kimura et al. 2018). The α-tocopherol and γ-tocopherol supplementation to embryos reduced the alcohol-induced neuronal damage (Miller et al. 2003).

A direct role for vitamin E in cognitive deficit is not proven. However, the regular supplementation of vitamin E is known to reduce the risk of developing amyotrophic lateral sclerosis (ALS) (Ascherio et al. 2005). It is suggested that vitamin E is better to prevent ALS than for treatment (Wang et al. 2011a). Vitamin E functions directly as an antioxidant. It prevents the increasing levels of lipid peroxidation (Blaner et al. 2021).

Vitamin B12 It is required for neuronal development and myelination. Vitamin B12 deficiency causes neural tube defect, stroke, and dementia. Deficiency also causes stunting of the brain development (Smith et al. 2018). Vitamin B12 is required for DNA synthesis and for RBC synthesis. It is required for motor neuron functioning. Vitamin B12 deficiency can affect the peripheral nervous system. It affects the spinal cord and nerves outside the brain. Its deficiency is a cause of cognitive impairment. In brain injury cases, vitamin B12 treatment improved microtubule stabilization, re-myelinization, and myelin reparation by inhibiting the ER stress-induced neuronal injury (Wu et al. 2021).

Vitamin C It plays a crucial role in the central nervous system. It is required for neurogenesis and differentiation (Jara et al. 2022). These processes are mostly occurring in the embryonic brain. However, in the adult brain neurogenesis is also affected by vitamin C (ascorbic acid) deficiency in subventricular zone. In prenatal rat, the ethanol and pentylenetetrazol-induced apoptosis in hippocampal neurons is prevented by prenatal treatment of vitamin C (Naseer et al. 2011). The neurodegeneration was triggered through the apoptotic pathway in the mitochondria. Vitamin C is important for the Ca^{2+}mediated regulation of neuronal processes such as growth, myelination, and neurotransmitter production (e.g. catecholamines). In stress condition, these processes are altered and vitamin C plays a major role as antioxidant (Zylinska et al. 2023). It is the main antioxidant in the CNS (Zylinska et al. 2023). Ascorbic acid (AA) is oxidizing to dehydroascorbic acid (DHA), thereby AA is depleted. In an oxidative stress milieu, neuronal survival is limited because DHA triggers non-apoptotic cell death (Ferrada et al. 2020). The neurotoxicity induced by glutamate by over-activation of glutamate receptors is a cause for many neurodegenerative diseases, and the neuroprotective effect of vitamin C is well-documented as an antioxidant (Shah et al. 2015). Vitamin C is effective in reducing glutamate-trigged phosphorylation of AMPK, which alleviated neuronal death. It also plays a role in epigenetic regulation of neural development. The functioning of enzymes such as the 10–11 translocation methylcytosine dioxygenases (TETs) and the Jumonji C domain-containing histone lysine demethylases required for oxidative demethylation is dependent on optimal supply of vitamin C (Coker et al. 2022). In the fetal midbrain, dopamine neuronal differentiation relies upon these epigenetics regulating enzymes and the process is enhanced by vitamin C (He et al. 2015).

Folic Acid Folic acid (FA) is necessary for proper neural tube development during embryogenesis. Maternal FA deficiency is associated with a higher incidence of neural tube defects. FA provides methyl groups for DNA synthesis. Its deficiency leads to abnormalities in DNA replication and ultimately results in mutations. FA can affect methylation of genes involved in the transcription either by methylating the promoter or the gene body (Ondičová et al. 2022). Animal experiments have clearly shown that folic acid supplementation prior to and during pregnancy reduced the risk of developing neural tube defects (Leung et al. 2013). However, certain cases of neural tube defects are resistant to folic acid supplementation. FA in one-carbon metabolism donates methyl (one-carbon group) for nucleotide (DNA and RNA) synthesis. Depending on the mutations in the folate metabolism pathway, purine or pyrimidine nucleotide precursors, or downstream product of folate can rescue the neural development. FA works as an antioxidant in hippocampal cell under stress conditions (Budni et al. 2013). It was shown that FA could restore antioxidant levels and reduce lipid peroxidation in hippocampal neurons. It is effective in reducing long-term memory loss after sepsis and death due to impaired permeability of blood-brain barrier (BBB) and increased oxidative stress in brain neurons (Novochadlo et al. 2021).

Vitamin A All trans-retinoic acid is a regulator of gene expression at transcriptional level. Retinoic acid response elements are in many genes. Vitamin A is an indirect antioxidant because it only transcriptionally regulates the expression of canonical antioxidant response genes in the body (Blaner et al. 2021). Since vitamin A indirectly acts as an antioxidant, it is not a bona fide antioxidant in the true sense (Blaner et al. 2021).

Zinc Zinc is required for the function of >300 enzymes under the catoreductase and carbonic anhydrase. Zinc in enzymes can directly involve catalysis. In other cases, it contributes to maintaining the structure and stability of protein. A direct link for zinc deficiency and cognitive functions is not well documented. However, autonomic nervous system disturbances were seen in fetuses of mothers who had zinc deficiency.

Iron It is the essential cofactor of hemoglobin. It binds molecular oxygen in hemoglobin and delivers it to the metabolizing cells. The redox reactions in electron transport chain translocate electrons and are finally accepted by the molecular oxygen. These steps are necessary for energy production. Thus, iron deficiency affects oxygen delivery and thereby energy metabolism. Iron deficiency is the most prevalent malnutrition seen worldwide. It is required for the development of fetal and infant brain. Iron deficiency causes anemia called iron deficiency anemia (IDA). It leads to metabolic dysfunction and ultimately causes growth retardation and stunting. Maternal deficiency of iron is associated with an increased incidence of structural anomalies of the brain and is associated with autism and schizophrenia in the offspring (Georgieff 2020). Iron deficiency affects processes such as monoamine neurotransmission, neuronal growth and differentiation, and myelination in adulthood. It also acutely compromised gene expression. The role of iron in neural development and cognitive functions is undisputed (Radlowski and Johnson 2013). Iron supplementation up to 12 weeks postpartum significantly improved the intellectual and fine motor functions in infants. It can influence intellectual, motor, and socioemotional development.

Iodine It is the main cause of preventable types of mental retardation. Its deficiency causes mental retardation in an estimated two billion patients. The iodine deficiency suffered in the first trimester of the fetus severely affects the brain development. The T3 required for fetus is supplied by the maternal T4. Thyroid hormone is critical for development and is necessary for the cerebellar development of the brain. Maternal deficiency of iodine causes developmental hypothyroidism, and it impaired neural development (Wang et al. 2011b). Hypothyroidism upregulated caveolin-1 expression and downregulated synatotagamin-1 expression. These are required for membrane trafficking and neural development. Thyroid hormones mediate critical cellular functions of the body to regulate metabolic process. It activates glycolytic and fatty acid oxidation pathways, enhances proliferation and maturation of cells, and is required for neurogenesis (Salazar et al. 2019). Thyroid hormones directly interact with nuclear receptors and regulate gene expression. It can also activate protein kinases and ion channels to affect metabolism. Hypothyroidism alters dendritic architecture, cell proliferation, and migration pattern and impairs synaptic signaling in the hippocampal neurons. The salt iodization program has significantly reduced the iodine deficiency in population; however, 30% are still at risk worldwide (Hatch-McChesney and Lieberman 2022).

Antioxidant Enzymes with Micro-Nutrients as the Co-Factors The antioxidant enzyme superoxide dismutase requires both copper (Cu) and zinc (Zn). SOD enzymatically converts highly reactive superoxide to less reactive oxygen and hydrogen peroxide. Superoxide dismutase also converts highly toxic superoxide into the lesser toxic hydrogen peroxide and dioxygen. This is a coupled oxidation-reduction reaction where one substance is oxidized and the other is reduced (Fukai and Ushio-Fukai 2011). The redox reaction is mediated by the active center, which has copper-zinc in the superoxide dismutase system, where copper catalyzes the reaction. In the first step, Cu^{2+} is reduced to Cu^{+}, and an electron is supplied by the superoxide molecule, O_2^{-}, which is thereby oxidized to O_2. In the second reaction, Cu^{+} is oxidized to Cu^{2+}, and the electron released by copper reduces a second superoxide molecule to hydrogen peroxide, H_2O_2.

Mammals have three different types of superoxide dismutases. The primary superoxide dismutase has a Cu:Zn core and is called SOD1. It mostly resides in the cytoplasm. The cofactor in SOD1 is copper, which is necessary for the catalytic activity of SOD1. The SOD with manganese (Mn) cofactor is called superoxide dismutase2 (SOD2). It is present in the mitochondrial matrix of the cell. In SOD2, Mn is the key for catalytic conversion of superoxide. The extracellular type SOD is Cu:Zn-containing superoxide dismutase, termed SOD3. It is mainly present in the extracellular matrix and on cell surfaces. A relatively small fraction of SOD3 is present in plasma and extracellular fluids. SOD1 and SOD3 have overlapping functions due to the presence of a similar cofactor copper (Fukai and Ushio-Fukai 2011). Thus, copper is a key cofactor required for functioning of the antioxidant system involving SOD1 and SOD3. On the other hand, manganese is equally important for the functioning of antioxidant system in the mitochondria. Mutation in *SOD1* is one of the reasons to cause amyotrophic lateral sclerosis (ALS) or Lou Gehrig's disease, a degenerative motor neuron disease affecting the myelin covering of the axon (Hilton et al. 2015).

17.3 Severe Malnutrition-Induced Oxidative Stress

Many nutrients such as vitamins A, C, and E and micronutrients such as zinc and selenium are good antioxidants. These are required for various biological processes in the cells. In addition, glutathione plays a major role in antioxidant systems in the body. The food consumed by the animals provides the essential antioxidants required by the body. Therefore, imbalanced nutrition takes a toll on these essential antioxidants in the body. In severe malnutrition, these nutrients will not be sufficiently reaching the cell compartments and put limits on antioxidant-mediated free radical scavenging systems. In addition to the nutrient antioxidants, enzymes such as SOD, catalase (CAT), and glutathione peroxidase (GPX) perform the scavenging of free radicals generated in the cells (Aly et al. 2014). Dietary deficiency of protein, selenium, and zinc leads to oxidative stress and cell damage. GPX activity in whole blood and the level of antioxidant vitamins (A, C, and E) and ceruloplasmin, copper, and selenium were reduced in plasma of children with protein-calorie deficiency (Ashour et al. 1999). In marasmic children, the glutathione levels and GPX activity were very low (Tatli et al. 2000). They also had a high concentration of malondialdehyde (MDA), an indicator of lipid peroxidation, in serum. Thus, protein-energy deficiency led to weakened antioxidant systems and increased oxidative stress in marasmic children. A similar study was reported from Turkey, where marasmic children had an impaired antioxidant system in the plasma (Catal et al. 2007). In children with kwashiorkor, the level of thiobarbituric acid reactive substances (TBARS) was very high. It is a marker of lipid peroxidation. These children had lower levels of glutathione (GSH) and GPX (Sive et al. 1993). Nigerian children with marasmus and kwashiorkor had lower levels of vitamins A and E compared to the controls (Becker et al. 1994). The free radical theory of kwashiorkor of Golden and Ramdath proposed that the clinical pathology seen in kwashiorkor may be the outcome of weakened antioxidant system unable to scavenge the overload of free radicals produced in the cells (Etukudo et al. 1999). Protein-energy malnutrition affects not only reduced the antioxidant scavenging system but also the electrolyte components such as sodium, potassium, and bicarbonate (Etukudo et al. 1999). Moreover, protein-calorie malnutrition has a counterproductive action on polyunsaturated fatty acid metabolism. It decreased the essential fatty acid (arachidonic acid) levels in plasma, whereas lipid peroxidation increased (Squali Houssaïni et al. 2001).

17.4 Oxidative Stress in Neurons

Aerobic respiration in animals requires oxygen for metabolism and energy production. ATP is produced by oxidative phosphorylation reaction in the mitochondrial membranes. The high demand of energy required for neurons in the brain is met by mitochondria. The mitochondria in neurons also contain monoamine oxidase, which enzymatically oxidizes biogenic amines. The end products of these processes are reactive oxygen species and reactive nitrogen species. They are main cause of neural cell damage seen in neurodegenerative diseases such as Alzheimer's disease (AD) and Parkinson's disease (PD). Oxidative stress increases cell-cell transfer of alpha-synuclein and is a trigger of PD and its progression (Musgrove et al. 2019).

Normally free radicals are formed from this reaction in the cells. In the brain, the metabolic rate is very high. It demands a constantly maintained homeostatic redox in neurons of the brain for survival (Cobb and Cole 2015). It is also required for maintaining the electrochemical gradients required in the neurons and for neurotransmitter release and membrane lipid stability. However, the brain has limited levels of antioxidants compared to other organs. Therefore, free radical scavenging is less efficient in neurons. It results in increased oxidative stress or reactive nitrogen species stress (OS/NS). The antioxidants such as vitamin E, minocycline, and resveratrol can scavenge free radicals generated in the neurons and thus render neuroprotection from free radicals. However, if the antioxidant system is overwhelmed by the free radicals the cell experiences redox imbalance. This triggers cellular damage and inflammatory pathways and ultimately result in the cell death. The first reaction to neuronal damage is displayed by microglia activation. Another pathway involved in neurodegeneration is dopamine metabolism. OS/NS has a significant role in lipid peroxidation (LPO) and fatty acid modifications in the nervous system. This activates microglia cells and triggers neuroinflammatory response. PUFAs such as linolenic acid and arachidonic acids are abundantly present in the neuronal cell membrane. The unsaturated double bonds in them are targets for oxidation by OS/NS. If antioxidants fail to scavenge these free radicals, the lipid membrane will be damaged and result in cell death. Accumulation of LPO products in neurons is a key cause for neurodegenerative diseases. LPO products trigger glial activation that often clogs the antioxidant defense system, and it also damages the organelle such as mitochondria. A prevailing diabetic condition or altered dopamine metabolism could aggravate neuronal inflammatory reaction. If the neuroinflammatory response sustains in neurons, the cellular defense gets overwhelmed. This leads to neurotoxicity and ultimately neuronal damage and brain disorders.

The ROS/NS creates an imbalance in redox potential homeostasis. The change in redox potential in neurons is mainly attributable to microglia activation (MGLA) and impaired dopamine metabolism. The microglia either salvages or clears the triggering agent in the neurons that can damage the cells. The triggers for MGLA are ROS/NS and cytokines released due to neuronal injury. Activated microglia induces inducible nitric oxide synthase (iNOS) expression and cytokine release. It ensues oxidative stress in the neurons. The final outcome will be neuroinflammation and neurotoxicity. MGLA could elicit the production of dopamine-quinones (DQ) instead of normal dopamine metabolite production. DQ creates a vicious cycle for MGLA activation (Kuhn et al. 2006). The dopamine metabolism is intricately connected with microglia homeostasis. Alterations in dopamine metabolism could lead to ROS generation. It can lead to nitration of proteins in dopamine metabolic pathway and apoptosis.

17.5 Stress-Induced Neuronal Disorders

Malnutrition is a definite factor causing oxidative stress in neurons. Malnutrition during embryogenesis heavily impacts neurogenesis because the number of neuronal division and neural connection formation are enormous in first 1000 days of life. Brain tissue has more oxidizable lipid, and it consumes a higher rate of oxygen. Both of these attributes make the brain highly prone to the generation of ROS and oxidative stress. The oxidative stress-induced neuronal damage early in life compromises the health of the neurons and their connections. Such damage leads to later-life neuronal disorders. It can develop into behavioral disorders such as autism (James et al. 2006) and neurodegenerative diseases such as AD (Huang et al. 2016) and PD (Dias et al. 2013).

17.5.1 Huntington Disease (HD)

Huntington's disease is an autosomal dominant disease with molecular defect in *HTT* gene. Mutants have abnormally increased tandem repeats of cytosine-adenine-guanine (CAG) in the gene. This results in having extra polyglutamines (PolyQ) in the mutant huntingtin protein. The structural deformity in the mutant HTT protein clogs the proteasomal degradation machinery, and the mutant protein fragments accumulate in cytoplasm. These fragments interact with other cytoplasmic proteins and interfere with calcium signaling (Zhang et al. 2008), which ultimately affects the mitochondrial function (Chen 2011). The general consensus is that mutant HTT accumulation in neurons causes oxidative stress and leads to neurodegeneration. The HD animal model created with 3-nitropropionic acid (3-NP, a mitochondrial disrupter) has increased ROS in neurons (Cho et al. 2013). The upregulated ROS was reduced with overexpression of SOD1. Importantly, many studies have shown that supplementation of antioxidants could reduce the impact of oxidative stress in HD animal models (Silva-Adaya et al. 2008). All this evidence indicated that oxidative stress is a culprit in HD.

17.5.2 Alzheimer's Disease (AD)

AD is a neurodegenerative disease with progressive disability of memory. One of the major hallmarks for AD is the accumulation of amyloid precursor protein (APP). Beta- and gamma-secretase-dependent cleavage of APP generates toxic extracellular fragments of β-amyloid peptide fragment. Beta-amyloid peptide is seen to increase phosphorylation of tau protein in the neurons. It creates neurofibrillary tangles inside the neurons. This affects the microtubule network in the neurons and ultimately destabilizes neuronal structure and axonal transport. Notably, β-amyloid peptide accumulation creates oxidative stress in the neurons, and it is believed to be the main cause for neurodegeneration. Poly (ADP-ribose) polymerase-1 (PARP1)-mediated enzymatic activity is increased in certain neurodegenerative diseases. It is required for DNA repair and is involved in apoptotic cell death. PARP1 is activated in oxidative stress in Alzheimer's disease model. Beta-amyloid peptide increases MDA, ROS, and protein carbonylation in synaptosomes. APP upregulation and/or β-amyloid peptide aggregation are seen in AD. It can create OS niche for LPO production. This could result in progressive neurodegeneration, again inducing further expression of APP. This in turn causes the constitutive production of PARP1 and reduction in antioxidant scavenging in neurons. Furthermore, choline availability is limited for acetylcholine biosynthesis. Taken together, the net outcome will be a prevailing neurotoxicity in neurons and gradual death of the cells. This at system level can result in cognitive deficits as seen in AD.

17.5.3 Parkinson's Disease (PD)

PD is a neurodegenerative disease, which impairs substantia nigra of midbrain. This brain area controls the movement and balance of the body. In PD, mainly dopaminergic neurons are damaged in substantia nigra. Dopamine is an inhibitory neurotransmitter required for suppressing certain unwanted (uncoordinated) neural firings in the motor neurons. In the damaged dopamine neurons, the accumulation of Lewy bodies is seen. Mutations in *Parkin* gene are a cause for monogenic PD. Parkin functions in proteasomal degradation pathway. Altered Parkin structure results in oxidative stress and cellular damage. Other than Parkin, PD progression can arise from alterations in DA transport and metabolism and PUFA oxidation. Neuronal damage in PD, like in other diseases, can activate MGLA. It can generate ROS. This in turn develops oxidative stress.

17.5.4 Multiple Sclerosis (MS)

Multiple sclerosis is an autoimmune disease of the nervous system. In MS, the myelination is degraded. Demyelination easily exposes neurons to the ROS and RNS. The correlation between oxidative stress, mitochondrial damage, and progression of MS is seen in patients. The presence of oxidized lipid (MDA) was seen more in damaged neurons. Oxidized DNA was also seen more at the site of neuronal injury, and it was mainly seen in oligodendrocyte nuclei (Haider et al. 2011). MDA and oxidized phospholipid oxidation are seen mainly in cytoplasm of the oligodendrocyte. These damages correlate with the upregulation of inflammatory markers such as CD3 T cells. MS patients indeed show higher levels of products of oxidative stress. In this aspect, MS shows a similar pathophysiology like AD and PD. It is shown that excessive production of reactive NO than MGLA is the main cause for neurodegeneration in MS (Rejdak et al. 2008).

17.5.5 Autism

The molecular cause of autism is not well understood. The ER stress, oxidative stress, and apoptosis are possible molecular basis for development of autism (Dong et al. 2018). Environmental factors are also a risk factor for autism. For example, exposure to bisphenol A (BPA) increased ROS generation and mitochondrial membrane potential, which were higher in autistic siblings than their normal counterpart (Kaur et al. 2014). Thus, BPA exposure increased oxidative stress and mitochondrial dysfunction in autistic patients. Many lines of evidence suggested an inefficient antioxidant scavenging system in autism patients. A weak antioxidant system in terms of superoxide dismutase, glutathione peroxidase, and catalase was reported in autism cases (Manivasagam et al. 2020). The level of biomarkers in antioxidant system in the body is a good indicator of oxidative stress, and it can predict the propensity for autism in children (El-Ansary et al. 2017). One of the studies showed that ROS species metabolism is abnormal in autistic patients (Meguid et al. 2017). For example, SOD2 was lower in autistic patients than in the healthy controls. Interestingly, the body fluids (blood and urine) of autistic children had metabolites such as short-chain fatty acids (SCFAs), indoles, and lipopolysaccharides (LPS), which are of bacterial origin. This is indicative of a role for microbiota-gut-brain-axis in autism (Srikantha and Mohajeri 2019), and more evidence in this regard is emerging (Liu et al. 2019). Thus, changes in the gut microbiota and metabolites can create a stress-prone condition and it can lead to neural disorders.

17.6 Conclusions

The importance of nutrition in development is undisputed. The requirement for protein-calorie-rich food during pregnancy is an established dictum. In early development, especially in the first 1000 days of life, the most rapidly growing and important organ in the body is the brain. Neural circuit development requires an enormous amount of nutrients, vitamins, and minerals in critical period of brain development. Neurons and axons demand a huge supply of lipids for their membrane elaboration. The lipid bilayer is a target for reactive oxygen and nitrogen species. In malnourished conditions, lack of antioxidants in food and deficiency of vitamins generate oxidative stress in neurons. This ultimately damages the cellular functions and leads to neurodegenerative diseases. To avoid such devastating health conditions early in development, a balanced diet with sufficient protein-calorie and fat supplement must be provided. The growing body of data showed the importance of PUFA in neural development and its requirement in critical windows of brain development. Many studies demonstrated that the best natural reserves of antioxidants are present in fruits and vegetables. Therefore, to improve the antioxidant defense system of the body, inclusion of more fruits and leafy vegetables in diet is necessary. The lack of micronutrients is a serious health problem even in developed countries. Therefore, strategies should be developed to focus on vulnerable communities where malnutrition is endemic, either in macronutrients like protein-calorie and fat or in micronutrient levels. Both types of malnutrition have a serious impact on the health and living standard of a person in a society. The neurodegenerative and psychiatric diseases are in rise globally. This will generate a huge burden on the healthcare system. On the other hand, the life span is increasing in societies with good healthcare systems. This can indirectly increase the number of patients with neurodegenerative disease. Since neural damage is often an irreparable event, it is essential to reduce the oxidative stress that causes neural damage. The best way to avoid oxidative stress is to nourish the body with an enriched supply of antioxidants containing natural products.

Acknowledgements The author acknowledges the ICMR-NIN intramural grant (No.ST/1/Intra-Proj-Allot/2022-223 20-FS04, 2022, dt 13-06-2022) support for the work.

References

Aly GS, Shaalan AH, Mattar MK, Ahmed HH, Zaki ME, Abdallah HR (2014) Oxidative stress status in nutritionally stunted children. Egypt Pediatr Assoc Gazette 62(1):28–33. https://doi.org/10.1016/j.epag.2014.02.003

Ascherio A, Weisskopf MG, O'Reilly EJ, Jacobs EJ, McCullough ML, Calle EE et al (2005) Vitamin E intake and risk of amyotrophic lateral sclerosis. Ann Neurol 57(1):104–110. https://doi.org/10.1002/ana.20316, PMID 15529299

Ashour MN, Salem SI, El-Gadban HM, Elwan NM, Basu TK (1999) Antioxidant status in children with protein-energy malnutrition (PEM) living in Cairo, Egypt. Eur J Clin Nutr 53(8):669–673. https://doi.org/10.1038/sj.ejcn.1600830, PMID 10477255

Aureli M, Grassi S, Prioni S, Sonnino S, Prinetti A (2015) Lipid membrane domains in the brain. Biochim Biophys Acta 1851(8):1006–1016. https://doi.org/10.1016/j.bbalip.2015.02.001, PMID 25677824

Becker K, Botticher D, Leichsenring M (1994) Antioxidant vitamins in malnourished Nigerian children. Int J Vitam Nutr Res 64:306–310

Belkind-Gerson J, Carreón-Rodríguez A, Contreras-Ochoa CO, Estrada-Mondaca S, Parra-Cabrera MS (2008) Fatty acids and neurodevelopment. J Pediatr Gastroenterol Nutr 47(Suppl 1):S7–S9. https://doi.org/10.1097/MPG.0b013e3181818e3f, PMID 18667917

Blaner WS, Shmarakov IO, Traber MG (2021) Vitamin A and vitamin E: will the real antioxidant please stand up? Annu Rev Nutr 41:105–131. https://doi.org/10.1146/annurev-nutr-082018-124228, PMID 34115520

Budni J, Zomkowski AD, Engel D, Santos DB, dos Santos AA, Moretti M et al (2013) Folic acid prevents depressive-like behavior and hippocampal antioxidant imbalance induced by restraint stress in mice. Exp Neurol 240:112–121. https://doi.org/10.1016/j.expneurol.2012.10.024, PMID 23142187

Catal F, Avci A, Karadag A, Alioglu B, Avci Z (2007) Oxidant and antioxidant status of Turkish marasmic children: a single center study. J Trace Elem Med Biol 21(2):108–112. https://doi.org/10.1016/j.jtemb.2007.01.003, PMID 17499150

Cerami C (2017) Iron nutriture of the fetus, neonate, infant, and child. Ann Nutr Metab 71(Suppl 3):8–14. https://doi.org/10.1159/000481447, PMID 29268254

Chen CM (2011) Mitochondrial dysfunction, metabolic deficits, and increased oxidative stress in Huntington's disease. Chang Gung Med J 34(2):135–152, PMID 21539755

Child CD (2007) The science of early childhood development (in brief). Harvard University, Cambridge

Cho KJ, Kim HW, Cheon SY, Lee JE, Kim GW (2013) Apoptosis signal-regulating kinase-1 aggravates ROS-mediated striatal degeneration in 3-nitropropionic acid-infused mice. Biochem Biophys Res Commun 441(2):280–285. https://doi.org/10.1016/j.bbrc.2013.08.103, PMID 24021285

Cobb CA, Cole MP (2015) Oxidative and nitrative stress in neurodegeneration. Neurobiol Dis 84:4–21. https://doi.org/10.1016/j.nbd.2015.04.020, PMID 26024962

Coker SJ, Smith-Díaz CC, Dyson RM, Vissers MCM, Berry MJ (2022) The epigenetic role of vitamin C in neurodevelopment. Int J Mol Sci 23(3):1208. https://doi.org/10.3390/ijms23031208, PMID 35163133

Cusick SE, Georgieff MK (2016) The role of nutrition in brain development: the golden opportunity of the "first 1000 days". J Pediatr 175:16–21. https://doi.org/10.1016/j.jpeds.2016.05.013, PMID 27266965

Delahaye F, Breton C, Risold PY, Enache M, Dutriez-Casteloot I, Laborie C et al (2008) Maternal perinatal undernutrition drastically reduces postnatal leptin surge and affects the development of arcuate nucleus proopiomelanocortin neurons in neonatal male rat pups. Endocrinology 149(2):470–475. https://doi.org/10.1210/en.2007-1263, PMID 18006626

Dias V, Junn E, Mouradian MM (2013) The role of oxidative stress in Parkinson's disease. J Parkinsons Dis 3(4):461–491. https://doi.org/10.3233/JPD-130230, PMID 24252804

Dong D, Zielke HR, Yeh D, Yang P (2018) Cellular stress and apoptosis contribute to the pathogenesis of autism spectrum disorder. Autism Res 11(7):1076–1090. https://doi.org/10.1002/aur.1966, PMID 29761862

El-Ansary A, Bjørklund G, Chirumbolo S, Alnakhli OM (2017) Predictive value of selected biomarkers related to metabolism and oxidative stress in children with autism spectrum disorder. Metab Brain Dis 32(4):1209–1221. https://doi.org/10.1007/s11011-017-0029-x, PMID 28497358

Etukudo MH, Agbedana EO, Akinyinka OO, Osifo BO (1999) Plasma electrolytes, total cholesterol, liver enzymes, and selected antioxidant status in protein energy malnutrition. Afr J Med Med Sci 28(1–2):81–85, PMID 12953993

Ferrada L, Barahona MJ, Salazar K, Vandenabeele P, Nualart F (2020) Vitamin C controls neuronal necroptosis under oxidative stress. Redox Biol 29:101408. https://doi.org/10.1016/j.redox.2019.101408, PMID 31926631

Frapin M, Guignard S, Meistermann D, Grit I, Moullé VS, Paillé V et al (2020) Maternal protein restriction in rats alters the expression of genes involved in mitochondrial metabolism and epitranscriptomics in fetal hypothalamus. Nutrients 12(5):1464. https://doi.org/10.3390/nu12051464, PMID 32438566

Fukai T, Ushio-Fukai M (2011) Superoxide dismutases: role in redox signaling, vascular function, and diseases. Antioxid Redox Signal 15(6):1583–1606. https://doi.org/10.1089/ars.2011.3999, PMID 21473702

Georgieff MK (2020) Iron deficiency in pregnancy. Am J Obstet Gynecol 223(4):516–524. https://doi.org/10.1016/j.ajog.2020.03.006, PMID 32184147

Gomes LC, Di Benedetto G, Scorrano L (2011) During autophagy mitochondria elongate, are spared from degradation and sustain cell viability. Nat Cell Biol 13(5):589–598. https://doi.org/10.1038/ncb2220, PMID 21478857

Gould JM, Smith PJ, Airey CJ, Mort EJ, Airey LE, Warricker FDM et al (2018) Mouse maternal protein restriction during preimplantation alone permanently alters brain neuron proportion and adult short-term memory. Proc Natl Acad Sci U S A 115(31):E7398–E7407. https://doi.org/10.1073/pnas.1721876115, PMID 29941596

Goyal MS, Raichle ME (2018) Glucose requirements of the developing human brain. J Pediatr Gastroenterol Nutr 66(Suppl 3):S46–S49. https://doi.org/10.1097/MPG.0000000000001875, PMID 29762377

Gunay A, Shin HH, Gozutok O, Gautam M, Ozdinler PH (2021) Importance of lipids for upper motor neuron health and disease. Semin Cell Dev Biol 112:92–104. https://doi.org/10.1016/j.semcdb.2020.11.004, PMID 33323321

Guy AT, Kamiguchi H (2021) Lipids as new players in axon guidance and circuit development. Curr Opin Neurobiol 66:22–29. https://doi.org/10.1016/j.conb.2020.09.003, PMID 33039927

Haider L, Fischer MT, Frischer JM, Bauer J, Höftberger R, Botond G et al (2011) Oxidative damage in multiple sclerosis lesions. Brain 134(7):1914–1924. https://doi.org/10.1093/brain/awr128, PMID 21653539

Hatch-McChesney A, Lieberman HR (2022) Iodine and iodine deficiency: a comprehensive review of a re-emerging issue. Nutrients 14(17):3474. https://doi.org/10.3390/nu14173474, PMID 36079737

He XB, Kim M, Kim SY, Yi SH, Rhee YH, Kim T et al (2015) Vitamin C facilitates dopamine neuron differentiation in fetal midbrain through TET1- and JMJD3-dependent epigenetic control manner. Stem Cells 33(4):1320–1332. https://doi.org/10.1002/stem.1932, PMID 25535150

Helland IB, Smith L, Saarem K, Saugstad OD, Drevon CA (2003) Maternal supplementation with very-long-chain n-3 fatty acids during pregnancy and lactation augments children's IQ at 4 years of age. Pediatrics 111(1):e39–e44. https://doi.org/10.1542/peds.111.1.e39, PMID 12509593

Hill MA(2020) Embryology timeline human development

Hilton JB, White AR, Crouch PJ (2015) Metal-deficient SOD1 in amyotrophic lateral sclerosis. J Mol Med (Berl) 93(5):481–487. https://doi.org/10.1007/s00109-015-1273-3, PMID 25754173

Huang WJ, Zhang X, Chen WW (2016) Role of oxidative stress in Alzheimer's disease. Biomed Rep 4(5):519–522. https://doi.org/10.3892/br.2016.630, PMID 27123241

James SJ, Melnyk S, Jernigan S, Cleves MA, Halsted CH, Wong DH et al (2006) Metabolic endophenotype and related genotypes are associated with oxidative stress in children with autism. Am J Med Genet B Neuropsychiatr Genet 141B(8):947–956. https://doi.org/10.1002/ajmg.b.30366, PMID 16917939

Jara N, Cifuentes M, Martínez F, González-Chavarría I, Salazar K, Ferrada L et al (2022) Vitamin C deficiency reduces neurogenesis and proliferation in the SVZ and lateral ventricle extensions of the young Guinea pig brain. Antioxidants (Basel) 11(10):2030. https://doi.org/10.3390/antiox11102030, PMID 36290753

Jiang Q (2014) Natural forms of vitamin E: metabolism, antioxidant, and anti-inflammatory activities and their role in disease prevention and therapy. Free Radic Biol Med 72:76–90. https://doi.org/10.1016/j.freeradbiomed.2014.03.035, PMID 24704972

Jiang Q, Im S, Wagner JG, Hernandez ML, Peden DB (2022) Gamma-tocopherol, a major form of vitamin E in diets: insights into antioxidant and anti-inflammatory effects, mechanisms, and roles in disease management. Free Radic Biol Med 178:347–359. https://doi.org/10.1016/j.freeradbiomed.2021.12.012, PMID 34896589

Kaur K, Chauhan V, Gu F, Chauhan A (2014) Bisphenol A induces oxidative stress and mitochondrial dysfunction in lymphoblasts from children with autism and unaffected siblings. Free Radic Biol Med 76:25–33. https://doi.org/10.1016/j.freeradbiomed.2014.07.030, PMID 25101517

Khacho M, Clark A, Svoboda DS, Azzi J, MacLaurin JG, Meghaizel C et al (2016) Mitochondrial dynamics impacts stem cell identity and fate decisions by regulating a nuclear transcriptional program. Cell Stem Cell 19(2):232–247. https://doi.org/10.1016/j.stem.2016.04.015, PMID 27237737

Khacho M, Harris R, Slack RS (2019) Mitochondria as central regulators of neural stem cell fate and cognitive function. Nat Rev Neurosci 20(1):34–48. https://doi.org/10.1038/s41583-018-0091-3, PMID 30464208

Kimura Y, Asa M, Urano Y, Saito Y, Nishikawa K, Noguchi N (2018) Tocopherol suppresses 24(S)-hydroxycholesterol-induced cell death via inhibition of CaMKII phosphorylation. Biochimie 153:203–209. https://doi.org/10.1016/j.biochi.2018.07.004, PMID 30244813

Kuhn DM, Francescutti-Verbeem DM, Thomas DM (2006) Dopamine quinones activate microglia and induce a neurotoxic gene expression profile: relationship to methamphetamine-induced nerve ending damage. Ann N Y Acad Sci 1074:31–41. https://doi.org/10.1196/annals.1369.003, PMID 17105901

Leung KY, De Castro SC, Savery D, Copp AJ, Greene ND (2013) Nucleotide precursors prevent folic acid-resistant neural tube defects in the mouse. Brain 136(9):2836–2841. https://doi.org/10.1093/brain/awt209, PMID 23935126

Liu S, Li E, Sun Z, Fu D, Duan G, Jiang M et al (2019) Altered gut microbiota and short chain fatty acids in Chinese children with autism spectrum disorder. Sci Rep 9(1):287. https://doi.org/10.1038/s41598-018-36430-z, PMID 30670726

Manivasagam T, Arunadevi S, Essa MM, SaravanaBabu C, Borah A, Thenmozhi AJ et al (2020) Role of oxidative stress and antioxidants in autism. Adv Neurobiol 24:193–206. https://doi.org/10.1007/978-3-030-30402-7_7, PMID 32006361

McCall KA, Huang C, Fierke CA (2000) Function and mechanism of zinc metalloenzymes. J Nutr 130(5S Suppl):1437S–1446S. https://doi.org/10.1093/jn/130.5.1437S, PMID 10801957

McCartney K, Phillips DA (2006) Blackwell handbook of early childhood development. Blackwell Publishing, Malden

Meguid NA, Ghozlan SAS, Mohamed MF, Ibrahim MK, Dawood RM, Din NGBE et al (2017) Expression of reactive oxygen species-related transcripts in Egyptian children with autism. Biomark Insights 12:1177271917691035. https://doi.org/10.1177/1177271917691035, PMID 28469396

Miller RR Jr, Olson BM, Rorick N, Wittingen AL, Bullock M (2003) Embryonic exposure to exogenous alpha- and gamma-tocopherol partially attenuates ethanol-induced changes in brain morphology and brain membrane fatty acid composition. Nutr Neurosci 6(4):201–212. https://doi.org/10.1080/1028415031000119329, PMID 12887137

Musgrove RE, Helwig M, Bae EJ, Aboutalebi H, Lee SJ, Ulusoy A et al (2019) Oxidative stress in vagal neurons promotes parkinsonian pathology and intercellular alpha-synuclein transfer. J Clin Invest 129(9):3738–3753. https://doi.org/10.1172/JCI127330, PMID 31194700

Naseer MI, Ullah N, Ullah I, Koh PO, Lee HY, Park MS et al (2011) Vitamin C protects against ethanol and PTZ-induced apoptotic neurodegeneration in prenatal rat hippocampal neurons. Synapse 65(7):562–571. https://doi.org/10.1002/syn.20875, PMID 20963815

Novochadlo M, Goldim MP, Bonfante S, Joaquim L, Mathias K, Metzker K et al (2021) Folic acid alleviates the blood brain barrier permeability and oxidative stress and prevents cognitive decline in sepsis-surviving rats. Microvasc Res 137:104193. https://doi.org/10.1016/j.mvr.2021.104193, PMID 34062190

Ondičová M, Irwin RE, Thursby SJ, Hilman L, Caffrey A, Cassidy T et al (2022) Folic acid intervention during pregnancy alters DNA methylation, affecting neural target genes through two distinct mechanisms. Clin Epigenetics 14(1):63. https://doi.org/10.1186/s13148-022-01282-y, PMID 35578268

Radlowski EC, Johnson RW (2013) Perinatal iron deficiency and neurocognitive development. Front Hum Neurosci 7:585. https://doi.org/10.3389/fnhum.2013.00585, PMID 24065908

Rejdak K, Petzold A, Stelmasiak Z, Giovannoni G (2008) Cerebrospinal fluid brain specific proteins in relation to nitric oxide metabolites during relapse of multiple sclerosis. Mult Scler 14(1):59–66. https://doi.org/10.1177/1352458507082061, PMID 17893112

Roy D, Tedeschi A (2021) The role of lipids, lipid metabolism and ectopic lipid accumulation in axon growth, regeneration and repair after CNS injury and disease. Cell 10(5):1078. https://doi.org/10.3390/cells10051078, PMID 34062747

Salazar P, Cisternas P, Martinez M, Inestrosa NC (2019) Hypothyroidism and cognitive disorders during development and adulthood: implications in the central nervous system. Mol Neurobiol 56(4):2952–2963. https://doi.org/10.1007/s12035-018-1270-y, PMID 30073507

Shah SA, Yoon GH, Kim HO, Kim MO (2015) Vitamin C neuroprotection against dose-dependent glutamate-induced neurodegeneration in the postnatal brain. Neurochem Res 40(5):875–884. https://doi.org/10.1007/s11064-015-1540-2, PMID 25701025

Silva-Adaya D, Pérez-De La Cruz V, Herrera-Mundo MN, Mendoza-Macedo K, Villeda-Hernández J, Binienda Z et al (2008) Excitotoxic damage, disrupted energy metabolism, and oxidative stress in the rat brain: antioxidant and neuroprotective effects of L-carnitine. J Neurochem 105(3):677–689. https://doi.org/10.1111/j.1471-4159.2007.05174.x, PMID 18194214

Sive AA, Subotzky EF, Malan H, Dempster WS, Heese HD (1993) Red blood cell antioxidant enzyme concentrations in kwashiorkor and marasmus. Ann Trop Paediatr 13(1):33–38. https://doi.org/10.1080/02724936.1993.11747622, PMID 7681643

Smith AD, Warren MJ, Refsum H (2018) Vitamin B(12). Adv Food Nutr Res 83:215–279. https://doi.org/10.1016/bs.afnr.2017.11.005, PMID 29477223

Squali Houssaïni FZ, Foulon T, Payen N, Iraqi MR, Arnaud J, Groslambert P (2001) Plasma fatty acid status in Moroccan children: increased lipid peroxidation and impaired polyunsaturated fatty acid metabolism in protein-calorie malnutrition. Biomed Pharmacother 55(3):155–162. https://doi.org/10.1016/s0753-3322(01)00041-5, PMID 11325213

Srikantha P, Mohajeri MH (2019) The possible role of the microbiota-gut-brain-axis in autism spectrum disorder. Int J Mol Sci 20(9):2115. https://doi.org/10.3390/ijms20092115, PMID 31035684

Tatli MM, Vural H, Koc A, Kosecik M, Atas A (2000) Altered anti-oxidant status and increased lipid peroxidation in marasmic children. Pediatr Int 42(3):289–292. https://doi.org/10.1046/j.1442-200x.2000.01217.x, PMID 10881588

Tsui-Pierchala BA, Encinas M, Milbrandt J, Johnson EM Jr (2002) Lipid rafts in neuronal signaling and function. Trends Neurosci 25(8):412–417. https://doi.org/10.1016/s0166-2236(02)02215-4, PMID 12127758

Veleri S (2021) Nutrients seminal role in the development and epigenetic reprogramming during the first 1000 days of life, epigenetics and metabolomics. Elsevier, Amsterdam, pp 265–277

Wang H, O'Reilly ÉJ, Weisskopf MG, Logroscino G, McCullough ML, Schatzkin A et al (2011a) Vitamin E intake and risk of amyotrophic lateral sclerosis: a pooled analysis of data from 5 prospective cohort studies. Am J Epidemiol 173(6):595–602. https://doi.org/10.1093/aje/kwq416, PMID 21335424

Wang Y, Zhong J, Wei W, Gong J, Dong J, Yu F et al (2011b) Developmental iodine deficiency and hypothyroidism impair neural development, upregulate caveolin-1, and downregulate synaptotagmin-1 in the rat cerebellum. Biol Trace Elem Res 144(1–3):1039–1049. https://doi.org/10.1007/s12011-011-9089-7, PMID 21611807

Wiggins RC, Fuller G, Enna SJ (1984) Undernutrition and the development of brain neurotransmitter systems. Life Sci 35(21):2085–2094. https://doi.org/10.1016/0024-3205(84)90507-1, PMID 6149444

Wu F, Xu K, Liu L, Zhang K, Xia L, Zhang M et al (2021) Corrigendum: vitamin B12 enhances nerve repair and improves functional recovery after traumatic brain injury by inhibiting ER stress-induced neuron injury. Front Pharmacol 12:598335. https://doi.org/10.3389/fphar.2021.598335, PMID 33912034

Zhang H, Li Q, Graham RK, Slow E, Hayden MR, Bezprozvanny I (2008) Full length mutant huntingtin is required for altered Ca2+ signaling and apoptosis of striatal neurons in the YAC mouse model of Huntington's disease. Neurobiol Dis 31(1):80–88. https://doi.org/10.1016/j.nbd.2008.03.010, PMID 18502655

Zylinska L, Lisek M, Guo F, Boczek T (2023) Vitamin C modes of action in calcium-involved signaling in the brain. Antioxidants (Basel). 12(2):231. https://doi.org/10.3390/antiox12020231, PMID 36829790

Oxidative Medicine and Novel Pharmacological Treatment Approaches in Liver Disease

18

Dinesh Mani Tripathi, Priyansha Singh, Akash Kumar Mourya, and Savneet Kaur

Abstract

The function of the liver is to carry out detoxification of xenobiotics, metabolism of amino acids and proteins, synthesis of bile from cholesterol, and storage of glucose as glycogen and its metabolism. As the xenobiotics are subjected to detoxification by microsomal enzymes of the liver that is the CYP450 family—they may also release free radicals and further cause reactive oxidative species stress. The antioxidative mechanisms also come into action to alleviate the increased oxidative stress caused due to free radicals, and the recent studies have focussed on novel approaches such as nutraceutical therapy, treatment with nanoparticles, repurposing of drugs, and with recent advancement in the coming of gene therapy. In this chapter, the emphasis lies on the pathophysiological mechanisms of liver diseases caused due to oxidative stress with utmost focus on the pharmacology of oxidative therapies studied so far for liver diseases and in the amelioration of various parameters of liver disease such as the increase in markers of liver disease such as alanine aminotransferase, aspartate aminotransferase, gamma, triglycerides, bilirubin, and the overall histological changes in the liver such as infiltration of inflammatory cells and hepatocyte necrosis.

Keywords

Oxidative stress · Liver damage · Antioxidative medicine

D. M. Tripathi (✉) · P. Singh · A. K. Mourya · S. Kaur
Department of Molecular and Cellular Medicine, Institute of Liver and Biliary Sciences, New Delhi, India
e-mail: dineshmanitripathi@gmail.com

R. Tulsawani, D. Vohora (eds.), *Adaptation under Stressful Environments through Biological Adjustments and Interventions*,
https://doi.org/10.1007/978-981-99-7652-2_18

18.1 Introduction

The liver is a major organ located in the right upper quadrant of the abdomen below the diaphragm, and it weighs around 1.2–1.5 kg. The liver performs numerous physiological processes namely macronutrient metabolism, blood volume regulation, immune system support, endocrine control of growth signalling pathways, lipid and cholesterol homeostasis, and the breakdown of xenobiotic compounds or drugs. Processing and metabolism of macronutrients provide the energy required to carry out the so mentioned processes and are among the most critical functions of the liver. Also, it stores glucose in the form of glycogen and also breaks down glycogen to glucose via gluconeogenesis, in response to fasting. The liver not only oxidises lipids, but can also package excess lipid for secretion to and storage in other tissues, especially adipose. Finally, protein and amino acid metabolism is also among the functions of liver as it is responsible for the majority of proteins secreted in the blood (whether based on mass or range of unique proteins), the processing of amino acids for energy, and disposal of nitrogenous waste from protein degradation in the form of urea metabolism. Also, since the liver is an accessory digestive organ, it synthesises bile from cholesterol, which on release from the bile duct emulsifies fat in the small intestine. Also, after the breakdown of haem in haemolysis, it is converted to conjugated bilirubin by the liver by uridine diphosphate glucuronyl transferase (UGT) enzyme and after that it is subjected to renal excretion (Willatts 1990). In this chapter, we are going to focus mainly on the role of free radicals and further the pharmacological and non-pharmacological management under the umbrella of oxidative medicines for liver diseases.

The liver metabolises a variety of compounds and most important being the drugs that result in the generation of free radicals (FR), hence increasing the reactive oxidative stress (ROS), which is induced by many factors but broadly by excess fat causing non-alcoholic fatty liver disease, and due to alcohol. Rest of other factors include lifestyle disorders such as metabolic disorders and bile duct obstruction causing excess bile build-up in liver parenchyma and infectious diseases such as hepatitis and schistosomiasis. Antioxidants, however, neutralise FR and keep the liver's oxidative/antioxidative balance in check. ROS is the term used to describe the condition when the liver's oxidative/antioxidative balance is disturbed. Oxidative stress triggers harmful liver processes that result in liver diseases. Antioxidants must be replenished in order to maintain homeostasis. Consuming naturally occurring compounds with antioxidant properties is one way to replenish antioxidants (Casas-Grajales and Muriel 2015).

18.2 The General Pathology of Oxidative Stress in Causing Liver Damage

While performing its essential functions of metabolism, liver is subjected to production and release of reactive oxidative species. Oxidative stress has been characterised as a co-occurring pathogenic process that initiates and advances liver injury.

Numerous risk factors, such as alcohol, drugs, environmental toxins, and radiation, can cause the liver to experience oxidative stress, which can lead to serious liver illnesses such as alcoholic liver disease and non-alcoholic steatohepatitis. During oxidative stress, three major processes occur—DNA damage, lipid peroxidation, and protein denaturation, which have the root cause in exogenous (e.g. alcohol, drugs, environmental toxins, and virus) and endogenous factors (e.g. obesity, metabolic disorders, and insulin resistance) (Li et al. 2015).

Alcohol/drugs (especially chemotherapeutic agents and infamously paracetamol)/toxins/β-oxidation on metabolism with microsomal enzyme CYP2E1 produces unstable free radicals, which react with lipids, proteins, and DNA, leading to further lipid peroxidation, and DNA mutations further induction of carcinogenicity and cause protein inactivation, which further inhibits the essential functioning enzymes of the cell, causing necrosis.

When there is an excess of either risk factors, there is an imbalance seen in the homeostasis of oxidative stress. The majority of chronic liver diseases (CLDs) are associated with oxidative stress (OS), which results in liver damage and loss of liver function. Preclinical models for fibrosis and cirrhosis have been found to have OS as a characteristic feature. In addition, ROS directly activate hepatic stellate cells and encourage the synthesis of pro-fibrogenic mediators from Kupffer cells and circulating inflammatory cells, which leads to the beginning of fibrosis. New molecular targets with therapeutic promise have been found as a result of advances in our understanding of the mechanisms behind fibrosis, allowing for more precise and individualised management of this condition.

The liver is a major organ that is subjected to ROS stress. The major cells in the liver that are injured by oxidative stress are its parenchymal cells. Parenchymal cells' mitochondrion, microsomes, and peroxisomes are able to produce ROS via controlling the expression of PPAR, which is mostly connected to the liver's fatty acid oxidation gene. Additionally, Kupffer cells, hepatic stellate cells, and endothelial cells may be more exposed to or susceptible to oxidative stress-related chemicals. Oxidative stress can cause Kupffer cells to generate a number of cytokines, including TNF-α, which could worsen inflammation and increase apoptosis. Hepatic stellate cells produce collagen and proliferate when their lipid peroxidation, which is brought on by oxidative stress, occurs (Li et al. 2015).

The perpetual parenchymal necrosis, changes in the composition of the ECM, the migration of immunological and/or inflammatory cells to the site of damage, and the activation of hepatic stellate cells (HSCs), macrophages, and Kupffer cells (KCs) are characteristics of the evolution of liver fibrosis (liver macrophages). Growth factors, cytokines, and chemokines, alterations in the structure and composition of the extracellular matrix (ECM), and reactive molecules resulting from oxidative stress (OS) have all been proposed to have a role in pathogenesis at the molecular level (Sánchez-Valle et al. 2012).

18.3 The Role of Defence Mechanisms in Oxidative Damage

ROS is endogenously balanced by the protective mechanisms of the body by the activation of nuclear factor erythroid 2-related factor 2 (Nrf2-Keap1-Cul3) complex, which transcribes to produce antioxidative enzymes such as superoxide dismutase (SOD), glutathione peroxidase (GPx)/reductase (GRed), glutathione transferase, thioredoxin reductase, catalase (CAT), haeme oxygenase, peroxiredoxin (Prx), and paraoxonase 1. The OS is mainly balanced by Nrf2-Keap1-Cul3 complex present in the cytoplasm (initially, it is inactive under normal conditions). On activation under excess free radicals causing oxidative stress, Nrf2 splits from Keap1-Cul3, which facilitates its translocation into the nucleus as it gets phosphorylated with the help of members of mitogen-activated protein kinases (MAPKs) such as extracellular signal-regulated kinase (ERK), p38 and c-June N-terminal kinase (JNK), phosphatidylinositide 3-kinases (PI3K), PKR-like endoplasmic reticulum kinase (PERK), and protein kinase C (PKC) and shows conformational change. These conformational changes prevent Nrf2 ubiquitination and proteasomal degradation, leading to nuclear translocation of Nrf2.

In the nucleus, Nrf2 heterodimerises with small Maf proteins and enhances its binding to antioxidant response element (ARE) to activate the transcription of Nrf2-dependent genes such as haem oxygenase-1 (HO-1), NADPH quinine oxido-reductase 1 (NQO1), aldo-ketoreductase (AKR), peroxiredoxin 1 (PRXD 1), glutathione-*S*-transferase (GST), glutathione peroxidase-1,2 (GPX-1,2), γ-glutamyl cysteine ligase-catalytic (γ-GCLC), γ-glutamyl cysteine ligase-modulatory (γ-GCLM), and superoxide dismutase (SOD).

It subsequently combines with AREs to trigger the transcription of endogenous protective genes, including antioxidant genes, phase II detoxification enzyme genes, and anti-inflammatory co-stimulating genes.

Nrf2 blocks the upregulation of tumour necrosis factor-α (TNF-α), interleukin-6 (IL-6), interleukin-17 (IL-17), interleukin-1β (IL-1β), monocyte chemo-attractant protein-1 (MCP-1), and macrophage inflammatory protein-2 (MIP-2) and inhibits the promoter activity of vascular cell adhesion molecule-1 (VCAM-1) whose synthesis is mediated by the transcription of NF-kB. Fibrosis is induced by reactive oxygen species (ROS) through a number of inflammatory mechanisms. Hepatocytes are susceptible to damage from alcohol, viruses, or other toxic stimuli, which cause the generation of ROS and damage-associated molecular patterns (DAMPs) that bind to Kupffer cells Toll-like receptor 4 (TLR4). The inflammatory response is then triggered by nuclear factor-B (NF-kB), which is activated by the TLR4/MyD88 pathway.

There is also a cross-talk observed between Nrf2 and TGFb1. In oxidative stress, Nrf2 inhibits TGFb1-mediated transcription of fibrotic genes such as a-SMA and collagen-1 (Gong and Yang 2020) (Fig. 18.1).

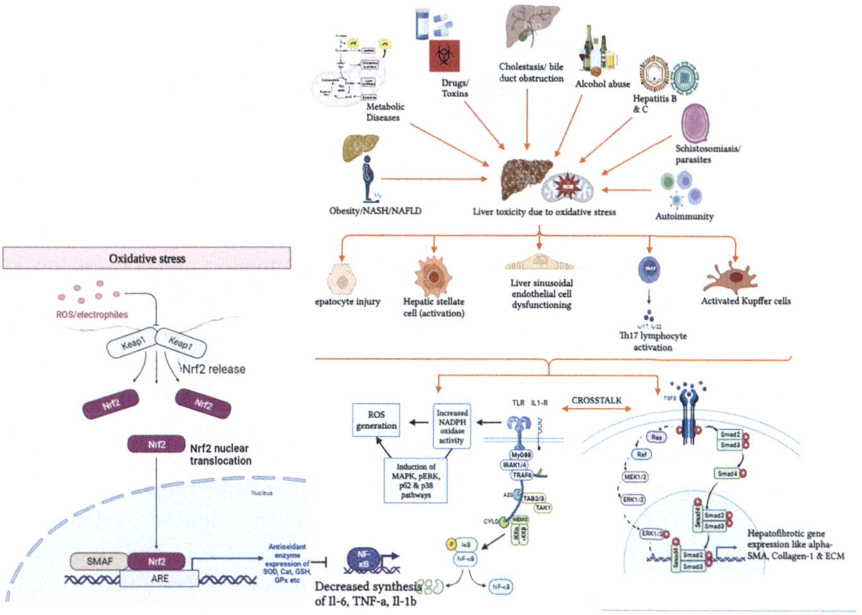

Fig. 18.1 Major pathway involved in antioxidative therapy for liver disease

18.4 Role of Antioxidants as Oxidative Medicines in Liver Damage

Now that we have seen that any risk factor whether exogenous or endogenous contributes to increased OS in liver parenchyma due to an imbalance in curbing of excess free radical species, the main focus is to treat liver disease at each stage by targeting the root cause at a molecular level. In this section, we are going to discuss the various therapeutic strategies for the said purpose.

(a) **Vitamin E/α-Tocopherol**—The management of NAFLD was found to require vitamin supplementation. Antioxidant-rich vitamins, such as vitamins E and C, have been shown to reduce ALT and AST serum activity, as well as lipoatrophy and lobular hepatitis without any effect on liver fibrosis. As a result, their supplementation should be thought of as a first-choice therapy in individuals with NASH. By lowering the intracellular mitochondrial membrane potential, raising levels of the anti-apoptotic protein BCL-2, and lowering levels of the pro-apoptotic proteins BAX and p53, it decreased the rates of apoptosis. Additionally, it was able to lessen the activity of caspase-8 and caspase-3 in the Fas/FasL apoptotic route and caspase-9 and cytochrome *c* in the mitochondrial apoptosis pathway. Vitamin E can lessen the inflammatory response in

NAFLD by blocking the nuclear localisation of NF-kB, lowering COX-2 production, and regulating the expression of cytokines such as TNF-α, IL-1, IL-2, IL-4, IL-6, and IL-8. By reducing hepatic lipid accumulation and peroxidation, vitamin E reduced the hepatic steatosis brought on by lipotoxicity. It was also found to decrease oxidative stress by reducing the amount of TBARS and increase the amount of glutathione and SOD activity. In diet-induced NASH, vitamin E reversed hyperinsulinemia, inhibited hepatic stellate cell activation, lowered inflammation, and avoided hepatic fibrosis, but it had no effect on insulin signalling. Notably, this anti-inflammatory impact was linked to active NF-kB and decreased MAPK (JNK/p38 MAPK) signalling. Additionally, vitamin E decreased the fraction of M1 macrophages and increased the proportion of M2 macrophages, which attenuated inflammation in NASH patients. The attenuation of JNK-mediated inflammatory signalling was linked to these benefits. For M1 macrophage polarisation during inflammation associated with obesity, JNK activity is necessary (Perumpail et al. 2018; Nagashimada and Ota 2019). It has intrinsic antioxidative properties because its chromanol ring can donate a hydrogen ion to scavenge lipid peroxyl radicals. Vitamin E raises superoxide dismutase (SOD), catalase (CAT), and glutathione peroxidase (GPx)—all essential antioxidant enzymes in the body that can split superoxide radicals into oxygen or hydrogen peroxide, which hence show its ROS scavenging activity (Perumpail et al. 2018). In animal models, vitamin E, or tocopherol, which is known to have antioxidant effects and to suppress peroxidation, has been shown to limit the expression of the TGF-1 gene in the liver and to prevent the development of liver fibrosis (Hasegawa et al. 2001). In clinical trials, the vitamin E supplements also curbed down the levels of the serum redox markers such as AGE and protein carbonyls (Podszun and Frank 2021).

(b) **Cannabinoid 2 Receptor Agonist**—CB2 receptors are abundantly expressed in Th17 lymphocytes, Kupffer cells, and hepatic stellate cells under fibrotic conditions, though in normal liver they are marginally expressed. The most often studied CB2 selective agonists in research are JWH-015/133, β-caryophyllene, HU-308/910, GW, schisandrin B, bisdemethoxycurcumin, and AM1241 (Ali et al. 2021). Hepatic stellate cell (HSC) activation and the inhibitory capacity of CB2 receptor agonists play a major role in the amelioration of liver fibrosis. Both their capacities to scavenge free radicals and inhibit lipid peroxidation were examined. It has been shown that CB2 receptor agonism achieved significant anti-lipid peroxidation activity in liver diseases. They demonstrated strong scavenging abilities against the hydroxyl radical and superoxide anion as well. Significant inhibition was seen in the activity of 5-lipoxygenase, an enzyme involved in fibrogenesis (Calleja et al. 2013). In alcoholic liver disease, it has been shown to have anti-inflammatory effects via upregulation of HO-1 in macrophages and stimulate the autophagy process via upregulating of HO-1 in macrophage that mediates the anti-inflammatory and antisteatogenic activities of CB2 receptors. In alcoholic liver disease, studied in CCl_4 or any other preclinical in vivo models of induced fibrosis, CB2 receptor agonist mediates HO-1 pathway. In alcoholic liver disease, studied in CCl_4 or any other preclinical in vivo models of induced fibrosis, CB2 receptor agonist mediates HO-1 pathway (Hashiesh et al. 2021). CB2 receptor agonist also reduced the complication of portal hypertension induced by liver cirrhosis. Also, a few CB2 receptor agonists have shown the facilitation of PPAR-γ-mediated anti-inflammatory actions and reduced the upregulation of NF-kB. Activated stellate cells were subjected to apoptotic cell death by the activation of death-inducing signalling cascade (DISC) by activation of caspase-3 and hence the markers for fibrogenesis such as α-SMA, Col1α1, Timp1, and MMP-1. Hepatic Kupffer cells on sensing endotoxins, the TLR4 receptor gets activated which releases NF-kB and thus activates the secretion of inflammatory mediators like TNF-α, IL-1β and IL-6 and upon treatment with CB2 receptor agonist suppresses the activation of NF-kB. CB2 receptor agonist also reduces LSEC inflammation by reducing P-selectin, VCAM-1, and ICAM-1 expression. CB2 receptor agonist also reduces the accumulation of fatty acids by reducing SREBP-1c and Fasn expression in hepatocyte cell lines (Lotersztajn et al. 2008; Calleja et al. 2013; Lee et al. 2015; Jha et al. 2021).

(c) **Dasatinib**—It is a potent and broad-spectrum tyrosine kinase second-generation oral multi-target tyrosine kinase inhibitor. Based on histopathological analysis, dasatinib significantly decreased TGF-b1 and PDGF gene expression, attenuated the mean score of fibrosis, relaxed the necro-inflammatory process, and decreased TNF-a gene and protein expression, and ALT and AST levels. Particularly with MDA, nitric oxide, and TNF-a, a remarkable restoration of oxidative and inflammatory indicators was observed. PDGFA, phosphorylated PDGFR, and IL-1, COX2, SREBP-1, iNOS, CD68, TGF-b1, and ASMA expression in the liver were dramatically reduced by dasatinib (Elsayed et al. 2021). The activation of HSCs is suggested to be affected by miRNAs via a number of signalling pathways, including the TGF-b/Smads and Wnt/b-catenin pathways. Dasatinib treatment decreased the expression of miRNA-17 and restored WiF-1 and Smad-7, which inhibited both Wnt/b-catenin and TGF-b/Smad signalling. Additionally, it increased miRNA-378, which caused Wnt-10 to decrease and facilitate suppress-activated HSCs. Dasatinib has been shown to inhibit Src kinase, which further inhibits PI3K/Aft pathway, and as a result, it inhibits activation of hepatic stellate cells and hence their proliferation, differentiation, and migration (Mohammadalipour et al. 2017; Zaafan and Abdelhamid 2022).

(d) **Farnesoid X Receptor Agonist**—FXR is a known target for treating liver disorders and associated complications (e.g. portal hypertension) For the treatment of non-alcoholic steatohepatitis (NASH), various FXR agonists are undergoing clinical trials, including obeticholic acid, cilofexor, nidufexor, tropifexor, and MET409 (Fiorucci et al. 2021). Activation of FXR also attenuates hepatic triglyceride accumulation. The liver, gastrointestinal system, and kidneys are the main organs where the farnesoid X receptor is expressed. It is a nuclear receptor superfamily member. FXR controls the transcription of target genes by heterodimerising with the retinoid X receptor a (RXRa;

NR2B1). It also binds to the FXR response element (FXRE). According to a prior study, p300 and SIRT1 dynamically control the acetylation of FXR, which limits FXR/RXR a heterodimerisation and FXR activity under normal circumstances. Acetylation of FXR is dynamically controlled by p300 and SIRT1 and inhibits FXR/RXR a heterodimerisation and its activity. Consuming ethanol decreased the level of the FXR/RXR a heterodimer and increased the level of acetylation of FXR in the liver in preclinical models of ALD. Treatment with an FXR agonist decreased liver inflammatory cell infiltration and the expression levels of inflammatory factors (TNF-α and Mcp-1) in the body. It has been shown that mice with ALD have much higher hepatic levels of TNF-α and Mcp-1 than mice that were pair-fed, both of which are crucial in the pathophysiology of alcohol-induced liver damage. Furthermore, it has been shown that activating FXR blocks the NF-kB signalling pathway and reduces the expression of pro-inflammatory genes in a preclinical model of non-alcoholic fatty liver disease. In preclinical models that were treated with ethanol, FXR agonists decreased CYP2E1 levels and decreased oxidative stress. After consuming alcohol, hepatic CYP2E1 is strongly stimulated; alcohol can maintain the enzyme by preventing its breakdown and promoting gene transcription. Lipid peroxidation and oxidative stress are brought on by the induction of CYP2E1, which has a role in liver injury (Wu et al. 2014). FXR ligands regulate PPAR gene expression, and FXR and PPAR ligands synergistically regulate the profibrogenic property of HSC. Peroxisomal proliferator-activated receptor (PPAR) expression is elevated by SHP when FXR is activated. HSC activation and the development of fibrosis are both influenced by PPAR. Through direct Smad3 binding and inhibition of TGF-induced CTGF and a-SMA production in smooth muscle cells, PPAR can control the TGF/Smad pathway. Inhibiting TGF synthesis during fibrosis in various tissues is possible with a number of drugs that activate PPAR. Although PPAR is downregulated during HSC activation, PPAR is upregulated during HSC quiescence. PPAR is implicated in HSC trans-differentiation and fibroblast transformation. The majority of PPAR agonists inhibit HSC proliferation and induce apoptosis in activated HSC to lessen hepatic fibrosis. The overexpression of type I collagen, hydroxyproline, and a-SMA can also be decreased by PPAR, which will prevent liver fibrosis (Stofan and Guo 2020; Monroy-Ramirez et al. 2021).

(e) **Taxifolin**—Taxifolin (TXN), a natural active molecule, has a pharmacological activity and therapeutic effects such as antioxidant, anti-inflammatory, anticancer, and antiviral actions. The abnormal serum levels of aminotransferases were controlled by TXN. Histological analysis revealed that TXN decreased collagen deposition, returned the damaged liver structure to its normal liver lobular shape, and decreased the infiltration of inflammatory cells in the liver. Alcohol-induced liver toxicity was alleviated by TXN by lowering oxidative stress injury, by the enhancement of GSH and SOD activity and inhibition of CYP2E1 overexpression. The PI3K/Akt signalling pathway was activated by TXN, which also boosted the expression of p-PI3K and p-Akt. Bax was also markedly suppressed, and Bcl-2 expression was improved. TXN therapy dramatically reduced the expression of caspase-3 in liver tissue. TXN was also found to decrease TGF-b1/Smad signalling pathway and also control HSC activation, hence decreasing fibrosis. TXN alleviates the synthesis of ECM via TGFb1-Smad and PI3K/Akt/mTOR pathway. Taxifolin improved the redox status of the liver tissues by modulating the redox-sensitive MAPK pathway (p38/c-Fos), as shown by a decrease in lipid peroxidation and protein oxidation and an increase in overall antioxidant capacity. During hepatocellular carcinoma, there is a prevalence of biomarkers such as hypoxia-induced factor 1 a (HIFa), vascular endothelial growth factor (VEGF), and protein kinase B (Salama and Kabel 2020; Ding et al. 2021; Liu et al. 2021).

(f) **Benzoquinone Derivatives**—On the assessment of their antioxidant activity and antifibrotic effects on activated hepatic stellate cells, benzoquinone derivatives such as 5-isopropyl-2-methyl-1,4-benzoquinone (TQ), 2-*tert*-butyl-1,4-benzoquinone (*t*Bu-Q), 2,5-dimethyl-*p*-benzoquinone (Dime-Q), and *p*-benzoquinone (Ph-Q) showed significantly decreased levels of liver injury markers and lipid peroxidation caused by excessive ROS, including aspartate aminotransferase (AST), alanine aminotransferase (ALT), and malondialdehyde (MDA). Treatment with benzoquinone derivatives also downregulated the mRNA and protein expression of liver fibrosis indicators such as collagen I, alpha-smooth muscle actin (α-SMA), and TIMP and dramatically reduced the deposition of extracellular matrix (ECM). By modifying the TLR4-CD14 signalling pathway to reduce the expression of procaspase-1 and IL-1 in cells, benzoquinone derivatives, which are ROS scavenging chemicals, also promoted apoptosis via a mitochondrial pathway by increasing the ratio of Bax/Bcl-2 and by activating caspase-3 (Miao et al. 2020).

(g) **Silymarin**—Silymarin is a naturally occurring substance that is obtained from the species *Silybum marianum*, also known as milk thistle. Because silymarin has the power to prevent free radicals from being formed during the metabolism of harmful chemicals such as ethanol, acetaminophen, and carbon tetrachloride, it has both hepatoprotective and antioxidant properties. It is well known that the production of free radicals harms cellular membranes and results in lipoperoxidation. Silymarin increases hepatic glutathione and may aid the liver's antioxidant defence. Silymarin's antifibrotic effect is principally brought on by its capacity to prevent hepatic stellate cells from developing into myofibroblasts by blocking fibrogenic pathways, such as those involved in cytoskeletal development, profibrogenic collagen, and electron transfer chains. In particular, silymarin inhibits NF-kB, downregulates TGF-ß1 mRNA, and suppresses activation of hepatic stellate cells. Also, silymarin extracts increase the expression of GSH, SOD, and GPx and decrease the abnormal levels of ALT, AST, and GGT in serum (Vargas-Mendoza et al. 2014; Kołota 2017; Tighe et al. 2020).

(h) **Hesperidin**—Citrus fruits such as oranges and lemons contain the phytoflavanone glycoside known as hesperidin (HSP). It has a wide range of pharmacological actions, including antioxidant, anti-inflammatory, anti-carcinogenic, hepatoprotective, and antidiabetic properties. The antioxidative effects of hesperidin were evaluated using preclinical rat fibrosis model induced by carbon tetrachloride, and hesperidin was administered in a dose range of 100–200 mg/kg/day. In histopathological studies, the liver tissues from the rats treated by HSP revealed normal hepatic architecture with a central vein and radiating cords of normal hepatocytes with central rounded vesicular nuclei. Blood sinusoids are separated from hepatic cords. On molecular analysis, it was revealed that hesperidin has proven to decrease serum levels of alanine transaminase (ALT), aspartate transaminase (AST), and bilirubin levels, which are markedly increased in hepatic fibrosis while increasing albumin levels, which are decreased. Also, the hepatic content of the most studied markers of hepatic fibrosis—TGF-β1, α-SMA, hydroxyproline, and oxidative stress markers such as TBARS, malondialdehyde (MDA), PCNA, and myeloperoxidase (MPO)—is reported to have been decreased to a level nearly comparable to that of the control group. Also, it drastically decreased microscopic hepatic damage. As an antioxidant, it has been shown to increase the expression of SOD, CAT, and GSH to counter oxidative stress so induced. HSP has been reported to even upregulate the genetic expression of antioxidant genes such as Nrf2, HO-1, and PPAR-γ and decrease apoptotic factors such as BAX and caspase-3. HSP facilitates the decreased expression of inflammatory markers like TNF-α, IL-6, IL-1β, iNOS, and MMP-9. In a randomised, double-blind, placebo-controlled clinical trial including patients with placebo ($n = 25$) and patients who were administered hesperidin ($n = 25$), anti-inflammatory effects were observed (Çetin et al. 2016; Abdel-sttar et al. 2017; Cheraghpour et al. 2019; Tabeshpour et al. 2020; Abd-Eltawab Tammam et al. 2022).

(i) **α-Lipoic acid**—Alpha-lipoic acid (ALA, 5-(1,2-dithiolan-3-yl) pentanoic acid) or ALA is a potent antioxidant with a good safety profile. ALA is a naturally occurring substance that can be obtained from a variety of dietary sources or produced endogenously. It is widely distributed throughout the cellular membranes of both prokaryotes and eukaryotes and is also known as thioctic acid. It participates in the metabolism of carbohydrates and fats endogenously, mostly through acting as a cofactor for several mitochondrial enzyme complexes. ALA markedly ameliorated liver cirrhosis, which was made evident with the help of positive histopathological alterations, reduced liver fibrosis, and decreased serum concentrations of ALT and AST along with HSC activity. Furthermore, ALA demonstrated the inhibition of the activation of the TGF-β/Smad3 pathway. Hepatocellular autophagy in oxidative liver diseases was suppressed by ALA. The administration of ALA additionally activated the AKT/mTOR pathway in cirrhotic liver. These results suggested that the inhibition of autophagy through the regulation of the AKT/mTOR pathway and the suppression of the TGF-β/Smad3 pathway may be a potential mechanism for the hepatoprotective effect of ALA against cirrhosis. ALA has also reportedly decreased the genetic expression of fibrotic markers such as α-SMA and Col1α1 and also the autophagy biomarkers such as LC3II, beclin-1, and p68. Alshammari et al. in 2022 conducted an experiment on liver cirrhosis induction by gold nanoparticles in rats, and ALA reversed all the hepatotoxic effects induced by AuNPs such as decreased serum levels of alanine aminotransferase (ALT) and gamma-glutamyltransferase (γ-GTT), and aspartate aminotransferase (AST), as well as liver levels of malonyldialdehyde (MDA) It also significantly enhanced hepatic levels of total glutathione (GSH), superoxide dismutase (SOD), and catalase (CAT) but decreased hepatic levels of IL-6 and TNF-α. It also upregulated mRNA levels of B-cell lymphoma 2 (Bcl-2) and haem oxygenase-1 (HO-1) and significantly decreased those of Bax and cleaved caspase-3. In addition, ALA reduced the total and nuclear levels of NF-κB p65 but facilitated the increase in the mRNA and total and nuclear protein levels of Nrf2 (Liu et al. 2019; El-Mancy et al. 2022).

(j) **Aloricumab**—Alcohol use modulates epigenetically the proprotein convertase subtilisin/kexin 9 (PCSK9) enzyme. The LDL receptor (LDL-R) is the target of PCSK9, which then modulates LDL-R production and recycling. PCSK9 is largely expressed in the liver and is crucial for the regulation of low-density lipoprotein cholesterol (LDL-C). Gain-of-function mutations in PCSK9, which produce autosomal dominant hypercholesterolemia, and loss-of-function mutations, which are connected to lower levels of LDL-C and low rates of cardiovascular disease, respectively, led to the early discovery of its function in lipid regulation. These findings have sped up the development of anti-PCSK9 medicines, resulting in novel strategies for significantly decreasing LDL-C. Treatment with the monoclonal antibody alirocumab significantly downregulates the expression of PCSK9 in the liver and then elevates LDL-R, which is accompanied by a reduction in the inflammation, oxidative stress, steatosis, and hepatocellular damage caused by alcohol. Hepatic mRNA expressions of SREBP-1, SREBP-2, and E2F1 are elevated in ALD, but alirocumab administration attenuated the alcohol-induced expressions of the transcription factors. Alcohol-induced lipid accumulation is reduced and lipid metabolism is regulated by PCSK9 inhibition. Alcohol consumption led to a rise in hepatic TG levels, which alirocumab therapy mitigated. It decreased the enhanced gene expression of fatty acid synthase (FAS), a lipogenic enzyme crucial to TG accumulation, in the alcohol-fed rat group. Additionally, alcohol exposure decreased the mRNA expression of the fatty acid catabolism regulators such as peroxisome proliferator-activated receptor alpha (PPAR) and carnitine palmitoyltransferase I (CPT1), whereas alirocumab treatment demonstrated a reduction. It reduced hepatic steatosis (multiple small vacuoles) and inflammation (infiltrating leukocytes) and blood ALT and AST levels and reduced the genetic expression of fibrosis markers, smooth muscle alpha-actin (ACTA2), alpha-1 type I collagen (COL1A1), matrix metalloproteinase-2 (MMP-2), and metallopeptidase inhibitor 1 (TIMP-1) along with the mRNA expression of pro-inflammatory cytokines (TNF-α, IL-1β, IL-22, IL-33, IL-17α, and IL-2), and chemokines (macrophage inflammatory protein (MIP)-2 and monocyte chemo-attractant protein (MCP)-1) (Lee et al. 2019).

(k) **Tanshinone**—Tanshinone IIA (TIIA), a fat-soluble extract of *Salvia miltiorrhiza*, possesses anti-inflammatory, antioxidant, anticancer, neuroprotective, and cardioprotective properties. The major mechanism of action of tanshinone is a decrease in TGF-1 mRNA and consequently protein expression, a decrease in TNF expression, a significant increase in SOD and GSH-Px levels, and a significant decrease in MDA levels. Significant reductions in Ang II, AT1R, VEGF, and HIF-1 levels were observed. Bax protein expression was considerably downregulated, but Bcl-2 protein expression was significantly upregulated. TIIA reduced the formation of ROS and MDA, increased the activity of T-SOD and GSH-PX, and possibly inhibited apoptosis and improved hepatic steatosis to reduce oxidative stress. Zhang et al. showed that cholestasis and hepatotoxicity could be attenuated by TIIA by nuclear upregulation of PXR, Cyp3a11, Cyp3a13, and Mdr1 in in vitro studies on HepG2 cell line. Sirtuin 1 (SIRT1)/protein kinase AMP and activated catalytic subunit alpha 1 (PRKAA1) pathways were both activated by the sodium sulphonate salt of TIIA. In case of orthotopic liver transplant, I/R injury to hepatocytes is a major complication involved—and hence, TIIA has come up as a potential therapeutic agent as reverses the injury by downregulation of HMGB1-TLR4-NF-kB pathway in hepatic keratinocytes and upregulation of PI3K-Akt pathway, and at micromolar concentrations, it can facilitate live regeneration by activating the canonical Wnt signalling pathway. Tan IIA increased the levels of calreticulin, caspase-12, and GADD153 in human HCC J5 (hepatocellular carcinoma cell line) cells to inhibit and induce apoptosis for the prevention of cancer. In case of lipid metabolism, tanshinone II A was found to be related to regulation of the intracellular molecular targets such as PPARα, CYP1A2, and MMP2 in the regulation of lipid metabolism (Ying et al. 2019; Li et al. 2020).

(l) **Matrix Metalloproteinases/Tissue Inhibitors of Metalloproteinase**—Matrix metalloproteinases, also known as TIMPs, are crucial enzymes that control the synthesis and breakdown of ECM. The primary ECM-degrading enzyme in the liver is MMP, and TIMPs are MMP's endogenous inhibitors. In activated HSCs, MMP2 and MMP14 are significantly expressed. MMPs weaken ECM in typical conditions. TIMPs, which are highly expressed in liver injury, however, block MMPs. Due to the imbalance between ECM deposition and degradation being disrupted, there is an excessive accumulation of ECM, which eventually results in liver fibrosis. Therefore, increasing MMP activity or decreasing TIMP activity is a beneficial approach to treating liver fibrosis. Halofuginone is a small-molecule quinolone derivative, which lowered the level of TIMPs, inhibited collagen expression, and alleviated hepatic fibrosis. An antifibrosis action was reported in the ethanol extract of *Fraxinus rhynchophylla* (FR(Et-OH)). By downregulating the expression of MMP2, MMP9, and TIMP1, FR(Et-OH) successfully reduced liver lesions and fibrous connective tissue growth. Type I and III collagens make up the majority of the ECM following liver injury, and type I collagen is the most abundant type of collagen in a fibrotic liver, therefore inhibiting type I or III collagen production and accumulation will be helpful in treating liver fibrosis (Tan et al. 2021).

(m) **Emricasan**—Emricasan is a pan-caspase inhibitor with anti-apoptotic and anti-inflammatory pathway, which was developed exclusively for the treatment of liver diseases. Caspase is a family of at least 11 human intracellular cysteines, which are involved in apoptosis, inflammation, and fibrosis. They are also involved in portal hypertension as they produce hemodynamically active pro-inflammatory micro-vesicles, which cause intrahepatic inflammation, vasoconstriction, and splanchnic vasodilation. Emricasan reportedly showed a decrease in HPVG in case of compensated cirrhosis and severe portal hypertension. Being a caspase inhibitor, it markedly decreases the activity of caspase-3, decreases overall hepatic cell death, and increases bile production. It also showed a reported decrease in serum markers of liver disease. There was a decrease in portal pressure and an improvement in HVR plus portal perfusion pressure and intrahepatic vascular tone, which ameliorated microcirculatory dysfunction. It decreased the expression of a-SMA, *p*-moesin, PDGFRβ, MMP (MMPs 2 and 9), TGF-β1, and TIMP (TIMPs 1 and 2) along with the decrease in the desmin content and a decrease in HSCs. On staining with Sirius red, it revealed the decrease in fibrotic lesions in the hepatic tissues. As it improved portal pressure, therefore at the root levels it was demonstrated by increase in endothelial fenestration, reduction in sinusoidal von Willebrand factor (vWF) expression, and improved NO availability, along with higher phospho-endothelial nitric oxide synthase (eNOS) expression and higher expression of cGMP content. Emricasan reportedly showed an attenuated expression of pro-inflammatory molecules (IL 1, IL-6, and TNF-α) and a reduction in CD68-positive cells. In human hepatocytes, there was an improved expression of HNF4a, slc22a1, and abcc3 and a high albumin and urea synthesis. Emricasan being a pan-caspase inhibitor showed improvement in hepatocytes, LSECs, HSCs, and Kupffer cells. There was also an increase in CYP3A4 activity signifying an improvement in the metabolic activity of hepatic microsomes (Garcia-Tsao et al. 2019; Gracia-Sancho et al. 2019; Mu et al. 2021).

(n) **Resveratrol**—Resveratrol is a natural polyphenolic flavonoid having anti-inflammatory, antioxidant, anti-neoplastic, and antifibrotic properties. Resveratrol showed a significant reduction in serum biomarkers of cirrhosis such as ALT and AST, hepatocyte necrosis, collagen deposition, and infiltration of inflammatory cells and significantly inhibited the accumulation of hydroxyproline. Resveratrol has demonstrated a reduction in liver fibrosis by considerably upregulating the number of Kupffer cells as it significantly enhanced the hepatic expression of F4/80 mRNA and has reportedly increased the transition of M1 (inflammatory phenotype) macrophage to M2 (anti-inflammatory phenotype) as there was an increase in the genetic expression markers of the anti-inflammatory phenotype such as Mrc1, Mrc2, Arg1, and CD163 and a suppression of the inflammatory phenotype by decreasing the expression of iNOS, TNF-a, and Mcp-1 (Yu et al. 2019). RSV primarily acts through the signalling pathways of AMPK/SIRT1/Nrf2, ERK/p38 (signal-regulated extracellular kinase), mitogen-activated protein kinase (MAPK), and phosphatase and tensin homolog/protein kinase B (PTEN/Akt), which promote the synthesis of antioxidant molecules and the expression of related genes involved in the biogenesis of mitochondrial energy. In steatosis, it had demonstrated an increase in superoxide dismutase (SOD), glutathione peroxidase (GPx), catalase) CAT), and a reduction in nitric oxide synthase (NOS) activity. In case of oxidative stress and inflammation, there was an activation of AMPK/SIRT1/Nrf2, ERK/p38, MAPK, and PTEN/Akt signalling pathways, inducing autophagy through an mTOR-dependent pathway and also inhibition of Nrf2 ubiquitination; promotion of the transcriptional functions of FoxOs results in induction of transcription of antioxidative enzymes. It has also shown a decrease in cirrhosis-induced portal pressure as an associated complication by reduction in MDA (malondialdehyde) levels and increase in GPx and SOD levels, as well as the inhibition of mRNA expression of inflammatory mediators, including inducible NO, tumour necrosis factor-α (TNF-α), and interleukin-1β (IL-1β). In cholestatic liver disease, RSV had decreased TIMP-1 and collagen Iα1 mRNA expression, mRNA expression levels of TNF-α and IL-6 mRNA, and the number of CD68-positive Kupffer cells recruited in the liver and increased Ki67-positive hepatocytes (Izzo et al. 2021).

(o) ***Bergenia ciliata***—*Bergenia ciliata*, an evergreen herb from the *Saxifragaceae* family, is revered as a miracle plant because of the variety of therapeutic uses it possesses (Latief et al. 2022). It reportedly possesses a plethora of pharmacological properties of antifungal, antiviral, anti-plasmodial, antibacterial, anti-inflammatory, anti-ulcer, and anticancer benefits. It comprises a wide variety of phytochemicals such as phenolic compounds, alcohols, terpenoids, and fatty acids, which contribute to its nutraceutical activity (Ahmad et al. 2018). The major hepatoprotective properties are shown by bergenin, gallic acid, gallicin, β-sitosterol, arbutin, catechin, afzelechin, linalool, limonene, pentanoic acid, α-terpineol, hexanoic acid, hexalactone, quercetin, protocatechuic acid, and traces of beta-caryophyllene, which upregulate the expression of anti-inflammatory and antioxidative molecules such as Bcl2 and PPAR-γ and downregulates the expression of biomarkers, which are pro-inflammatory such as Bax, NF-kB, TNF-α, IL-1β, and IL-6 and also decreases the expression of fibrogenic molecules such as Col1α1, Timp1, and α-SMA, which synergistically reduce ROS-induced apoptosis, liver inflammation, and NAFLD (Latief et al. 2022). Latief et al. in 2023 reported positive results of formulation of phytochemical constituents of bergenia as liposomes due to its enhanced bioavailability and reduced clearance. Xenobiotics absorbed by reticuloendothelial system (RES) are inhibited when they are formulated as nanocarriers such as liposomes, thus increasing their bioavailability, especially for hydrophobic molecules, and they also improve patient compliance, efficacy, product stability, solubility, and also the therapeutic index. When the extracts are formulated as liposomes, they reverted back the serum LFT markers in a better way as compared to the free extract, thus reducing oxidative stress and hepatic inflammation (Latief et al. 2023).

(p) ***Lobularia maritima*** **heteropolysaccharide**—Sweet alyssum or sweet Alison is the common name for *Lobularia maritima* (*Alyssum maritimum, Brassicaceae*). It is a halophyte endemic to Tunisia. The novel heteropolysaccharides found in the extracts are composed of mainly glucose, galactose, and xylose as their monomers. As per the in vitro analysis, the ROS scavenging activity of LmPS was reportedly increased to 90% and is even 20% more than catechin. LmPS extract has also reportedly decreased liver function markers such as ALT, AST, ALP, and LDH in serum and MDA levels in the liver, and an increase in the activities of antioxidative enzymes such as SOD, CAT, and GPx. LmPS has also demonstrated an ameliorative effect on cirrhosis by decreasing mediators of inflammation such as TNF-α, IL-10, and TGF-β1 and a significant difference in infiltration of neutrophils in comparison with the healthy liver in preclinical studies. Sulodexide also decreased the DNA smear and migration seen in hepatocytes in case of fibrosis (Ben Hsouna et al. 2022).

(q) **Sulodexide**—Sulodexide (SDX) is a heparinoid derivative compound. Sulodexide is a heparinoid compound extracted and purified from porcine intestinal mucosa, comprising 80% electrophoretically fast-moving heparin and 20% dermatan sulphate. Sulodexide has demonstrated therapeutic effects on the vascular system, including anti-inflammatory, antithrombotic, pro-fibrinolytic, vasoregulatory, and endothelial protective effects. Huang et al. in 2023 reported the hepatoprotective actions of SLD in TAA and DDC-induced hepatic fibrosis and cholestasis in mice. It also reduced liver weight and body weight ratios, and on staining, it was revealed that ECM deposits were decreased in TAA or DDC-induced mice treated with SDX. Also, it was found to down regulate the genetic expression of α-SMA and Col1α1, Acta2, and also reduced hepatic hydroxyproline and collagen content in fibrosis. It was reported that SDX reduced the fibrosis-associated LSEC capillarisation as it reduced the increased expression of CD34, CD31, and laminin and maintained eNOS expression in LSEC dysfunction up to normal levels. SDX reduced genetic expression of fibrosis-related genes as it downregulated the TGF-b signalling pathway and ECM receptor signalling and also reduced SMAD2 phosphorylation levels. It was observed that SDX blocked LSEC endothelial-mesenchymal transition as it again downregulated the expression of Col1a1, vimentin, Acta2, and Sm22 and in in vitro analysis showed the reduction in capillarisation genetic markers, EDN1 and CD34 in primary cultured LSECs, and hence reversed the effects of decreased fenestration and increased capillarisation both in vivo and in vitro (Huang et al. 2023).

(r) **Cerium Oxide Nanoparticles/Nanoceria**—Nanoceria has excellent biomedical properties in healing diseases caused due to oxidative stress as they switch between the valences +3 to +4. The nanoceria acts as a self-regenerating catalyst, which helps in exhibiting its superior antioxidant effects and hence decreasing inflammatory response. The nanosized particles increase its targeted drug delivery properties (its main targets were liver and spleen) and decreased adverse effects. Also, it was reported as nontoxic in in vitro in hepatic cell lines of HepG2, WRL68, and RAW264.7 and in vivo at therapeutic doses. They have demonstrated a drastic decrease in ALT, AST, GGT, and LDH activity and decreased concentrations of serum glutamic oxaloacetic transaminase (SGOT), serum glutamic pyruvic transaminase (SGPT), alkaline phosphatase (ALP), total bilirubin, and triglycerides and also hepatic injury score. It also showed improvement in portal hypertension, mean arterial hypertension, and splanchnic perfusion pressure. The levels of infiltrated CD68-positive cells and viable cells and genetic expression of α-SMA and caspase-3 could be significantly decreased in hepatic tissues. A marked decrease in genetic expression of inflammatory cytokines (TNF-α, IL-1β, IL-6, IL-17, COX-2, p65-NF-kB, and iNOS), oxidative messengers (Epx, Ncf1, and Ncf2), and endoplasmic reticulum messengers (Atf3 and Hspa5) related to stress signalling pathways was observed and also a decrease in other markers of hepatic oxidative stress such as hydroxyproline, TOS, MDA, and nitrite along with an increase in antioxidant markers such as CAT, SOD, TAC, and GSH. Nanoceria demonstrated a positive antifibrotic effect as it abrogated the genetic effect of TGF-b1, a-SMA, TIMP1, LOXL2 TIMP2, MMP-2, N-cadherin, fibronectin, Col1a1, and Snail. It also reduced the levels of ET-1, the main vasoconstrictor involved in portal hypertension, and decreased the transition of Kupffer cells into M1 pro-inflammatory phenotype (Karakoti et al. 2007; Oró et al. 2016; Casals et al. 2020; Stephen Inbaraj and Chen 2020; Abbasi et al. 2021; Godugu et al. 2023).

(s) **Liraglutide:** Liraglutide is a glucagon-like peptide (GLP-1) agonist developed by Novo Nordisk for type 2 diabetes mellitus (T2DM), which can be used to treat NAFLD (Shan et al. 2022). In type 2 diabetes mellitus, there is an insulin resistance, which increases the free fatty acid and triglyceride amount in liposomes of hepatocytes, which in excess amounts causes lipotoxicity and stimulates β-oxidation of FFA, which releases oxidative free radicals, and hence causes NASH which worsens to cause cirrhosis. Liraglutide was found to improve metabolism and reduce intracellular lipid accumulation in hepatocytes. It was reported to decrease the increased fasting blood glucose levels, insulin resistance (HOMA-IR), and serum levels of ALT, AST, and FFA in preclinical models of high-fat diet-induced diabetes and liver steatosis. Liraglutide upregulated the expression of insulin receptor substrate 2 (IRS2), insulin receptor (InsR), PI3K, p85, and Akt and decreased the expression of a-SMA, which was increased in case of NASH (Yang et al. 2019). Liraglutide was also found to reduce the expression of receptor for advanced glycation end products (RAGE), which was activated due to processed food-derived advanced glycation end products, which further activate downstream NADPH activity, which further activates ROS and upregulates the genetic expression of TGF-b signals by MAPK signalling. NADPH is formed from MOX1, NOX1, and NOX2 and NOX4 whose expression leads to NASH by NOX2 in preclinical murine models and NOX1 in human NASH patients. It was reported that LG reduced fibrogenesis and collagen. In in vitro analysis in AML12 cell line (murine hepatocyte cell line) was subjected to treatment with palmitic acid (PA) and H_2O_2 to mimic lipid peroxidation to increase ROS and p-JNK and p-p38-MAPK, and with liraglutide treatment, RAGE and NOX2 expressions were decreased and increased cell viability as it upregulated the expression of Bcl-2 and downregulated the expression of Bax. Additionally, it also reduced the expression of TGF-β-pSmad2 complex and inflammatory markers such as TNF-α, IL-1β, and IL-6 (Ji et al. 2022). In NAFLD, ACE/Ang II/AT1R axis gets activated in which Ang II promotes inflammation, oxidative stress, fibrosis, and vascular changes via the activation of AT1R, while liraglutide activates the antagonistic pathway of ACE2/angiotensin1-7/Mas axis. ACE and AT1R gene expression is activated in NAFLD, while ACE2 and Mas genetic expression was also high in preclinical mouse models of HFD-induced NASH as a compensatory and protective mechanism. Liraglutide showed a decrease in liver triglyceride content/TG concentration, NF-kB mRNA expression, IL-1b mRNA expression and concentration, G6P and PEPCK (gluconeogenesis-related genes) mRNA expression, and an increase in CPT-1a and ACOX1 mRNA (fatty acid oxidation-related genes) expression and increase in ACE2 protein expression and in Ang1-7. GLP-1 receptor agonist has been shown to reduce hepatocyte apoptosis and ER stress by inhibiting the activation of NATCH, LRR, and PYD domains of NALP3 inflammasome and NF-kB. Other mechanisms may also include an increased production of cyclic adenosine monophosphate (cAMP) and increased activation of an AMP-activated protein kinase (AMPK)-dependent pathway in hepatocytes, hence increasing fatty acid oxidation and decreasing in lipogenesis, and/or an increase in hepatic insulin signalling and sensitivity with GLP-1 and subsequent improvement of the hepatic glucose metabolism. Liraglutide by activating the SIRT1/AMPK pathway regulates lipid synthesis and metabolism and inhibits the de novo synthesis of fatty acids. Early intervention with liraglutide can reduce blood glucose, inhibit fat synthesis, and reduce insulin resistance, inflammation, fibrosis, and oxidative stress damage, thereby alleviating NAFLD, which may be related to the activation of SIRT1/AMPK and its downstream genes (Shan et al. 2022). In palmitic acid-induced steatosis in HepG2 cell line, RAS activity was regulated by PI3K/Akt pathway (Yang et al. 2020). It has shown pro-apoptotic properties for HepG2 cells in case of hepatocellular carcinoma in a dose-dependent manner by activation of JNK signalling pathway (Li et al. 2019). In HFD rats and HepG2 cells treated with palmitate fatty acid, liraglutide increased the expression of autophagy markers such as LC3, beclin1, and Atg7. Liraglutide-induced autophagy increased with the deceased expression of LC3, beclin1, and Atg7 was inhibited by the AMPK inhibitor in vitro. In addition, liraglutide therapy increased AMPK levels, decreased p-mTOR expression, and increased TSC1 (He et al. 2020). Liraglutide significantly improved hepatic vascular resistance and endothelial functioning as it decreased the genetic expression of a-SMA, PDGRFβ, and Col1a1 and as a result decreased the contractility of HSCs, which plays a major role in portal hypertension, as it reduced portal pressure and HVR and increased fenestrae and porosity (De Mesquita et al. 2017).

(t) **Gene therapy**—The only known effective treatment for end-stage liver fibrosis is liver transplant, although it has numerous drawbacks, including challenges in locating a liver donor, immunological rejection, and poor prognosis. These issues are anticipated to be resolved by gene therapy techniques such as antisense oligonucleotide chains, RNA intervention with the help of siRNA, and decoy oligonucleotides, but the study of gene therapy as an effective mechanism in controlling liver disease is still subjected to preclinical studies. In HSC-T6 cells, siRNA-mediated direct knockout of TGF-b1 drastically lowered the production of a-SMA and type I collagen and thus had an antifibrotic effect. Histone deacetylase 2 (HDAC2) is a protein that is upregulated and is present in fibrotic liver tissues or HSCs treated with TGF-1. In HSC-T6 cells exposed to TGF-1, siRNA used to inhibit HDAC2 expression resulted in a reduction in the expression of COL1a1 and α-SMA. As a result of downregulating the Wnt/b-catenin signal pathway, inhibiting the proliferation of HSC-T6 cells, and inducing apoptosis, b-catenin siRNA decreased collagen synthesis and b-catenin expression in HSCs in a time-dependent manner, halting the progression of hepatic fibrosis implying that β-catenin siRNA offers a novel approach to the management of liver fibrosis.

(u) The liver function of preclinical mouse models of hepatic fibrosis was considerably enhanced by miR-101, a short non-coding microRNA that controls the MAPK response. By suppressing the levels of α-SMA and COL1a1, lowering the accumulation of ECM components, and blocking the PI3K/AKT/mTOR signal pathway, miR-101 significantly reduced the damage to the liver parenchyma and delayed liver fibrosis. In a preclinical mouse model of hepatic fibrosis, miR-29b, a short non-coding RNA that is downregulated in fibrotic liver tissues and in primary activated HSCs, significantly inhibited the production of Smad3 in HSC line LX1 and decreased the expression of α-SMA and type I collagen. Additionally, miR-29b significantly slowed the progression of liver fibrosis by reducing HSC activation and inhibiting the apoptosis of HSCs brought on by the PI3K/AKT pathway (Tan et al. 2021) (Fig. 18.2).

18.5 Conclusion and Future Perspectives

Liver cirrhosis is in Western nations, and liver cirrhosis is the fourth leading cause of adult mortality, with portal hypertension consequences accounting for the majority of deaths. Development of precise diagnostic techniques for early diagnosis, successful etiologic treatment, improved pharmacological therapy for portal hypertension, and successful therapies for end-stage liver failure are necessary to lower mortality (Berzigotti 2017).

Here, in this chapter we have gone through the various mechanisms with which antioxidants can act at various aspects at the molecular level and treat liver cirrhosis and associated complications. Although antioxidative therapy remains one of the safest treatment options due to high safety and therapeutic index, research still remains going on in improving the drug delivery through its formulation as targeted drug delivery system with the help of nano-biomaterials and at genetic level by gene therapy with the help of miRNA and RNAi or siRNA.

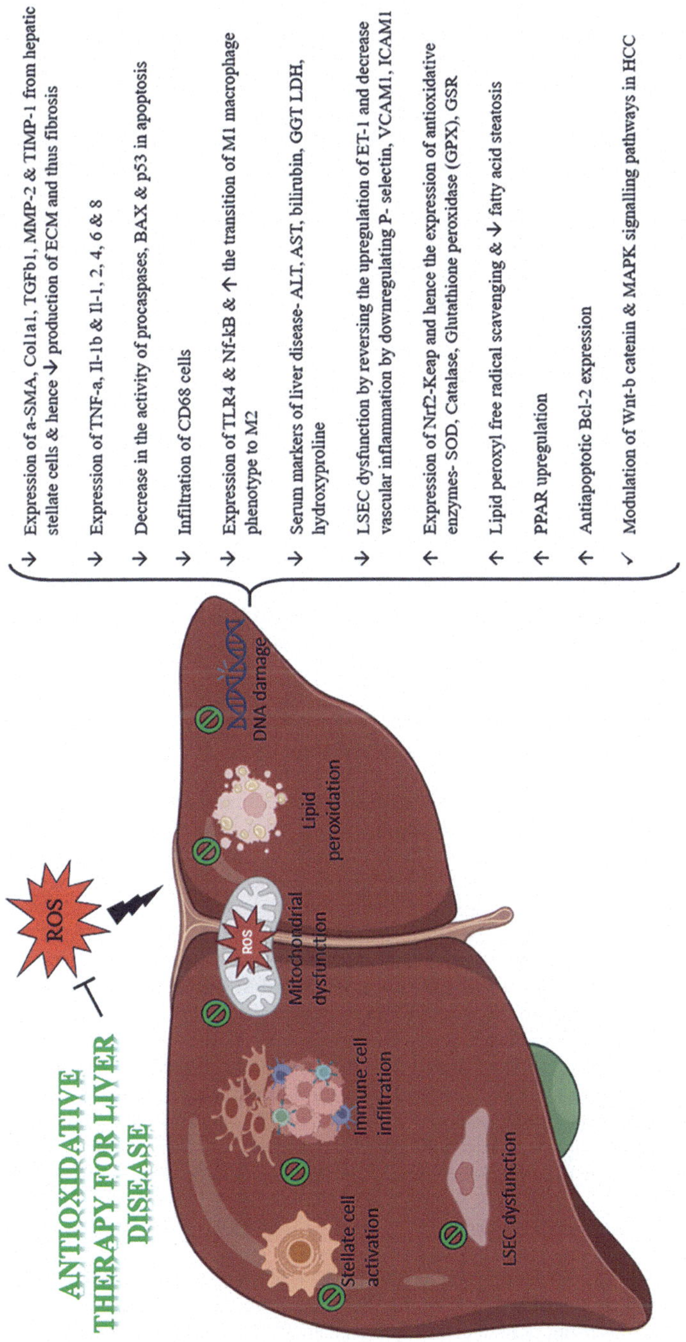

Fig. 18.2 A flowchart depicting the positive effects of antioxidative therapy for liver disease caused due to increased oxidative stress

References

Abbasi E, Vafaei SA, Naseri N, Darini A, Azandaryani MT, Ara FK et al (2021) Protective effects of cerium oxide nanoparticles in non-alcoholic fatty liver disease (NAFLD) and carbon tetrachloride-induced liver damage in rats: study on intestine and liver. Metabol Open 12: 100151. https://doi.org/10.1016/j.metop.2021.100151, PMID 34870139

Abdel-sttar AR, Khalaf MM, Aboyoussef AM, Abosaif AA (2017) Ameliorative effect of hesperidin on carbon tetrachloride induced liver fibrosis in rats. Int J Pharm Pharm Sci 9(7):45. https://doi.org/10.22159/ijpps.2017v9i7.17611

Abd-Eltawab Tammam A, Khalaf AAA, Zaki AR, Mansour Khalifa M, Ibrahim MA, Mekkawy AM et al (2022) Hesperidin protects rats' liver and kidney from oxidative damage and physiological disruption induced by nickel oxide nanoparticles. Front Physiol 13:912625. https://doi.org/10.3389/fphys.2022.912625, PMID 36338490, PMCID PMC9626958

Ahmad M, Butt MA, Zhang G, Sultana S, Tariq A, Zafar M (2018) Bergenia ciliata: a comprehensive review of its traditional uses, phytochemistry, pharmacology and safety. Biomed Pharmacother 97:708–721. https://doi.org/10.1016/j.biopha.2017.10.141, PMID 29102914

Ali AM, El-Tawil OS, Al-Mokaddem AK, Abd El-Rahman SS (2021) Promoted inhibition of TLR4/miR-155/NFkB p65 signaling by cannabinoid receptor 2 agonist (AM1241), aborts inflammation and progress of hepatic fibrosis induced by thioacetamide. Chem Biol Interact 336:109398. https://doi.org/10.1016/j.cbi.2021.109398, PMID 33503444

Alshammari GM, Abdelhalim MA, Al-Ayed MS, Al-Harbi LN, Yahya MA (2022) The protective effect of α-lipoic acid against gold nanoparticles (AuNPs)-mediated liver damage is associated with upregulating Nrf2 and suppressing NF-κB. Nutrients 14(16):3327. https://doi.org/10.3390/nu14163327, PMID 36014833, PMCID PMC9414933

Ben Hsouna A, Hfaiedh M, Ben Slima S, Romdhane WB, Akacha BB, Bouterra MT et al (2022) Antioxidant and hepatoprotective effects of novel heteropolysaccharide isolated from Lobularia maritima on CCl4-induced liver injury in rats. Food Sci Nutr 10(7):2271–2284. https://doi.org/10.1002/fsn3.2836, PMID 35844920

Berzigotti A (2017) Advances and challenges in cirrhosis and portal hypertension. BMC Med 15(1):200. https://doi.org/10.1186/s12916-017-0966-6, PMID 29121925

Calleja MA, Vieites JM, Montero-Meléndez T, Torres MI, Faus MJ, Gil A et al (2013) The antioxidant effect of β-caryophyllene protects rat liver from carbon tetrachloride-induced fibrosis by inhibiting hepatic stellate cell activation. Br J Nutr 109(3):394–401. https://doi.org/10.1017/S0007114512001298, PMID 22717234

Casals E, Zeng M, Parra-Robert M, Fernández-Varo G, Morales-Ruiz M, Jiménez W et al (2020) Cerium oxide nanoparticles: advances in biodistribution, toxicity, and preclinical exploration. Small 16(20):e1907322. https://doi.org/10.1002/smll.201907322, PMID 32329572

Casas-Grajales S, Muriel P (2015) Antioxidants in liver health. World J Gastrointest Pharmacol Ther 6(3):59–72. https://doi.org/10.4292/wjgpt.v6.i3.59, PMID 26261734

Çetin A, Çiftçi O, Otlu A (2016) Protective effect of hesperidin on oxidative and histological liver damage following carbon tetrachloride administration in Wistar rats. Arch Med Sci 12(3): 486–493. https://doi.org/10.5114/aoms.2015.49484, PMID 27279838

Cheraghpour M, Imani H, Ommi S, Alavian SM, Karimi-Shahrbabak E, Hedayati M et al (2019) Hesperidin improves hepatic steatosis, hepatic enzymes, and metabolic and inflammatory parameters in patients with nonalcoholic fatty liver disease: a randomized, placebo-controlled, double-blind clinical trial. Phytother Res 33(8):2118–2125. https://doi.org/10.1002/ptr.6406, PMID 31264313

De Mesquita FC, Guixé-Muntet S, Fernández-Iglesias A, Maeso-Díaz R, Vila S, Hide D et al (2017) Liraglutide improves liver microvascular dysfunction in cirrhosis: evidence from translational studies. Sci Rep 7(1):3255. https://doi.org/10.1038/s41598-017-02866-y, PMID 28607430

Ding C, Zhao Y, Chen X, Zheng Y, Liu W, Liu X (2021) Taxifolin, a novel food, attenuates acute alcohol-induced liver injury in mice through regulating the NF-κB-mediated inflammation and PI3K/Akt signalling pathways. Pharm Biol 59(1):868–879. https://doi.org/10.1080/13880209.2021.1942504, PMID 34225578

El-Mancy EM, Elsherbini DMA, Al-Serwi RH, El-Sherbiny M, Ahmed Shaker G, Abdel-Moneim AH et al (2022) α-Lipoic acid protects against cyclosporine A-induced hepatic toxicity in rats: effect on oxidative stress, inflammation, and apoptosis. Toxics 10(8):442. https://doi.org/10.3390/toxics10080442, PMID 36006121

Elsayed HRH, El-Nablaway M, Othman BH, Abdalla AM, El Nashar EM, Abd-Elmonem MM et al (2021) Can dasatinib ameliorate the hepatic changes, induced by long term Western diet, in mice? Ann Anat 234:151626. https://doi.org/10.1016/j.aanat.2020.151626, PMID 33144268

Fiorucci S, Biagioli M, Baldoni M, Ricci P, Sepe V, Zampella A et al (2021) The identification of farnesoid X receptor modulators as treatment options for nonalcoholic fatty liver disease. Expert Opin Drug Discov 16(10):1193–1208. https://doi.org/10.1080/17460441.2021.1916465, PMID 33849361

Garcia-Tsao G, Fuchs M, Shiffman M, Borg BB, Pyrsopoulos N, Shetty K et al (2019) Emricasan (IDN-6556) lowers portal pressure in patients with compensated cirrhosis and severe portal hypertension. Hepatology 69(2):717–728. https://doi.org/10.1002/hep.30199, PMID 30063802

Godugu C, Khurana A, Saifi MA (2023) Rare earth cerium oxide nanoparticles attenuated liver fibrosis in bile duct ligation mice model. J Trace Elem Med Biol 75:127102. https://doi.org/10.1016/j.jtemb.2022.127102, PMID 36423438

Gong Y, Yang Y (2020) Activation of Nrf2/AREs-mediated antioxidant signalling, and suppression of profibrotic TGF-β1/Smad3 pathway: a promising therapeutic strategy for hepatic fibrosis—a review. Life Sci 256:117909. https://doi.org/10.1016/j.lfs.2020.117909, PMID 32512009

Gracia-Sancho J, Manicardi N, Ortega-Ribera M, Maeso-Díaz R, Guixé-Muntet S, Fernández-Iglesias A et al (2019) Emricasan ameliorates portal hypertension and liver fibrosis in cirrhotic rats through a hepatocyte-mediated paracrine mechanism. Hepatol Commun 3(7):987–1000. https://doi.org/10.1002/hep4.1360, PMID 31304452

Hasegawa T, Yoneda M, Nakamura K, Makino I, Terano A (2001) Plasma transforming growth factor-β1 level and efficacy of α-tocopherol in patients with non-alcoholic steatohepatitis: a pilot study. Aliment Pharmacol Ther 15(10):1667–1672. https://doi.org/10.1046/j.1365-2036.2001.01083.x, PMID 11564008

Hashiesh HM, Sharma C, Goyal SN, Jha NK, Ojha S (2021) Pharmacological properties, therapeutic potential and molecular mechanisms of JWH133, a CB2 receptor-selective agonist. Front Pharmacol 12:702675. https://doi.org/10.3389/fphar.2021.702675, PMID 34393784

He Y, Ao N, Yang J, Wang X, Jin S, Du J (2020) The preventive effect of liraglutide on the lipotoxic liver injury via increasing autophagy. Ann Hepatol 19(1):44–52. https://doi.org/10.1016/j.aohep.2019.06.023, PMID 31787541

Huang R, Deng J, Zhu CP, Liu SQ, Cui YL, Chen F et al (2023) Sulodexide attenuates liver fibrosis in mice by restoration of differentiated liver sinusoidal endothelial cell. Biomed Pharmacother 160:114396. https://doi.org/10.1016/j.biopha.2023.114396, PMID 36791568

Izzo C, Annunziata M, Melara G, Sciorio R, Dallio M, Masarone M et al (2021) The role of resveratrol in liver disease: a comprehensive review from in vitro to clinical trials. Nutrients 13(3):1–23. https://doi.org/10.3390/nu13030933, PMID 33805795

Jha NK, Sharma C, Meeran MFN, Jha SK, Dwivedi VD, Gupta PK et al (2021) Cannabinoid Type-2 receptor agonist, JWH133 may be a possible candidate for targeting infection, inflammation, and immunity in COVID-19. Immunology 1(3):285–304. https://doi.org/10.3390/immuno1030020

Ji J, Feng M, Huang Y, Niu X (2022) Liraglutide inhibits receptor for advanced glycation end products (RAGE)/reduced form of nicotinamide-adenine dinucleotide phosphate (NAPDH) signaling to ameliorate non-alcoholic fatty liver disease (NAFLD) in vivo and vitro. Bioengineered 13(3):5091–5102. https://doi.org/10.1080/21655979.2022.2036902, PMID 351 64657

Karakoti AS, Kuchibhatla SVNT, Babu KS, Seal S (2007) Direct synthesis of nanoceria in aqueous polyhydroxyl solutions. J Phys Chem C 111(46):17232–17240. https://doi.org/10.1021/jp076164k

Kołota A (2017) The possibilities of silymarin extract use in alcoholic liver disease—a literature review. Alcohol Drug Addict 30(2):137–154. https://doi.org/10.5114/ain.2017.70290
Latief U, Tung GK, Singh H, Per TS, Jain SK (2022) Bergenia ciliata as a future candidate for liver diseases: a concise review. J Basic Appl Zool 83(1):17. https://doi.org/10.1186/s41936-022-00282-x
Latief U, Kaur M, Dar SH, Thakur S, Per TS, Tung GK et al (2023) Preparation and characterisation of liposomes of Bergenia ciliata extract and evaluation of their hepatoprotective activity. J Pharm Sci 112(1):328–335. https://doi.org/10.1016/j.xphs.2022.07.013, PMID 35872024
Lee PJ, Woo SJ, Jee JG, Sung SH, Kim HP (2015) Bisdemethoxycurcumin induces apoptosis in activated hepatic stellate cells via cannabinoid receptor 2. Molecules 20(1):1277–1292. https://doi.org/10.3390/molecules20011277, PMID 25594342
Lee JS, Mukhopadhyay P, Matyas C, Trojnar E, Paloczi J, Yang YR et al (2019) PCSK9 inhibition as a novel therapeutic target for alcoholic liver disease. Sci Rep 9(1):17167. https://doi.org/10.1038/s41598-019-53603-6, PMID 31748600
Li S, Tan HY, Wang N, Zhang ZJ, Lao L, Wong CW et al (2015) The role of oxidative stress and antioxidants in liver diseases. Int J Mol Sci 16(11):26087–26124. https://doi.org/10.3390/ijms161125942, PMID 26540040
Li Q, Xue AY, Li ZL, Yin Z (2019) Liraglutide promotes apoptosis of HepG2 cells by activating JNK signaling pathway. Eur Rev Med Pharmacol Sci 23(8):3520–3526. https://doi.org/10.26355/eurrev_201904_17719, PMID 31081108
Li Z, Zou J, Cao D, Ma X (2020) Pharmacological basis of tanshinone and new insights into tanshinone as a multitarget natural product for multifaceted diseases. Biomed Pharmacother 130:110599. https://doi.org/10.1016/j.biopha.2020.110599, PMID 33236719
Liu G, Liu J, Pian L, Gui S, Lu B (2019) α-Lipoic acid protects against carbon tetrachloride-induced liver cirrhosis through the suppression of the TGF-β/Smad3 pathway and autophagy. Mol Med Rep 19(2):841–850. https://doi.org/10.3892/mmr.2018.9719, PMID 30535447
Liu X, Liu W, Ding C, Zhao Y, Chen X, Ling D et al (2021) Taxifolin, extracted from waste larix olgensis roots, attenuates CCl4-induced liver fibrosis by regulating the PI3K/AKT/mTOR and TGF-β1/Smads signaling pathways. Drug Des Dev Ther 15:871–887. https://doi.org/10.2147/DDDT.S281369, PMID 33664566
Lotersztajn S, Teixeira-Clerc F, Julien B, Deveaux V, Ichigotani Y, Manin S et al (2008) CB2 receptors as new therapeutic targets for liver diseases. Br J Pharmacol 153(2):286–289. https://doi.org/10.1038/sj.bjp.0707511, PMID 17952109
Miao Y, Wu Y, Jin Y, Lei M, Nan J, Wu X (2020) Benzoquinone derivatives with antioxidant activity inhibit activated hepatic stellate cells and attenuate liver fibrosis in TAA-induced mice. Chem Biol Interact 317:108945. https://doi.org/10.1016/j.cbi.2020.108945, PMID 31935363
Mohammadalipour A, Karimi J, Khodadadi I, Solgi G, Hashemnia M, Sheikh N et al (2017) Dasatinib prevent hepatic fibrosis induced by carbon tetrachloride (CCl4) via anti-inflammatory and antioxidant mechanism. Immunopharmacol Immunotoxicol 39(1):19–27. https://doi.org/10.1080/08923973.2016.1263860, PMID 27908221
Monroy-Ramirez HC, Galicia-Moreno M, Sandoval-Rodriguez A, Meza-Rios A, Santos A, Armendariz-Borunda J (2021) PPARs as metabolic sensors and therapeutic targets in liver diseases. Int J Mol Sci 22(15):8298. https://doi.org/10.3390/ijms22158298, PMID 34361064
Mu LY, Li S, Tang L, Li R (2021) Efficacy and safety of emricasan in liver cirrhosis and/or fibrosis. Clinics 76(17):e2409. https://doi.org/10.6061/CLINICS/2021/E2409
Nagashimada M, Ota T (2019) Role of vitamin E in nonalcoholic fatty liver disease. IUBMB Life 71(4):516–522. https://doi.org/10.1002/iub.1991, PMID 30592129
Oró D, Yudina T, Fernández-Varo G, Casals E, Reichenbach V, Casals G et al (2016) Cerium oxide nanoparticles reduce steatosis, portal hypertension and display anti-inflammatory properties in rats with liver fibrosis. J Hepatol 64(3):691–698. https://doi.org/10.1016/j.jhep.2015.10.020, PMID 26519601
Perumpail BJ, Li AA, John N, Sallam S, Shah ND, Kwong W et al (2018) The role of vitamin E in the treatment of NAFLD. Diseases 6(4):86. https://doi.org/10.3390/diseases6040086, PMID 30249972

Podszun MC, Frank J (2021) Impact of vitamin E on redox biomarkers in non-alcoholic fatty liver disease. Redox Biol 42:101937. https://doi.org/10.1016/j.redox.2021.101937, PMID 33773953

Salama SA, Kabel AM (2020) Taxifolin ameliorates iron overload-induced hepatocellular injury: modulating PI3K/AKT and p38 MAPK signaling, inflammatory response, and hepatocellular regeneration. Chem Biol Interact 330:109230. https://doi.org/10.1016/j.cbi.2020.109230, PMID 32828744

Sánchez-Valle V, Chávez-Tapia NC, Uribe M, Méndez-Sánchez N (2012) Role of oxidative stress and molecular changes in liver fibrosis: a review. Curr Med Chem 19(28):4850–4860. https://doi.org/10.2174/092986712803341520, PMID 22709007

Shan L, Wang F, Zhai D, Meng X, Liu J, Lv X (2022) New drugs for hepatic fibrosis. Front Pharmacol 13:874408. https://doi.org/10.3389/fphar.2022.874408, PMID 35770089

Stephen Inbaraj B, Chen BH (2020) An overview on recent in vivo biological application of cerium oxide nanoparticles. Asian J Pharm Sci 15(5):558–575. https://doi.org/10.1016/j.ajps.2019.10.005, PMID 33193860

Stofan M, Guo GL (2020) Bile acids and FXR: novel targets for liver diseases. Front Med (Lausanne) 7:544. https://doi.org/10.3389/fmed.2020.00544, PMID 33015098

Tabeshpour J, Hosseinzadeh H, Hashemzaei M, Karimi G (2020) A review of the hepatoprotective effects of hesperidin, a flavanon glycoside in citrus fruits, against natural and chemical toxicities. Daru 28(1):305–317. https://doi.org/10.1007/s40199-020-00344-x, PMID 32277430

Tan Z, Sun H, Xue T, Gan C, Liu H, Xie Y et al (2021) Liver fibrosis: therapeutic targets and advances in drug therapy. Front Cell Dev Biol 9:730176. https://doi.org/10.3389/fcell.2021.730176, PMID 34621747

Tighe SP, Akhtar D, Iqbal U, Ahmed A (2020) Chronic liver disease and silymarin: a biochemical and clinical review. J Clin Transl Hepatol 8(4):454–458. https://doi.org/10.14218/JCTH.2020.00012, PMID 33447529

Vargas-Mendoza N, Madrigal-Santillán E, Morales-González A, Esquivel-Soto J, Esquivel-Chirino C, García-Luna Y, González-Rubio M et al (2014) Hepatoprotective effect of silymarin. World J Hepatol 6(3):144–149. https://doi.org/10.4254/wjh.v6.i3.144, PMID 24672644

Willatts SM (1990) The liver. Curr Anaesth Crit Care 1(4):195. https://doi.org/10.1016/0953-7112(90)90001-I

Wu W, Zhu B, Peng X, Zhou M, Jia D, Gu J (2014) Activation of farnesoid X receptor attenuates hepatic injury in a murine model of alcoholic liver disease. Biochem Biophys Res Commun 443(1):68–73. https://doi.org/10.1016/j.bbrc.2013.11.057, PMID 24269813

Yang P, Liang Y, Luo Y, Li Z, Wen Y, Shen J et al (2019) Liraglutide ameliorates nonalcoholic fatty liver disease in diabetic mice via the IRS2/PI3K/Akt signaling pathway. Diabetes Metab Syndr Obes 12:1013–1021. https://doi.org/10.2147/DMSO.S206867, PMID 31308717

Yang M, Ma X, Xuan X, Deng H, Chen Q, Yuan L (2020) Liraglutide attenuates non-alcoholic fatty liver disease in mice by regulating the local renin-angiotensin system. Front Pharmacol 11:432. https://doi.org/10.3389/fphar.2020.00432, PMID 32322207

Ying Q, Teng Y, Zhang J, Cai Z, Xue Z (2019) Therapeutic effect of tanshinone IIA on liver fibrosis and the possible mechanism: a preclinical meta-analysis. Evid Based Complement Alternat Med 2019:7514046. https://doi.org/10.1155/2019/7514046, PMID 31915451

Yu B, Qin SY, Hu BL, Qin QY, Jiang HX, Luo W (2019) Resveratrol improves CCl4-induced liver fibrosis in mouse by upregulating endogenous IL-10 to reprogramme macrophages phenotype from M(LPS) to M(IL-4). Biomed Pharmacother 117(6):109110. https://doi.org/10.1016/j.biopha.2019.109110, PMID 31252263

Zaafan MA, Abdelhamid AM (2022) Dasatinib ameliorates thioacetamide-induced liver fibrosis: modulation of miR-378 and miR-17 and their linked Wnt/β-catenin and TGF-β/smads pathways. J Enzyme Inhib Med Chem 37(1):118–124. https://doi.org/10.1080/14756366.2021.1995379, PMID 34894966

Sleep, Brain, and Stress

19

Bhanuteja Thondala, Harsh Pawar, and Usha Panjwani

Abstract

The brain regulates adaptations to the body driven by behavioral states like waking and environment. These physiological changes support adaptive behaviors including movement and dealing with unpleasant circumstances and discrete stimuli like noise, crowding, hunger, extreme heat or cold, and other safety risks. Sleep is an indispensable phenomenon of life. The biological function of sleep is preserved evolutionarily across species. The importance of sleep in normal physical and mental function is well known and researched in both human studies and animal models. Modern society is impacted by deleterious consequences of sleep deprivation due to lifestyle, often at variance with the natural zeitgeber. Both acute and chronic sleep deprivations lead to serious neurobehavioral effects. This chapter looks at recent trends in nonpharmacological approaches to improve sleep quality, both proven and in the research domain, its advantages, and challenges in adverse conditions. It also traces the history of research on sleep deprivation. The regulation of sleep function by circadian and homeostatic mechanisms still eludes scientists. The neural correlates, neurotransmitters, neuromodulators, pharmacological interventions for improving sleep quality, and adverse effects, along with some

B. Thondala · H. Pawar
Soldier Performance Vertical, Defence Institute of Physiology and Allied Sciences (DIPAS), Defence Research and Development Organisation (DRDO), Delhi, India

U. Panjwani (✉)
Soldier Performance Vertical, Defence Institute of Physiology and Allied Sciences (DIPAS), Defence Research and Development Organisation (DRDO), Delhi, India

Manav Rachna International Institute of Research and Studies, Faridabad, Haryana, India
e-mail: ushapanjwani6@gmail.com

R. Tulsawani, D. Vohora (eds.), *Adaptation under Stressful Environments through Biological Adjustments and Interventions*,
https://doi.org/10.1007/978-981-99-7652-2_19

of the well-studied active components of plants to improve mental health are discussed.

Keywords

Sleep quality · Sleep deprivation · Adaptogens · High-altitude hypoxia · Neurotransmitters · Pharmacological management

19.1 Introduction

19.1.1 History of Sleep Research

Sleep research originated with an emphasis on decoding the function of sleep by depriving animals of sleep. In contrast to the history of research in other medical sciences, the history of sleep research is limited. In 1894, Marie De Manaceine found that sleep deprivation was fatal to young and adult dogs. George Thomas White Patrick and J. Allen Gilbert conducted preliminary experiments on sleep deprivation in humans in 1896. In 1898, Lamberto Daddi showed that neuronal degeneration occurred in the brains of dogs after total sleep deprivation. Cesare Agostini observed "progressive exhaustion of psychic activity" in dogs subjected to total sleep deprivation. Researchers at the Walter Reed Army Institute of Research documented that acute sleep loss disrupts sleep-deprived males' performance. William C. Dement performed the foremost REM sleep deprivation experiment and also showed REM sleep rebound. Allan Rechtschaffen reported that rats subjected to sleep deprivation showed increased energy expenditure, impaired thermoregulation, and REM rebound.

Adequate sleep is essential for optimum health as it promotes many physiological processes, such as memory consolidation, clearance of brain metabolites, and regeneration of nervous, immune, skeletal, and muscular systems (Fernández-Moriano et al. 2017). Over the past few decades, industrialization and the fast-paced urbanized world gradually posed unobtrusive complications, raising concerns for operations requiring high human cognitive performance. This brings us to the debate over sleep need and homeostasis in modern humans.

19.1.2 Adaptogens

Adaptogens are a class of herbal, nutritional, medicinal, and therapeutic agents that support the resilience, adaptation, and survival of living things under stress. An adaptogen should be harmless and only slightly impair an organism's physiological processes. Adaptogens also increase resistance to harmful influences from a wide range of physical, chemical, and biological factors and have a normalizing effect regardless of the direction of the pathologic changes (Brekhman and Dardymov 1969). Because of their pharmacologically pleiotropic effects on the neuroendocrine and immunological systems, adaptogens have long been used to treat a variety of ailments. They display a biphasic dosage response, meaning that at low doses, they mimic moderate stress by activating the adaptive stress response signaling pathways, which help the body to deal with more intense stress (Panossian et al. 2021). However, the precise mechanism of adaptogen activity is still debatable despite the diverse identified mechanisms. Adaptogens are thought to function as "metabolic regulators" at the cellular as well as organism levels which boost an organism's ability to adapt to external stressors and avoid noxious effects to the organism by such stressors (Panossian et al. 1999). They promote neuroendocrine function and homeostasis by altering the levels of plasma cortisol, neuropeptide Y (NPY) (Panossian et al. 2012), and other stress-related mediators, such as membrane-bound G-protein receptors, heat shock proteins 70, and nitric oxide, and are involved in the regulation of the hypothalamic–pituitary–adrenal (HPA) axis (Panossian 2013). They are also reported to alter gene expression of important mediators of intracellular communication involved in stress-induced signal transduction pathways, such as G-protein signaling phosphatidylinositol, phospholipase C pathways, stress-activated protein kinase JNK (MAPK-9), Hsp70 (stress-protective effect on the cellular level), phosphatidylinositol, and phospholipase C pathways. Another most commonly reported mechanism of their activity is their ROS-countering abilities which are common to most adaptogens due to the presence of polyflavonoids. Environmental stressors generally cause higher levels of ROS production at cellular levels and produce debilitating effects through oxidative stress (Fernández-Moriano et al. 2017). Antioxidant dietary supplements, such as certain vitamins, lessen the negative effects of particulate pollutants as there is no definite prevention. Unsaturated fatty acids in fish oil and olive oil have the effect of reducing cardiovascular effects to brief exposure to air pollutants. In particular, it was shown that vitamins C and E protect against oxidative damage from exposure to O3 and PM2.5 (Tong 2016).

19.1.3 Neurological Manifestations

The human brain perceives and evaluates any stress, as well as plays important role in behavioral and physiological responses to the stressor, which makes the brain, the primary organ controlling stress and adaptation response. These responses encourage adaptation or allostasis, but they can also make a significant contribution to pathophysiology when they are overused and deregulated called allostatic overload. In response to stresses and other noxious stimuli, the human brain exhibits amazing structural and functional plasticity, including neuronal and dendritic remodeling and reserves, and synaptic alterations. Neural circuits underlying the vital process of cognition, memory, mood, anxiety, and depression are finely balanced with the levels of certain neurotransmitters that can be disrupted by stress which in turn affects the behavioral state and normal physiology of the body. Stress-induced changes such as autonomic, metabolic, neuroendocrine, and immunological and their consecutive pathophysiology are short-term and generally require no alteration;

however, persistent anomalies in behavior or physiological function are required to intervene either pharmacologically or nonpharmacologically. Adaptive responses or associated pathological changes due to stressors could be epigenetic which are long-lasting and need to be prevented. Brain plasticity is an important process that could be harnessed therapeutically to prevent or treat such neurophysiologic manifestations (McEwen 2017).

19.1.4 High-Altitude Hypoxia and Brain

Nonacclimatized people who are exposed to elevations exceeding 3000 m (9,843 ft.) for periods ranging from a few hours or days may experience severe effects from high-altitude surroundings. Nearly 20% of the body's total oxygen intake is thought to be used up by the brain. Thus, high altitudes have a significant impact on the brain. Headache, disorientation, exhaustion, dizziness, shortness of breath, decreased appetite, nausea, interrupted sleep, and general malaise are common symptoms of ascending to a higher altitude. Within a few hours, moderate hypoxia causes significant changes in physiological and psychological characteristics (Bahrke and Shukitt-Hale 1993). Climbing to high altitude causes a drop in blood oxygenation, which lowers the flow of oxygen to the brain and rest of the body. With time, the body makes a range of physiological adaptations and changes to match the level of oxygen, at least in part. This process is known as acclimatization overall. Altitude exposure causes detrimental changes in behavior as well as a reduction in psychological capacity. Chronic high-altitude exposure impairs cognitive functions like attentiveness, memory, decision, emotion, and especially on spatial memory (Wang et al. 2014). Cognitive impairment is the result of decreased cerebral autoregulation, vascular changes, microglial activation, oxidative stress, neuroinflammation, and consecutive degeneration (Wilson et al. 2009). Increased PGE2 levels and altered COX 1- and COX 2-driven neuroinflammatory pathophysiology have also been linked to hypoxia exposure (Chauhan et al. 2019). Chronic exposure to moderate hypoxia caused temporary vascular leaks in the blood–brain barrier's blood vessels associated with the aggregation and activation of microglia near damaged vasculature (Halder and Milner 2020).

19.1.5 High-Altitude Cerebral Edema

Most severe and potentially deadly condition in AMS is termed as High-Altitude Cerebral Edema (HACE) characterized by confusion, ataxia, fever, fatigue, rapid heartbeat, altered mental status, and loss of consciousness. Though HACE is the least prevalent type of altitude sickness, but to the unavailability of treatment, it can develop quickly to coma and death due to brain herniation within 24 h and hence considered as the final form of AMS (Jensen and Vincent 2022). It can occur at any altitude above (2500 m), however not common below 4000–5000 m with almost 0.5%–1% individuals suffering HACE. Typically, HACE manifests after 48 h with or without symptoms like anorexia, lassitude, headache, nausea, and insomnia (Gallagher and Hackett 2004). The mechanism behind the development of HACE is not fully understood; however, certain neurohormonal ailments such as NO, ROS, inflammatory cytokines, VEGF, neurotransmitters, and alternation in cerebral hemodynamics due to dilatation of cerebral arteries and overperfusion leading to increased intracranial pressure and capillary leakage could be responsible. Other researchers suggest individual intracranial space available for edema in the individual as a possible cause behind the susceptibility for HACE in a few individuals called the "tight-fit" hypothesis (Ross 1985). Increased intracranial pressure, as well as overload volume, are still the most preferred hypotheses of HACE development. The Na +/K+ ATPase pump dysregulation due to hypoxia stress-induced free radicals, causes astrocyte swelling due to osmotic imbalances and could cause cytotoxic edema. Corner stone of the treatment success relies on the immediate descent of the affected individual by at least 1000 m or more until symptoms subside without giving extra exertion to the individual. A portable hyperbaric chamber and supplemental oxygen (2–4 L/min) are alternatives till descent from altitude is not possible. The pharmacological treatment approach involves dexamethasone 8 mg administration via oral, IM, or IV and a subsequent dose of 4 mg every 6 h. About 0.15 mg/kg dexamethasone per 6 h should be administered to pediatric individuals. Dexamethasone which is primarily a glucocorticoid receptor agonist known for its potent anti-inflammatory activity is the first line of treatment as a well prophylactic option in the case of AMS susceptible individuals; however, exact mechanism of its beneficial effect is unclear in the case of HACE. The proposed mechanism involved inhibition of the inflammatory signaling cascade, mitigating vascular permeability, improving antioxidant balance, water channels like aquaporin-4 channel (AQP4), and blockade of the sympathetic outflow. Acetazolamide is another reputed first-line therapeutic as well as a prophylactic approach given 250 mg oral twice daily in case of severe AMS and HACE. Acetazolamide primarily inhibits carbonic anhydrase enzyme which results in bicarbonate excretion or bicarbonaturia and shifts equilibrium toward metabolic acidosis and mitigates the respiratory alkalosis resulting from hypoxia-induced hyperventilation allowing chemoreceptors to respond hypoxia much efficiently. Contraindication to acetazolamide involves individuals previously allergic to sulfa drugs. In addition, Ibuprofen (NSAID) is said to be preventive for AMS, albeit there are few studies supporting this and risk of GI bleeding in hypoxic environments (Xiong et al. 2017). Budesonide, ginkgo Biloba, chewed coca leaves, and acetaminophen are among more possible substitutes that have been researched. Since there is currently insufficient data in the literature to support their effectiveness, they are not advised for use in HACE prevention (Zelmanovich et al. 2022). With little evidence, sildenafil, tadalafil, and theophylline may also aid in the therapy of HACE (Wilson et al. 2009). In rat models of AMS, tetrahydrocurcumin and ganglioside (GM1) are also said to be beneficial in high-altitude cerebral edema (Zelmanovich et al. 2022).

19.1.6 Extreme Cold and Brain

Extreme cold environment triggers the body's temperature regulation system and induces shivering, vasoconstriction, and reduction of sweating are the basic reactions to a cold stimulus, all of which help to maintain a normal core body temperature of 37 °C. Neural feedback centers located in the hypothalamus, spinal cord, skin, and visceral layer regulate these responses. Hypothalamus induces shivering by stimulating the sympathetic nervous system and primary motor center. Evidence of an effect of cold on response times is also somewhat inconsistent. Several studies have demonstrated that response times during cognitive tasks are slowed when subjects are exposed to cold air or water. Exposure to acute cold stress affects the performance on matching to sample tests of working memory functions. There has been a significant decline in memory recall but not recognition, according to certain research, while some others reported that even 50 min of exposure to 4 °C, may lead to greater arousal, as shown by decreased evoked potential latencies and reaction times. Extensive memory impairments and depressive behavior linked to hypothermic cold exposures have been seen in animal research. An increase in depressive symptoms and reductions in cognitive function have been observed in clinically healthy adults throughout the winter months, according to the studies of polar expeditions and employees in Antarctic research stations known as "winter over syndrome" (Palinkas 2001). Certain dietary and pharmaceutical therapies, in particular, may help to avoid cognitive decline brought on by colds. Tyrosine, a nutritional precursor of catecholamines, appears to boost catecholaminergic function and avoid cold-induced declines in cognitive performance, according to studies on both humans and animals. Tyrosine has been shown by Gibson and Wurtman to speed up the rate of norepinephrine (NE) synthesis in the CNS of rats following acute ambient cold exposure of 4 °C in comparison to rats not given tyrosine and cold stressed. Rauch and Lieberman reported that pretreatment with 400 mg/kg L-tyrosine can counteract the behavioral depression brought on by hypothermia (Rauch and Lieberman 1990). Compared to saline, doses of 100 and 200 mg/kg L-tyrosine greatly improved overall matching accuracy, although they did not entirely counteract the effects of cold exposure. Daily administration of 50 mg levothyroxine mitigates disorientation and fatigue and improves performance on a delayed matching-to-sample test, since alteration in thyroid hormone levels is quite common in Antarctica known as "Polar T3 syndrome" (Reed et al. 1990).

19.1.7 Extreme Heat and Brain

Elevated temperatures, whether of environmental (i.e., heat stroke) or body-related origin (i.e., infectious or noninfectious), do have an impact on numerous organ systems of the human body and may even be fatal. The central nervous system is particularly susceptible to hyperthermia, especially if it is protracted or extreme. Symptoms of heat-related conditions include heat exhaustion, life-threatening heat stroke, and heat syncope/cramps. A heat stroke typically results in aphasia, dysphagia, cerebellar ataxia, and cognitive impairment. The neurological and cognitive effects of temperature on the brain may be seen immediately after the incident, continue for a considerable amount of time afterward, or even last forever. Even modest and transient hyperthermia may have an adverse effect on cognitive function. This could be permanent in some circumstances. It has been demonstrated that acutely negative effects of hyperthermia on cognition, memory, and information processing. Hyperthermia may have a more significant impact on some cognitive functions than others (Walter and Carraretto 2016). Heat stress causes a wide range of pathophysiological reactions and significantly alters the structure of the brain. Both experimental animals and humans have shown cognitive impairment after being exposed to high temperatures. Heat stress has been linked to memory loss, neuronal death, and reduced adult neurogenesis. It can activate glial cells and induce inflammatory chemicals in the hippocampus (Lee et al. 2015). Animal studies also reported impaired nigrostriatal dopaminergic signaling due to repeated exposure to heat (Kim et al. 2013).

19.1.8 Environmental Radiations and Neurophysiology

Multiple ionizing and nonionizing radiations, both of natural and man-made origin are present around us in the environment. Exposure to ultraviolet radiation, cosmic radiations that are substantially more prevalent at high altitudes, and various ionizing radiations like radon gas, are a few prominent examples. Professionals like astronauts will unavoidably face greater health hazards as they spend more time in space because of their exposure to solar particle events and galactic cosmic rays. Crew members of spacecraft exposed to radiation may face serious risks later, when slowly manifesting negative consequences may finally become obvious. Numerous researchers have demonstrated that cells in the hippocampus DG's neurogenic zone are very sensitive to deep space relevant low doses of radiation and that such changes are linked to hippocampal-dependent cognitive deficits. Dendritic density, synaptic plasticity, and reduced neurogenesis are all reported to be affected by simulated radiation exposure (Davis et al. 2021).

19.1.9 Environmental Pollutants and Neurological Functioning

Pollution is a growing issue for developing as well wealthy nations as it widely impacts public health. Numerous health hazards are reported including cancers, autism, hearing and vision loss, memory problems, developmental disorders which are associated with pollution. Gaseous pollutants are readily absorbable from lungs and can cause serious neurological side effects. Brain is quite immunologically privileged organ; however, recent researches are showing that particulate pollutants can even cross blood–brain barrier to reach brain depending on their sizes. PM2.5 is the most studied particulate pollutant that was reported to be associated with distinct gray matter changes in the frontal, temporal, occipital, and parietal lobe which are hemisphere-specific, when exposed in childhood (Cserbik et al. 2020). Exposure to fine PM (PM 2.5 m) has been linked to an increased incidence of dementia and Alzheimer's disease, according to various epidemiological studies (Jung et al. 2015). The association between air pollution and brain dysfunction has been linked to a number of biological mechanisms, including systemic inflammation and oxidative stress. Studies also reported that leading cause of stroke is NO_2, the main component of air pollution. Additionally, ischemic stroke risk was considerably enhanced by brief exposure to SO_2, NO_2, and PM10 (Atkinson et al. 2013). Another study reported that the brain's corpus callosum region is dramatically reduced in late pregnancy with PM exposure, and a reduction in CC size of 50 mm^3 results in noticeably higher hyperactivity subscores proving its association with ADHD. Potential routes for air pollutants to cross the blood–brain barrier include the olfactory tract, the gastrointestinal system, and the vagus nerve. Although the molecular mechanisms by which fine particulate impacts brain problems are unknown, it has been claimed that the two main mechanisms by which air pollution exerts its damaging systemic and CNS effects are oxidative stress and inflammatory responses. Increases in a number of proinflammatory cytokines like IL-6, IL-10, and TNF especially in the olfactory bulb and hippocampus, were observed (Steenhof et al. 2011).

Why it is important to study the effects of sleep loss on our brain at adverse conditions?

Insufficient sleep and its medical consequences are often neglected or go unrecognized among the global population. Sleep-related disorders and comorbid medical conditions are enormously prevalent that are infrequently diagnosed. Some of the prevalent sleep-related disorders categorized by ICSD-2 are:

Insomnias, sleep-related breathing disorders like obstructive sleep apnea, hypersomnias of central origin like narcolepsy, Kleine–Levin syndrome, circadian rhythm sleep disorders, parasomnias, sleep-related movement disorders, other psychiatric/behavioral disorders frequently encountered in the differential diagnosis of sleep disorders. Emphasis on the deteriorating effects of sleep deprivation reflects the disruption of higher-order cognitive domains like attentional functions, working memory, and long-term memory. Acute sleep deprivation and insomnia exhibit distinct morphological changes of gray matter volume (GMV). Sleep deprivation has accumulative negative effects on brain morphology and advanced cognitive function. The altered GMV may provide the neurobiological basis for attention and memory impairments following sleep loss (Dai et al. 2018).

Performance on attentional tasks deteriorates in a dose-dependent manner with the amount of accrued time awake, owing to increasing sleep pressure (Borbély 1982; Belenky et al. 2003; Van Dongen et al. 2003). Impairments in such tasks are referred to as "micro sleeps" or "lapses" meaning ineffective response in the error of omission (Krause et al. 2017). The dorsolateral prefrontal cortex, intraparietal sulcus, and thalamus are well studied to show disrupted fMRI signals due to sleep loss (Chuah et al. 2010). Neural networks involved in working memory and brain regions associated with attentional tasks are overlapped and are shown to be impaired by sleep deprivation (Turner et al. 2007; Drummond et al. 2012). Studies suggest that brain regions associated with reward and valuation (subcortex) are hypersensitized due to acute sleep deprivation (Tufik 1981; Venkatraman et al. 2009; Volkow et al. 2012; Mullin et al. 2013). Emotional discrimination and expression is another significant virtue to be considered in context to sleep and sleep loss. Studies reported that sleep-deprived individuals exhibit a generalized inflation in emotional vulnerability, along with a disrupted ability to distinguish emotions of negative valence (Goldstein and Walker 2014). The antidepressive effect of sleep deprivation could be explained by the elevated levels of serotonin, tryptophan, and taurine in the brain (Davies et al. 2014). It was shown that norepinephrine and dopamine levels are elevated in frontal cortex, brain stem, cerebellum, and pons and medulla posthypoxic exposure. In contrast to this, serotonin concentration decreased in the frontal cortex and pons, and medulla after 14 days of hypobaric hypoxia. The concentration of tyrosine hydroxylase in locus coeruleus (LC) was also elevated but choline acetyl transferase and glutamic acid decarboxylase (GAD) decreased in laterodorsal-tegmentum and pedunculopontine-tegmentum (Ray et al. 2011).

One of the consequences of sleep deprivation addressed is elevated anxiety. Caffeine and modafinil significantly ameliorated anxious behaviors. Anxiety, proinflammatory cytokines, and activated microglial cell count are positively correlated during sleep deprivation (Wadhwa et al. 2018). Recognition memory and synaptic protein impairment in the hippocampus during sleep deprivation are also rescued by caffeine and modafinil. Spatial memory and adult neurogenesis are deteriorated due to sleep deprivation (Wadhwa et al. 2017, 2019). Adenosine A1R antagonist 8-CPT suppresses the sleep deprivation-induced impairment in spatial memory and adult neurogenesis by regulating BDNF levels in the hippocampus (Chauhan et al. 2016). Sleep architecture is modulated at high altitude (3000 m), like diminished S3 and S4 sleep stages, and rise in Stage 2 of NREM sleep, unaltered REM sleep, a decreased subjective state of sleep quality. At 4300 m, sleep architecture is similarly influenced but at higher magnitude on the second night of staged induction, than reported at 3500 m (Thakur et al. 2012). The long-term efficacy of such interventions could only be understood by executing sustained research emphasized that on both advantages and disadvantages, which is what we seem to lack. In addition to the DMN, a genetic and physiological framework of other intrinsically connected networks should be studied to better understand the impacts of sleep deprivation and its association with circadian phenotype. Metacognitive beliefs and sleep quality were studied to be the mediators of morningness and eveningness to have positive and negative affect respectively.

19.1.10 Endogenous Mechanisms Involved in Sleep Regulation

Sleep is regulated by two important physiological processes: Homeostatic sleep regulation and circadian rhythm to maintain the sleep/wake cycle. Homeostasis maintains all the variables involved in the sleep at a certain threshold levels whereas the circadian regulation changes the magnitude of each variable for optimal functioning. Sleep deprivation directly leads to the increase of homeostatic sleep pressure. Research shows that knockout of several critical clock genes (Period: Per1–3, Bmal1, Cry1, Cry2, Clock, and Npas2) leads to the impairment of both the circadian and homeostatic processes (Albrecht 2002). Double knockout of Cry1 and Cry2 and Per1, Per2 leads to the phenotype (enhanced NREM sleep, sleep consolidation, and EEG SWA) similar to increased homeostatic sleep pressure which is a clear correlation that clock genes are play a critical role in sleep homeostasis (Wisor et al. 2008). Sleep pressure fluctuations influencing the phase-shifting potential of circadian clock are evident from previous studies. It could be possible by the reduction in SCN neurons activity with respect to the diminished phase-shifting potential of circadian clock in response to light. Serotonin concentration is increased in brain following short 4–6 h sleep deprivation (Asikainen et al. 1995, 1997; Grossman et al. 2000). Increased serotonin concentration reduces the SCN neuronal activity (Yu et al. 2001). EEG SWA is a significant marker of homeostatic sleep pressure during NREM sleep. After sleep deprivation, increase in slow-wave activity (SWA) led to reduction in SCN neuronal activity which reverted back to normal levels after the decrease in SWA during recovery sleep. We can address the question that how do sleep loss is compensated endogenously by starting to focus on the relationship between circadian and homeostatic processes. These two mechanisms are known to be involved in sleep regulation; however, their interdependence is least studied. Does the sleep homeostasis is influenced by the circadian clock or vice versa? Recent fMRI data suggest that activity in cortical regions is impacted by the combined action of sleep pressure and circadian rhythm whereas subcortical regions are mainly influenced by the endogenous circadian process and not by the influence of sleep pressure increased by sleep deprivation (Muto et al. 2016). More comprehensive data are required to ascertain the endogenous sleep regulatory mechanisms influencing sleep pressure, sleep timing. *Valeriana wallichi* aqueous root extract was found to be improving sleep quality by modulating concentrations of monoamines in cortex and brainstem (Sahu et al. 2012). Although changes in sleep pressure tend to have a direct effect on the phase-shifting ability of the circadian clock, it has not yet been studied that the expression of clock genes in the SCN is altered under the influence of sleep deprivation (Curie et al. 2015). This is supported by the finding that the process of the circadian clock does not alter sleep deprivation per se in hamsters (Mistlberger et al. 2002) and only slight changes (Van Diepen et al. 2014) have been found in mice. Moreover, the magnitude of human circadian clock output activity markers, such as melatonin release or core body temperature, does not alter under the impact of accumulated sleep pressure (Birchler-Pedross et al. 2009). The impact of homeostatic sleep pressure on the central circadian pacemaker may therefore be on the input or output of the clock, likely altering the feedback to the phase-shifting zeitgeber, but not specifically on the expression of the clock gene and the clock phase or time. The most common misconception rooted into modern-day society is that sleep debt during the week could be repaid by sleeping extra hours during the weekend. Research suggests that professionals who are deprived of sleep chronically during night shifts can utilize 30-min nap as a countermeasure to cognitive decline and other event-related potentials involved in working memory (Panjwani et al. 2010).

Meditation has also been studied as an intervention to improve the deterioration caused by sleep deprivation (Chatterjee et al. 2012). According to a recent study reported in *Current Biology*, which debunks this myth and reckons that ad libitum weekend recovery sleep does not reverse the negative consequences induced by the sleep loss (Depner et al. 2019). Comprehensive mechanisms involved in the interaction of circadian and homeostatic processes are lacking and further research is required to decode the endogenous circuits for counteracting sleep loss.

19.1.11 Neurotransmitters Involved in Onset and Maintenance of Sleep

Diverse chemicals function together to regulate sleep in our body. Gamma-aminobutyric acid (GABA) and histamine are released in the forebrain and hypothalamus. Suppression of the neuronal activity in the cortex and thalamus by higher GABA and lower histamine levels leads to the onset of NREM sleep. The sleep–wake cycle is another vital virtue which is impacted by neurotransmitters secreted by reticular activating system (RAS) neurons, such as norepinephrine, acetylcholine, and serotonin. These neurotransmitters maintain wake state and decline during REM sleep. The hypothalamus releases another important neuropeptide orexin that contributes to maintaining arousal. Melatonin synchronizes the body with the environment's light–dark cycle, thus maintaining the diurnal rhythm, increasing during the night and decreasing during the day. Several important pharmacotherapeutic drugs approved by FDA target those neurons which are releasing the discussed neurotransmitters regulating the sleep, wakefulness, and circadian rhythm.

Why do we need alternative nonpharmacological interventions for improving sleep quality?

Economical, nonpharmacological therapies potentially easing the management of the diseases are prudent in any context of health. These therapies are essential in not only improving the quality of life, but also may aid in minimal healthcare visits and unplanned hospitalizations. Some of the most commonly prescribed drugs like zolpidem (Ambiem) and eszopiclone (Lunesta) slow down brain function by indiscriminately targeting receptors for GABA, an important neurotransmitter involved in cognition. Benzodiazepines are reported to have some of the side effects like psychomotor problems, memory impairments, paradoxical inhibition, depression, and teratogenic effects in pregnant women (Longo and Johnson 2000). Prominent effects are observed when the drug concentrations are high; however, considering short-acting compounds as an alternative due to dosage issue is also tricky. Unfortunately, shorter-acting compounds are inefficient in maintaining sleep quality throughout the night (Weinling et al. 2006).

19.1.12 Relevance of Circadian Phenotype on Physiological Effects of Sleep Deprivation

Circadian phenotype of an individual reflects strong synchronization between endogenous biological clock and exogenous environmental zeitgebers or cues that influence sleep/wake cycles. Endogenous circadian oscillator is suprachiasmatic nucleus (SCN) of the hypothalamus integrating environmental cues such as light exposure and it also regulates the same in peripheral tissues. Some of the circadian rhythm sleep disorders are advanced sleep phase disorder (ASPD), delayed sleep wake phase disorder (DSWPD). Comprehensive understanding of an individual's circadian phenotype is critical in the treatment of disorders related to circadian rhythm impairment and physiological effects caused by sleep loss. Individual differences in circadian clock based on their diurnal preference are classified as morningness/eveningness types, also referred to larks/owls respectively. Studies have evidently suggested that the individuals with late diurnal preference are susceptible to negative health impacts. Societal time has emerged as a serious modern-day concern that cannot be ignored. It is the desynchronization of the biological clock with natural zeitgebers and aligning ourselves with new time that fits our work/study schedules. Maximum susceptibility of individuals to sleep deprivation could be explained by the impairment in the synchronized activity of intrinsic functional networks, often called as intrinsic connected networks (ICNs). One such most investigated ICN is default mode network (DMN) critical in varied cognitive domains, is highly vulnerable to sleep loss. Studies have reported that functional connectivity is diminished in primary nodes of DMN in late phenotypes as compared to their early counterparts. But these studies have limitations like functional connectivity analysis was done during clock time 08:00–20:00 instead of adhering to the individual's internal biological clock. In healthy adults, regular exercise (combination of resistance training (RT) and body-weight functional mobility exercises with limited rest intervals) improves objective sleep outcomes such as sleep efficiency, slow-wave sleep, total sleep time, and latency to sleep onset while also improving subjective sleep quality. Chronic insomnia is a condition with a broad definition and a significant clinical heterogeneity among affected patients, its pathophysiology possibly comprehends a diversity of mechanisms (Romero et al. 2017). Some of the most prevalent and well-studied nonpharmacological interventions like Cognitive Behavioral Therapy (Table 19.1) for improving sleep quality at adverse conditions are discussed below.

19.1.13 Deep Brain Stimulation

Deep brain stimulation (DBS) is an interventional surgical technique in which some brain regions are implanted with electrodes. These electrodes, or leads, produce electrical impulses that regulate brain activity that is abnormal. The electrical impulses will also compensate for the chemical imbalances that cause different conditions within the brain. Brain area stimulation is managed by a programmable

Table 19.1 Components of cognitive behavioral therapy to improve sleep quality

S.NO	Behavioral technique	Description
1	Sleep restriction therapy	Process of eliminating naps, forcefully staying up beyond normal bedtime. This decreases the time spent lying in bed awake. It develops a deeper association between bed and sleep instead of bed and lying awake
2	Stimulus control therapy	It identifies and changes sleep habits that disrupt better sleep
3	Improving your sleep environment and sleep hygiene	The sleeping area should be dim, silent, cool, and relaxed, so blackout curtains, earplugs or a noise-reducing sound machine might be recommended by your therapist
4	Remaining passively awake	It is also known as "paradoxical intention." Since deep thoughts of failing to fall asleep generate anxiety that increases wake time. Taking control over such thoughts and making no voluntary effort to sleep may, paradoxically, enhance sleep quality
5	Yoga and relaxation training	Practices like yoga, mindfulness meditation, progressive muscle relaxation, and breathing exercises may improve sleep quality
6	Biofeedback	Usage of heart rate, breathing, and muscle tension sensors that monitor basic physiological functions. Biofeedback helps you to understand and regulate the responsiveness of the body to anxiety that alters sleep patterns
7	Hypnosis	The hypnotherapist helps to change negative thought patterns promoting restful sleep

generator in the upper chest that is mounted under the skin. This technique is used when treatments are no longer successful in preserving a reasonable quality of life for patients. Medical conditions that are treated using DBS are dystonia, epilepsy, obsessive-compulsive disorder, Parkinson's disease. In this review, we emphasize on studies involving the execution of DBS to improve sleep disturbances in Parkinson's disease. Brain regions that are targeted for DBS are subthalamic nucleus (STN), globus pallidus internus (GPi), ventral intermediate nucleus (VIM), and pedunculopontine nucleus (PPN.) But emerging evidence suggests that STN-DBS exerts positive effect on numerous sleep parameters such as sleep efficiency, total sleep time, and wakefulness after sleep onset. A significant increase of 155% was reported by Monaca et al., in the duration of deep slow-wave and cumulative REM sleep time after STN-DBS. But they found only mild motor improvement. In patients with PD, DBS therapy at either site tends to enhance objective and subjective sleep parameters. Improvement in motor and some nonmotor nocturnal symptoms results in an increase in overall sleep time by up to 1 h, as well as a decrease in sleep fragmentation (Vgontzas et al. 2004; Odekerken et al. 2016). Consequently, the causes of sleep disruptions appear complicated and potentially multifactorial. Motor and nonmotor PD symptoms, sleep–wake function dysregulation, and comorbidities such as breathing complications linked to sleep will often play a role. Therefore, various factors contribute in enhancing the quality of sleep. Studies suggested that the improvement in quality of sleep could be achieved by augmentation of nocturnal mobility. However, Monaca et al. constructed experimental evidence on only a mild motor enhancement, but a significant improvement in sleep. This leads to conclusion that sleep/wakefulness regulatory centers are crucial despite motor enhancement is accomplished. Recent evidence suggests that sleep onset and maintenance insomnia improved less convincingly. Together, this may indicate that subthalamic nucleus deep brain stimulation (STN-DBS) exerts direct effect on sleep–wake regulation centers or could also be on nocturnal motor and NMS.

19.1.14 Transcutaneous Electrical Nerve Stimulation

Transcutaneous electrical nerve stimulation (TENS) is a procedure that uses electrical current at low voltage to alleviate pain. How does it work? Two theories have been proposed; nerve cells that inhibit the transmission of pain signals are stimulated by the electric current, altering your pain perception. The other theory is that the level of endorphins, the normal pain-killing hormone of the body, is enhanced by this nerve stimulation. In patients with chronic insomnia who were on medications, low-frequency electrical stimulation had multiple beneficial effects. Treatment response predictor variables were daytime sleepiness and depressive and anxious behavior, and diminishing occipital delta power was correlated with relieving symptoms of insomnia. In individuals with chronic low-back pain, routine TENS enhanced self-reported and objective sleep metrics. The improved group had longer overall sleep time and fewer PLMs contrasted to the unimproved group. Sleep in chronic low-back pain can be a significant outcome for TENS potency (Gozani 2019). In patients with obstructive sleep apnea, TENS management of the upper airway dilator muscles has led to the reduction of the Apnea–Hypopnea Index (Steier et al. 2011; Jauregui et al. 2016).

19.1.15 Resistance Training

Alterations in energy dissipation, rise in body temperature levels, and easing of musculoskeletal pain are some of the possible mechanisms by which resistance training could improve sleep quality. Studies reported that higher body temperature caused by resistance training prior to sleep aids in increasing slow-wave sleep (SWS). Investigations on the acute effects of timing of resistance training on sleep pattern and quality are insufficient to arrive at a comprehensive conclusion. Jessica R et al. reported that no statistical impact was found to impact sleep architecture, but involving resistance training in our lifestyle leads to improvement of sleep quality. RT sessions at certain times of the day can achieve sleep objectives, such as shortening the time to fall asleep or decreasing the amount of time spent awake during the night. The critical analysis of RT in hemodialysis patients as a nonpharmacological technique resulted in an increase in overall sleep time, sleep latency, nocturnal waking, and sleep efficiency. In addition, RT improved the redox and inflammatory profile, possibly resulting from elevated NO bioavailability and lowered ADMA concentration. Improving NO and ADMA can play a crucial role in improving the quality of sleep and health-related variables in patients with CKD.

19.1.16 Pharmacological Interventions

Since the dawn of human civilization, we have been surrounded by environmental stressors like intense heat waves, intense cold temperatures, noise, and space radiation. However, as human civilization has advanced, so have various man-made stresses like pollutants, noise, hazardous technology, and geopolitical disturbances like wars. Life on earth is equipped with an elegant process of adaptation to these stressors on Earth; there is an intricate mechanism developed by life forms for adaptation to these stresses. Although humans have an effective adapting mechanism, harsh environments nonetheless provide significant challenges to the human body and can even alter normal physiological processes. Adaptation is crucial to human existence. Every system in the human body is impacted by such severe environmental stress, but the brain is particularly vulnerable. Numerous studies have demonstrated that environmental stresses could impact the cognitive capabilities of humans. Extreme heat and high-altitude hypoxia are well documented to cause neuroinflammatory alterations, cognitive problems, and disturbed sleep patterns. Proinflammatory changes made in the human brain can be readily triggered by space microgravity, radiation, noise, and other particulate and gaseous

Table 19.2 Pharmacological management of acute mountain sickness

S. No	Drugs	Mechanism	Treatment
1	Mild AMS		
	Paracetamol, ibuprofen	NSAID (blocking COX-1,2), inhibiting release of inflammatory mediators	Headache at high altitude
	Prochlorperazine, promethazine, cinnarizine	D2 antagonist Antihistamine H1 antagonist Antihistamine H1 antagonist	Nausea, dizziness, vertigo
2	Severe AMS/ high-altitude cerebral edema		
	Dexamethasone Acetazolamide	Glucocorticoid receptor agonist Carbonic anhydrase inhibitor	Management of AMS and cerebral edema
		Rapid descent altitude, supplemental oxygen, and portable hyperbaric chamber alongside drug treatment	
3	Severe AMS/high-altitude pulmonary edema		
	Nifedipine	Calcium channel blocker	Management of severe AMS and pulmonary edema
		Rapid descent altitude, supplemental oxygen, and portable hyperbaric chamber alongside drug treatment	

Table 19.3 List of plants and their active components to improve mental health under stressful conditions

Plant	Common name	Active constituents	Effects
Nardostachysjatamansi (Valerianaceae)	Jatamansi	Jatamansic acid, ketones jatamansone, and nardostachone	Brain tonic and helps to improve memory and brain functions by preventing cell damage due to its antioxidant property. It also calms down the brain and manages anxiety as well as insomnia
Canscoradecussta (Gentianaceae)	Shankhpushpi	Shankhapushpine, convolamine, convoline, convolidine, convolvine, confoline, convosine	Helps in enhancing concentration, learning capabilities, mental fatigue, insomnia, stress, anxiety, depression, etc
Bacopamonniera (Scrophulariaceae)	Brahmi	Alkaloids brahmine, herpestine and nicotine, hersaponin, apigenin, D-mannitol, monnierasides I-III, plantainoside B, and cucurbitacin; alkaloids brahmine, herpestine, and nicotine	Are memory improvement, epilepsy, insomnia, and anxiolytic
Ecliptaerecta (Asteraceae)	Bhringraj	Ecliptine, wedelolactone, wedelic acid, apigenin, luteolin, b-amyrin	Antioxidants and flavonoids present in it improves the memory capacity, focus, concentration, calmness, alertness of an individual
Acorus calamus (Acoraceae)	Vacha	Methyleugenol, cis-methylisoeugenol, β-asarone, geranylacetate, β-farnesene, shyobunone, epishyobunone, and isoshyobunon	Prevents inflammation, thereby improving behavioral changes, memory as well as mental performance

contaminants. Even cytokines and hormonal changes have been linked to these stresses, to which the brain is quite responsive. Some of the existing pharmacological interventions for the management of acute mountain sickness are listed in Table 19.2. Active components of plants being studied to improve mental health in adverse conditions are listed below (Tables 19.2 and 19.3).

References

Albrecht U (2002) Invited review: regulation of mammalian circadian clock genes. J Appl Physiol (1985) 92(3):1348–1355. https://doi.org/10.1152/japplphysiol.00759.2001, PMID 11842077

Asikainen M, Deboer T, Porkka-Heiskanen T, Stenberg D, Tobler I (1995) Sleep deprivation increases brain serotonin turnover in the Djungarian hamster. Neurosci Lett 198(1):21–24. https://doi.org/10.1016/0304-3940(95)11953-t, PMID 8570087

Asikainen M, Toppila J, Alanko L, Ward DJ, Stenberg D, Porkka-Heiskanen T (1997) Sleep deprivation increases brain serotonin turnover in the rat. Neuroreport 8(7):1577–1582. https://doi.org/10.1097/00001756-199705060-00006, PMID 9189895

Atkinson RW, Carey IM, Kent AJ, Van Staa TP, Anderson HR, Cook DG (2013) Long-term exposure to outdoor air pollution and incidence of cardiovascular diseases. Epidemiology 24(1): 44–53. https://doi.org/10.1097/EDE.0b013e318276ccb8, PMID 23222514

Bahrke MS, Shukitt-Hale B (1993) Effects of altitude on mood, behaviour and cognitive functioning. A review. Sports Med 16(2):97–125. https://doi.org/10.2165/00007256-199316020-00003, PMID 8378672

Belenky G, Wesensten NJ, Thorne DR, Thomas ML, Sing HC, Redmond DP et al (2003) Patterns of performance degradation and restoration during sleep restriction and subsequent recovery: a sleep dose-response study. J Sleep Res 12(1):1–12. https://doi.org/10.1046/j.1365-2869.2003.00337.x, PMID 12603781

Birchler-Pedross A, Schröder CM, Münch M, Knoblauch V, Blatter K, Schnitzler-Sack C et al (2009) Subjective well-being is modulated by circadian phase, sleep pressure, age, and gender. J Biol Rhythm 24(3):232–242. https://doi.org/10.1177/0748730409335546, PMID 19465700

Borbély AA (1982) A two process model of sleep regulation. Hum Neurobiol 1(3):195–204, PMID 7185792

Brekhman II, Dardymov IV (1969) New substances of plant origin which increase nonspecific resistance. Annu Rev Pharmacol 9:419–430. https://doi.org/10.1146/annurev.pa.09.040169.002223, PMID 4892434

Chatterjee A, Ray K, Panjwani U, Thakur L, Anand JP (2012) Meditation as an intervention for cognitive disturbances following total sleep deprivation. Indian J Med Res 136(6):1031–1038, PMID 23391801

Chauhan G, Ray K, Sahu S, Roy K, Jain V, Wadhwa M et al (2016) Adenosine A1 receptor antagonist mitigates deleterious effects of sleep deprivation on adult neurogenesis and spatial reference memory in rats. Neuroscience 337:107–116. https://doi.org/10.1016/j.neuroscience.2016.09.007, PMID 27623393

Chauhan G, Roy K, Kumar G, Kumari P, Alam S, Kishore K et al (2019) Distinct influence of COX-1 and COX-2 on neuroinflammatory response and associated cognitive deficits during high altitude hypoxia. Neuropharmacology 146:138–148. https://doi.org/10.1016/j.neuropharm.2018.11.026, PMID 30476507

Chuah LY, Dolcos F, Chen AK, Zheng H, Parimal S, Chee MW (2010) Sleep deprivation and interference by emotional distracters. Sleep 33(10):1305–1313. https://doi.org/10.1093/sleep/33.10.1305, PMID 21061852

Cserbik D, Chen JC, Mcconnell R, Berhane K, Er S, Schwartz J et al (2020) Fine particulate matter exposure during childhood relates to hemispheric-specific differences in brain structure. Environ Int 143:105933. https://doi.org/10.1016/j.envint.2020.105933, PMID 32659528

Curie T, Maret S, Emmenegger Y, Franken P (2015) In vivo imaging of the central and peripheral effects of sleep deprivation and suprachiasmatic nuclei lesion on PERIOD-2 protein in mice. Sleep 38(9):1381–1394. https://doi.org/10.5665/sleep.4974, PMID 25581923

Dai XJ, Jiang J, Zhang Z, Nie X, Bx L, Pei L et al (2018) Plasticity and susceptibility of brain morphometry alterations to insufficient sleep. Front Psychiatr 9:266. https://doi.org/10.3389/fpsyt.2018.00266, PMID 29997530

Davies SK, Ang JE, Vl R, Holmes B, Mann A, Robertson FP et al (2014) Effect of sleep deprivation on the human metabolome. Proc Natl Acad Sci U S A 111(29):10761–10766. https://doi.org/10.1073/pnas.1402663111, PMID 25002497

Davis CM, Allen AR, Bowles DE (2021) Consequences of space radiation on the brain and cardiovascular system. J Environ Sci Health C Toxicol Carcinog 39(2):180–218. https://doi.org/10.1080/26896583.2021.1891825, PMID 33902387

Depner CM, Melanson EL, Eckel RH, Snell-Bergeon JK, Perreault L, Bergman BC et al (2019) Ad libitum weekend recovery sleep fails to prevent metabolic dysregulation during a repeating pattern of insufficient sleep and weekend recovery sleep. Curr Biol 29(6):957–967.e4. https://doi.org/10.1016/j.cub.2019.01.069, PMID 30827911

Drummond SP, Anderson DE, Straus LD, Vogel EK, Perez VB (2012) The effects of two types of sleep deprivation on visual working memory capacity and filtering efficiency. PLoS One 7(4): e35653. https://doi.org/10.1371/journal.pone.0035653, PMID 22530064

Fernández-Moriano C, González-Burgos E, Iglesias I, Lozano R, Gómez-Serranillos MP (2017) Evaluation of the adaptogenic potential exerted by ginsenosides Rb1 and Rg1 against oxidative stress-mediated neurotoxicity in an in vitro neuronal model. PLoS One 12(8):e0182933. https://doi.org/10.1371/journal.pone.0182933, PMID 28813475

Gallagher SA, Hackett PH (2004) High-altitude illness. Emerg Med Clin North Am 22(2):329–355. https://doi.org/10.1016/j.emc.2004.02.001, PMID 15163571, viii

Goldstein AN, Walker MP (2014) The role of sleep in emotional brain function. Annu Rev Clin Psychol 10:679–708. https://doi.org/10.1146/annurev-clinpsy-032813-153716, PMID 24499013

Gozani SN (2019) Remote analgesic effects of conventional transcutaneous electrical nerve stimulation: a scientific and clinical review with a focus on chronic pain. J Pain Res 12:3185–3201. https://doi.org/10.2147/JPR.S226600, PMID 31819603

Grossman G, Mistlberger RE, Antle MC, Ehlen JC, Glass JD (2000) Sleep deprivation stimulates serotonin release in the suprachiasmatic nucleus. Neuroreport 11(9):1929–1932. https://doi.org/10.1097/00001756-200006260-00024, PMID 10884045

Halder SK, Milner R (2020) Mild hypoxia triggers transient blood-brain barrier disruption: a fundamental protective role for microglia. Acta Neuropathol Commun 8(1):175. https://doi.org/10.1186/s40478-020-01051-z, PMID 33115539

Jauregui JJ, Cherian JJ, Gwam CU, Chughtai M, Mistry JB, Elmallah RK et al (2016) A meta-analysis of transcutaneous electrical nerve stimulation for chronic low back pain. Surg Technol Int 28:296–302, PMID 27042787

Jensen JD, Vincent AL (2022) High-altitude cerebral edema. In: Statpearls. Statpearls Publishing. Copyright, Statpearls Publishing LLC, Treasure Island (FL)

Jung CR, Lin YT, Hwang BF (2015) Ozone, particulate matter, and newly diagnosed Alzheimer's disease: a population-based cohort study in Taiwan. J Alzheimers Dis 44(2):573–584. https://doi.org/10.3233/JAD-140855, PMID 25310992

Kim HG, Kim TM, Park G, Lee TH, Oh MS (2013) Repeated heat exposure impairs nigrostriatal dopaminergic neurons in mice. Biol Pharm Bull 36(10):1556–1561. https://doi.org/10.1248/bpb.b13-00268, PMID 23842035

Krause AJ, Simon EB, Mander BA, Greer SM, Saletin JM, Goldstein-Piekarski AN et al (2017) The sleep-deprived human brain. Nat Rev Neurosci 18(7):404–418. https://doi.org/10.1038/nrn.2017.55, PMID 28515433

Lee W, Moon M, Hg K, Lee TH, Oh MS (2015) Heat stress-induced memory impairment is associated with neuroinflammation in mice. J Neuroinflammation 12:102. https://doi.org/10.1186/s12974-015-0324-6, PMID 26001832

Longo LP, Johnson B (2000) Addiction: Part I. Benzodiazepines—side effects, abuse risk and alternatives. Am Fam Physician 61(7):2121–2128, PMID 10779253

McEwen BS (2017) Neurobiological and systemic effects of chronic stress. Chronic Stress (Thousand Oaks) 1:1. https://doi.org/10.1177/2470547017692328, PMID 28856337

Mistlberger RE, Belcourt J, Antle MC (2002) Circadian clock resetting by sleep deprivation without exercise in Syrian hamsters: dark pulses revisited. J Biol Rhythm 17(3):227–237. https://doi.org/10.1177/07430402017003006, PMID 12054194

Mullin BC, Phillips ML, Siegle GJ, Buysse DJ, Forbes EE, Franzen PL (2013) Sleep deprivation amplifies striatal activation to monetary reward. Psychol Med 43(10):2215–2225. https://doi.org/10.1017/S0033291712002875, PMID 23286303

Muto V, Jaspar M, Meyer C, Kussé C, Chellappa SL, Degueldre C et al (2016) Local modulation of human brain responses by circadian rhythmicity and sleep debt. Science 353(6300):687–690. https://doi.org/10.1126/science.aad2993, PMID 27516598

Odekerken VJ, Boel JA, Schmand BA, De Haan RJ, Figee M, Van Den Munckhof P et al (2016) Gpi vs STN deep brain stimulation for Parkinson disease: three-year follow-up. Neurology 86(8):755–761. https://doi.org/10.1212/WNL.0000000000002401, PMID 26819458

Palinkas LA (2001) Mental and cognitive performance in the cold. Int J Circumpolar Health 60(3): 430–439. https://doi.org/10.1080/22423982.2001.12113048, PMID 11590885

Panjwani U, Ray K, Chatterjee A, Bhaumik S, Kumar S (2010) Electrophysiological correlates of cognition improve with nap during sleep deprivation. Eur J Appl Physiol 108(3):549–556. https://doi.org/10.1007/s00421-009-1222-3, PMID 19865829

Panossian AG (2013) Adaptogens in mental and behavioral disorders. Psychiatr Clin North Am 36(1):49–64. https://doi.org/10.1016/j.psc.2012.12.005, PMID 23538076

Panossian AG, Efferth T, Shikov AN, Pozharitskaya ON, Kuchta K, Mukherjee PK et al (2021) Evolution of the adaptogenic concept from traditional use to medical systems: pharmacology of stress- and aging-related diseases. Med Res Rev 41(1):630–703. https://doi.org/10.1002/med.21743, PMID 33103257

Panossian A, Wikman G, Kaur P, Asea A (2012) Adaptogens stimulate neuropeptide Y and hsp72 expression and release in neuroglia cells. Front Neurosci 6:6. https://doi.org/10.3389/fnins.2012.00006, PMID 22347152

Panossian A, Wikman G, Wagner H, Plant adaptogens. III. (1999) Earlier and more recent aspects and concepts on their mode of action. Phytomedicine 6(4):287–300. https://doi.org/10.1016/S0944-7113(99)80023-3, PMID 10589450

Rauch TM, Lieberman HR (1990) Tyrosine pretreatment reverses hypothermia-induced behavioral depression. Brain Res Bull 24(1):147–150. https://doi.org/10.1016/0361-9230(90)90299-f, PMID 2310942

Ray K, Dutta A, Panjwani U, Thakur L, Anand JP, Kumar S (2011) Hypobaric hypoxia modulates brain biogenic amines and disturbs sleep architecture. Neurochem Int 58(1):112–118. https://doi.org/10.1016/j.neuint.2010.11.003, PMID 21075155

Reed HL, Silverman ED, Shakir KM, Dons R, Burman KD, O'brian JT (1990) Changes in serum triiodothyronine (T3) kinetics after prolonged Antarctic residence: the polar T3 syndrome. J Clin Endocrinol Metab 70(4):965–974. https://doi.org/10.1210/jcem-70-4-965, PMID 2318952

Romero K, Goparaju B, Russo K, Westover MB, Bianchi MT (2017) Alternative remedies for insomnia: a proposed method for personalized therapeutic trials. Nat Sci Sleep 9:97–108. https://doi.org/10.2147/NSS.S128095, PMID 28360539

Ross RT (1985) The random nature of cerebral mountain sickness. Lancet 1(8435):990–991. https://doi.org/10.1016/s0140-6736(85)91771-4, PMID 2859454

Sahu S, Ray K, Yogendra Kumar MS, Gupta S, Kauser H, Kumar S et al (2012) Valeriana wallichii root extract improves sleep quality and modulates brain monoamine level in rats. Phytomedicine 19(10):924–929. https://doi.org/10.1016/j.phymed.2012.05.005, PMID 22766307

Steenhof M, Gosens I, Strak M, Godri KJ, Hoek G, Cassee FR et al (2011) In vitro toxicity of particulate matter (PM) collected at different sites in the Netherlands is associated with PM composition, size fraction and oxidative potential—the RAPTES project. Part Fibre Toxicol 8: 26. https://doi.org/10.1186/1743-8977-8-26, PMID 21888644

Steier J, Seymour J, Rafferty GF, Jolley CJ, Solomon E, Luo Y et al (2011) Continuous transcutaneous submental electrical stimulation in obstructive sleep apnea: a feasibility study. Chest 140(4):998–1007. https://doi.org/10.1378/chest.10-2614, PMID 21454399

Thakur L, Anand JP, Malhotra AS, Sharma YK, Panjwani U (2012) Sleep architecture at 4300 m altitude in a sample of Indian lowlanders. Indian J Physiol Pharmacol 56(4):295–300. PMID 23781648

Tong H (2016) Dietary and pharmacological intervention to mitigate the cardiopulmonary effects of air pollution toxicity. Biochim Biophys Acta 1860(12):2891–2898. https://doi.org/10.1016/j.bbagen.2016.05.014, PMID 27189803

Tufik S (1981) Changes of response to dopaminergic drugs in rats submitted to REM-sleep deprivation. Psychopharmacol (Berl) 72(3):257–260. https://doi.org/10.1007/BF00431826, PMID 6111823

Turner TH, Drummond SP, Salamat JS, Brown GG (2007) Effects of 42 hr of total sleep deprivation on component processes of verbal working memory. Neuropsychology 21(6):787–795. https://doi.org/10.1037/0894-4105.21.6.787, PMID 17983292

Van Diepen HC, Lucassen EA, Yasenkov R, Groenen I, Ijzerman AP, Meijer JH et al (2014) Caffeine increases light responsiveness of the mouse circadian pacemaker. Eur J Neurosci 40(10):3504–3511. https://doi.org/10.1111/ejn.12715, PMID 25196050

Van Dongen HP, Maislin G, Mullington JM, Dinges DF (2003) The cumulative cost of additional wakefulness: dose-response effects on neurobehavioral functions and sleep physiology from chronic sleep restriction and total sleep deprivation. Sleep 26(2):117–126. https://doi.org/10.1093/sleep/26.2.117, PMID 12683469

Venkatraman V, Payne JW, Bettman JR, Luce MF, Huettel SA (2009) Separate neural mechanisms underlie choices and strategic preferences in risky decision making. Neuron 62(4):593–602. https://doi.org/10.1016/j.neuron.2009.04.007, PMID 19477159

Vgontzas AN, Zoumakis E, Bixler EO, Lin HM, Follett H, Kales A et al (2004) Adverse effects of modest sleep restriction on sleepiness, performance, and inflammatory cytokines. J Clin Endocrinol Metab 89(5):2119–2126. https://doi.org/10.1210/jc.2003-031562, PMID 15126529

Volkow ND, Tomasi D, Wang GJ, Telang F, Fowler JS, Logan J et al (2012) Evidence that sleep deprivation downregulates dopamine D2R in ventral striatum in the human brain. J Neurosci 32(19):6711–6717. https://doi.org/10.1523/JNEUROSCI.0045-12.2012, PMID 22573693

Wadhwa M, Chauhan G, Roy K, Sahu S, Deep S, Jain V et al (2018) Caffeine and modafinil ameliorate the neuroinflammation and anxious behavior in rats during sleep deprivation by inhibiting the microglia activation. Front Cell Neurosci 12:49. https://doi.org/10.3389/fncel.2018.00049, PMID 29599709

Wadhwa M, Kumari P, Chauhan G, Roy K, Alam S, Kishore K et al (2017) Sleep deprivation induces spatial memory impairment by altered hippocampus neuroinflammatory responses and glial cells activation in rats. J Neuroimmunol 312:38–48. https://doi.org/10.1016/j.jneuroim.2017.09.003, PMID 28912034

Wadhwa M, Prabhakar A, Anand JP, Ray K, Prasad D, Kumar B et al (2019) Complement activation sustains neuroinflammation and deteriorates adult neurogenesis and spatial memory impairment in rat hippocampus following sleep deprivation. Brain Behav Immun 82:129–144. https://doi.org/10.1016/j.bbi.2019.08.004, PMID 31408672

Walter EJ, Carraretto M (2016) The neurological and cognitive consequences of hyperthermia. Crit Care 20(1):199. https://doi.org/10.1186/s13054-016-1376-4, PMID 27411704

Wang Y, Ma H, Fu S, Guo S, Yang X, Luo P et al (2014) Long-term exposure to high altitude affects voluntary spatial attention at early and late processing stages. Sci Rep 4(1):4443. https://doi.org/10.1038/srep04443

Weinling E, Mcdougall S, Andre F, Bianchetti G, Dubruc C (2006) Pharmacokinetic profile of a new modified release formulation of zolpidem designed to improve sleep maintenance. Fundam Clin Pharmacol 20(4):397–403. https://doi.org/10.1111/j.1472-8206.2006.00415.x, PMID 16867025

Wilson MH, Newman S, Imray CH (2009) The cerebral effects of ascent to high altitudes. Lancet Neurol 8(2):175–191. https://doi.org/10.1016/S1474-4422(09)70014-6, PMID 19161909

Wisor JP, Pasumarthi RK, Gerashchenko D, Thompson CL, Pathak S, Sancar A et al (2008) Sleep deprivation effects on circadian clock gene expression in the cerebral cortex parallel electroencephalographic differences among mouse strains. J Neurosci 28(28):7193–7201. https://doi.org/10.1523/JNEUROSCI.1150-08.2008, PMID 18614689

Xiong J, Lu H, Wang R, Jia Z (2017) Efficacy of ibuprofen on prevention of high altitude headache: a systematic review and meta-analysis. PLoS One 12(6):e0179788. https://doi.org/10.1371/journal.pone.0179788, PMID 28632763

Yu GD, Liu YL, Jiang XH, Guo SY, Zhang HQ, Yin QZ et al (2001) The inhibitory effect of serotonin on the spontaneous discharge of suprachiasmatic neurons in hypothalamic slice is mediated by 5-HT(7) receptor. Brain Res Bull 54(4):395–398. https://doi.org/10.1016/s0361-9230(00)00462-7, PMID 11306191

Zelmanovich R, Pierre K, Felisma P, Cole D, Goldman M, Lucke-Wold B (2022) High-altitude cerebral edema: improving treatment options. Biologics (Basel) 2(1):81–91. https://doi.org/10.3390/biologics2010007, PMID 35425940

20 Senolytics in Brain Disorders: A Novel Pharmacological Approach to Control Cellular Senescence and Oxidative Stress

Faraha Ahmed and Razia Khanam

Abstract

Cellular senescence is considered to play an important role in age-related disorders in humans. Various studies indicate that brain cells also undergo cellular senescence via multiple mechanisms such as metabolic dysfunction, inflammation, oxidative stress, and alterations of proteins, thereby contributing to neuronal loss in age-related neurodegenerative and other disorders. Hence, suppressing the burden of cell senescence via pharmacological agents known as senolytics could offer great promise for the management of such conditions. There are reports that various senolytics acting via various mechanisms have been studied in animal models of CNS disorders and in human postmortem brain tissues. These agents such as dasatinib, quercetin, and fisetin have shown some positive results in decreasing the load of cell senescence in a few CNS disorders. In this review, the impact of cell senescence in various brain disorders and the studies done on senolytics so far on CNS disorders have been reviewed.

Keywords

Cell senescence · Senolytics · Dasatinib · Quercetin · Fisetin

F. Ahmed
Department of Pharmacology, School of Pharmaceutical Education and Research, Jamia Hamdard, New Delhi, India

R. Khanam (✉)
Department of Biomedical Sciences, College of Medicine, Gulf Medical University, Ajman, United Arab Emirates
e-mail: drrazia@gmu.ac.ae

R. Tulsawani, D. Vohora (eds.), *Adaptation under Stressful Environments through Biological Adjustments and Interventions*,
https://doi.org/10.1007/978-981-99-7652-2_20

20.1 Introduction

In 1961, human subcultured cells showed replicative and irreversible arrest of the cell cycle (Hickson et al. 2019). This phenomenon was termed senescence which is derived from the Latin word, Senex, which means "growing old," Cellular senescence has been shown to have both beneficial and deleterious effects on organisms (Kang et al. 2015). The cellular response may differ depending on the intensity and type of stress and cell type, which may range from repair to cell death (Sapieha and Mallette 2018). It is not only a response to stress, but several stimuli such as radiation, chemotherapeutic agents, and oxidative stress impose cellular senescence in cells. Telomere damage, metabolic dysfunction, alterations of proteins, inflammation, stress, oncogenes, nucleotide damage, microbial damage, and mitogens also drive a cell into senescence (Mohamad Kamal et al. 2020). Similar to cellular fate, senescence also involves differentiation, proliferation, necrosis, and apoptosis of cells. Senescent cells release growth factors and cytokines and are metabolically active cells that play a critical role in tissue homeostasis, embryogenesis, tissue patterning, and wound healing. Senescence modifies gene expression, alters the function of cell organelles, modifies histones, and hence leads to significant metabolic and morphological alterations (Calcinotto et al. 2019; Ogrodnik et al. 2019; Ogrodnik 2021). Chronic stress can cause senescent cells to accumulate. When they accumulate, they are capable to cause destruction of tissue, dysfunction of progenitor and stem cells, and inhibition of immune system along with inflammation (both local and systemic). Elevation in number of senescent cells contributes toward several pathological conditions such as type 2 diabetes mellitus, liver diseases, atherosclerosis, osteoarthritis, and idiopathic pulmonary fibrosis. Thus, senescence is favorable for acute stress-induced cellular damage but has deleterious effects on chronic stress-induced disorders (Ogrodnik et al. 2019; Wissler Gerdes et al. 2020). Besides arrest of cell cycle, senescent cells also undergo functional, morphological, and biochemical changes (Sedelnikova et al. 2004; Bernadotte et al. 2016). The cells appear larger in size, cell survival signaling pathways get activated, and formation of heterochromatin foci occurs along with increase in expression of growth factors, pro-inflammatory cytokines, and metalloproteinases (Si et al. 2021).

Human body's natural defense mechanism identifies senescent cells through helper T-cells and is then removed by natural killer cells, lymphocytes, and macrophages (Han et al. 2003; Akbar and Henson 2011; Kowald et al. 2020). Multiple transcription factors like p53/CDKN1A (cyclin-dependent kinase inhibitor) (p21), CDKN2A (p16)/RB (retinoblastoma) (Shay et al. 1991; Hernandez-Segura et al. 2018; Casella et al. 2019), and p16^{INK4a} (Robles and Adami 1998; Calcinotto et al. 2019), p21^{CIP1}, p27, p19^{ARF}, p53, and PAI-1 are tumor suppressors that have considerable role in cellular senescence (Calcinotto and Alimonti 2017). p53 regulates the expression of p21 transcriptionally (El-Deiry et al. 1993). There is upregulation of p21 and p16 and increased expressions of CDKN in senescent cells (Stein et al. 1999). Further, elevated activity of senescence-associated β-galactosidase (SA-β gal^{+}) at pH 6 is observed during cellular senescence (Hernandez-Segura et al. 2018; Casella et al. 2019). GATA4 (a protein-coding

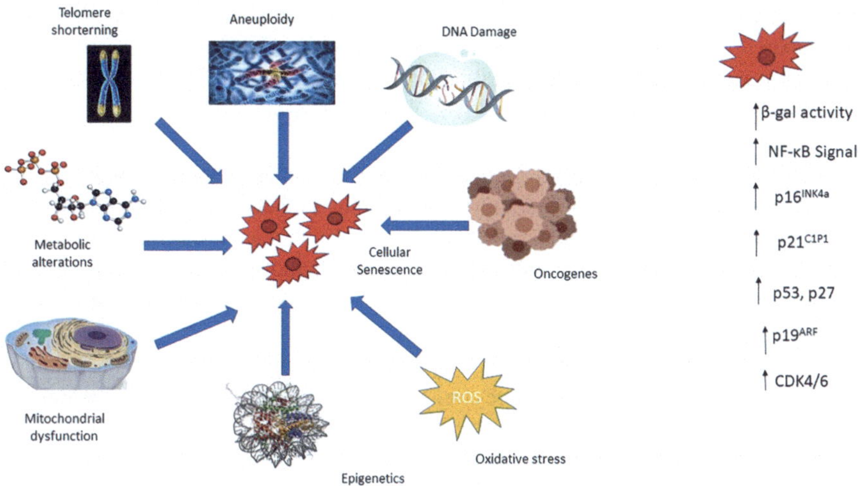

Fig. 20.1 Factors and mechanisms leading to cellular senescence. Various factors such as telomere shortening, aneuploidy, damage of DNA, development of oncogenes, development of reactive oxygen species (ROS) through oxidative stress, dysfunctional epigenetics, mitochondrial dysfunction, and metabolic alterations can drive cell into senescence. There is activation of beta-galactosidase activity that activates $p53/p21^{CIP1}$ and $p16^{INK4a}$/RB networks. There is upregulation of NF-κB (nuclear factor kappa B) and cyclin-dependent kinase (CDK) functions

gene), a zinc finger transcription factor, is another important regulator of senescence. It acts as a link between DNA damage and autophagy that leads to inflammation and senescence by activating NF-κB through IL1A and TRAF3IP2 and serves several roles in multiple organs. It functions independently of p53 and $p16^{INK4a}$ (Kang et al. 2015). Nuclear factor kappa B (NF-κB) pathway has been reported to promote senescence-associated secretory phenotype (SASP)-related cellular senescence. Shao et al. also revealed that Nrf-2 is another crucial antioxidant transcription factor (Shao et al. 2021). Markers of SASP-induced senescence include telomere-linked DNA damage foci, elevated response of lysosomal SA-β gal^+, senescence-linked heterochromatin foci, and increased level of $p21^{CIP1}$ and $p16^{INK4a}$ (Ben-Porath and Weinberg 2005; Krishnamurthy et al. 2006). Figure 20.1 describes the various factors responsible for cellular senescence.

According to Costa and team, deficiency of paraoxonase 2 (PON2) leads to mitochondrial dysfunction. PON2 is found in several dopaminergic regions such as substantia nigra and substantia striatum; brain stem is an antioxidant and anti-inflammatory protein that binds to coenzyme Q10 in mitochondrial complex (Costa et al. 2014). Accumulation of senescent cells accelerates with process of normal aging in animals and humans. Permanent cell cycle arrest by senescent cells is regulated by p53-p21-RB and $p16^{INK4A}$. Initially, cell cycle is arrested due to higher levels of p53 that upregulates CDKN(p21), and arrest is mediated by cyclin-dependent kinase 4 (CDK4) and cyclin-dependent kinase 6 (CDK6) which leads to hypo-phosphorylation of RB protein and arrests S phase of cell cycle. Apart from

NF-κB-SASP, other markers of cellular senescence include cyclic GMP-AMP synthase (cGAS), p38 mitogen-activated protein kinases (p38MAPK), stimulator of interferon genes (STING), mammalian target of rapamycin (mTOR), and NOTCH signaling pathways (Kritsilis et al. 2018; Watanabe et al. 2017; González-Gualda et al. 2021).

SASP is developed by most cells and consists of chemokines, growth factors, procoagulant factors, pro-inflammatory chemokines such as interleukins (IL-1β and IL-6), proteases bioactive lipids, miRNAs, micro-vesicles, and nucleotides (Xu et al. 2018; Martel et al. 2020). In addition to this, some chemotherapeutic agents such as doxorubicin and etoposide, exposure to ionizing radiation, and oxidative stress also contribute to senescence (Zhang et al. 2021).

There are SCAPs (senescent cell anti-apoptotic pathways) that inhibit the death of senescent cells in human pre-adipocytes. SCAPs pathways include ephrins/dependence receptors, PI3Kδ/Akt, Bcl-2/B-cell, p53/FOXO4a/p21/serpine, HIF-1α, and HSP-90 pathway (Wissler Gerdes et al. 2020).

Progressive research to identify drugs that can selectively eliminate senescent cells without affecting non-senescent cells leads to the development of senolytics. These are the class of drugs that help in removal or suppression of senescence of cells pharmacologically by inducing apoptosis of senescent cells. Senolytics do not exhibit their intrinsic activity through any biochemical pathway or receptors or enzymes rather they directly affect senescent cells by critically targeting them. Approach toward the process of senolysis seems promising for good health, age-related disorders, and prolonged life (Wissler Gerdes et al. 2020; Zhang et al. 2021). Senolytics promote apoptosis of senescent cells through SASP. Quercetin (Q) and dasatinib (D) are potent senolytics that suppress the senescent cell anti-apoptotic pathway network. Q is more selective as it targets Bcl-2 family, HIF-1α, PI3K, and p21-elated anti-apoptotic pathways (Hickson et al. 2019). Figure 20.2 presents the overview of mechanism of senolytics action.

There are several other drugs/compounds that exhibit senolytic or senotherapeutic activities in several disorders. The first-generation senolytics described in 2015 include dasatinib (kinase inhibitor) and quercetin. After that, many classes of senolytics have been described like natural compounds (fisetin, piperlongumine, curcumin, ouabain), Bcl-2 family inhibitors/Bcl-2 homology 3 (BH3) mimetics (navitoclax/ABT263, ABT-737, A1331852 and A1155463, UBX1325), inhibitors of MDM2/p53 interaction (UBX0101, P5091), Hsp90 inhibitors (geldanamycin, tanespimycin, alvespimycin/17-DMAG), p53-binding inhibitors (FOXO4-DRI), and histone deacetylase (HDAC) inhibitors (Panobinostat) (Kim and Kim 2019; Lagoumtzi and Chondrogianni 2021).

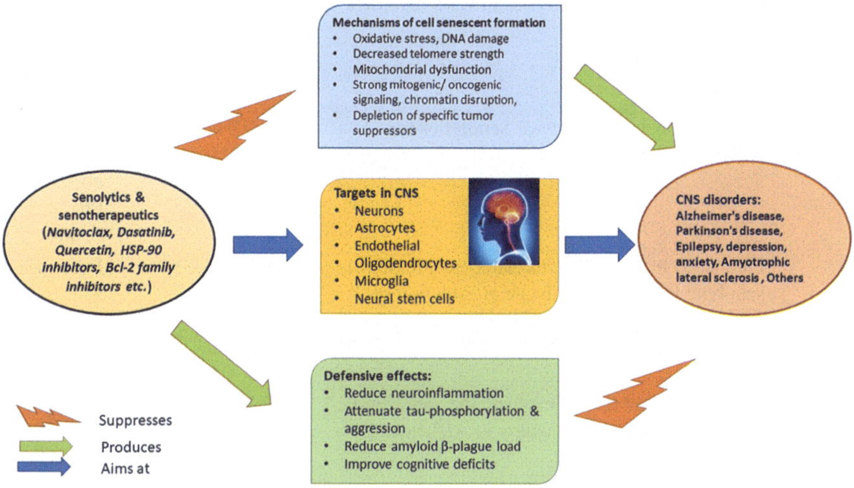

Fig. 20.2 Overview of senolytics on mechanisms inducing cell senescence and their effects on CNS disorders

20.2 The Role of Cell Senescence and Senolytics in CNS Disorders

Cellular senescence is associated with several CNS disorders. Preclinical and clinical research studies to substantiate their role in therapy of diseases have been conducted to substantiate their role.

Table 20.1 enlists various clinical trials regarding the role of senolytics in CNS disorders.

20.2.1 Anxiety

Psychological stress has a direct relation with increased oxidative stress, lower telomerase activity, and reduced telomere length; hence, the patients with depression, anxiety, or stress disorders have shorter telomere length than healthy controls (Wang et al. 2017). Few studies have reported that antianxiety and antidepressant drug therapy shows neuroprotective effect which is related to the reversal of aging-induced hallmarks like telomere length (Zhou et al. 2011; Perna et al. 2016).

Nociceptive opioid receptor agonist MCOPPB (1-[1-(1-methylcyclooctyl)-4-piperidinyl]-2-[(3R)-3-piperidinyl]-1H-benzimidazole), a potent anxiolytic, recently demonstrated its senolytic potential. It increased lipid storage in male C57/BL6 mice and *E. elegance* worm indicating longevity. Additionally, it showed tissue-specific cellular senescence in peripheral tissues of mice. Further, it significantly decreased

Table 20.1 Clinical trials on senolytics in CNS disorders

S. No.	Official title	Clinical trial registration number	Drugs and phases
1	An open-label pilot study of Senolytics for Alzheimer's disease	NCT04785300	Dasatinib + quercetin Phase 1, phase 2
2	Pilot study to investigate the safety and feasibility of senolytic therapy to modulate progression of Alzheimer's disease (SToMP-AD)	NCT04063124	Dasatinib + quercetin Phase 1, phase 2
3	Phase II clinical trial to evaluate the safety and feasibility of senolytic therapy in Alzheimer's disease	NCT04685590	Dasatinib + quercetin Phase 2
4	A multi-domain lifestyle intervention protocol to detect changes in retinal amyloid among individuals with mild cognitive impairment	NCT02741804	Fisetin + others Phase NA
5	Impact of the combined treatment of curcumin and resveratrol liposomed polyphenols with G04CB02 on the clinical improvement of ALS patients	NCT04654689	Curcumin Phase 2
6	A phase 2, prospective, multicenter, randomized, double-masked, active-controlled study to assess the safety, tolerability, and evidence of activity of a repeat intravitreal injection of UBX1325 in patients with neovascular AMD	NCT05275205	UBX1325 Phase 2

the number of senescent cells in DOX-treated Huh-7 and HepG2 cells (Raffaele et al. 2022).

Metabolic disorder like obesity is also related to augmented senescent cell burden and neuropsychiatric disorders like anxiety and depression (Gariepy et al. 2010; Hryhorczuk et al. 2013; Palmer et al. 2022). To evaluate these effects, Ogrodnik and team proved that dasatinib and quercetin treatment improved high-fat diet (HFD)-induced obesity, anxiety, and senescence in 8-month-old C57Bl/6 mice. As accumulation of lipids is crucial for senescence, they investigated the role of glial senescent cells in obesity-associated neurogenesis and anxiety-like behavior in vivo. Elevated plus maze model was used to evaluate the anxiety-like behavior in mice. Perigonadal adipose tissue was measured as a senescent marker. It was revealed that removal of senescent cells reduced the obesity-associated anxiety-like behavior in the animal model. HFD-treated mice displayed proximity of senescent $Plin2^+$ glial cells that are also expressed marker of neurons and neuronal precursor cells, doublecortin (Dcx). Senolytic therapy also upregulated Dcx, $CD133^+$, and $Nestin^+$ that are markers of immature neurons, ependymal cells, and neural precursor cells, respectively (Ogrodnik et al. 2019).

20.2.2 Depression

Clearing of senescent cells could provide new avenues for stress and depressive illness. To prove this theory, Lin and co-workers studied SA-β-gal activity of active senescent cells in vivo in C57BL/6 male mice. Behavioral despair and depression in animals were studied with forced swim test and chronic unpredictable stress (CUS) model. CUS was employed for 1 month with different stressors such as cage tilting, sleep deprivation, food/water deprivation, rapid light and dark cycles, restraints, removal of bedding, and water at the bottom of the cages. Results showed that CUS-treated mice exhibited higher levels of $p16^{INK4a}$ levels indicating cellular senescence. The outcome also revealed that dasatinib and quercetin treatment decreased the load of SA-β-gal-type cells and expression of $p16^{INK4a}$ in mice hippocampal region (Lin et al. 2021).

Effect of fisetin was evaluated in X-ray (10 Gy)-induced depression in female MRL/lpr systemic lupus erythematosus (SLE) and haplotype female MRl/MPJ mice. Oral administration of fisetin decreased the proximity of senescent cells in CA3 region, part of hippocampus associated with depression and presence of SA-β-gal^+ cells. Further, transcription of SASP genes at the hippocampal region in SLE mouse and attenuation of depression-like behavior in MRL/lpr mouse model was seen as well (Saito et al. 2021).

20.2.3 Alzheimer's Disease

Neurodegenerative disorders like Alzheimer's disease (AD) are outcome of imbalanced autophagy and lysosomal dysfunction. Cellular senescence also contributes toward pathology of AD and ascertain age, susceptibility, and rate of progression of disease (Saez-Atienzar and Masliah 2020). Major mechanisms of cellular senescence involved in neurodegeneration includes chronic inflammation, dysfunctional regenerative capacities of brain cerebral hypoperfusion leading to functional loss altered blood brain barrier and erratic re-entry of differentiated post-mitotic neurons. Higher expression of key regulators of cell cycle dysfunction and cyclin-dependent kinase inhibitor $p21^{CIP1}$ and p53 along with raised activity of p38MAPK activity is also found. Additionally, elevated levels of TNF-α, interleukins (IL-6 and IL-1β), transforming growth factor-β (TGF-β), and γH2AX were found in hippocampus of AD patients (Kritsilis et al. 2018).

Curraise et al. investigated the role of oral fisetin (25 mg/kg/day) on learning and memory in double transgenic APPswe/PS1dE9 AD mice (3–12 months old). The results showed that fisetin elevated extracellular signal-regulated kinase (ERK) phosphorylation and reduced protein carbonylation suggesting its role in oxidative stress. Further, it also decreased the expression of p25 (cyclin-dependent kinase 5 activator) in brains of mice, which is responsible for neurodegeneration and neuroinflammation (Currais et al. 2014).

Fisetin treatment remarkably suppressed phosphorylation of tau protein, levels of BACE-1, and Aβ accumulation in $A\beta_{1-42}$ (3 μL/5 min/mouse)-treated mice. Fisetin also reversed the $A\beta_{1-42}$-induced elevated synaptic proteins such as SYN, SNAP-25, SNAP-23, post-synaptic density protein 95 (PSD-95), phospho-CREB (p-CREB), phospho-GLuR1 (p-GluR1), and phospho-calcium/calmodulin-stimulated protein kinase II (p-CAMKII). Additionally, it also increased levels ofp-Akt, p-PI3K, and p-GSK3β in $A\beta_{1-42}$ mice and suppressed inflammation in neurons and apoptotic neurodegeneration (Ahmad et al. 2017).

Zhang et al. evaluated the impact of Q (50 m/kg) and D (12 mg/kg) for 9 days on APP/PS1 mouse model after oral administration. Since, Aβ plaque is responsible for neuritic degeneration, local inflammation, and lysosomal dysfunction. Hence, they also quantified Aβ plaque and associated oligodendrocyte progenitor cells (OPC) as they were accumulated in the regions of brain in AD patients. The results showed that senescent cells in AD patients and Aβ mouse model of AD-expressed OPCs transform into pro-inflammatory phenotypes of senescent cells and accumulation of Aβ plaque also induces cellular senescence. Senolytic treatment cleared the p16 cellular expression of OPC cells in the mouse model and decreased the microglial activation and expression of pro-inflammatory cytokines as well. Senolytic treatment also improved the patterns of learning and enhanced memory in APP/PS1 AD mouse model affirming its substantial role in cellular senescence responsible for AD (Zhang et al. 2019).

Fatt et al. reported that administration of navitoclax (ABT-263), in vivo in middle-aged mice, caused ablation of senescent cells at hippocampus after a month therapy. This in turn rapidly increased the expression of neural precursor cells at the hippocampus. Further, after a year of therapy, there was an increase in SA-β-Gal-positive in the sub-granular zone by 3 times. Additionally, navitoclax also exhibited an improvement in spatial learning and memory when tested through the Morris water maze (Fatt et al. 2022). Navitoclax also reversed the cell senescence in irradiation (5 Gy)-induced senescence in brain astrocytes of transgenic p16-3MR mice (Yabluchanskiy et al. 2020).

Studies suggested that progressive telomere shortening is related to biological aging and age-related morbidities, weakness, and mortality (Whittemore et al. 2019; Zhu et al. 2019). Further, it is also shown to activate cellular senescence along with inflammation and mitochondria (Zhu et al. 2019). Metformin via multiple mechanisms is shown to reduce telomere shortening in diabetic patients and also encourage protective feedback mechanisms on the maintenance of telomere in addition to the improvement in glucose tolerance and insulin sensitivity (Kulkarni et al. 2020).

20.2.4 Parkinson's Disease (PD)

The damage of neurons from the substantia nigra and the build-up of ubiquitinated alpha synuclein along with Lewy bodies and Lewy neuritis in cytoplasmic inclusion and within neurites, respectively, are the characteristic features of PD (Kritsilis et al. 2018). Many studies documented involvement of cellular senescence in the pathogenesis of neurodegenerative disorders including PD (Lee et al. 2021). The expression of p16^{INK4a} mRNA was increased in brain samples of PD patients. Senescent cell markers are also present within astrocytes of brain tissues in postmortem PD samples. These samples exhibited elevated levels of pro-inflammatory SASP factors, including IL-6, IL-8, and IL-α (Chinta et al. 2018).

Paraquat, a toxic compound capable of inducing PD, can also stimulate cellular senescence and generate the production of SASP factors in astrocytes. This action leads to neuronal cell death. When these senescent astrocytes were removed in in vitro study, it reversed the paraquat-induced PD (Chinta et al. 2018).

In the pathogenesis of PD, mitochondrial hyperactivity has been reported as an early event. Approaches that decrease mitochondrial respiration may serve as new strategy for the management of PD. In this regard, antidiabetic drug metformin seems to be a promising agent. It improves motor functions and neuronal viability by decreasing mitochondrial respiration (Mor et al. 2020). Hence, reducing mitochondrial respiration could be employed as a strategy to explore new treatment approach in PD.

Quercetin has ability to bind with α-synuclein through a covalent bond (Singh et al. 2020). The neuroprotective and anti-PD roles of quercetin were observed in primary cultured rat cortical cells against oxidative neuronal injury and on 6-hydroxydomaine-induced rat model, respectively (Haleagrahara et al. 2011; Bakoyiannis et al. 2019). The mechanisms suggested for neuroprotective effect of quercetin involves activation of sirtuins (SIRT1) and induction of autophagy (Kida and Goligorsky 2016). Quercetin alters the cellular senescence via Nrf2 pathway as it suppresses SASP-induced cellular senescence (Stein et al. 1999). Along with induction of Nrf2-ARE (antioxidant response element), it also induces PON2 (Acıkara et al. 2022).

20.2.5 Epilepsy

Among the various biomarkers indicated for cell senescence, neurodegeneration and aging involve altered plasma Aβ levels, epigenetic stress, tumor suppressor loss, and shortening of telomere length. These factors are also associated with vascular, metabolic, and psychological risk factors for neurodegenerative disorders (Si et al. 2021; Innes et al. 2018). Telomeres are essential for the maintenance of genomic integrity. Its length is considered a biological marker for stress and cellular aging. Interestingly, a significant shorter telomere in the peripheral cell leukocytes of treatment-resistant mesial temporal lobe epilepsy (mTLE) has also been reported (Miranda et al. 2020). More such studies can help in designing drugs targeting telomeres. This would be effective for the treatment of resistant cases of mTLE and other types of epilepsies. This might ultimately open the window of opportunities in exploring novel therapeutic avenues against epilepsy and in cognitive decline.

Studies performed to discover the options that can suppress the epileptic seizure symptoms by modulating cellular and molecular changes such as oxidative stress, neuroinflammation, apoptosis, and mTOR pathway inhibition have shown metformin as a potential candidate in this aspect. The changes observed during the different phases of epileptic disease like neuronal loss are also reported to be modified by metformin (Yimer et al. 2019; Nandini et al. 2019).

There are no reports of dasatinib effects on epilepsy; however, quercetin has been reported recently to inhibit KA-induced epilepsy by microglia cell inactivation and the production of NF-κB, TNF-α, and IL-1β in mice (Wu et al. 2020). On the other hand, Xie et al. (2022) recently suggested that quercetin effectively protects against seizure-induced neuron death in vivo and in vitro and improves cognitive function impairment via the multiple pathways including SIRT1/Nrf2/SLC7A11/GPX4 pathway (Xie et al. 2022).

20.2.6 Amyotrophic Lateral Sclerosis

The hallmark effects of senescence have also been found to be associated with amyotrophic lateral sclerosis (ALS), and the role of removing senescent cells or restoring senescent cells back to normal function has been tried as a strategy for producing a disease-modifying ALS therapy (Maximova et al. 2021). In the experimental animal models and in postmortem human tissues of ALS, the glial senescence has been reported (Trias et al. 2019; Sivasubramanian et al. 2020). Senolytics affecting heat shock protein 90 (Hsp90) (17-DMAG), PI3K inhibition and DNA intercalation (Q), Akt inhibition (D), non-selective Bcl-2 family inhibitions (navitoclax), etc., have been studied for their role in removal of senescent cells and in the models of CNS disorders (Maximova et al. 2021). It has been observed that current senolytics targets such as Hsp90 and PI3K inhibition may risk degenerating motor neurons. On the contrary, inhibiting Bcl-2 family members may provide neuroprotection both directly and indirectly, via glial senolysis. Exploration of alternative strategies that are selectively upregulated in senescent glia in ALS are required to be identified that could be targeted without compromising motor neurons. If such strategies work out, the slowing down of ALS at early stage would be possible and may eventually extend lifespan of such patients.

20.2.7 Miscellaneous

Agents like navitoclax and BCL-X_L inhibitors like A1331852 and A1155463 have also been reported recently to possess senolytic activity (Zhu et al. 2015). Navitoclax modulates the anti-apoptotic pathways of BCL2 and BCL-XL, which are upregulated in senescent cells. The role of navitoclax as a senolytic agent has been suggested in delivering functional advantages in mice after brain radiation (Zhu et al. 2015, 2016, 2017). While BCL-2 inhibitors are associated with a number of adverse effects, A1331852 and A1155463 are relatively selective BCL-X_L inhibitors. They appear less likely to cause neutropenia than navitoclax, making them possibly better contenders for subsequent translation into clinical applications (Zhu et al. 2016, 2017).

Piperlongumine (PL) is another natural product that is reported to have multiple effects including anti-tumor activity. Wang and co-workers showed that it killed senescent human WI-38 fibroblasts preferentially when senescence was induced by different methods such as ionizing radiation, replicative exhaustion, or ectopic expression of the oncogene *Ras* (Wang et al. 2016). It killed senescent cells alone by inducing apoptosis. It further exhibited synergistic effects with navitoclax. The studies demonstrated PL as a novel lead for developing senolytic agents, but further studies are warranted to develop piperlongumine as an anti-aging agent.

Curcumin is known to suppress acute and chronic inflammation via multiple mechanisms including ROS scavenging, stimulating antioxidant defense actions, and reducing lipid peroxidation. It has also been reported to possess anticancer effects and to inhibit cell growth in tumor cell lines such as SGC7901, HL60, and Bel7402 in a dose-dependent manner (Cui et al. 2006). Following the treatment with curcumin in the telomerase-positive human glioblastoma and medulloblastoma cell lines, inhibition of telomerase activity has also been reported (Khaw et al. 2013).

The details of dasatinib, quercetin, fisetin, curcumin, and UBX1325 are available for trials on CNS disorders such as Alzheimer's disease, cognitive impairment, ALS, and neovascular age-related macular degeneration. For the remaining senolytics, the clinical trials are mainly ongoing/completed for osteoarthritis of knee, skin diseases, lymphomas, diabetic macular edema, etc. (Kirkland and Tchkonia 2020; Raffaele and Vinciguerra 2022).

20.3 Conclusion

A variety of complex mechanisms such as mitochondrial dysfunction, neuroinflammation, oxidative load, and various other homeostatic mechanisms that are otherwise compromised during cell senescence are able to initiate the neurodegeneration and other mechanisms leading to various disorders. Senolytics seem to re-establish progenitor dysfunction, reduce tissue inflammation, and improve age- and disease-related metabolic dysfunction. Hence, they are likely to delay, prevent, or suppress multiple age-related diseases and chronic conditions and improve health and lifespan in experimental animals. Therefore, it is predicted that they may also lead to interventions for humans that delay, prevent, or treat senescence- and age-related conditions. Though the effects and role of senolytics in peripheral chronic disorders such as diabetes, aging, osteoporosis, cardiovascular diseases, pulmonary fibrosis, and cancer are reported, only a few agents have been evaluated in CNS disorders. Some of them are reported to possess serious adverse effects such as navitoclax producing hematological toxicity, hence is not completely safe. Strategies like combination of agents with distinct mechanism of action may help in showing synergistic effects. Recent therapeutic approaches based on specific drug delivery strategies and use of novel biomarkers are also being required to be developed to increase the specificity and reduce the side effects of senolytics. Clinical trials are required to assess the safety, efficacy, tolerability, and specific targets of these drugs before being available for routine clinical use. It is also predicted that most of the senolytic agents target particular types of senescent cells, based on their origin or mechanisms by which senescence was induced. Another future approach may also be to explore methods of senescence prevention, which would require further detailed research. There is still a long way to go, but senolytics might come up a completely new range of drugs targeting novel mechanism and may become therapeutic agents to serve treatment-resistant/refractory disorders.

Funding No funding to disclose.

Conflict of Interest The authors have no conflicts of interest to declare that are relevant to the content of this article.

References

Acıkara OB, Karatoprak GŞ, Yücel Ç, Akkol EK, Sobarzo-Sánchez E, Khayatkashani M et al (2022) A critical analysis of quercetin as the attractive target for the treatment of Parkinson's disease. CNS Neurol Disord Drug Targets 21(9):795–817. https://doi.org/10.2174/1871527320666211206122407

Ahmad A, Ali T, Park HY, Badshah H, Rehman SU, Kim MO (2017) Neuroprotective effect of fisetin against amyloid-beta-induced cognitive/synaptic dysfunction, neuroinflammation, and neurodegeneration in adult mice. Mol Neurobiol 54(3):2269–2285. https://doi.org/10.1007/s12035-016-9795-4

Akbar AN, Henson SM (2011) Are senescence and exhaustion intertwined or unrelated processes that compromise immunity? Nat Rev Immunol 11(4):289–295. https://doi.org/10.1038/nri2959

Bakoyiannis I, Daskalopoulou A, Pergialiotis V, Perrea D (2019) Phytochemicals and cognitive health: are flavonoids doing the trick? Biomed Pharmacother 109:1488–1497. https://doi.org/10.1016/j.biopha.2018.10.086

Ben-Porath I, Weinberg RA (2005) The signals and pathways activating cellular senescence. Int J Biochem Cell Biol 37(5):961–976. https://doi.org/10.1016/j.biocel.2004.10.013

Bernadotte A, Mikhelson VM, Spivak IM (2016) Markers of cellular senescence. Telomere shortening as a marker of cellular senescence. Aging (Albany, NY) 8(1):3–11. https://doi.org/10.18632/aging.100871

Calcinotto A, Alimonti A (2017) Aging tumour cells to cure cancer: "pro-senescence" therapy for cancer. Swiss Med Wkly 147:w14367. https://doi.org/10.57187/smw.2017.14367

Calcinotto A, Kohli J, Zagato E, Pellegrini L, Demaria M, Alimonti A (2019) Cellular senescence: aging, cancer, and injury. Physiol Rev 99(2):1047–1078. https://doi.org/10.1152/physrev.00020.2018

Casella G, Munk R, Kim KM, Piao Y, De S, Abdelmohsen K et al (2019) Transcriptome signature of cellular senescence. Nucleic Acids Res 47(14):7294–7305. https://doi.org/10.1093/nar/gkz555

Chinta SJ, Woods G, Demaria M, Rane A, Zou Y, McQuade A et al (2018) Cellular senescence is induced by the environmental neurotoxin paraquat and contributes to neuropathology linked to Parkinson's disease. Cell Rep 22(4):930–940. https://doi.org/10.1016/j.celrep.2017.12.092

Costa LG, De Laat R, Dao K, Pellacani C, Cole TB, Furlong CE (2014) Paraoxonase-2 (PON2) in brain and its potential role in neuroprotection. Neurotoxicology 43:3–9. https://doi.org/10.1016/j.neuro.2013.08.011

Cui SX, Qu XJ, Xie YY, Zhou L, Nakata M, Makuuchi M et al (2006) Curcumin inhibits telomerase activity in human cancer cell lines. Int J Mol Med 18(2):227–231. https://doi.org/10.3892/ijmm.18.2.227

Currais A, Prior M, Dargusch R, Armando A, Ehren J, Schubert D et al (2014) Modulation of p25 and inflammatory pathways by fisetin maintains cognitive function in Alzheimer's disease transgenic mice. Aging Cell 13(2):379–390. https://doi.org/10.1111/acel.12185

El-Deiry WS, Tokino T, Velculescu VE, Levy DB, Parsons R, Trent JM et al (1993) WAF1, a potential mediator of p53 tumor suppression. Cell 75(4):817–825. https://doi.org/10.1016/0092-8674(93)90500-p

Fatt MP, Tran LM, Vetere G, Storer MA, Simonetta JV, Miller FD et al (2022) Restoration of hippocampal neural precursor function by ablation of senescent cells in the aging stem cell niche. Stem Cell Rep 17(2):259–275. https://doi.org/10.1016/j.stemcr.2021.12.010, /pmc/articles/PMC8828532

Gariepy G, Nitka D, Schmitz N (2010) The association between obesity and anxiety disorders in the population: a systematic review and meta-analysis. Int J Obes 34(3):407–419. https://doi.org/10.1038/ijo.2009.252

González-Gualda E, Baker AG, Fruk L, Muñoz-Espín D (2021) A guide to assessing cellular senescence in vitro and in vivo. FEBS J 288(1):56–80. https://doi.org/10.1111/febs.15570

Haleagrahara N, Siew CJ, Mitra NK, Kumari M (2011) Neuroprotective effect of bioflavonoid quercetin in 6-hydroxydopamine-induced oxidative stress biomarkers in the rat striatum. Neurosci Lett 500(2):139–143. https://doi.org/10.1016/j.neulet.2011.06.021

Han S, Yang K, Ozen Z, Peng W, Marinova E, Kelsoe G et al (2003) Enhanced differentiation of splenic plasma cells but diminished long-lived high-affinity bone marrow plasma cells in aged mice. J Immunol 170(3):1267–1273. https://doi.org/10.4049/jimmunol.170.3.1267

Hernandez-Segura A, Nehme J, Demaria M (2018) Hallmarks of cellular senescence. Trends Cell Biol 28(6):436–453. https://doi.org/10.1016/j.tcb.2018.02.001

Hickson LTJ, Langhi Prata LGP, Bobart SA, Evans TK, Giorgadze N, Hashmi SK et al (2019) Senolytics decrease senescent cells in humans: preliminary report from a clinical trial of dasatinib plus quercetin in individuals with diabetic kidney disease. EBioMedicine 47:446–456; [cited Sep 8 2022]. https://pubmed.ncbi.nlm.nih.gov/31542391/

Hryhorczuk C, Sharma S, Fulton SE (2013) Metabolic disturbances connecting obesity and depression. Front Neurosci 7(7):177. https://doi.org/10.3389/fnins.2013.00177

Innes KE, Selfe TK, Brundage K, Montgomery C, Wen S, Kandati S et al (2018) Effects of meditation and music-listening on blood biomarkers of cellular aging and Alzheimer's disease in adults with subjective cognitive decline: an exploratory randomized clinical trial. J Alzheimers Dis 66(3):947–970. https://doi.org/10.3233/JAD-180164

Kang C, Xu Q, Martin TD, Li MZ, Demaria M, Aron L et al (2015) The DNA damage response induces inflammation and senescence by inhibiting autophagy of GATA4. Science 349(6255):aaa5612. https://doi.org/10.1126/science.aaa5612

Khaw AK, Hande MP, Kalthur G, Hande MP (2013) Curcumin inhibits telomerase and induces telomere shortening and apoptosis in brain tumour cells. J Cell Biochem 114(6):1257–1270. https://doi.org/10.1002/jcb.24466

Kida Y, Goligorsky MS (2016) Sirtuins, cell senescence, and vascular aging. Can J Cardiol 32(5):634–641. https://doi.org/10.1016/j.cjca.2015.11.022

Kim EC, Kim JR (2019) Senotherapeutics: emerging strategy for healthy aging and age-related disease. BMB Rep 52(1):47–55. https://doi.org/10.5483/BMBRep.2019.52.1.293

Kirkland JL, Tchkonia T (2020) Senolytic drugs: from discovery to translation. J Intern Med 288(5):518–536. https://doi.org/10.1111/joim.13141

Kowald A, Passos JF, Kirkwood TBL (2020) On the evolution of cellular senescence. Aging Cell 19(12):e13270. https://doi.org/10.1111/acel.13270

Krishnamurthy J, Ramsey MR, Ligon KL, Torrice C, Koh A, Bonner-Weir S et al (2006) p16INK4a induces an age-dependent decline in islet regenerative potential. Nature 443(7110):453–457. https://doi.org/10.1038/nature05092

Kritsilis M, Rizou SV, Koutsoudaki PN, Evangelou K, Gorgoulis VG, Papadopoulos D (2018) Ageing, cellular senescence and neurodegenerative disease. Int J Mol Sci 19(10):2937. https://doi.org/10.3390/ijms19102937

Kulkarni AS, Gubbi S, Barzilai N (2020) Benefits of metformin in attenuating the hallmarks of aging. Cell Metab 32(1):15–30. https://doi.org/10.1016/j.cmet.2020.04.001
Lagoumtzi SM, Chondrogianni N (2021) Senolytics and senomorphics: natural and synthetic therapeutics in the treatment of aging and chronic diseases. Free Radic Biol Med 171:169–190. https://doi.org/10.1016/j.freeradbiomed.2021.05.003
Lee S, Wang EY, Steinberg AB, Walton CC, Chinta SJ, Andersen JK (2021) A guide to senolytic intervention in neurodegenerative disease. Mech Ageing Dev 200:111585. https://doi.org/10.1016/j.mad.2021.111585
Lin YF, Wang LY, Chen CS, Li CC, Hsiao YH (2021) Cellular senescence as a driver of cognitive decline triggered by chronic unpredictable stress. Neurobiol Stress 15:100341. https://doi.org/10.1016/j.ynstr.2021.100341
Martel J, Ojcius DM, Wu CY, Peng HH, Voisin L, Perfettini JL et al (2020) Emerging use of senolytics and senomorphics against aging and chronic diseases. Med Res Rev 40(6):2114–2131. https://doi.org/10.1002/med.21702
Maximova A, Werry EL, Kassiou M (2021) Senolytics: a novel strategy for neuroprotection in ALS? Int J Mol Sci 22(21):12078. https://doi.org/10.3390/ijms222112078
Miranda DM, Rosa DV, Costa BS, Nicolau NF, Magno LAV, de Paula JJ et al (2020) Telomere shortening in patients with drug-resistant epilepsy. Epilepsy Res 166:106427. https://doi.org/10.1016/j.eplepsyres.2020.106427
Mohamad Kamal NS, Safuan S, Shamsuddin S, Foroozandeh P (2020) Aging of the cells: insight into cellular senescence and detection methods. Eur J Cell Biol 99(6):151108. https://doi.org/10.1016/j.ejcb.2020.151108
Mor DE, Sohrabi S, Kaletsky R, Keyes W, Tartici A, Kalia V et al (2020) Metformin rescues Parkinson's disease phenotypes caused by hyperactive mitochondria. Proc Natl Acad Sci U S A 117(42):26438–26447. https://doi.org/10.1073/pnas.2009838117
Nandini HS, Paudel YN, Krishna KL (2019) Envisioning the neuroprotective effect of metformin in experimental epilepsy: a portrait of molecular crosstalk. Life Sci 233:116686. https://doi.org/10.1016/j.lfs.2019.116686
Ogrodnik M (2021) Cellular aging beyond cellular senescence: markers of senescence prior to cell cycle arrest in vitro and in vivo. Aging Cell 20(4):e13338. https://doi.org/10.1111/acel.13338
Ogrodnik M, Zhu Y, Langhi LGP, Tchkonia T, Krüger P, Fielder E et al (2019) Obesity-induced cellular senescence drives anxiety and impairs neurogenesis. Cell Metab 29(5):1233. https://doi.org/10.1016/j.cmet.2019.01.013
Palmer AK, Jensen MD, Tchkonia T, Kirkland JL (2022) In: Serrano M, Muñoz-Espin D (eds) Senescence in obesity: causes and consequences. Academic Press, pp 289–308
Perna G, Iannone G, Alciati A, Caldirola D (2016) Are anxiety disorders associated with accelerated aging? A focus on neuroprogression. Neural Plast 2016:8457612. https://doi.org/10.1155/2016/8457612
Raffaele M, Kovacovicova K, Biagini T, Lo Re O, Frohlich J, Giallongo S et al (2022) Nociceptin/orphanin FQ opioid receptor (NOP) selective ligand MCOPPB links anxiolytic and senolytic effects. GeroScience 44(1):463–483. https://doi.org/10.1007/s11357-021-00487-y
Raffaele M, Vinciguerra M (2022) The costs and benefits of senotherapeutics for human health. Lancet Healthy Longev 3(1):e67–e77. https://doi.org/10.1016/S2666-7568(21)00300-7
Robles SJ, Adami GR (1998) Agents that cause DNA double strand breaks lead to p16INK4a enrichment and the premature senescence of normal fibroblasts. Oncogene 16(9):1113–1123. https://doi.org/10.1038/sj.onc.1201862
Saez-Atienzar S, Masliah E (2020) Cellular senescence and Alzheimer disease: the egg and the chicken scenario. Nat Rev Neurosci 21(8):433–444. https://doi.org/10.1038/s41583-020-0325-z
Saito Y, Miyajima M, Yamamoto S, Sato T, Miura N, Fujimiya M et al (2021) Accumulation of senescent neural cells in murine lupus with depression-like behavior. Front Immunol 12:692321. https://doi.org/10.3389/fimmu.2021.692321

Sapieha P, Mallette FA (2018) Cellular senescence in postmitotic cells: beyond growth arrest. Trends Cell Biol 28(8):595–607. https://doi.org/10.1016/j.tcb.2018.03.003

Sedelnikova OA, Horikawa I, Zimonjic DB, Popescu NC, Bonner WM, Barrett JC (2004) Senescing human cells and ageing mice accumulate DNA lesions with unrepairable double-strand breaks. Nat Cell Biol 6(2):168–170. https://doi.org/10.1038/ncb1095

Shao Z, Wang B, Shi Y, Xie C, Huang C, Chen B et al (2021) Senolytic agent quercetin ameliorates intervertebral disc degeneration via the Nrf2/NF-κB axis. Osteoarthr Cartil 29(3):413–422. https://doi.org/10.1016/j.joca.2020.11.006

Shay JW, Pereira-Smith OM, Wright WE (1991) A role for both RB and p53 in the regulation of human cellular senescence. Exp Cell Res 196(1):33–39. https://doi.org/10.1016/0014-4827(91)90453-2

Si Z, Sun L, Wang X (2021) Evidence and perspectives of cell senescence in neurodegenerative diseases. Biomed Pharmacother 137:111327. https://doi.org/10.1016/j.biopha.2021.111327

Singh A, Tripathi P, Yadawa AK, Singh S (2020) Promising polyphenols in Parkinson's disease therapeutics. Neurochem Res 45(8):1731–1745. https://doi.org/10.1007/s11064-020-03058-3

Sivasubramanian MK, Monteiro R, Balasubramanian P, Subramanian M (2020) Oxidative stress-induced senescence alters glutamate transporter expression in human brainstem astrocytes. FASEB J 34(S1):1. https://doi.org/10.1096/fasebj.2020.34.s1.06566

Stein GH, Drullinger LF, Soulard A, Dulić V (1999) Differential roles for cyclin-dependent kinase inhibitors p21 and p16 in the mechanisms of senescence and differentiation in human fibroblasts. Mol Cell Biol 19(3):2109–2117. https://doi.org/10.1128/MCB.19.3.2109

Trias E, Beilby PR, Kovacs M, Ibarburu S, Varela V, Barreto-Núñez R et al (2019) Emergence of microglia bearing senescence markers during paralysis progression in a rat model of inherited ALS. Front Aging Neurosci 11:42. https://doi.org/10.3389/fnagi.2019.00042

Wang Y, Chang J, Liu X, Zhang X, Zhang S, Zhang X et al (2016) Discovery of piperlongumine as a potential novel lead for the development of senolytic agents. Aging (Albany NY) 8(11): 2915–2926. https://doi.org/10.18632/aging.101100, /pmc/articles/PMC5191878

Wang X, Sundquist K, Hedelius A, Palmér K, Memon AA, Sundquist J (2017) Leukocyte telomere length and depression, anxiety and stress and adjustment disorders in primary health care patients. BMC Psychiatry 17(1):148. https://doi.org/10.1186/s12888-017-1308-0

Watanabe S, Kawamoto S, Ohtani N, Hara E (2017) Impact of senescence-associated secretory phenotype and its potential as a therapeutic target for senescence-associated diseases. Cancer Sci 108(4):563–569. https://doi.org/10.1111/cas.13184

Whittemore K, Vera E, Martínez-Nevado E, Sanpera C, Blasco MA (2019) Telomere shortening rate predicts species life span. Proc Natl Acad Sci U S A 116(30):15122–15127. https://doi.org/10.1073/pnas.1902452116

Wissler Gerdes EO, Zhu Y, Tchkonia T, Kirkland JL (2020) Discovery, development, and future application of senolytics: theories and predictions. FEBS J 287(12):2418–2427. https://doi.org/10.1111/febs.15264

Wu D, Zheng Z, Fan S, Wen X, Han X, Wang S et al (2020) Ameliorating effect of quercetin on epilepsy by inhibition of inflammation in glial cells. Exp Ther Med 20(2):854–859. https://doi.org/10.3892/etm.2020.8742

Xie R, Zhao W, Lowe S, Bentley R, Hu G, Mei H et al (2022) Quercetin alleviates kainic acid-induced seizure by inhibiting the Nrf2-mediated ferroptosis pathway. Free Radic Biol Med 191: 212–226. https://doi.org/10.1016/j.freeradbiomed.2022.09.001

Xu M, Pirtskhalava T, Farr JN, Weigand BM, Palmer AK, Weivoda MM et al (2018) Senolytics improve physical function and increase lifespan in old age. Nat Med 24(8):1246–1256. https://doi.org/10.1038/s41591-018-0092-9

Yabluchanskiy A, Tarantini S, Balasubramanian P, Kiss T, Csipo T, Fülöp GA et al (2020) Pharmacological or genetic depletion of senescent astrocytes prevents whole brain irradiation–induced impairment of neurovascular coupling responses protecting cognitive function in mice. GeroScience 42(2):409–428. https://doi.org/10.1007/s11357-020-00154-8, /pmc/articles/PMC7205933

Yimer EM, Surur A, Wondafrash DZ, Gebre AK (2019) The effect of metformin in experimentally induced animal models of epileptic seizure. Behav Neurol 2019:6234758. https://doi.org/10.1155/2019/6234758

Zhang P, Kishimoto Y, Grammatikakis I, Gottimukkala K, Cutler RG, Zhang S et al (2019) Senolytic therapy alleviates Aβ-associated oligodendrocyte progenitor cell senescence and cognitive deficits in an Alzheimer's disease model. Nat Neurosci 22(5):719–728. https://doi.org/10.1038/s41593-019-0372-9

Zhang L, Pitcher LE, Prahalad V, Niedernhofer LJ, Robbins PD (2021) Recent advances in the discovery of senolytics. Mech Ageing Dev 200:111587. https://doi.org/10.1016/j.mad.2021.111587

Zhou QG, Hu Y, Wu DL, Zhu LJ, Chen C, Jin X et al (2011) Hippocampal telomerase is involved in the modulation of depressive behaviors. J Neurosci 31(34):12258–12269. https://doi.org/10.1523/JNEUROSCI.0805-11.2011

Zhu Y, Doornebal EJ, Pirtskhalava T, Giorgadze N, Wentworth M, Fuhrmann-Stroissnigg H et al (2017) New agents that target senescent cells: the flavone, fisetin, and the BCL-X L inhibitors, A1331852 and A1155463. Aging (Albany NY) 9(3):955–963. https://doi.org/10.18632/aging.101202

Zhu Y, Liu X, Ding X, Wang F, Geng X (2019) Telomere and its role in the aging pathways: telomere shortening, cell senescence and mitochondria dysfunction. Biogerontology 20(1):1–16. https://doi.org/10.1007/s10522-018-9769-1

Zhu Y, Tchkonia T, Fuhrmann-Stroissnigg H, Dai HM, Ling YY, Stout MB et al (2016) Identification of a novel senolytic agent, navitoclax, targeting the Bcl-2 family of anti-apoptotic factors. Aging Cell 15(3):428–435. https://doi.org/10.1111/acel.12445

Zhu Y, Tchkonia T, Pirtskhalava T, Gower AC, Ding H, Giorgadze N et al (2015) The Achilles' heel of senescent cells: from transcriptome to senolytic drugs. Aging Cell 14(4):644–658. https://doi.org/10.1111/acel.12344

Part III

Stress Management Through Dietary Modifications and Nutritional Interventions

Prevention of Pro-inflammatory Markers By Medicinal Mushrooms Under Stressful Environment

21

Rekha Maithani, Ronika Yadav, Ratnika Sharma, and Rajkumar Tulsawani

Abstract

Hypoxia and inflammation are vicious cycles and often difficult to point out which signal activates first, and often these signals reinforce in a feedback loop. Inflammation significantly contributes to high-altitude sickness by increasing pro-inflammatory cytokines that participate in high-altitude associated pulmonary hypertension, pulmonary edema, and cerebral edema. Two medicinal mushrooms growing at high altitude, i.e., *Ganoderma lucidum* and *Cordyceps sinensis,* play vital role in diffusing cytokine storm observed upon high-altitude exposure, and these fungus exhibit strong anti-inflammatory properties.

Keywords

Inflammation · Anti-inflammatory · *Ganoderma lucidum* · *Cordyceps sinensis* · Immune system

21.1 Introduction

21.1.1 Inflammation

Generally, inflammation is body's defensive response to infections and tissue injury. However, emerging knowledge indicates that inflammation has a role beyond healing response, and chronic inflammation is associated with disorders and diseases

R. Maithani · R. Yadav · R. Sharma · R. Tulsawani (✉)
Defence Institute of Physiology and Allied Sciences (DIPAS), Defence Research and Development Organization (DRDO), Delhi, India
e-mail: rktulsawani@yahoo.com

R. Tulsawani, D. Vohora (eds.), *Adaptation under Stressful Environments through Biological Adjustments and Interventions*,
https://doi.org/10.1007/978-981-99-7652-2_21

such as arthritis, cardiovascular diseases, diabetes, asthma, cancer, and high-altitude sickness (Furman et al. 2019). The acute inflammatory response to infections, stressors, and trauma may be short term with high-grade magnitude which results in removal of stress response, healing, and repair of affected tissues; however, persistent low-grade inflammatory response leads to tissues damage or lifestyle-related metabolic disorders which may initiate collateral damage (Furman et al. 2019).

Cellular injury, hypoxia, infections, and oxidative stress lead to elevation of pro-inflammatory cytokines such as tumor necrosis factor-α (TNF-α) and interleukin IL1-β and interleukin-6 (IL-6) to initiate sequence of defensive events including recruitment of immune cells and vasodilation at the site of infection or cell injury. Pro-inflammatory cytokines are markers of inflammation (Dinarello 2000) that arises locally at the site of infection or injury and later in circulation, and such cytokine outburst is typically termed cytokine storm (Ferrara et al. 1993; Lange et al. 2012; Czepiel et al. 2014). The fanatic generation of cytokines has negative impact on host and causes clinical manifestations including edema, hemorrhage, organ failure, and even death. These pro-inflammatory cytokines are key regulators of acute inflammation (Chousterman et al. 2017) and chronic inflammation (Turner et al. 2014). Furman and authors, highlighted that acute inflammation may be represented by elevated bio-markers such as TNF-α, IL-6, IL1-β, and CRP, but chronic inflammation may be salient or with absence of classic or standard markers (Furman et al. 2019).

21.1.2 Hypobaric Hypoxia and Inflammation

Hypobaric hypoxia is unique scenario resulting in hypoxia and inflammation that coexist and stabilize hypoxia-inducible factor-1α (HIF-1α) and various inflammatory molecules including nuclear factor kappa B (Eltzschig and Carmeliet 2011; Shay and Simon 2012; Patel et al. 2017; El Alam et al. 2022). It is reported that at high-altitude inflammatory process is rapidly activated in human subjects and animal models (Maston et al. 2018; Hartmann et al. 2000) and results in lung and brain inflammation (fluid retention) with the surge in inflammatory cell infiltration (macrophages and neutrophils) and pro-cytokines (IL1-β, IL-6, TNF-α). The cytokine storm is counteracted with anti-inflammatory mediators such as IL-10 in the body which augment levels of cytokine inhibitors and block synthesis of pro-inflammatory cytokines and adhesion. However, if still the balance is tilted toward cytokine storm and body response fails, there is need for pharmacological interventions such as anti-inflammatory drugs/therapeutic regimes that must be able to block cytokine storm. The treatments of inflammation targeting pro-inflammatory cytokines such as TNF and IL1β have proved ineffective with antagonist of pro-inflammatory mediators. In this chapter, two medicinal mushrooms, *Ganoderma lucidum* and *Cordyceps sinensis*, growing under wild environment at high altitude with strong anti-inflammatory activity are discussed.

21.1.3 Inflammation and High-Altitude Illness

An increase in pro-inflammatory mediator(s) under hypoxia microenvironment is causative of high-altitude illnesses such as acute mountain sickness, high-altitude pulmonary, and cerebral edema (Pham et al. 2021).

At high altitude, reduced partial pressure generate free radicals in the body and disturbs redox homeostasis resulting in oxidative stress which in turn results in inflammation. Oxidative stress is casually connected with higher nuclear factor kappa B (NFk-β) and leads to pulmonary edema. HIF-1α produces TNF-α in macrophages which in turn stabilize and activate HIF-1α in feedback loop which is mediated by NFk-β (Jung et al. 2003; Zhou et al. 2003; Frede et al. 2006; Hayden and Ghosh 2008; Imtiyaz and Simon 2010; Pham et al. 2021). High-altitude-induced inflammation contributes to onset and progression of high-altitude pulmonary and cerebral edema, HAPE and HACE (Gallagher and Hackett 2004; Mehta et al. 2008; Luks et al. 2017).

It is documented that there is role of inflammation in the development of cerebral edema at high altitude (Bailey et al. 2009; Song et al. 2016; Zhou et al. 2017). It is well documented that pro-inflammatory cytokines are found elevated in the cortex of hypoxic brain. Moreover, LPS pretreatment induces inflammatory response in circulation that finally results in cerebral edema (Song et al. 2016; Zhou et al. 2017). Existence of hypoxia and inflammation together causes increased pro-inflammatory cytokines and increased blood vessel permeability resulting in vasogenic edema (Song et al. 2016; Zhou et al. 2017).

Pulmonary vasoconstrictions at high altitude cause pulmonary hypertension and along with persistent hypoxic stress result in vascular remodeling (increased muscularization) and aggravate pulmonary artery pressures (Groves et al. 1987; Wilkins et al. 2015; Arias-Stella and Saldana 1963; Hislop and Reid 1976). Persistent hypoxia produces a pro-inflammatory cytokine in pulmonary artery walls (Burke et al. 2009), and in response, resident fibroblasts, immune cells, and progenitor cells in the vascular adventitia release additional inflammatory mediators and growth factors contributing to increased muscularization (Stenmark et al. 2013). This inflammatory microenvironment recruits more inflammatory cells (Burke et al. 2009). Ascend to altitude activates pulmonary vasculature response including hypoxic pulmonary vasoconstriction, altered capillary permeability, and alveolar fluid accumulation (Talbot et al. 2005; Dunham-Snary et al. 2017; Brito et al. 2020; Swenson 2020; Sydykov et al. 2021). HAPE is initially a non-inflammatory in nature and hydrostatic pulmonary edema (Swenson et al. 2002), but later it becomes inflammatory response in progression and pathology of HAPE. It is observed that there is an increase in inflammatory mediators and chemokines such as IL1-β, IL-6, IL-8, TNF-α, and IL1RA in bronchoalveolar fluid at later stages of the HAPE (Kubo et al. 1998).

21.2 Anti-inflammatory Activity of *Ganoderma lucidum*

Despite there is ambiguity in the number of Ganoderma species present worldwide, various reports available show over 300 species distributed globally and about 15 species in India (Bhosle et al. 2010; Costa-Rezende et al. 2020; Petruzzello 2022). Among these species, *Ganoderma lucidum* has been widely researched and utilized in Asian Traditional Medicine, especially for treatment of many diseases and enhancing quality of life. *Ganoderma lucidum* (Fr.) Karst, which belongs to Ganodermataceae (Basidiomycetes), has worldwide distribution, in both tropical and temperate geographical regions, including Asia, Africa, Europe, and North and South America (Ziegenbein et al. 2006). *Ganoderma lucidum* grows on stumps of deciduous trees as parasite/saprophytes. It contains phytochemicals, which are attributed to various pharmacological effects against stressors, infections, injury and illness, including ganoderic acids, lactones, organic germanium, polysaccharides, rare minerals, and triterpenoids (Min et al. 1998). These phytoconstituents possessed by *Ganoderma lucidum* improve blood circulation, discard toxins, eliminate fatigue, positive energy balance, and augment immune system. There are numerous pharmacological activities reported in the *Ganoderma lucidum* including anti-inflammatory activity, immunological disorders, controlling infection, and anti-stress and anti-oxidative potential (Boh et al. 2007; Gao et al. 2004; Liu et al. 2009; Sliva 2006; Yeung et al. 2004; Sharma et al. 2019a, b). *Ganoderma lucidum* has been shown to be effective in renal and cerebral ischemia–reperfusion injury infectious diseases including bronchitis and viral hepatitis (Zhang et al. 2014; Geng et al. 2020). *Ganoderma lucidum* has been extensively used and reported for anti-inflammatory activity (Yoon et al. 2013), and this property is assigned due to presence of many bioactive molecules including polysaccharide and triterpenes (Dudhgaonkar et al. 2009, Joseph et al. 2011; Jiang et al. 2008; Zhang et al. 2018; Fu et al. 2019; Su et al. 2020; Koo et al. 2021; Cör et al. 2022; Wen et al. 2022). Hasnat and authors, documented *Ganoderma lucidum* grown on germinated brown rice prevented rise of pro-inflammatory cytokines including IL1-β, IL-6, TNF-α, NFk-β in macrophages treated with lipopolysaccharide suggesting its anti-inflammatory property (Hasnat et al. 2015).

Further, Sheena et al. (2003) reported that methanol and ethyl acetate extract has anti-inflammatory activity in mouse carrageenan and formalin acute and chronic inflammation model. Similarly, anti-inflammatory activity was reported in polysaccharides isolated from *Ganoderma lucidum* using similar carrageenan and formalin paw edema inflammation model, and activity was in doses ranging from 25 to 100 mg/kg body weight (Joesph et al. 2011). The strong anti-inflammatory activity of *Ganoderma lucidum* polysaccharides has been inhibition of higher pro-inflammatory cytokines levels of interleukin-1β (IL-1β) and interleukin-6 (IL-6) in response to induction by lipopolysaccharide (Wen et al. 2022). Dudhgaonkar and authors (2009) demonstrated that triterpenes extracted from *Ganoderma lucidum* suppressed the surge of inflammatory cytokines such as COX 2, IL-6, and TNF-α in murine RAW264.7 cells grown in lipopolysaccharide micro-environment. Many other studies have demonstrated anti-inflammatory activity of polysaccharides triterpenes and triterpenoids isolated from *Ganoderma lucidum* (Cai et al. 2017; Su et al. 2020; Koo et al. 2021; Cör et al. 2022).

Tulsawani and group documented that anti-inflammatory activity of *Ganoderma lucidum* against high-altitude extreme stressor, i.e., hypobaric hypoxia and adaptogenic potential (Sharma et al. 2019a) reported that exposure to hypobaric hypoxia in experimental animals resulted in higher pro-inflammatory cytokines such as IL-6, NFk-β, and TNF-α in circulation, and the surge in cytokines can be prevented with treatment of *Ganoderma lucidum* indicating its anti-inflammatory role. Aqueous extract of *Ganoderma lucidum* has been reported to deactivate hypobaric hypoxia-induced overexpression NF-κB which is casually associated with inflammatory response by stabilizing HIF-1α/NF-κB axis, thereby reducing neuronal inflammation and vasogenic cerebral edema (Sharma and Tulsawani 2020). In another study, rat muscle exposed to extreme hypobaric hypoxia and supplemented with aqueous extract of *Ganoderma lucidum* prevented overproduction of pro-inflammatory cytokines such as IL-6 and TNF-α (Jatwani and Tulsawani 2021). Overall, available literature supports anti-inflammatory potential of *Ganoderma lucidum* by regulating cytokine cycle.

21.3 Anti-inflammatory Activity of *Cordyceps sinensis*

Medicinal mushrooms are traditionally utilized in Asia as alternative therapies to cure various ailments including inflammatory diseases and infections. One of these medicinal mushrooms, *Cordyceps sinensis,* commonly referred to as caterpillar fungus, belongs to family Ophiocordycipitaceae and grows at high-altitude zones and widely used in Asian Traditional Medicinal System for enhancing vitality, longevity, and endurance (Zhu et al. 1998). There are numerous reports citing that the Genus Cordyceps is widely distributed and represented 400–750 species worldwide (Sung et al. 2007; Olatunji et al. 2018; Das et al. 2021), but *Cordyceps sinensis* has been widely researched. The health benefits of *Cordyceps sinensis* include anti-inflammatory, anti-oxidant, anti-stress, and immune-modulatory properties (Kuo et al. 1996, 2001; Wasser and Weis 1999; Li et al. 2001; Shahed et al. 2001; Won and Park 2005; Kim et al. 2006; Liu et al. 2007; Belwal et al. 2019; Chen et al. 2020; Das et al. 2021), and it is believed that observed pharmacological effects are due to the presence of wide array of bioactive compounds in *Cordyceps sinensis* such as cordycepin, polysaccharides, phenols, flavonoids, nucleoside, sterol, protein, amino acid, and polypeptide (Wasser and Weis 1999; Kuo et al. 2001; Li et al. 2001; Liu et al. 2015).

Cordyceps sinensis and isolated bioactiFve components possess strong anti-inflammatory activity (Kuo et al. 1996; Shahed et al. 2001; Won and Park 2005; Kim et al. 2006; Liu et al. 2007; Zhou et al. 2008; Jeong et al. 2010; Zhang et al. 2011; Yang et al. 2011; Qian et al. 2012; Chen et al. 2020; Das et al. 2021). It is reported that anti-inflammatory activity of *Cordyceps sinensis* is mediated via regulation pro-inflammatory markers such as IL-1B, IL-6, TNF-α, cyclooxygenase 2, and NF-κB and has been scientifically proved in various experimental models and is well documented (Kuo et al. 1996; Shahed et al. 2001; Won and Park 2005; Kim et al. 2006; Liu et al. 2007; Wang et al. 2018; Chen et al. 2020; Das et al. 2021). Besides, polysaccharides present in *Cordyceps sinensis* are biologically active and linked to numerous pharmacological effects including anti-inflammatory activity (Zhang et al. 2005; Kim 2010; Huang et al. 2018; Chen et al. 2010; Wang et al. 2014; Chang et al. 2015). *Cordyceps sinensis* polysaccharides are complex monosaccharide composition with galactose, glucose, and mannose and represent anti-inflammatory activity (Yan et al. 2014; Yuan et al. 2022; Li et al. 2021). The inflammatory response can be prevented with the treatment of *C. sinensis* polysaccharides through inhibition of pro-inflammatory cytokines TNF-α, IL-1β, NF-κB, and cyclooxygenase 2 using inflammation inducers including lipopolysaccharide and cyclophosphamide (Wang et al. 2015; Sheu et al. 2018; Li et al. 2020; Fan et al. 2018).

Cordycepin bioactive compound of *Cordyceps sinensis* suppressed inflammatory markers including TNF-α, cyclooxygenase 2, and NF-κB translocation in macrophages suggesting its utility in inflammation-associated illness (Kim et al. 2006; Jeong et al. 2010; Ashraf et al. 2020). Moreover, Cordycepin and *Cordyceps sinensis* regulate the functions of human immune cells by modulating cytokines (Kuo et al. 1996; Zhou et al. 2008; Zhang et al. 2011).

Many other bioactive compounds present in *Cordyceps sinensis* such as cordymine (Qian et al. 2012; Chen et al. 2013), adenosine (Shashidhar et al. 2013), cerevisterol (Wang et al. 2019), and d-mannitol (Hu et al. 2015) regulate pro-inflammatory cytokine including IL-1β, TNF-α, cox 2, IL-1β, and NF-kβ (Yang et al. 2011; Qian et al. 2012).

Moreover, *Cordyceps sinensis* is reported to possess anti-inflammatory activity under stressful environment. It is documented to prevent inflammatory response in experimental animals subjected to ischemia–reperfusion stress and modulate cytokines such as adhesion molecule (ICAM-1), cyclooxygenase 2, interleukin-1β, nuclear factor kappa B, and tumor necrosis factor-α in brain (Liu et al. 2011). It is well reported that *Cordyceps sinensis* induce tolerance against hypoxia in human lung epithelial cells by diffusing pro-inflammatory cytokine surge (Singh et al. 2013). Further, *Cordyceps sinensis* aqueous extract prevents the surge in pro-inflammatory cytokines to exert adaptive response under simulated hypobaric hypoxia (Sharma and Tulsawani 2022). Overall, *Cordyceps sinensis* regulate pro-inflammatory cytokines to exert strong anti-inflammatory response and have adaptive role under stressful environment.

21.4 Conclusions

Ganoderma lucidum and *Cordyceps sinensis* are two important medicinal mushrooms enhancing vitality, longevity, and endurance and are rich source of bioactive compounds and energy-providing metabolites. These medicinal mushrooms deactivate cytokine storm by preventing pro-inflammatory cytokines by virtue of its anti-inflammatory nature and even depict anti-inflammatory activity under extreme stressful environment.

References

Arias-Stella J, Saldana M (1963) The terminal portion of the pulmonary arterial tree in people native to high altitudes. Circulation 28(5):915–925. https://doi.org/10.1161/01.CIR.28.5.915

Ashraf SA, Elkhalifa AEO, Siddiqui AJ, Patel M, Awadelkareem AM, Snoussi M et al (2020) Cordycepin for health and wellbeing: a potent bioactive metabolite of an Entomopathogenic cordyceps medicinal fungus and its nutraceutical and therapeutic potential. Molecules 25(12): 2735. https://doi.org/10.3390/molecules25122735

Bailey DM, Bärtsch P, Knauth M, Baumgartner RW (2009) Emerging concepts in acute mountain sickness and high-altitude cerebral edema: from the molecular to the morphological. Cell Mol Life Sci 66(22):3583–3594. https://doi.org/10.1007/s00018-009-0145-9

Belwal T, Bhatt ID, Kashyap D, Sak K, Tuli HS, Pathak R et al (2019) Ophio *Cordyceps sinensis*. In: Nonvitamin and nonmineral nutritional supplements. Academic Press, pp 527–537

Bhosle S, Ranadive K, Bapat G, Garad S, Deshpande G, Vaidya J (2010) Taxonomy and diversity of Ganoderma from the Western parts of Maharashtra (India). Mycosphere 1:249–262

Boh B, Berovic M, Zhang J, Zhi-Bin L (2007) Ganoderma lucidum and its pharmaceutically active compounds. Biotechnol Annu Rev 13:265–301. https://doi.org/10.1016/S1387-2656(07)13010-6. PMID: 17875480

Brito J, Siques P, Pena E (2020) Long-term chronic intermittent hypoxia: a particular form of chronic high-altitude pulmonary hypertension. Pulm Circ 10(1 Suppl):5–12. https://doi.org/10.1177/2045894020934625. PMID: 33110494; PMCID: PMC7557688

Burke DL, Frid MG, Kunrath CL, Karoor V, Anwar A, Wagner BD et al (2009) Sustained hypoxia promotes the development of a pulmonary artery-specific chronic inflammatory microenvironment. Am J Physiol Lung Cell Mol Physiol 297(2):L238–L250. https://doi.org/10.1152/ajplung.90591.2008

Cai Q, Li Y, Pei G (2017) Polysaccharides from *Ganoderma lucidum* attenuate microglia-mediated neuroinflammation and modulate microglial phagocytosis and behavioural response. J Neuroinflammation 14(1):63. https://doi.org/10.1186/s12974-017-0839-0

Chang Y, Hsu WH, Lu WJ, Jayakumar T, Liao JC, Lin MJ et al (2015) Inhibitory mechanisms of CME-1, a novel polysaccharide from the mycelia of *Cordyceps sinensis*, in platelet activation. Curr Pharm Biotechnol 16(5):451–461. https://doi.org/10.2174/1389201016666150303152237

Chen J, Chan WM, Leung HY, Leong PK, Yan CTM, Ko KM (2020) Anti-inflammatory effects of a *Cordyceps sinensis* mycelium culture extract (Cs-4) on rodent models of allergic rhinitis and asthma. Molecules 25(18):4051. https://doi.org/10.3390/molecules25184051

Chen PX, Wang S, Nie S, Marcone M (2013) Properties of *Cordyceps sinensis*: a review. J Funct Foods 5(2):550–569. https://doi.org/10.1016/j.jff.2013.01.034

Chen W, Zhang W, Shen W, Wang K (2010) Effects of the acid polysaccharide fraction isolated from a cultivated *Cordyceps sinensis* on macrophages in vitro. Cell Immunol 262(1):69–74. https://doi.org/10.1016/j.cellimm.2010.01.001

Chousterman BG, Swirski FK, Weber GF (2017) Cytokine storm and sepsis disease pathogenesis. Semin Immunopathol 39(5):517–528. https://doi.org/10.1007/s00281-017-0639-8; Abstract| CrossRef Full Text|Google Scholar

Cör AD, Knez Ž, Knez MM (2022) Antioxidant, antibacterial, antitumor, antifungal, antiviral, anti-inflammatory, and nevro-protective activity of *Ganoderma lucidum*: an overview. Front Pharmacol 13:934982. https://doi.org/10.3389/fphar.2022.934982, PMID 35935849, PMCID PMC9353308

Costa-Rezende DH, Robleo GL, Drechsler-Santos ER, Glen M, Gates G, de Madrignac BR, et al (2020) Taxonomy and phylogeny of polypores with ganodermatoid

Czepiel J, Biesiada G, Brzozowski T, Ptak-Belowska A, Perucki W, Birczynska M et al (2014) The role of local and systemic cytokines in patients infected with Clostridium difficile. J Physiol Pharmacol 65(5):695–703

Das G, Shin HS, Leyva-Gómez G, Prado-Audelo ML, Cortes H, Singh YD et al (2021) Cordyceps spp.: a review on its immune-stimulatory and other biological potentials. Front Pharmacol 11: 2250

Dinarello CA (2000) Proinflammatory cytokines. Chest 118(2):503–508. https://doi.org/10.1378/chest.118.2.503

Dudhgaonkar S, Thyagarajan A, Sliva D (2009) Suppression of the inflammatory response by triterpenes isolated from the mushroom *Ganoderma lucidum*. Int Immunopharmacol 9(11): 1272–1280. https://doi.org/10.1016/j.intimp.2009.07.011

Dunham-Snary KJ, Wu D, Sykes EA, Thakrar A, Parlow LRG, Mewburn JD, Parlow JL, Archer SL (2016) Hypoxic pulmonary vasoconstriction: from molecular mechanisms to medicine. Chest 151(1):181–192. https://doi.org/10.1016/j.chest.2016.09.001. Epub 2016 Sep 16. PMID: 27645688; PMCID: PMC5310129

El Alam S, Pena E, Aguilera D, Siques P, Brito J (2022) Inflammation in pulmonary hypertension and edema induced by hypobaric hypoxia exposure. Int J Mol Sci 23(20):12656. https://doi.org/10.3390/ijms232012656, PMID 36293512, PMCID PMC9604159

Eltzschig HK, Carmeliet P (2011) Hypoxia and inflammation. N Engl J Med 364(7):656–665. https://doi.org/10.1056/NEJMra0910283

Fan S, Huang X, Wang S, Li C, Zhang Z, Xie M et al (2018) Combinatorial usage of fungal polysaccharides from *Cordyceps sinensis* and Ganoderma atrum ameliorate drug-induced liver injury in mice. Food Chem Toxicol 119:66–72. https://doi.org/10.1016/j.fct.2018.05.027

Ferrara JL, Abhyankar S, Gilliland DG (1993) Cytokine storm of graft-versus-host disease: a critical effector role for interleukin-1. Transplant Proc 25(1 Pt 2):1216–1217

Frede S, Stockmann C, Freitag P, Fandrey J (2006) Bacterial lipopolysaccharide induces HIF-1 activation in human monocytes via p44/42 MAPK and NF-κB. Biochem J 396(3):517–527. https://doi.org/10.1042/BJ20051839

Fu Y, Shi L, Ding K (2019) Structure elucidation and anti-tumor activity in vivo of a polysaccharide from spores of *Ganoderma lucidum* (Fr.) karst. Int J Biol Macromol 141:693–699. https://doi.org/10.1016/j.ijbiomac.2019.09.046

Furman D, Campisi J, Verdin E, Carrera-Bastos P, Targ S, Franceschi C et al (2019) Chronic inflammation in the etiology of disease across the life span. Nat Med 25(12):1822–1832. https://doi.org/10.1038/s41591-019-0675-0, PMID 31806905, PMCID PMC7147972

Gallagher SA, Hackett PH (2004) High-altitude illness. Emerg Med Clin North Am 22(2):329–355. https://doi.org/10.1016/j.emc.2004.02.001

Gao Y, Chan E, Zhou S (2004) Immunomodulating activities of Ganoderma a mushroom with medicinal properties. Food Rev Int 20(2):123–161. https://doi.org/10.1081/FRI-120037158

Geng X, Zhong D, Su L, Lin Z, Yang B (2020) Preventive and therapeutic effect of *Ganoderma lucidum* on kidney injuries and diseases. Adv Pharmacol 87:257–276. https://doi.org/10.1016/bs.apha.2019.10.003

Groves BM, Reeves JT, Sutton JR, Wagner PD, Cymerman A, Malconian MK et al (1987) Operation EVEREST II: elevated high-altitude pulmonary resistance unresponsive to oxygen. J Appl Physiol (1985) 63(2):521–530. https://doi.org/10.1152/jappl.1987.63.2.521

Hartmann G, Tschöp M, Fischer R, Bidlingmaier C, Riepl R, Tschöp K et al (2000) High altitude increases circulating interleukin-6, interleukin-1 receptor antagonist and C-reactive protein. Cytokine 12(3):246–252. https://doi.org/10.1006/cyto.1999.0533, PMID 10704252

Hasnat MA, Pervin M, Cha KM, Kim SK, Lim BO (2015) Anti-inflammatory activity on mice of extract of *Ganoderma lucidum* grown on rice via modulation of MAPK and NF-κB pathways. Phytochemistry 114:125–136. https://doi.org/10.1016/j.phytochem.2014.10.019

Hayden MS, Ghosh S (2008) Shared principles in NF-κB signaling. Cell 132(3):344–362. https://doi.org/10.1016/j.cell.2008.01.020

Hislop A, Reid L (1976) New findings in pulmonary arteries of rats with hypoxia induced pulmonary hypertension. Br J Exp Pathol 57(5):542–554

Hu H, Xiao L, Zheng B, Wei X, Ellis A, Liu YM (2015) Identification of chemical markers in *Cordyceps sinensis* by HPLC-MS/MS. Anal Bioanal Chem 407(26):8059–8066. https://doi.org/10.1007/s00216-015-8978-6

Huang D, Meran S, Nie SP, Midgley A, Junqiao W, Cui SW et al (2018) *Cordyceps sinensis*: anti-fibrotic and inflammatory effects of a cultured polysaccharide extract. Bioact Carbohydr Diet Fibre 14:2–8. https://doi.org/10.1016/j.bcdf.2017.07.012, ISSN 2212-6198

Imtiyaz HZ, Simon MC (2010) Hypoxia-inducible factors as essential regulators of inflammation. Curr Top Microbiol Immunol 345:105–120. https://doi.org/10.1007/82_2010_74, PMID 20517715, PMCID PMC3144567

Jatwani A, Tulsawani R (2021) *Ganoderma lucidum* induces myogenesis markers to avert damage to skeletal muscles in rats exposed to hypobaric hypoxia. High Alt Med Biol. https://doi.org/10.1089/ham.2020.0172

Jeong JW, Jin CY, Kim GY, Lee JD, Park C, Kim GD, Kim WJ, Jung WK, Seo SK, Choi IW, Choi YH (2010) Anti-inflammatory effects of cordycepin via suppression of inflammatory mediators in BV2 microglial cells. Int Immunopharmacol 10(12):1580–1586. https://doi.org/10.1016/j.intimp.2010.09.011. Epub 2010 Oct 14. PMID: 20937401

Jiang J, Grieb B, Thyagarajan A, Sliva D (2008) Ganoderic acids suppress growth and invasive behavior of breast cancer cells by modulating AP-1 and NF-κB signaling. Int J Mol Med 21(5):577–584

Joseph S, Sabulal B, George V, Antony KR, Janardhanan KK (2011) Antitumor and anti-inflammatory activities of polysaccharides isolated from Ganoderma lucidum. Acta Pharma 61(3):335–342. https://doi.org/10.2478/v10007-011-0030-6. PMID: 21945912

Jung Y, Isaacs JS, Lee S, Trepel J, Liu ZG, Neckers L (2003) Hypoxia-inducible factor induction by tumour necrosis factor in normoxic cells requires receptor-interacting protein-dependent nuclear factor κB activation. Biochem J 370(3):1011–1017. https://doi.org/10.1042/BJ20021279

Kim SD (2010) Isolation, structure and cholesterol esterase inhibitory activity of a polysaccharide, PS-A, from *Cordyceps sinensis*. J Korean Soc Appl Biol Chem 53(6):784–789. https://doi.org/10.3839/jksabc.2010.118

Kim HG, Shrestha B, Lim SY, Yoon DH, Chang WC, Shin DJ et al (2006) Cordycepin inhibits lipopolysaccharide-induced inflammation by the suppression of NF-kappaB through Akt and p38 inhibition in RAW 264.7 macrophage cells. Eur J Pharmacol 545(2–3):192–199. https://doi.org/10.1016/j.ejphar.2006.06.047

Koo MH, Chae HJ, Lee JH, Suh SS, Youn UJ (2021) Antiinflammatory lanostane triterpenoids from *Ganoderma lucidum*. Nat Prod Res 35(22):4295–4302. https://doi.org/10.1080/14786419.2019.1705815

Kubo K, Hanaoka M, Hayano T, Miyahara T, Hachiya T, Hayasaka M et al (1998) Inflammatory cytokines in BAL fluid and pulmonary hemodynamics in high-altitude pulmonary edema. Respir Physiol 111(3):301–310. https://doi.org/10.1016/s0034-5687(98)00006-1

Kuo YC, Tsai WJ, Shiao MS, Chen CF, Lin CY (1996) *Cordyceps sinensis* as an immunomodulatory agent. Am J Chin Med 24(2):111–125. https://doi.org/10.1142/S0192415X96000165

Kuo YC, Tsai WJ, Wang JY, Chang SC, Lin CY, Shiao MS (2001) Regulation of bronchoalveolar lavage fluids cell function by the immunomodulatory agents from Cordyceps sinensis. Life Sci 68(9):1067–1082. https://doi.org/10.1016/s0024-3205(00)01011-0. PMID: 11212870

Lange M, Nakano Y, Traber DL, Hamahata A, Traber LD, Enkhbaatar P (2012) Time course of the inflammatory and oxidative stress response to pulmonary infection in mice. Exp Lung Res 38(3):157–163. https://doi.org/10.3109/01902148.2012.663453

Li SP, Li P, Dong TT, Tsim KW (2011) Anti-oxidation activity of different types of natural Cordyceps sinensis and cultured Cordyceps mycelia. Phytomedicine 8(3):207–212. https://doi.org/10.1078/0944-7113-00030. PMID: 11417914

Li LQ, Song AX, Wong WT, Wu JY (2021) Isolation and assessment of a highly-active anti-inflammatory exopolysaccharide from mycelial fermentation of a medicinal fungus Cs-HK1. Int

J Mol Sci 22(5):2450. https://doi.org/10.3390/ijms22052450, PMID 33671052, PMCID PMC7957654
Li LQ, Song AX, Yin JY, Siu KC, Wong WT, Wu JY (2020) Anti-inflammation activity of exopolysaccharides produced by a medicinal fungus *Cordyceps sinensis* Cs-HK1 in cell and animal models. Int J Biol Macromol 149:1042–1050. https://doi.org/10.1016/j.ijbiomac.2020.02.022
Liu CQ, Zheng AJ, Li DJ, Cao QS (2007) Improvement of immunity in mice by ultra-fine pulverization of Cordyceps militaris. Jiangsu J Agric Sci 1:63–66
Liu YW, Gao JL, Guan J, Qian ZM, Feng K, Li SP (2009) Evaluation of antiproliferative activities and action mechanisms of extracts from two species of Ganoderma on tumor cell lines. J Agric Food Chem 57(8):3087–3093. https://doi.org/10.1021/jf900011f. PMID: 19368349
Liu Z, Li P, Zhao D, Tang H, Guo J (2011) Anti-inflammation effects of *Cordyceps sinensis* mycelium in focal cerebral ischemic injury rats. Inflammation 34(6):639–644. https://doi.org/10.1007/s10753-010-9273-5
Liu Y, Wang J, Wang W, Zhang H, Zhang X, Han C (2015) The chemical constituents and pharmacological actions of *Cordyceps sinensis*. Evid Based Complement Alternat Med 2015: 575063. https://doi.org/10.1155/2015/575063, PMCID. PMC4415478, 1
Luks AM, Swenson ER, Bärtsch P (2017) Acute high-altitude sickness. Eur Respir Rev 26(143): 160096. https://doi.org/10.1183/16000617.0096-2016
Maston LD, Jones DT, Giermakowska W, Resta TC, Ramiro-Diaz J, Howard TA et al (2018) Interleukin-6 trans-signaling contributes to chronic hypoxia-induced pulmonary hypertension. Pulm Circ 8(3):2045894018780734. https://doi.org/10.1177/2045894018780734, (PMC Free article). PMID 29767573
Mehta SR, Chawla A, Kashyap AS (2008) Acute mountain sickness, high altitude cerebral oedema, high altitude pulmonary oedema: the current concepts. Med J Armed Forces India 64(2): 149–153. https://doi.org/10.1016/S0377-1237(08)80062-7
Min BS, Nakamura N, Miyashiro H, Bae KW, Hattori M (1998) Triterpenes from the spores of Ganoderma lucidum and their inhibitory activity against HIV-1 protease. Chem Pharm Bull (Tokyo) 46(10):1607–1612. https://doi.org/10.1248/cpb.46.1607. PMID: 9810695
Olatunji OJ, Tang J, Tola A, Auberon F, Oluwaniyi O, Ouyang Z (2018) The genus Cordyceps: an extensive review of its traditional uses, phytochemistry and pharmacology. Fitoterapia 129: 293–316. https://doi.org/10.1016/j.fitote.2018.05.010
Patel H, Zaghloul N, Lin K, Liu SF, Miller EJ, Ahmed M (2017) Hypoxia-induced activation of specific members of the NF-kB family and its relevance to pulmonary vascular remodeling. Int J Biochem Cell Biol 92:141–147. https://doi.org/10.1016/j.biocel.2017.09.022, PMID 28987523
Petruzzello M (2022) Ganoderma. In: Encyclopaedia Britannica; [cited Jun 29 2023]. https://www.britannica.com/science/Ganoderma
Pham K, Parikh K, Heinrich EC (2021) Hypoxia and inflammation: insights from high-altitude physiology. Front Physiol 12:676782. https://doi.org/10.3389/fphys.2021.676782, PMID 34122145, PMCID PMC8188852
Qian GM, Pan GF, Guo JY (2012) Anti-inflammatory and antinociceptive effects of cordymin, a peptide purified from the medicinal mushroom Cordyceps sinensis. Nat Prod Res 26(24):2358–2362. https://doi.org/10.1080/14786419.2012.658800. Epub 2012 Feb 21. PMID: 22348255
Shahed AR, Kim SI, Shoskes DA (2001) Down-regulation of apoptotic and inflammatory genes by *Cordyceps sinensis* extract in rat kidney following ischemia/reperfusion. Transplant Proc 33(6): 2986–2987. https://doi.org/10.1016/s0041-1345(01)02282-5
Sharma C, Bhardwaj N, Sharma A, Tuli HS, Batra P, Beniwal V et al (2019b) Bioactive metabolites of *Ganoderma lucidum*: factors, mechanism and broad spectrum therapeutic potential. J Herb Med 17:100268
Sharma P, Tulsawani R (2020) *Ganoderma lucidum* aqueous extract prevents hypobaric hypoxia induced memory deficit by modulating neurotransmission, neuroplasticity and maintaining redox homeostasis. Sci Rep 10(1):8944. https://doi.org/10.1038/s41598-020-65812-5, PMID 32488040, PMCID PMC7265456

Sharma P, Tulsawani R (2022) Efficacy of aqueous extract of Chinese caterpillar mushroom Ophio *Cordyceps sinensis* (Ascomycota) against simulated altitude stress and subacute toxicity studies. Int J Med Mushrooms 24(11):21–34. https://doi.org/10.1615/IntJMedMushrooms.2022045194
Sharma P, Tulsawani R, Agrawal U (2019a) Pharmacological effects of *Ganoderma lucidum* extract against high-altitude stressors and its subchronic toxicity assessment. J Food Biochem 43(12):e13081. https://doi.org/10.1111/jfbc.13081
Shashidhar MG, Giridhar P, Udaya Sankar K, Manohar B (2013) Bioactive principles from *Cordyceps sinensis*: a potent food supplement–a review. J Funct Foods 5(3):1013–1030. https://doi.org/10.1016/j.jff.2013.04.018
Shay JE, Simon MC (2012) Hypoxia-inducible factors: crosstalk between inflammation and metabolism. Semin Cell Dev Biol 23(4):389–394. https://doi.org/10.1016/j.semcdb.2012.04.004, PMID 22525300
Sheena N, Ajith TA, Janardhanan KK (2003) Anti-inflammatory and anti[1]nociceptive activities of *Ganoderma lucidum* occurring in South India. Pharm Biol 41(4):301–304. https://doi.org/10.1076/phbi.41.4.301.15677
Sheu JR, Chen ZC, Hsu MJ, Wang SH, Jung KW, Wu WF et al (2018) CME-1, a novel polysaccharide, suppresses iNOS expression in lipopolysaccharide-stimulated macrophages through ceramide-initiated protein phosphatase 2A activation. J Cell Mol Med 22(2):999–1013. https://doi.org/10.1111/jcmm.13424, PMID 29214724, PMCID PMC5783865
Sliva D (2006) Ganoderma lucidum in cancer research. Leuk Res 30(7):767–768. https://doi.org/10.1016/j.leukres.2005.12.015
Singh M, Tulsawani R, Koganti P, Chauhan A, Manickam M, Misra K (2013) *Cordyceps sinensis* increases hypoxia tolerance by inducing heme oxygenase-1 and metallothionein via Nrf2 activation in human lung epithelial cells. Biomed Res Int 2013:569206. https://doi.org/10.1155/2013/569206, PMID 24063008, PMCID PMC3770031, 1
Song TT, Bi YH, Gao YQ, Huang R, Hao K, Xu G et al (2016) Systemic pro-inflammatory response facilitates the development of cerebral edema during short hypoxia. J Neuroinflammation 13(1):63. https://doi.org/10.1186/s12974-016-0528-4
Stenmark KR, Yeager ME, El Kasmi KC, Nozik-Grayck E, Gerasimovskaya EV, Li M et al (2013) The adventitia: essential regulator of vascular wall structure and function. Annu Rev Physiol 75:23–47. https://doi.org/10.1146/annurev-physiol-030212-183802
Su HG, Peng XR, Shi QQ, Huang YJ, Zhou L, Qiu MH (2020) Lanostane triterpenoids with anti-inflammatory activities from *Ganoderma lucidum*. Phytochemistry 173:112256. https://doi.org/10.1016/j.phytochem.2019.112256
Sung GH, Hywel-Jones NL, Sung JM, Luangsa-Ard JJ, Shrestha B, Spatafora JW (2007) Phylogenetic classification of Cordyceps and the clavicipitaceous fungi. Stud Mycol 57:5–59. https://doi.org/10.3114/sim.2007.57.01, PMID 18490993, PMCID PMC2104736
Swenson ER (2020) Early hours in the development of high-altitude pulmonary edema: time course and mechanisms. J Appl Physiol 128(6):1539–1546. https://doi.org/10.1152/japplphysiol.00824.2019. Epub 2020 Mar 26. PMID: 32213112
Swenson ER, Maggiorini M, Mongovin S, Gibbs JS, Greve I, Mairbäurl H, Bärtsch P (2002) Pathogenesis of high-altitude pulmonary edema: inflammation is not an etiologic factor. JAMA 287(17):2228–2235. https://doi.org/10.1001/jama.287.17.2228. Erratum in: JAMA 2002 Sep 4;288(9):1064. PMID: 11980523
Sydykov A, Mamazhakypov A, Maripov A, Kosanovic D, Weissmann N, Ghofrani HA, Sarybaev AS, Schermuly RT (2021) Pulmonary hypertension in acute and chronic high altitude maladaptation disorders. Int J Environ Res Public Health 18(4):1692. https://doi.org/10.3390/ijerph18041692. PMID: 33578749; PMCID: PMC7916528
Talbot NP, Balanos GM, Dorrington KL, Robbins PA (2004) Two temporal components within the human pulmonary vascular response to approximately 2 h of isocapnic hypoxia. J Appl Physiol 98(3):1125–1139. https://doi.org/10.1152/japplphysiol.00903.2004. Epub 2004 Nov 12. PMID: 15542574

Turner MD, Nedjai B, Hurst T, Pennington DJ (2014) Cytokines and chemokines: at the crossroads of cell signalling and inflammatory disease. Biochim Biophys Acta 1843(11):2563–2582. https://doi.org/10.1016/j.bbamcr.2014.05.014

Wang C, Hou XX, Rui HL, Li LJ, Zhao J, Yang M et al (2018) Artificially cultivated Ophio *Cordyceps sinensis* alleviates diabetic nephropathy and its podocyte injury via inhibiting P2X7R expression and NLRP3 inflammasome activation. J Diabetes Res 2018:1390418. https://doi.org/10.1155/2018/1390418

Wang M, Kornsakulkarn J, Srichomthong K, Feng T, Liu JK, Isaka M et al (2019) Antimicrobial anthraquinones from cultures of the ant pathogenic fungus Cordyceps morakotii BCC 56811. J Antibiot (Tokyo) 72(3):141–147. https://doi.org/10.1038/s41429-018-0135-y

Wang Y, Liu D, Zhao H, Jiang H, Luo C, Wang M et al (2014) *Cordyceps sinensis* polysaccharide CPS-2 protects human mesangial cells from PDGF-BB-induced proliferation through the PDGF/ERK and TGF-β1/Smad pathways. Mol Cell Endocrinol 382(2):979–988. https://doi.org/10.1016/j.mce.2013.11.018

Wang Y, Wang Y, Liu D, Wang W, Zhao H, Wang M et al (2015) *Cordyceps sinensis* polysaccharide inhibits PDGF-BB-induced inflammation and ROS production in human mesangial cells. Carbohydr Polym 125:135–145. https://doi.org/10.1016/j.carbpol.2015.02.012

Wasser SP, Weis AL (1999) Medicinal properties of substances occurring in higher basidiomycetes mushrooms: current perspectives (review). Int J Med Mushrooms 1(1):31–62. https://doi.org/10.1615/IntJMedMushrooms.v1.i1.30

Wen L, Sheng Z, Wang J, Jiang Y, Yang B (2022) Structure of water-soluble polysaccharides in spore of *Ganoderma lucidum* and their anti-inflammatory activity. Food Chem 373(A):131374. https://doi.org/10.1016/j.foodchem.2021.131374

Wilkins MR, Ghofrani HA, Weissmann N, Aldashev A, Zhao L (2015) Pathophysiology and treatment of high-altitude pulmonary vascular disease. Circulation 131(6):582–590. https://doi.org/10.1161/CIRCULATIONAHA.114.006977

Won SY, Park EH (2005) Anti-inflammatory and related pharmacological activities of cultured mycelia and fruiting bodies of Cordyceps militaris. J Ethnopharmacol 96(3):555–561. https://doi.org/10.1016/j.jep.2004.10.009

Yan JK, Wang WQ, Wu JY (2014) Recent advances in Cordyceps sinensis polysaccharides: mycelial fermentation, isolation, structure, and bioactivities: a review. J Funct Foods 6:33–47. https://doi.org/10.1016/j.jff.2013.11.024, PMID 32362940, PMCID PMC7185505

Yang ML, Kuo PC, Hwang TL, Wu TS (2011) Anti-inflammatory principles from Cordyceps sinensis. J Nat Prod 74:1996–2000. https://doi.org/10.1021/np100902f

Yeung HW, Qing-Yi L, Qifeng Z, Go VLW (2004) Chemical and biochemical basis of the potential anti-tumor properties of Ganoderma lucidum. Curr Top Nutr Res 2:67–77

Yoon HM, Jang KJ, Han MS, Jeong JW, Kim GY, Lee JH et al (2013) *Ganoderma lucidum* ethanol extract inhibits the inflammatory response by suppressing the NF-κB and toll-like receptor pathways in lipopolysaccharide-stimulated BV2 microglial cells. Exp Ther Med 5(3):957–963. https://doi.org/10.3892/etm.2013.895, PMID 23408713, PMCID PMC3570243

Yuan Q, Xie F, Tan J, Yuan Y, Mei H, Zheng Y et al (2022) Extraction, structure and pharmacological effects of the polysaccharides from *Cordyceps sinensis*: a review. J Funct Foods 89: 104909. https://doi.org/10.1016/j.jff.2021.104909

Zhang J, Yu Y, Zhang Z, Ding Y, Dai X, Li Y (2011) Effect of polysaccharide from cultured Cordyceps sinensis on immune function and anti-oxidation activity of mice exposed to 60Co. Int Immunopharmacol 11:2251–2257. https://doi.org/10.1016/j.intimp.2011.09.019

Zhang W, Tao J, Yang X, Yang Z, Zhang L, Liu H, Wu K, Wu J (2014) Antiviral effects of two Ganoderma lucidum triterpenoids against enterovirus 71 infection. Biochem Biophys Res Commun 449(3):307–312. https://doi.org/10.1016/j.bbrc.2014.05.019. Epub 2014 May 15. PMID: 24845570

Zhang K, Liu Y, Zhao X, Tang Q, Dernedde J, Zhang J et al (2018) Anti-inflammatory properties of GLPss58, a sulfated polysaccharide from *Ganoderma lucidum*. Int J Biol Macromol 107 (A):486–493. https://doi.org/10.1016/j.ijbiomac.2017.09.015

Zhang W, Yang J, Chen J, Hou Y, Han X (2005) Immunomodulatory and antitumour effects of an exopolysaccharide fraction from cultivated *Cordyceps sinensis* (Chinese caterpillar fungus) on tumour-bearing mice. Biotechnol Appl Biochem 42(1):9–15. https://doi.org/10.1042/BA20040183

Zhou Y, Huang X, Zhao T, Qiao M, Zhao X, Zhao M et al (2017) Hypoxia augments LPS-induced inflammation and triggers high altitude cerebral edema in mice. Brain Behav Immun 64:266–275. https://doi.org/10.1016/j.bbi.2017.04.013

Zhou J, Schmid T, Brüne B (2003) Tumor necrosis factor-alpha causes accumulation of a ubiquitinated form of hypoxia inducible factor-1alpha through a nuclear factor-kappaB-dependent pathway. Mol Biol Cell 14(6):2216–2225. https://doi.org/10.1091/mbc.e02-09-0598

Zhou X, Luo L, Dressel W, Shadier G, Krumbiegel D, Schmidtke P, Zepp F, Meyer CU (2008) Cordycepin is an immunoregulatory active ingredient of Cordyceps sinensis. Am J Chin Med 36:967–980. https://doi.org/10.1142/S0192415X08006387

Zhu JS, Halpern GM, Jones K (1998) The scientific rediscovery of an ancient Chinese herbal medicine: *Cordyceps sinensis*: part I. J Altern Complement Med 4(3):289–303. https://doi.org/10.1089/acm.1998.4.3-289

Ziegenbein FC, Hanssen HP, König WA (2005) Secondary metabolites from Ganoderma lucidum and Spongiporus leucomallellus. Phytochemistry 67(2):202–211. https://doi.org/10.1016/j.phytochem.2005.10.025. Epub 2005 Dec 13. PMID: 16356517

Nutraceuticals in Alleviating Aging

22

Ratnika Sharma and Rajkumar Tulsawani

Abstract

Aging is a physiological process characterized by the loss of biological functions such as senescence and is associated with degenerative diseases. Reactive oxygen (ROS) and nitrogen (RNS) species accumulated over time cause an age-associated loss in bodily functions, and this forms the basis of the free radical theory of aging. Aging affects an organism at physical, physiological, cellular, and molecular levels that manifest in degenerative diseases. There exists a positive relationship between aging and nutraceuticals, and it has been of great research interest worldwide. A significant number of studies have highlighted the bioactivities of medicinal mushrooms—Cordyceps and Ganoderma that contribute to preventing aging and age-related disorders. Here, we discuss the different factors of aging and the role of nutraceuticals in combating them.

Keywords

Aging · Free radical · Nutraceuticals · Cordyceps · Ganoderma

22.1 Introduction

Aging is a physiological process that every living organism experiences in their lifetime. Globally, an increase in life expectancy has been observed over a span of 50 years, and therefore, subsequent changes are witnessed in the mortality pattern in

R. Sharma · R. Tulsawani (✉)
Defence Institute of Physiology and Allied Sciences (DIPAS), Defence Research and Development Organization (DRDO), Delhi, India
e-mail: rktulsawani@yahoo.com

R. Tulsawani, D. Vohora (eds.), *Adaptation under Stressful Environments through Biological Adjustments and Interventions*,
https://doi.org/10.1007/978-981-99-7652-2_22

the countries of different socioeconomic development. With advancing age, deleterious effects occur on the biological systems that affect their proper functioning and cause degenerative diseases such as Alzheimer's disease, cancer, cardiovascular disease, and type 2 diabetes and cause vulnerability to death. Molecular damage like senescence, impaired intercellular contact, and loss of mitochondrial homeostasis occur that alter the structure and functions of cells and tissues (Singh et al. 2019; Partridge et al. 2018).

Furthermore, there is an associated risk factor of multimorbidity; i.e., at any particular given time, a person has more than one disorder. This multimorbidity may arise due to the interaction of different disorders, one disorder being impacted by the treatment of another one and between treatments of one disorder with another one (Beard et al. 2016).

Another important factor associated with the multidimensional dynamics between chronic diseases, physiological changes, and multimorbidity generates a complex health state that cannot be described using specific disease characteristics. These are called "geriatric syndromes." Frailty is one such syndrome that involves a progressive deterioration of physiological health with aging that results in increased susceptibility to various stressors and may ultimately have adverse outcomes (Beard et al. 2016). Frailty is described as different from normal aging, and it increases the vulnerability of older individuals to stressors. It thus results in decreased strength and endurance along with a decline in nutritional, physiological, and cognitive functions (Klein and Fain 2022).

22.2 Characteristics of Aging

Aging effects can be described in the following terms (Zhang et al. 2015):

22.2.1 Physical Effects

The initial effect of aging on an organism can be observed as physical deterioration and increased susceptibility to age-associated diseases. There are concomitant changes that occur at cellular level such as in number and composition that occur during aging. In age-related diseases, cellular loss is observed due to depletion in stem cells, except in cancer where there is indefinite growth of cells.

22.2.2 Physiological Effects

Body systems such as inflammation, nutrient sensing deregulation, and endocrine dysfunction regulate the physiological effects associated with aging. Inflammation plays an important role in clearing pathogens and healing wounded tissues during normal physiological conditions. However, chronic inflammation during aging occurs that leads to diseases. For example, inflammatory pathway of nuclear factor (NF)-κβ is involved in insulin resistance and neurodegenerative diseases, and therefore, inhibiting it helps in attenuating the onset of these diseases (Li et al. 2012a; Zhang et al. 2013). Similarly, deregulation of insulin/insulin-like growth factors and muscle-derived cytokines such as myokines is witnessed during aging (Demontis et al. 2014). Hence, dietary restriction and physical activity play an important role in maintaining health and longevity.

22.2.3 Changes at Cellular Level

In aging, several cellular malfunctions occur. For example, there is cellular senescence and an increase in reactive oxygen species occurs due to electron leakage from the mitochondrial respiratory chain which causes the oxidative stress in age-associated diseases. Also, there is an increase in endoplasmic reticulum stress that deregulates the protein folding along with defects in autophagy–lysosome and ubiquitin–proteasome systems causing proteotoxic stress.

22.2.4 Changes in Biomolecules

Biomolecules suffer malfunction with age due to disruption of cellular homeostasis and environmental factors. During aging, DNA damage occurs that causes mutations along with changes in the telomere length. Epigenetic modifications such as histone acetylation, histone modifications, and DNA methylation are also observed. Additionally, chaperone functions and regulation of proteolysis are affected during aging which further causes the mutations and protein folding errors.

22.3 Free Radicals and Aging

Aging is associated with many deleterious effects that are linked with inflammation, DNA mutations, and increased oxidative stress. One of the most prominent theories of aging was "Free radical theory of Aging" given by Dr. Harman in 1956 (Harman 1956). According to his theory, with age, there is an increased accumulation of free radicals that cause damage to the biomolecules and pathological disorders leading to cell senescence and organismal aging.

Free radicals are highly reactive, unstable molecules with one or more unpaired electrons. These free radicals can either accept or donate an electron to other molecules and thus may act as oxidants or reductants. They can be produced endogenously in the body as a byproduct of metabolic reactions such as inflammatory responses, phagocytosis, and/or mitochondrial electron transport and can be caused due to exogenous sources such as drugs, xenobiotics, radiations, and environmental pollutants. Different types of free radicals are ROS including superoxide radicals, hydroxyl radicals, peroxyl radicals, alkoxyl radicals, and RNS such as nitric oxide and nitrogen dioxide (Dhalaria et al. 2020).

These free radicals can be regulated via enzymatic or nonenzymatic antioxidants. Within the human body, enzymatic antioxidants function to maintain the balance between free radicals and antioxidants. Within the cells and tissues, enzymatic antioxidants are catalase, glutathione peroxidase, and superoxide dismutase (SOD). In addition, nutraceuticals, i.e., natural foods and their compounds, can act as antioxidants by neutralizing the ROS and RNS (Afanas'ev 2010; Dhalaria et al. 2020).

ROS signaling plays a significant role during normal physiological processes. However, during aging, genes such as Sirtuin, Klotho, FOXO3a, and p66shc interact with protein kinases such as Akt kinases and regulate ROS signaling (Afanas'ev 2010). Besides Akt kinases, other kinases also play an important role in these signaling pathways. For example, in their study on rats of age groups 6, 27 and 33 months, Wu et al. (2009) observed an increased blood glucose and superoxide along with phosphorylation of soleus p38-MAPK and ERK 1/2.

Silent information regulator Sirtuin has been reported to prevent the overproduction of ROS in conjunction with FOXO3a. Sirt1 has been shown to be considerably upregulated at low or moderate stress conditions (Alcendor et al. 2007), whereas Sirt3 and FOXO1 have been suggested to be involved in the mitochondrial signaling survival cascade.

FOXO3a is a transcription factor belonging to the O subclass of the forkhead family. It has been demonstrated to be activated due to ROS which further results in cellular senescence. Miyauchi et al. (2004) also reported that inhibition of FOXO3a and activation of Akt kinase caused senescence in human endothelial cells through a p53/p21-dependent pathway.

Similarly, Klotho protein may help in suppressing aging by inhibiting ROS formation. It has also been demonstrated to induce the nuclear translocation of FOXO3a where it binds to the manganese superoxide dismutase (MnSOD) promoter and further prevents ROS generation (Yamamoto et al. 2005).

22.4 Nutraceuticals Against Aging

Due to the rise in novel infectious agents, associated treatment costs, and increased proportion of the elderly population in the world, there is accelerated interest in diet-based health management. Nutrition plays an important role in maintaining the individual's health and preventing chronic diseases.

Nutraceuticals are natural foods that provide physiological benefits in addition to being a source of nutrition through diet (Sharma and Padwad 2020). Different bioactive components of these nutraceuticals have been reported to have positive effects on the health of the elderly (Kalache et al. 2019). These bioactive components may exert an antiaging effect by their antioxidant activity and thus counteracting oxidative stress. Exogenous antioxidants may function by following mechanisms such as scavenging free radicals, decreasing ROS production, and/or repairing oxidized membranes (Simioni et al. 2018). Since nutraceuticals can help in ameliorating excessive ROS burden, the elderly can benefit from the intake of antioxidants and can prevent oxidative stress that accompanies the aging process. Additionally, cellular senescence and immunosenescence that impair the immune system during aging may be counteracted by the immunotherapeutic potential of nutraceuticals (Sharma and Padwad 2020).

Nutritional factors may affect health by mediating epigenetic mechanisms. Epigenetic modifications, for example, DNA methylation, histone modifications, and noncoding microRNAs (miRNAs), assist the response to internal and external factors interacting with nutritional factors (Ribarič 2012).

22.5 Antiaging Effects of Cordyceps

Cordyceps is a fungus belonging to the family Ascomycetes which commonly grows at an altitude of 14,000 ft. in the Indian Himalayan regions, China, Nepal, and Tibet. It is a highly valued mushroom, and different species such as *C. sinensis*, *C. militaris*, *C. taii*, and *C. sobolifera* from this genus have been studied for their therapeutic value. For many years, this fungus has been utilized as a traditional medicine for renal problems, hyperglycemia, and respiratory disorders. *C. sinensis* has been used to treat bacterial and fungal infections, cardiovascular diseases, cancers, oxidative stress, and aging (Das et al. 2021). These biological activities may be attributed to bioactive compounds existing in *C. sinensis* such as cordycepin, fatty acids, and flavonoids, nucleotide, polysaccharides, proteins, and sterols. Aging is linked to oxidative stress, Ji et al. (2009) reported in their study antiaging effects of water extract prepared from *C. sinensis*. D-galactose-induced aging caused symptoms such as reduced antioxidant enzymes, memory dysfunction, and decreased sexual function, and *C. sinensis* extract was able to improve these symptoms in a dose-dependent manner. Table 22.1 compiles the antiaging properties of different extracts or components of Cordyceps.

22.6 Antiaging Effects of Ganoderma

Ganoderma, otherwise referred to as Lingzhi/Reishi, is a basidiomycete white rot fungus. It is frequently seen to grow in temperate and subtropical climatic regions including Asia, Africa, and Europe. *G. lucidum* has been traditionally used as an herbal medicine to promote health. It has been reported to have nonvolatile components such as fiber (59%), fat, (3–5%) protein (7–8%), carbohydrate (26–28%), and ash (1.8%). However, the major ingredients that contribute to its pharmacological effects are polysaccharides, peptidoglycans, and triterpenes found in mycelia, fruiting bodies, and spores.

Several studies point toward the antiaging or antiaging-related roles of the extracts and the bioactive components derived from these extracts. Sudheesh et al. (2010) reported in their study on aged mice, a decrease in ROS through the

Table 22.1 Antiaging properties of different extracts or components of Cordyceps

Extracts/Bioactive components	Function	Reference
Cordycepin	Anti-inflammatory Immunomodulatory Antiaging	Ashraf et al. (2020), Kopalli et al. (2019)
Adenosine	Antioxidant Anti-inflammatory	Jiang et al. (2011), Shin et al. (2009)
Exopolysaccharide fraction	Antioxidant Anti-inflammatory, immunomodulatory	Huang et al. (2013), Li et al. (2020)
Acid polysaccharides from *C. sinensis*	Antioxidant, immunomodulatory	Chen et al. (2012), Li et al. (2003)
C. militaris polysaccharide	Antioxidant	Zhang et al. (2020)
CP2-c2-S2 polysaccharide from *C. militaris*	Antiaging, immunomodulatory	Liu et al. (2016)
Fumosoroseain A from *C. fumosorosea*	Antiaging	Wei et al. (2022)
Water soluble polysaccharide CPS50-I from *C. gunnii*	Antiaging	Zhu et al. (2011)
Water soluble polysaccharide from *C. taii*	Antioxidant, antiaging	Xiao et al. (2012)

antioxidant activity of ethanol extract of *G. lucidum*. Also, it could help in increasing the activity of important antioxidant enzymes for instance catalase, glutathione peroxidase, and MnSOD in the mitochondria of the heart, liver, and brain in aged mice. Yoon et al. (2013) also highlighted the anti-inflammatory activity of ethanol extract by inhibition of NF-κβ and toll-like receptors in lipopolysaccharide-stimulated BV2 microglial cells. Similarly, the aqueous extract of *G. lucidum* improved stress tolerance and increased the longevity of *Caenorhabditis elegans* via the germline signaling pathway (Cuong et al. 2019), whereas polysaccharide derived from *G. lucidum* prolonged the lifespan of these nematodes by upregulating in Daf-16 expression through TIR-1 receptor and MAPK pathway (Chuang et al. 2009). Table 22.2 compiles the antiaging properties of different extracts or components of Ganoderma.

22.7 Conclusion

Nutraceuticals may exert antiaging effects mainly through antioxidant, free radical scavenging, immunomodulation, and anti-inflammatory activities. Aging is a complex process that has different impacts on different biological scales—physical, physiological, cellular, and molecular levels. These levels may help in identifying better strategies to counterbalance the stress and thus delay aging. Nutraceuticals could help alleviate these stressors and provide a way of healthy aging in organisms.

Table 22.2 Antiaging properties of different extracts or components of Ganoderma

Extracts/bioactive components	Function	Reference
Reishi polysaccharide fraction 3 (RF3)	Immunomodulatory effect Lifespan extension	Chen et al. (2004), Chuang et al. (2009)
G. lucidum polysaccharide I, II, III, IV	Antioxidant	Shi et al. (2013)
G. lucidum polysaccharide B	Immunomodulatory effect	Lei and Lin (1993), Mingchun et al. (2001)
G. lucidum polysaccharide peptide	Antioxidant, immunomodulatory effect	Cao and Lin (2004)
Ganodermasides A, B, C, D	Antiaging	Weng et al. (2010), Weng et al. (2011)
RLZ8 recombinant *G. lucidum* protein	Antiaging	Liu et al. (2018)
Ethanol extract of *G. lucidum*	Antioxidant, antiaging	Sudheesh et al. (2010), Ajith et al. (2009)
Water extract of *G. lucidum*	Antiaging	Cuong et al. (2019)
G. atrum polysaccharide PSG-1	Antioxidant, immunomodulation, antiaging	Li et al. (2012b), Yu et al. (2015)
G. capense polysaccharides	Antioxidant, antiaging	Huang et al. (2015), Yan et al. (2013)

References

Afanas'ev I (2010) Signaling and damaging functions of free radicals in aging—free radical theory, hormesis, and TOR. Aging Dis 1(2):75–88

Ajith TA, Sudheesh NP, Roshny D, Abishek G, Janardhanan KK (2009) Effect of Ganoderma lucidum on the activities of mitochondrial dehydrogenases and complex I and II of electron transport chain in the brain of aged rats. Exp Gerontol 44(3):219–223. https://doi.org/10.1016/j.exger.2008.11.002

Alcendor RR, Gao S, Zhai P, Zablocki D, Holle E, Yu X et al (2007) Sirt1 regulates aging and resistance to oxidative stress in the heart. Circ Res 100(10):1512–1521. https://doi.org/10.1161/01.RES.0000267723.65696.4a

Ashraf SA, Elkhalifa AEO, Siddiqui AJ, Patel M, Awadelkareem AM, Snoussi M et al (2020) Cordycepin for health and wellbeing: a potent bioactive metabolite of an entomopathogenic medicinal fungus Cordyceps with its nutraceutical and therapeutic potential. Molecules 25(12): 2735. https://doi.org/10.3390/molecules25122735

Beard JR, Officer A, De Carvalho IA, Sadana R, Pot AM, Michel JP et al (2016) The world report on ageing and health: a policy framework for healthy ageing. Lancet 387(10033):2145–2154. https://doi.org/10.1016/S0140-6736(15)00516-4

Cao QZ, Lin ZB (2004) Antitumor and anti-angiogenic activity of Ganoderma lucidum polysaccharides peptide. Acta Pharmacol Sin 25(6):833–838

Chen HS, Tsai YF, Lin S, Lin CC, Khoo KH, Lin CH et al (2004) Studies on the immuno-modulating and anti-tumor activities of Ganoderma lucidum (Reishi) polysaccharides. Bioorg Med Chem 12(21):5595–5601. https://doi.org/10.1016/j.bmc.2004.08.003

Chen W, Yuan F, Wang K, Song D, Zhang W (2012) Modulatory effects of the acid polysaccharide fraction from one of anamorph of Cordyceps sinensis on Ana-1 cells. J Ethnopharmacol 142(3): 739–745. https://doi.org/10.1016/j.jep.2012.05.055

Chuang MH, Chiou SH, Huang CH, Yang WB, Wong CH (2009) The lifespan-promoting effect of acetic acid and Reishi polysaccharide. Bioorg Med Chem 17(22):7831–7840. https://doi.org/10.1016/j.bmc.2009.09.002

Cuong VT, Chen W, Shi J, Zhang M, Yang H, Wang N et al (2019) The anti-oxidation and anti-aging effects of Ganoderma lucidum in Caenorhabditis elegans. Exp Gerontol 117:99–105. https://doi.org/10.1016/j.exger.2018.11.016

Das G, Shin HS, Leyva-Gómez G, Prado-Audelo ML, Cortes H, Singh YD et al (2021) Cordyceps spp.: a review on its immune-stimulatory and other biological potentials. Front Pharmacol 11: 2250

Demontis F, Patel VK, Swindell WR, Perrimon N (2014) Intertissue control of the nucleolus via a myokine-dependent longevity pathway. Cell Rep 7(5):1481–1494. https://doi.org/10.1016/j.celrep.2014.05.001

Dhalaria R, Verma R, Kumar D, Puri S, Tapwal A, Kumar V et al (2020) Bioactive compounds of edible fruits with their anti-aging properties: a comprehensive review to prolong human life. Antioxidants (Basel) 9(11):1123. https://doi.org/10.3390/antiox9111123

Harman D (1956) Aging: a theory based on free radical and radiation chemistry. J Gerontol 11(3): 298–300. https://doi.org/10.1093/geronj/11.3.298

Huang Y, Li N, Wan JB, Zhang D, Yan C (2015) Structural characterization and antioxidant activity of a novel heteropolysaccharide from the submerged fermentation mycelia of Ganoderma capense. Carbohydr Polym 134:752–760. https://doi.org/10.1016/j.carbpol.2015.08.067

Huang QL, Siu KC, Wang WQ, Cheung YC, Wu JY (2013) Fractionation, characterization and antioxidant activity of exopolysaccharides from fermentation broth of a Cordyceps sinensis fungus. Process Biochem 48(2):380–386. https://doi.org/10.1016/j.procbio.2013.01.001

Ji DB, Ye J, Li CL, Wang YH, Zhao J, Cai SQ (2009) Antiaging effect of Cordyceps sinensis extract. Phytotherapy research: an international journal devoted to pharmacological and toxicological evaluation of natural product derivatives. Phytother Res 23(1):116–122. https://doi.org/10.1002/ptr.2576

Jiang Y, Wong JH, Fu M, Ng TB, Liu ZK, Wang CR et al (2011) Isolation of adenosine, iso-sinensetin and dimethylguanosine with antioxidant and HIV-1 protease inhibiting activities from fruiting bodies of Cordyceps militaris. Phytomedicine 15((18)(2–3)):189–193. https://doi.org/10.1016/j.phymed.2010.04.010

Kalache A, De Hoogh AI, Howlett SE, Kennedy B, Eggersdorfer M, Marsman DS et al (2019) Nutrition interventions for healthy ageing across the lifespan: a conference report. Eur J Nutr 58 (Suppl 1):1. https://doi.org/10.1007/s00394-019-02027-z

Klein AS, Fain MJ (2022) Geriatric syndromes. Reichel's care elderly. Clin Aspects Aging 21:81

Kopalli SR, Cha KM, Lee SH, Hwang SY, Lee YJ, Koppula S et al (2019) Cordycepin, an active constituent of nutrient powerhouse and potential medicinal mushroom Cordyceps militaris Linn., ameliorates age-related testicular dysfunction in rats. Nutrients 11(4):906. https://doi.org/10.3390/nu11040906

Lei LS, Lin ZB (1993) Effects of Ganoderma polysaccharides on the activity of DNA polymerase alpha of splenocytes and immune function in aged mice. Yao Xue Xue Bao 28(8):577–582

Li WJ, Nie SP, Peng XP, Liu XZ, Li C, Chen Y et al (2012b) Ganoderma atrum polysaccharide improves age-related oxidative stress and immune impairment in mice. J Agric Food Chem 60(6):1413–1418. https://doi.org/10.1021/jf204748a

Li LQ, Song AX, Yin JY, Siu KC, Wong WT, Wu JY (2020) Anti-inflammation activity of exopolysaccharides produced by a medicinal fungus Cordyceps sinensis Cs-HK1 in cell and animal models. Int J Biol Macromol 149:1042–1050. https://doi.org/10.1016/j.ijbiomac.2020.02.022

Li J, Tang Y, Cai D (2012a) IKKβ/NF-κB disrupts adult hypothalamic neural stem cells to mediate a neurodegenerative mechanism of dietary obesity and pre-diabetes. Nat Cell Biol 14(10): 999–1012. https://doi.org/10.1038/ncb2562

Li SP, Zhao KJ, Ji ZN, Song ZH, Dong TT, Lo CK et al (2003) A polysaccharide isolated from Cordyceps sinensis, a traditional Chinese medicine, protects PC12 cells against hydrogen

peroxide-induced injury. Life Sci 73(19):2503–2513. https://doi.org/10.1016/s0024-3205(03)00652-0
Liu X, Huang Y, Chen Y, Cao Y (2016) Partial structural characterization, as well as immunomodulatory and anti-aging activities of CP2-c2-s2 polysaccharide from Cordyceps militaris. RSC Adv 6(106):104094–104103. https://doi.org/10.1039/C6RA23612J
Liu K, Liu YY, Sun F, Ren LQ (2018) Antiaging effect of recombinant Ganoderma lucidum immunomodulatory protein on aging model mice. Chin Older J 38:5539–5540
Mingchun L, Linsheng L, Qingbiao W, Dongsheng L, Ziming X, Shuqin Y et al (2001) Effects of Ganoderma lucidum polysaccharides on inositol trisphosphate and diacylglycerol in murine T cells. Zhongguo yao xue za zhi (Zhongguo yao xue hui: 1989) 36(8):526–528
Miyauchi H, Minamino T, Tateno K, Kunieda T, Toko H, Komuro I (2004) Akt negatively regulates the in vitro lifespan of human endothelial cells via a p53/p21-dependent pathway. EMBO J 23(1):212–220. https://doi.org/10.1038/sj.emboj.7600045
Partridge L, Deelen J, Slagboom PE (2018) Facing up to the global challenges of ageing. Nature 561(7721):45–56. https://doi.org/10.1038/s41586-018-0457-8
Ribarič SS (2012) Diet and aging. Oxidative Med Cell Longev 2012:741468. https://doi.org/10.1155/2012/741468
Sharma R, Padwad Y (2020) Nutraceuticals-based immunotherapeutic concepts and opportunities for the mitigation of cellular senescence and aging: a narrative review. Ageing Res Rev 63:101141. https://doi.org/10.1016/j.arr.2020.101141
Shi M, Zhang Z, Yang Y (2013) Antioxidant and immunoregulatory activity of Ganoderma lucidum polysaccharide (GLP). Carbohydr Polym 95(1):200–206. https://doi.org/10.1016/j.carbpol.2013.02.081
Shin S, Moon S, Park Y, Kwon J, Lee S, Lee CK et al (2009) Role of cordycepin and adenosine on the phenotypic switch of macrophages via induced anti-inflammatory cytokines. Immune Netw 9(6):255–264. https://doi.org/10.4110/in.2009.9.6.255
Simioni C, Zauli G, Martelli AM, Vitale M, Sacchetti G, Gonelli A et al (2018) Oxidative stress: role of physical exercise and antioxidant nutraceuticals in adulthood and aging. Oncotarget 9(24):17181–17198. https://doi.org/10.18632/oncotarget.24729
Singh PP, Demmitt BA, Nath RD, Brunet A (2019) The genetics of aging: a vertebrate perspective. Cell 177(1):200–220. https://doi.org/10.1016/j.cell.2019.02.038
Sudheesh NP, Ajith TA, Ramnath V, Janardhanan KK (2010) Therapeutic potential of Ganoderma lucidum (Fr.) P. Karst. Against the declined antioxidant status in the mitochondria of post-mitotic tissues of aged mice. Clin Nutr 29(3):406–412. https://doi.org/10.1016/j.clnu.2009.12.003
Wei J, Zhou X, Dong M, Yang L, Zhao C, Lu R et al (2022) Metabolites and novel compounds with anti-microbial or antiaging activities from Cordyceps fumosorosea. AMB Express 12(1):40. https://doi.org/10.1186/s13568-022-01379-w
Weng Y, Lu J, Xiang L, Matsuura A, Zhang Y, Huang Q et al (2011) Two new antiaging ergosterols from spores of the medicinal mushroom Ganoderma lucidum. Biosci Biotechnol Biochem 75(4):800–803
Weng Y, Xiang L, Matsuura A, Zhang Y, Huang Q, Qi J et al (2010) Two novel antiaging ergosterols from spores of a medicinal mushroom Ganoderma lucidum on yeast via UTH1 gene. Bioorg Med Chem 18(3):999–1002
Wu M, Desai DH, Kakarla SK, Katta A, Paturi S, Gutta AK et al (2009) Acetaminophen prevents aging-associated hyperglycemia in aged rats: effect of aging-associated hyperactivation of p38-MAPK and ERK1/2. Diabetes Metab Res Rev 25(3):279–286. https://doi.org/10.1002/dmrr.932
Xiao JH, Xiao DM, Chen DX, Xiao Y, Liang ZQ, Zhong JJ (2012) Polysaccharides from the medicinal mushroom Cordyceps taii show antioxidant and immunoenhancing activities in a D-galactose-induced aging mouse model. Evid Based Complement Alternat Med 2012:273435. https://doi.org/10.1155/2012/273435

Yamamoto M, Clark JD, Pastor JV, Gurnani P, Nandi A, Kurosu H et al (2005) Regulation of oxidative stress by the anti-aging hormone klotho. J Biol Chem 280(45):38029–38034. https://doi.org/10.1074/jbc.M509039200

Yan C, Kong F, Zhang D, Cui J (2013) Anti-glycated and antiradical activities in vitro of polysaccharides from Ganoderma capense. Pharmacogn Mag 9(33):23–27. https://doi.org/10.4103/0973-1296.108132

Yoon HM, Jang KJ, Han MS, Jeong JW, Kim GY, Lee JH et al (2013) Ganoderma lucidum ethanol extract inhibits the inflammatory response by suppressing the NF-κB and toll-like receptor pathways in lipopolysaccharide-stimulated BV2 microglial cells. Exp Ther Med 5(3):957–963. https://doi.org/10.3892/etm.2013.895

Yu Q, Nie SP, Wang JQ, Huang DF, Li WJ, Xie MY (2015) Signaling pathway involved in the immunomodulatory effect of Ganoderma atrum polysaccharide in spleen lymphocytes. J Agric Food Chem 63(10):2734–2740. https://doi.org/10.1021/acs.jafc.5b00028

Zhang R, Chen HZ, Liu DP (2015) The four layers of aging. Cell Syst 1(3):180–186. https://doi.org/10.1016/j.cels.2015.09.002

Zhang G, Li J, Purkayastha S, Tang Y, Zhang H, Yin Y et al (2013) Hypothalamic programming of systemic ageing involving IKK-β, NF-κB and GnRH. Nature 497(7448):211–216. https://doi.org/10.1038/nature12143

Zhang Y, Zeng Y, Cui Y, Liu H, Dong C, Sun Y (2020) Structural characterization, antioxidant and immunomodulatory activities of a neutral polysaccharide from Cordyceps militaris cultivated on hull-less barley. Carbohydr Polym 235:115969. https://doi.org/10.1016/j.carbpol.2020.115969

Zhu ZY, Si CL, Zhong YR, Zhu CM, Zhou JP, Liu AJ et al (2011) The purification and antioxidative activities in d-galactose-induced aging mice of a water-soluble polysaccharide from Cordyceps gunnii (berk.) berk. Mycelium. J Food Biochem 35(1):303–322. https://doi.org/10.1111/j.1745-4514.2010.00383.x

Heat Stress and Dehydration 23

Pooja Chaudhary, Abhishek Bharadwaj, Rajinder Kumar Gupta, Sachin Kumar, and Shubhangi Sharma

Abstract

Global warming and its related issues have become a topic of concern in recent times. The most severe problem associated with increased environmental temperature is the higher heat load on the body. Passive heat exposure as such has many debilitating effects such as prickly heat rash, heat syncope, and other heat illnesses resulting in altered body functioning, mainly in elderly people and children. Working in a hot environment further increases heat load which can result in increased core body temperature (T_c) and can have severe outcomes. Thermoregulation is the homeostatic mechanism to maintain the T_c within a narrow range which is optimal for body functioning. Under a hot environment, sweat evaporation becomes the major thermoregulatory effector to reduce Tc. Loss of water and electrolytes in sweat, if not replenished, leads to dehydration which can be detrimental to human performance and survival. As temperatures are rising all around the globe, it is important to understand the challenges imposed on the body by dehydration and the importance of remaining hydrated. This book chapter has been conceptualized and designed to provide the necessary information about thermoregulation, core body temperature, heat illnesses, dehydration, and its consequences.

P. Chaudhary (✉) · A. Bharadwaj · R. K. Gupta · S. Kumar
Defence Institute of Physiology and Allied Sciences (DIPAS), Defence Research and Development Organization (DRDO), Delhi, India
e-mail: poojachaudhary155@gmail.com

S. Sharma
Defence Institute of Physiology and Allied Sciences (DIPAS), Defence Research and Development Organization (DRDO), Delhi, India

Amity Institute of Biotechnology, Amity University, Noida, Uttar Pradesh, India

R. Tulsawani, D. Vohora (eds.), *Adaptation under Stressful Environments through Biological Adjustments and Interventions*,
https://doi.org/10.1007/978-981-99-7652-2_23

Keywords

Dehydration · Heat stress · Thermoregulation

23.1 Introduction

With the increase in global warming, the frequency and severity of heat illnesses and heat-mediated mortality are on the rise worldwide (Romanello et al. 2021). Heat illnesses include heat rash, cramps, syncope, exhaustion, and stroke. Heat rashes are red-colored skin rashes usually caused by excessive sweating. These rashes are usually itchy and prickly. Heat cramps are painful muscular cramps or spasms caused by excessive loss of salt with water during sweating. Heat syncope is fainting or feeling dizziness after prolonged standing in a hot environment. This is due to transiently reduced blood supply to the brain. Another form of heat illness is heat exhaustion which is defined as the body's inability to continue working under stressful conditions. It usually occurs due to extreme sweating without fluid replacement resulting in increased core body temperature. If not treated in time, heat exhaustion may lead to heat stroke. Heat stroke is characterized by central nervous system dysfunction such as delirium and coma. Heat stroke is an outcome of the body's failed thermoregulation resulting in >40.5 °C core body temperature. While heat rash, cramps, and syncope are milder forms of heat illnesses, heat exhaustion and heat stroke are more severe forms that can result in death (Howe and Boden 2007). Heat illnesses result from heat accumulation in the body. An increase in the core body temperature (T_c) is an indicator of the accumulation of heat stress (Rossi 2014). The various factors leading to excessive body heat are metabolic heat production, exercise, and exposure to high ambient temperatures. Cellular metabolism generates heat from the oxidation of food constituents and is an important source of heat production in the body. Exercising muscles increase the metabolism by 5 to 15 times the resting rate, thereby contributing significantly to body heat generation (Sawka et al. 1993). Performing physical activity in high ambient temperatures further increases the heat load.

23.2 Heat Loss Mechanisms

Under normal conditions, the human core body temperature is maintained at a constant value of about 37 °C by hypothalamic thermoregulatory center of the autonomic nervous system. The anterior hypothalamic–preoptic area contains cold and heat-sensitive neurons that are called central thermoreceptors. Peripheral thermoreceptors are present in the skin. Both central and peripheral thermoreceptors sense the high temperature and activate hypothalamic response which, in turn, decrease body heat production and increase heat loss to the environment. Blood vessels are present profusely beneath the skin layer. Hypothalamic regulatory center senses high core temperatures and sends signals for blood vessel dilation in order to

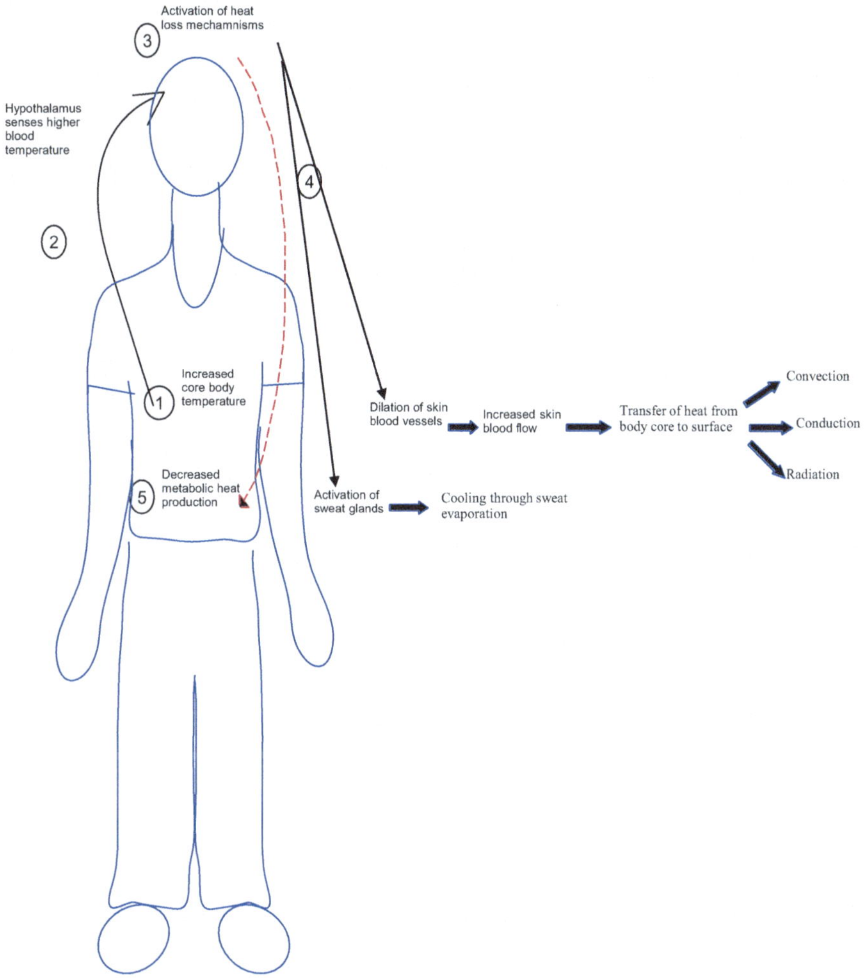

Fig. 23.1 Heat loss mechanisms of the body

circulate increased blood volume through them (Fig. 23.1). Heat is transferred from the core of the body to the periphery via increased skin blood flow from where it can be removed through radiation, convection, conduction, and evaporation. Radiation is described as heat loss in the form of infrared radiation from the skin surface to the environment. Conduction is the loss of heat through physical contact with any colder surface. Convection is the process of transfer of heat from the body to air or water moving across the skin's surface. Sweat on the skin's surface vaporizes and carries heat along with it in a process called evaporation. Radiation, conduction, and convection play important roles as heat loss mechanisms when the ambient environmental temperature is lower than the skin temperature. However, when the ambient environmental temperature is higher than the skin temperature, dilation of cutaneous

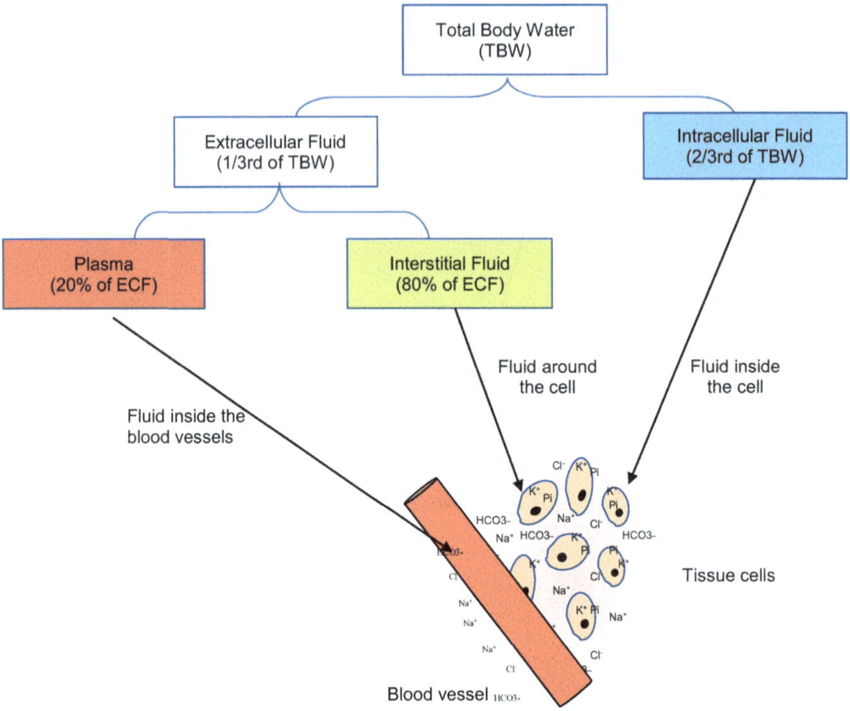

Fig. 23.2 Total body water distribution in different compartments

vasculature and sweat evaporation becomes the major thermoregulatory effector to reduce T_c (Osilla et al. 2022). Along with the dilation of blood vessels, the hypothalamus also sends signals to sweat glands to secrete more sweat (Fig. 23.1) in a hot environment.

23.3 Total Body Water: Fluid–Electrolyte Balance

As sweating is the primary mechanism of heat loss from body under hot environment, the total body water content (TBW) plays an important role in maintaining thermal homeostasis in the body. On an average, the TBW in a human body is about 60% of the body weight. As depicted in Fig. 23.2, the two main components of TBW are extracellular fluid (ECF) and intracellular fluid (ICF). ECF is the fluid outside the cells, and ICF is the fluid inside the cells. About one-third of TBW is present in the form of ECF, and the rest is present in the form of ICF. ECF is further divided into interstitial fluid (80% of ECF) and plasma (20% of ECF). A water-permeable cell membrane separates these fluid compartments, which allows continuous fluid exchange between the segments (Armstrong 2005). The ionic composition of intravascular and interstitial compartment is almost similar; sodium (Na+) is the major cation, and the major anions are chloride and bicarbonate. However, the major cation in ICF is potassium (K+), and major anions are inorganic phosphates (Muhammad et al. 2019). These ions form the electrolytes concentration of the different segments. The osmolality (concentration of solutes in the solvent) of different water compartments is usually the same, thus maintaining the water–electrolyte balance among them. Sweat loss during exercise in hot environment disturbs the normal water–electrolyte balance resulting in movement of water between these segments (Périard et al. 2021).

23.4 Thirst and Hormone-Mediated Regulation of Fluid–Electrolyte Balance

Thirst and hormonal mechanisms maintain the TBW and electrolyte balance within a narrow range (Bossingham et al. 2005). Under hot environment, sweat loss results in hypertonic (increased osmolality)–hypovolemic (decreased volume) dehydration (Sawka et al. 2015). Both water and solutes/electrolytes are lost in sweat; however, the loss of water is more than the electrolytes. Thus, plasma volume depletion caused by excessive sweating results in a proportionally decreased blood pressure and increased plasma osmolality (Sawka et al. 1985). Decreased plasma volume, increased plasma osmolality, and decrease in blood pressure are activators of thirst sensation and secretion of anti-diuretic hormone (ADH) or vasopressin (McNeil-Masuka and Boyer 2022).

The most important signal for thirst is blood osmolality. As little as 1% increase in plasma osmolality is sufficient to stimulate thirst (Leib et al. 2016). Neurons in the forebrain's subfornical organ (SFO) and organum vasculosum of the lamina terminalis (OVLT) sense alterations in blood osmolality and generate thirst response (McKinley and Johnson 2004).

Hypovolemia (decreased plasma volume) and the resulting hypotension (decreased blood pressure) also stimulate thirst (Arai et al. 2013; Leib et al. 2016). In addition to thirst sensation, the osmoreceptors also send signals to release ADH from the posterior pituitary gland. This hormone primarily affects the ability of the kidney to reabsorb water by inducing the expression of water transport proteins called aquaporins in the late distal tubule and collecting duct to increase water reabsorption (Boone and Deen 2008). Hypovolemia and hypotension are sensed by the baroreceptors in the kidney to release the enzyme renin that starts the hormone–enzyme cascade known as the renin–angiotensin–aldosterone system (RAAS pathway) to produce angiotensin II, which further stimulates ADH secretion (Thornton 2010) (Fig. 23.3).

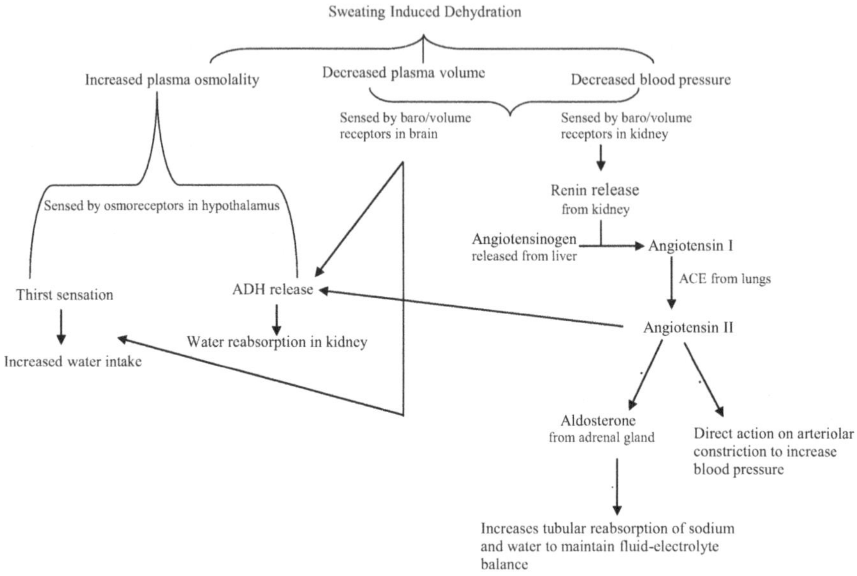

Fig. 23.3 Body's response to dehydration

23.5 Consequences of Dehydration

Dehydration impacts human health and performance in a negative way. The most severe consequences of dehydration are failed thermoregulation and cardiovascular strain. As discussed earlier, evaporative sweating is the major thermoregulatory effector under hot environment. Dehydration decreased water availability reduces sweat rate thereby hampering effective thermoregulation. Failed thermoregulation can result in multiple organ dysfunction which if not identified and treated urgently can result in high mortality rate and permanent neurological damage (Ninan et al. 2020). Dehydration-induced hypovolemia also amplifies the cardiac load by decreasing stroke volume and cardiac output. Thus, dehydration during intense and prolonged whole-body exercise in the heat also results in cardiac stress (Watanabe et al. 2020).

23.6 Summary

Heat stress and its related illnesses are on the rise around the globe due to global warming. Though some heat illnesses are mild and easily avoidable/reversible such as heat rash, heat cramps, and heat syncope, the more severe forms of heat illnesses such as exhaustion and stroke can have severe implications on sustainability and survival. Dehydration has been considered to be one of the most common factors leading to heat illnesses (Casa et al. 2000; Coris et al. 2004). Dehydration is an outcome of loss of body water with inadequate fluid replenishment. Sweat loss while working in a hot environment usually results in dehydration if the fluid is not replenished immediately. Sweat-induced dehydration is often categorized as hypovolemic (decreased fluid volume) and hyperosmotic (increased electrolyte concentration). The body's fluid–electrolyte balancing mechanisms come into action on receiving dehydration signals such as decreased plasma volume, increased osmotic concentration, and decreased blood pressure. Inadequate fluid and electrolyte replenishment results in failed compensatory mechanisms causing further dehydration. Thus, with continuous activity in a hot environment, the heat load and subsequently the T_c of the body keep on increasing, while the evaporative cooling diminishes due to increased fluid loss in the sweat. Therefore, fluid replenishment is the most important management measure for counteracting heat stress.

References

Arai S, Stotts N, Puntillo K (2013) Thirst in critically ill patients: from physiology to sensation. Am J Crit Care 22(4):328–335. https://doi.org/10.4037/ajcc2013533

Armstrong LE (2005) Hydration assessment techniques. Nutr Rev 63(6 Pt 2):S40–S54. https://doi.org/10.1111/j.1753-4887.2005.tb00153.x

Boone M, Deen PMT (2008) Physiology and pathophysiology of the vasopressin-regulated renal water reabsorption. Pflugers Arch 456(6):1005–1024. https://doi.org/10.1007/s00424-008-0498-1

Bossingham MJ, Carnell NS, Campbell WW (2005) Water balance, hydration status, and fat-free mass hydration in younger and older adults. Am J Clin Nutr 81(6):1342–1350. https://doi.org/10.1093/ajcn/81.6.1342

Casa DJ, Armstrong LE, Hillman SK, Montain SJ, Reiff RV, Rich BS et al (2000) National Athletic Trainers' association position statement: fluid replacement for athletes. J Athl Train 35(2):212–224

Coris EE, Ramirez AM, Van Durme DJ (2004) Heat illness in athletes. The dangerous combination of heat, humidity and exercise. Sports Med 34(1):9–16. https://doi.org/10.2165/00007256-200434010-00002

Howe AS, Boden BP (2007) Heat-related illness in athletes. Am J Sports Med 35(8):1384–1395. https://doi.org/10.1177/0363546507305013

Leib DE, Zimmerman CA, Knight ZA (2016) Thirst. Curr Biol 26(24):R1260–R1265. https://doi.org/10.1016/j.cub.2016.11.019

McKinley MJ, Johnson AK (2004) He physiological regulation of thirst and fluid intake. Physiological regulation of thirst and fluid intake. News Physiol Sci 19:1–6. https://doi.org/10.1152/nips.01470.2003

McNeil-Masuka J, Boyer TJ (2022) Insensible fluid loss. In: StatPearls, p 2023

Muhammad A, Mohamad MA, Hania A, Ghaffar A (2019) Dehydration and volume depletion: how to handle the misconceptions. World J Nephrol 8(1):23–32

Ninan GA, Gunasekaran K, Jayakaran JAJ, Johnson J, Abhilash KPP, Pichamuthu K et al (2020) Heat-related illness—clinical profile and predictors of outcome from a healthcare center in South India. J Family Med Prim Care 9(8):4210–4215. https://doi.org/10.4103/jfmpc.jfmpc_690_20

Osilla EV, Marsidi JL, Sharma S (2022) Physiology, temperature regulation. StatPearls Publishing, Treasure Island, (FL), p 2023; http://www.ncbi.nlm.nih.gov/books/NBK507838

Périard JD, Eijsvogels TMH, Daanen HAM (2021) Exercise under heat stress:thermoregulation, hydration, performance implications, and mitigation strategies. Physiol Rev 101(4):1873–1979. https://doi.org/10.1152/physrev.00038.2020

Romanello M, McGushin A, Napoli CD, Drummond P, Hughes N, Jamart L (2021) The 2021 report of the lancet countdown on health and climate change: code red for a healthy future. Lancet 398(10311):1619–1662

Rossi R (2014) Clothing for protection against heat and flames. Managing thermal stress. Woodhead Publishing Series in Textiles; 2014:70–89

Sawka MN, Cheuvront SN, Kenefick RW (2015) Hypohydration and human performance: impact of environment and physiological mechanisms. Sports Med 45(Suppl 1):S51–S60. https://doi.org/10.1007/s40279-015-0395-7

Sawka MN, Young AJ, Francesconi RP, Muza SR, Pandolf KB (1985) Thermoregulatory and blood responses during exercise at graded hypohydration levels. J Appl Physiol (1985) 59(5): 1394–1401. https://doi.org/10.1152/jappl.1985.59.5.1394

Sawka MN, Wenger B, Young AJ, Pandolf KB (1993) Institute of Medicine (US) Committee on Military Nutrition Research. In: Marriott BM (ed) Nutritional needs in hot environments: applications for military personnel in field operations. National Academies Press (US), Washington (DC). 3, Physiological Responses to Exercise in the Heat

Thornton SN (2010) Thirst and hydration: physiology and consequences of dysfunction. Physiol Behav 100(1):15–21. https://doi.org/10.1016/j.physbeh.2010.02.026

Watanabe K, Stöhr EJ, Akiyama K, Watanabe S, González-Alonso J (2020) Dehydration reduces stroke volume and cardiac output during exercise because of impaired cardiac filling and venous return, not left ventricular function. Physiol Rep 8(11):e14433. https://doi.org/10.14814/phy2.14433

Management of Stress Through Dietary Modifications

24

Rifat Ara Begum, Emran Kabir Chowdhury, and Hossain Uddin Shekhar

Abstract

Stress and depression are silent killers in the present day. Due to lifestyle, workload, social factors, and unhealthy food consumption, every person is at risk of developing such mental disorders and being unproductive. These mental conditions are not curable by the available medications; moreover, the medications have side effects which can develop obesity, diabetes, and cardio-vascular diseases in the long run. Luckily, we can find a solution in our food. Some nutrients such as—vitamins (niacin, folate), minerals (magnesium, zinc) and ω-3 long-chain polyunsaturated fatty acids (LCPUFAs) are well studied to have a prominent role in the maintenance of healthy brain function. These nutrients cannot cure depression, but they can prevent and protect mental health in the long run. We suggest dietary modification and including such nutrient-rich foods into the diet could be the best precaution and care we can do for our mental health. The traditional Mediterranean diet is the best we can adapt to as it is equipped with all foods (seeds, vegetables, fruits, fish, and seafood) which are rich sources of the aforementioned nutrients.

Keywords

Stress · Depression · Nutrition · EPA · DHA · Microbiota

R. A. Begum · E. K. Chowdhury · H. U. Shekhar (✉)
Department of Biochemistry and Molecular Biology, University of Dhaka, Dhaka, Bangladesh
e-mail: shekhardu@hotmail.com

R. Tulsawani, D. Vohora (eds.), *Adaptation under Stressful Environments through Biological Adjustments and Interventions*,
https://doi.org/10.1007/978-981-99-7652-2_24

24.1 Introduction

Nutritious food not only helps maintain our physique, but also contributes to maintaining our mental health. A healthy diet can boost brain function and improve mood, but research also suggests that deficiencies in certain nutrients may contribute to or exacerbate conditions such as depression, bipolar disorder, schizophrenia, and obsessive-compulsive disorder (OCD) (Murray et al. 1996; Rao et al. 2008). This chapter aims to discuss stress, mental health, and the link between a healthy diet and stress management.

24.2 Stress and Its Effects on the Human Body

Stress is a condition that may arise when the body faces any disturbances in equilibrium caused by any particular stimuli. Humans have a very complex psycho-physiological integrated system, so responding to such stimuli is a complex process. People vary in the degree and intensity of their responses to such stimuli; hence, they may have various degrees or levels of stress and different coping abilities both mentally and physically (Gonzalez and Miranda-Massari 2014). Stress and stressors (stress-causing stimuli) are now a common problem, as modern lifestyle has created several triggers, such as financial pressures, political and religious aspects, social issues and status, extreme population, environmental pollution, and most of all food industry. Food industry now provides main staple foods that are additional stressors, containing refined carbohydrates, excessive animal fats, artificial colors, preservatives, and sweeteners.

Stress affects a person in three major ways. The primary response is a surge or increase of adrenaline, which is known to cause the flight-or-fight response, which a person can withstand. Secondary response is typically a temporary resistance mechanism that a person's body can activate in order to deal with a stressful situation. The final stage appears when the person's body gets exhausted. At this level, the body cannot cope with the stress and also develops some chronic conditions like high blood pressure, cardiovascular disease, a weak immune system (developing susceptibility to allergies and infections), skin irritations, pain and stiffness of the neck, shoulder and back, breathing difficulties and even infertility (Fig. 24.1, Carlson 2004).

There have been many studies to predict the relationship between stress and depression (Kessler et al. 1997; Mazure 1998; Monroe and Hadjiyannakis 2002; Tennant 2002; Paykel 2003). The studies revealed acute and chronic stressors, which are usually different life events and elements of society, that lead to the development of depressive symptoms. In the present day, depression is a very serious psychopathological condition that causes the patient to be unable to experience pleasure, sadness, insomnia, hypersomnia, retardation, lack of focus or concentration, feeling helpless, and, in the long run, may develop suicidal tendencies (Ali and Hendawy 2018). Aside from social and environmental factors, other factors such as changes in hormone or neurotransmitter levels or oxidative stress can cause stress and

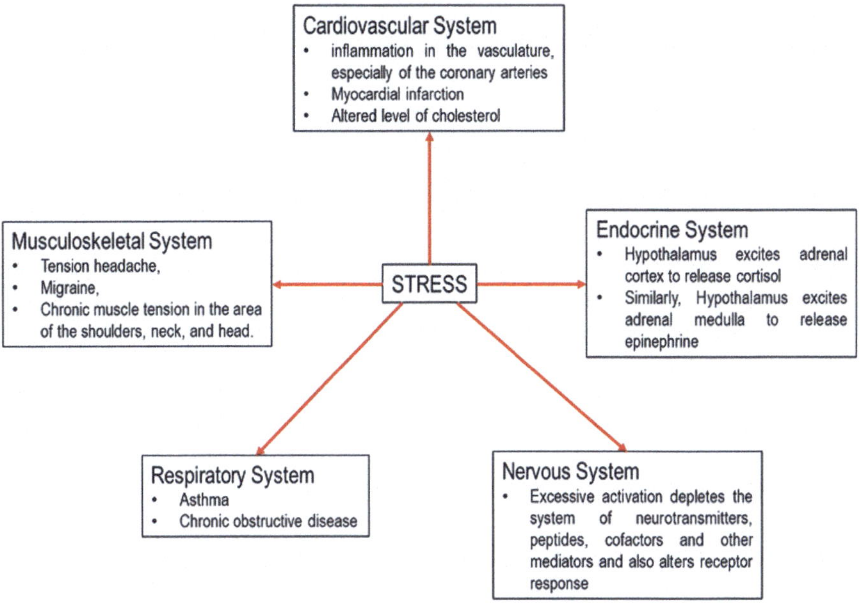

Fig. 24.1 Stress affects different major tissue systems of human body. (Adapted from Gonzalez and Miranda-Massari (2014))

depression (Ali and Hendawy 2018). Antidepressants are frequently used to treat depression (Dale et al. 2015; Chávez-Castillo et al. 2019), but they can cause side effects (e.g., obesity, diabetes, cardiovascular disease, addiction, and so on) (Ali and Hendawy 2018).

24.3 Relationship Between Diet and Stress

A research area called "nutritional psychiatry" has emerged due to the immense importance of adequate nutrients in maintaining healthy brain functions (Jacka 2017). This discipline focuses on dietary nutrients and their effects on mood alterations—mainly depressive disorders and anxiety disorders (Sanchez-Villegas and Martínez-González 2013; Opie et al. 2015; Lutz et al. 2021). People who are stressed generally require a more nutrition-rich diet, whereas people who are stressed frequently choose "comfort" foods that are not sufficiently nutritious, developing unhealthy eating habits that worsen the stress situation. Stressed people often opt for foods containing high sugar and fat, also coffee is included in the list that contains high caffeine. Caffeine increases the release of cortisol and adrenaline hormones and dopamine neurotransmitters, which develop restlessness while resting is a must-do for the stressed person. Cortisol can increase appetite for foods rich in sugar and/or carbohydrate and fat, which can disturb normal metabolic function and hence enhance fat deposition in the body, leading to the development of diabetes, obesity, and cardiovascular diseases (Gonzalez and Miranda-Massari 2014) in the long run.

24.4 Nutrients Related to Stress Control

The human body and brain require macro- and micronutrients from food to maintain physical and mental health (Harbottle 2019; Gómez-Pinilla 2008; Goyal et al. 2018; Moore et al. 2018). Inadequate nutrition may cause alterations in neurotransmitter levels, causing mood alterations and the onset of depression (Jacka et al. 2010). Different nutrients were studied for their relevance in preventing the development of stress and depression in individuals. Miura and co-researchers have shown the significance of tryptophan metabolism in the maintenance of balanced mental health and any impairment in that pathway may develop depressive disorders (Miura et al. 2008). Shaw and the group showed that tryptophan and its structural analog 5-hydroxytryptophan can be used therapeutically for the treatment of depression (Shaw et al. 2002). Multiple studies have confirmed that folic acid, vitamin B6, and vitamin B12 coordinately play crucial roles in maintaining mental health and preventing the early onset of depression (Sachdev et al. 2005; Sánchez-Villegas et al. 2009; Skarupski et al. 2010). Even vitamin D deficiency has been found to be linked to the development of depression (Cuomo et al. 2017). Micronutrients like magnesium and zinc have been found to play potential roles in improving brain function and reducing the risk of depression (Lai et al. 2012; Yary and Aazami 2012; Derom et al. 2013). A semi-essential sulfur-containing amino acid namely taurine, and its analogs have been reported to be extremely useful against neurodevelopmental disorders, including depression (Jakaria et al. 2019). The essential polyunsaturated fatty acids, particularly omega-3 fatty acids, are critical in preventing depression and the development of any mental disorder (McNamara and Carlson 2006; Hibbeln 2009; Payne 2010). Several phenolic compounds such as resveratrol (3,4,5-trihydroxy-trans-stilbene) have been extensively studied for their various roles in protecting neurons (Xu et al. 2010; Bastianetto et al. 2015; Kuršvietienė et al. 2016). More research is being done on various other nutrients for their role in stress prevention and mental health improvement.

24.5 Roles of the Gut Microbiome in Stress and Depression

Some recent studies have discovered a link between major depressive disorders and the gut microbiome. The gut microbiome is the collection of trillions of microbes living in the human intestine that help with food digestion and conduct many other functions (Tannock 2017; Cheung et al. 2019). The gut microbiome is considered an "organ" itself as it is integrated into the functions of the host (O'Hara and Shanahan 2006). The gut microbiome is known to influence the nervous system's hypothalamus-pituitary-adrenal axis (Ait-Belgnaoui et al. 2012), the secretion of a familiar neurotransmitter, aminobutyric acid (GABA) (Digaitiene et al. 2012) and some short-chain fatty acids (SCFAs), which can influence the immune system (43–48). The gut-brain axis (a bidirectional system transferring impulses between CNS and gastrointestinal tract) not only is related to the function of the digestive system, but also is involved with inflammation of the gut, eating disorders, chronic abdominal pains, and stress-related gastrointestinal disorders, such as –irritable bowel syndrome (IBS) or inflammatory bowel disorder (IBD), etc. (Mayer 2011; Kennedy et al. 2012; Dinan and Cryan 2013). All of the connections point to the investigation of the link between the gut microbiome and major depressive disorders.

Studies related to gut inflammation and gut microbiome have revealed that such inflammation can lead to depression (Maes and Leunis 2008; Maes et al. 2012, 2013). Guida and co-researchers showed that antibiotic-induced disruption of the microbiome (a.k.a. dysbiosis) can develop depression-like behavior in mice, which can be reversed by administration of probiotic Lactobacillus casei DG (Guida et al. 2018). Interestingly, gut microbiome remodeling studies showed that gut microbiota from humans suffering from chronic depression were transferred to microbiome-deficient rodents that developed anxiety and depression-like symptoms in those rodents, compared to the control rodent group which received microbiota from healthy human controls (Kelly et al. 2016; Zheng et al. 2016).

Kwak and co-researchers detected Bifidobacterium genera of gut bacteria which are able to produce vitamin B2 (riboflavin), vitamin B3 (niacin), and vitamin B9 (folate) (Kwak et al. 2016). Many studies have shown that a low folate level is associated with depression and that people suffering from depression usually have low folate levels (Gilbody et al. 2007; Beydoun et al. 2010). Most of the human gut microbes are thiamin auxotroph (i.e., they need to intake thiamin for survival) (Paerl et al. 2018) and the deficiency of thiamin can also cause a depression-like condition in older adults (Zhang et al. 2013). Apart from the micronutrients, gut microbiomes are also known for the production of neurotransmitters and their precursors, such as –GABA (Barrett et al. 2012; Yunes et al. 2016), tryptamine (Williams et al. 2014), norepinephrine, dopamine (Wall et al. 2014), and serotonin (O'Mahony et al. 2015; Yunes et al. 2016).

Very few such studies have been conducted on humans directly (Hsiao et al. 2013; Mi et al. 2015); however, the results are promising and thus bring the gut microbiome to the attention of researchers as a possible target for the treatment of major depressive disorders. The gut microbiome can be controlled by a planned and healthy diet.

24.6 Dietary Modification and Stress Control

As stress- and depression-related mental conditions are not completely curable by known medications, it could be a better option to modify lifestyle and food habits. The Western diet is usually rich in saturated fats and sugar, which are very unhealthy. Unfortunately, the majority of the US and UK citizens are now used to such unhealthy diets (Kant 2000; Nielsen et al. 2002; Akbaraly et al. 2009). Rachelle

Table24.1 Antidepressant food scores (AFSs) of common foods from both plant and animal sources. (Data from LaChance and Ramsey 2018)

Vegetables	48%
Organ meats	25%
Fruits	20%
Seafood	16%
Legumes	8%
Grains	5%
Nuts and seeds	5%
Dairy	3%

S. Opie and co-researchers published a recommended diet plan to prevent depression in 2017 (Opie et al. 2017). This five-step diet plan suggests that:

1. Following any of the 'traditional' dietary patterns (such as the Mediterranean, Norwegian, or Japanese diet).
2. Increasing consumption of fruits, vegetables, legumes, wholegrain cereals, nuts, and seed (rich in vitamins, minerals and phenolics).
3. Eating foods high in omega-3 polyunsaturated fatty acids, such as shellfish and seafood.
4. Replacing unhealthy fast foods and fried foods with nutritional foods.
5. Controlling the intake of processed-foods, commercial bakery goods, and sweets.

The "Supporting the Modification of Lifestyle in Lowered Emotional States" (SMILES) trial was the first controlled trial of dietary modification to treat depression and the outcome of administration of a modified Mediterranean diet was very promising (O'Neil et al. 2013). As shown in Table 24.1, LaChance and colleagues successfully studied various animal and plant food sources and assigned them an Antidepressant Food Score (AFS)—the first system to nutritionally profile food in order to plan a diet. Foods with a higher AFS are beneficial in the prevention of depression (LaChance and Ramsey 2018).

Among all the traditional diets, the one most studied and applied for the treatment purpose of depression patients is the Mediterranean diet (MedD). Initial modification to any diet according to MedD can be the addition of foods rich in ω-3 long-chain polyunsaturated fatty acids (LCPUFA) will be a good start. The most commonly found ω-3 LCPUFA are α-linolenic acid [ALA, 18: 3n-3], eicosapentaenoic acid [EPA, 20: 5n-3] and docosahexaenoic acid [DHA, 22: 6n-3]. Seeds and nuts are rich in ALA; whereas, fatty fish and seafood are rich in EPA and DHA –the MedD is a holy grail as it is enriched with nutrient-rich seeds, nuts, fish, and seafood (Hibbeln and Salem Jr 1995; Hibbeln 1998; Estruch 2010). These ω-3 LCPUFAs can be incorporated into the neuronal membranes (e.g., synaptosomes and myelin), especially DHA is rich in the membranes of synaptic, mitochondrial, and endoplasmic membranes (Logan 2004; Beltz et al. 2007). EPA and DHA are involved in synapse synthesis, synaptic responses, actions of ion channels in the synapses, controlling both dopaminergic and serotonin-mediated pathways etc., and hence have an immense role in maintaining healthy brain function (Lutz and Durand 1994;

Kawakita et al. 2006; Uauy and Dangour 2006; Innis 2007; Dyall 2015). There have been several studies on clinical trials of ω-3 LCPUFA on patients with chronic depression, these fatty acids can show preventive effects and improve the cognitive process (Stoll et al. 1999; Colangelo et al. 2009; Lucas et al. 2011; Sublette et al. 2011; Muldoon et al. 2014; Pusceddu et al. 2016; Hamazaki et al. 2018). Apart from the fatty acids, eating fresh vegetables and fruits can be a rich source of vitamin B, vitamin C, minerals, and phenolics, which are also known to be associated with maintaining healthy brain functions. Reduction in caffeine intake can improve the sleep pattern and release of relaxing hormones significantly. Minerals and some vitamins can also control the gut microbial flora which is essential for healthy brain functions. Antioxidants may be an important addition to the diet for maintaining a healthy nervous system, possibly by preventing oxidative damage and stress on neurons; however, the direct function of antioxidants remains unknown. Daily exercises or physical activities could be an added bonus to a modified nutrition-rich diet as they are known to improve mood.

24.7 Concluding Remarks

This chapter discusses the nutrients that can maintain healthy brain function and can prevent the development of stress-induced depressive disorders. Adaptation to the most familiar Mediterranean diet could be a life-saving step because it makes a diet nutrition-rich and is enough to prevent any damage to the nervous system. We lack proper medical treatment for depressive disorders, and dietary modification could be our best hope to stay healthy and functional for a long time.

References

Ait-Belgnaoui A et al (2012) Prevention of gut leakiness by a probiotic treatment leads to attenuated HPA response to an acute psychological stress in rats. Psychoneuroendocrinology 37(11):1885–1895. https://doi.org/10.1016/j.psyneuen.2012.03.024

Akbaraly TN, Brunner EJ, Ferrie JE, Marmot MG, Kivimaki M, Singh-Manoux A (2009) Dietary pattern and depressive symptoms in middle age. Br J Psychiatry 195(5):408–413. https://doi.org/10.1192/bjp.bp.108.058925. PMID 19880930

Ali AM, Hendawy AO (2018) So, antidepressant drugs have serious adverse effects, but what are the alternatives. Nov Appro Drug Des Dev 4(3):555636

Barrett E, Ross RP, O'Toole PW, Fitzgerald GF, Stanton C (2012) γ-Aminobutyric acid production by culturable bacteria from the human intestine. J Appl Microbiol 113(2):411–417. https://doi.org/10.1111/j.1365-2672.2012.05344.x, PMID 22612585

Bastianetto S, Ménard C, Quirion R (2015) Neuroprotective action of resveratrol. Biochim Biophys Acta 1852(6):1195–1201. https://doi.org/10.1016/j.bbadis.2014.09.011, PMID 25281824

Beltz BS, Tlusty MF, Benton JL, Sandeman DC (2007) Omega-3 fatty acids upregulate adult neurogenesis. Neurosci Lett 415(2):154–158. https://doi.org/10.1016/j.neulet.2007.01.010, PMID 17240063

Beydoun MA et al (2010) Serum folate, vitamin B-12 and homocysteine and their association with depressive symptoms among US adults. Psychosom Med 72(9):862

Carlson NR (2004) Physiology of behavior. Pearson higher, 8th edn. Allyn & Bacon, New York, NY

Chávez-Castillo M et al (2019) Depression as a neuroendocrine disorder: emerging neuropsychopharmacological approaches beyond monoamines. Adv Pharmacol Sci 2019:7943481

Cheung SG, Goldenthal AR, Uhlemann AC, Mann JJ, Miller JM, Sublette ME (2019) Systematic review of gut microbiota and major depression. Front Psych 10:34. https://doi.org/10.3389/fpsyt.2019.00034, PMID 30804820

Colangelo LA, He K, Whooley MA, Daviglus ML, Liu K (2009) Higher dietary intake of long-chain ω-3 polyunsaturated fatty acids is inversely associated with depressive symptoms in women. Nutrition 25(10):1011–1019. https://doi.org/10.1016/j.nut.2008.12.008, PMID 19195841

Cuomo A, Giordano N, Goracci A, Fagiolini A (2017) Depression and vitamin D deficiency: causality, assessment, and clinical practice implications. Neuropsychiatry 07(5):606–614. https://doi.org/10.4172/Neuropsychiatry.1000255

Dale E, Bang-Andersen B, Sánchez C (2015) Emerging mechanisms and treatments for depression beyond SSRIs and SNRIs. Biochem Pharmacol 95(2):81–97

Derom ML, Sayón-Orea C, Martínez-Ortega JM, Martínez-González MA (2013) Magnesium and depression: a systematic review. Nutr Neurosci 16(5):191–206. https://doi.org/10.1179/1476830512Y.0000000044, PMID 23321048

Digaitiene A et al (2012) Lactic acid bacteria isolated from rye sourdoughs produce bacteriocin-like inhibitory substances active against Bacillus subtilis and fungi. J Appl Microbiol 112(4):732–742

Dinan TG, Cryan JF (2013) Melancholic microbes: a link between gut microbiota and depression? Neurogastroenterol Motil 25(9):713–719. https://doi.org/10.1111/nmo.12198, PMID 23910373

Dyall SC (2015) Long-chain omega-3 fatty acids and the brain: a review of the independent and shared effects of EPA, DPA and DHA. Front Aging Neurosci 7:52. https://doi.org/10.3389/fnagi.2015.00052, PMID 25954194

Estruch R (2010) Anti-inflammatory effects of the mediterranean diet: the experience of the PREDIMED study. Proc Nutr Soc 69(3):333–340. https://doi.org/10.1017/S0029665110001539, PMID 20515519

Gilbody S, Lightfoot T, Sheldon T (2007) Is low folate a risk factor for depression? A meta-analysis and exploration of heterogeneity. J Epidemiol Community Health 61(7):631–637

Gómez-Pinilla F (2008) Brain foods: the effects of nutrients on brain function. Nat Rev Neurosci 9 (7):568–578

Gonzalez MJ, Miranda-Massari JR (2014) Diet and stress. Psychiatr Clin N Am 37(4):579–589. https://doi.org/10.1016/j.psc.2014.08.004

Goyal MS, Iannotti LL, Raichle ME (2018) Brain nutrition: a life span approach. Annu Rev Nutr 38:381–399

Guida F, Turco F, Iannotta M, De Gregorio D, Palumbo I, Sarnelli G et al (2018) Antibiotic-induced microbiota perturbation causes gut endocannabinoidome changes, hippocampal neuroglial reorganization and depression in mice. Brain Behav Immun 67:230–245. https://doi.org/10.1016/j.bbi.2017.09.001, PMID 28890155

Hamazaki K, Takamori A, Tsuchida A, Kigawa M, Tanaka T, Ito M et al (2018) Dietary intake of fish and n-3 polyunsaturated fatty acids and risks of perinatal depression: the Japan Environment and Children's Study (JECS). J Psychiatr Res 98:9–16. https://doi.org/10.1016/j.jpsychires.2017.11.013, PMID 29253720

Harbottle L (2019) The effect of nutrition on older people's mental health. Br J Community Nurs 24 (Sup7):S12–S16

Hibbeln JR (1998) Fish consumption and major depression. Lancet 351(9110):1213. https://doi.org/10.1016/S0140-6736(05)79168-6, PMID 9643729

Hibbeln JR (2009) Depression, suicide and deficiencies of omega–3 essential fatty acids in modern diets. World Rev Nutr Diet 99:17–30. https://doi.org/10.1159/000192992, PMID 19136836

Hibbeln JR, Salem N Jr (1995) Dietary polyunsaturated fatty acids and depression: when cholesterol does not satisfy. Am J Clin Nutr 62(1):1–9. https://doi.org/10.1093/ajcn/62.1.1, PMID 7598049

Hsiao EY, McBride SW, Hsien S, Sharon G, Hyde ER, McCue T et al (2013) Microbiota modulate behavioral and physiological abnormalities associated with neurodevelopmental disorders. Cell 155(7):1451–1463. https://doi.org/10.1016/j.cell.2013.11.024, PMID 24315484

Innis SM (2007) Dietary (n-3) fatty acids and brain development. J Nutr 137(4):855–859. https://doi.org/10.1093/jn/137.4.855, PMID 17374644

Jacka FN (2017) Nutritional psychiatry: where to next? EBiomedicine 17:24–29. https://doi.org/10.1016/j.ebiom.2017.02.020, PMID 28242200

Jacka FN, Pasco JA, Mykletun A, Williams LJ, Hodge AM, O'Reilly SL et al (2010) Association of western and traditional diets with depression and anxiety in women. Am J Psychiatry 167(3): 305–311. https://doi.org/10.1176/appi.ajp.2009.09060881, PMID 20048020

Jakaria M, Azam S, Haque ME, Jo SH, Uddin MS, Kim IS et al (2019) Taurine and its analogs in neurological disorders: focus on therapeutic potential and molecular mechanisms. Redox Biol 24:101223. https://doi.org/10.1016/j.redox.2019.101223, PMID 31141786

Kant AK (2000) Consumption of energy-dense, nutrient-poor foods by adult Americans: nutritional and health implications. The third National Health and Nutrition Examination Survey, 1988–1994. Am J Clin Nutr 72(4):929–936. https://doi.org/10.1093/ajcn/72.4.929, PMID 11010933

Kawakita E, Hashimoto M, Shido O (2006) Docosahexaenoic acid promotes neurogenesis in vitro and in vivo. Neuroscience 139(3):991–997. https://doi.org/10.1016/j.neuroscience.2006.01.021, PMID 16527422

Kelly JR, Borre Y, Patterson E, El Aidy S, Deane J et al (2016) Transferring the blues: depression-associated gut microbiota induces neurobehavioural changes in the rat. J Psychiatr Res 82:109–118. https://doi.org/10.1016/j.jpsychires.2016.07.019, PMID 27491067

Kennedy PJ, Clarke G, Quigley EM, Groeger JA, Dinan TG, Cryan JF (2012) Gut memories: towards a cognitive neurobiology of irritable bowel syndrome. Neurosci Biobehav Rev 36(1): 310–340. https://doi.org/10.1016/j.neubiorev.2011.07.001, PMID 21777613

Kessler RC, Davis CG, Kendler KS (1997) Childhood adversity and adult psychiatric disorder in the US national comorbidity survey. Psychol Med 27(5):1101–1119. https://doi.org/10.1017/s0033291797005588, PMID 9300515

Kuršvietienė L, Stanevičienė I, Mongirdienė A, Bernatonienė J (2016) Multiplicity of effects and health benefits of resveratrol. Medicina (Kaunas) 52(3):148–155. https://doi.org/10.1016/j.medici.2016.03.003, PMID 27496184

Kwak MJ, Kwon SK, Yoon JK, Song JY, Seo JG, Chung MJ et al (2016) Evolutionary architecture of the infant-adapted group of Bifidobacterium species associated with the probiotic function. Syst Appl Microbiol 39(7):429–439. https://doi.org/10.1016/j.syapm.2016.07.004, PMID 27524178

LaChance LR, Ramsey D (2018) Antidepressant foods: an evidence-based nutrient profiling system for depression. World J Psychiatry 8(3):97–104. https://doi.org/10.5498/wjp.v8.i3.97

Lai J, Moxey A, Nowak G, Vashum K, Bailey K, McEvoy M (2012) The efficacy of zinc supplementation in depression: systematic review of randomised controlled trials. J Affect Disord 136(1–2):e31–e39. https://doi.org/10.1016/j.jad.2011.06.022, PMID 21798601

Logan AC (2004) Omega-3 fatty acids and major depression: a primer for the mental health professional. Lipids Health Dis 3(1):25. https://doi.org/10.1186/1476-511X-3-25, PMID 15535884

Lucas M, Mirzaei F, O'Reilly EJ, Pan A, Willett WC, Kawachi I et al (2011) Dietary intake of n−3 and n−6 fatty acids and the risk of clinical depression in women: a 10-y prospective follow-up study. Am J Clin Nutr 93(6):1337–1343. https://doi.org/10.3945/ajcn.111.011817, PMID 21471279

Lutz M, Durand G (1994) Influence of dietary lipids on fatty acid composition of nervous membranes (myelin and synaptosomes) in rats. Nutr Res 14(9):1365–1373. https://doi.org/10.1016/S0271-5317(05)80296-9

Lutz M, Vargas C, Stojanova J, Arancibia M (2021) Diet and depressive disorders. Arch Clin Psychiatr (S Paulo) 48:117–122

Maes M, Kubera M, Leunis JC, Berk M (2012) Increased IgA and IgM responses against gut commensals in chronic depression: further evidence for increased bacterial translocation or leaky gut. J Affect Disord 141(1):55–62. https://doi.org/10.1016/j.jad.2012.02.023, PMID 22410503

Maes M, Kubera M, Leunis JC, Berk M, Geffard M, Bosmans E (2013) In depression, bacterial translocation may drive inflammatory responses, oxidative and nitrosative stress (O&NS), and autoimmune responses directed against O&NS-damaged neoepitopes. Acta Psychiatr Scand 127(5):344–354. https://doi.org/10.1111/j.1600-0447.2012.01908.x, PMID 22900942

Maes M, Leunis JC (2008) Normalization of leaky gut in chronic fatigue syndrome (CFS) is accompanied by a clinical improvement: effects of age, duration of illness and the translocation of LPS from gram-negative bacteria. Neuro Endocrinol Lett 29(6):902–910, PMID 19112401

Mayer EA (2011) Gut feelings: the emerging biology of gut–brain communication. Nat Rev Neurosci 12(8):453–466. https://doi.org/10.1038/nrn3071, PMID 21750565

Mazure CM (1998) Life stressors as risk factors in depression. Clin Psychol Sci Pract 5(3):291–313. https://doi.org/10.1111/j.1468-2850.1998.tb00151.x

McNamara RK, Carlson SE (2006) Role of omega-3 fatty acids in brain development and function: potential implications for the pathogenesis and prevention of psychopathology. Prostaglandins Leukot Essent Fat Acids 75(4–5):329–349

Mi GL, Zhao L, Qiao DD, Kang WQ, Tang MQ, Xu JK (2015) Effectiveness of lactobacillus reuteri in infantile colic and colicky induced maternal depression: a prospective single blind randomized trial. Antonie Van Leeuwenhoek 107(6):1547–1553. https://doi.org/10.1007/s10482-015-0448-9, PMID 25876529

Miura H, Ozaki N, Sawada M, Isobe K, Ohta T, Nagatsu T (2008) A link between stress and depression: shifts in the balance between the kynurenine and serotonin pathways of tryptophan metabolism and the etiology and pathophysiology of depression. Stress 11(3):198–209. https://doi.org/10.1080/10253890701754068, PMID 18465467

Monroe SM, Hadjiyannakis K (2002) The social environment and depression: focusing on severe life stress. In: Gotlib IH, Hammen CL (eds) Handbook of depression. Guilford Press, New York, NY, pp 314–340

Moore K et al (2018) Diet, nutrition and the ageing brain: current evidence and new directions. Proc Nutr Soc 77(2):152–163

Muldoon MF, Ryan CM, Yao JK, Conklin SM, Manuck SB (2014) Long-chain omega-3 fatty acids and optimization of cognitive performance. Mil Med 179(11 Suppl):95–105. https://doi.org/10.7205/MILMED-D-14-00168, PMID 25373092

Murray CJ, Lopez AD, World Health Organization (1996) The global burden of disease: a comprehensive assessment of mortality and disability from diseases, injuries, and risk factors in 1990 and projected to 2020: summary. World Health Organization

Nielsen SJ, Siega-Riz AM, Popkin BM (2002) Trends in energy intake in US between 1977 and 1996: similar shifts seen across age groups. Obes Res 10(5):370–378. https://doi.org/10.1038/oby.2002.51, PMID 12006636

O'Hara AM, Shanahan F (2006) The gut flora as a forgotten organ. EMBO Rep 7(7):688–693. https://doi.org/10.1038/sj.embor.7400731, PMID 16819463

O'Mahony SM, Clarke G, Borre YE, Dinan TG, Cryan JF (2015) Serotonin, tryptophan metabolism and the brain-gut-microbiome axis. Behav Brain Res 277:32–48. https://doi.org/10.1016/j.bbr.2014.07.027, PMID 25078296

O'Neil A, Berk M, Itsiopoulos C, Castle D, Opie R, Pizzinga J et al (2013) A randomised, controlled trial of a dietary intervention for adults with major depression (the 'SMILES' trial): study protocol. BMC Psychiatry 13(1):1–7

Opie RS, Itsiopoulos C, Parletta N, Sanchez-Villegas A, Akbaraly TN, Ruusunen A et al (2017) Dietary recommendations for the prevention of depression. Nutr Neurosci 20(3):161–171. https://doi.org/10.1179/1476830515Y.0000000043, PMID 26317148

Opie RS, O'Neil A, Itsiopoulos C, Jacka FN (2015) The impact of whole-of-diet interventions on depression and anxiety: a systematic review of randomised controlled trials. Public Health Nutr 18(11):2074–2093. https://doi.org/10.1017/S1368980014002614, PMID 25465596

Paerl RW, Bertrand EM, Rowland E, Schatt P, Mehiri M, Niehaus TD et al (2018) Carboxythiazole is a key microbial nutrient currency and critical component of thiamin biosynthesis. Sci Rep 8(1):5940. https://doi.org/10.1038/s41598-018-24321-2, PMID 29654239

Paykel ES (2003) Life events and affective disorders. Acta Psychiatr Scand Suppl 108(418):61–66. https://doi.org/10.1034/j.1600-0447.108.s418.13.x, PMID 12956817

Payne ME (2010) Nutrition and late-life depression: etiological considerations Depression: a significant problem for older adults. Aging Health 6(1):133–143. Available at: https://www.ncbi.nlm.nih.gov/pmc/articles/PMC2841351/pdf/nihms180649.pdf

Pusceddu MM, Kelly P, Stanton C, Cryan JF, Dinan TG (2016) N-3 polyunsaturated fatty acids through the lifespan: implication for psychopathology. Int J Neuropsychopharmacol 19(12): pyw078. https://doi.org/10.1093/ijnp/pyw078, PMID 27608809

Rao TS, Asha MR, Ramesh BN, Rao KS (2008) Understanding nutrition, depression and mental illnesses. Indian J Psychiatry 50(2):77–82. https://doi.org/10.4103/0019-5545.42391, PMID 19742217

Sachdev PS, Parslow RA, Lux ORA, Salonikas C, Wen WEI, Naidoo D et al (2005) Relationship of homocysteine, folic acid and vitamin B12 with depression in a middle-aged community sample. Psychol Med 35(4):529–538. https://doi.org/10.1017/s0033291704003721, PMID 15856723

Sánchez-Villegas A, Doreste J, Schlatter J, Pla J, Bes-Rastrollo M, Martínez-González MA (2009) Association between folate, vitamin B6 and vitamin B12 intake and depression in the SUN cohort study. J Hum Nutr Diet 22(2):122–133. https://doi.org/10.1111/j.1365-277X.2008.00931.x, PMID 19175490

Sanchez-Villegas A, Martínez-González MA (2013) Diet, a new target to prevent depression? BMC Med 11(1):3. https://doi.org/10.1186/1741-7015-11-3, PMID 23286788

Shaw K, Turner J, Del Mar C (2002) Are tryptophan and 5-hydroxytryptophan effective treatments for depression? A meta-analysis. Aust N Z J Psychiatry 36(4):488–491. https://doi.org/10.1046/j.1440-1614.2002.01046.x, PMID 12169147

Skarupski KA, Tangney C, Li H, Ouyang B, Evans DA, Morris MC (2010) Longitudinal association of vitamin B-6, folate, and vitamin B-12 with depressive symptoms among older adults over time. Am J Clin Nutr 92(2):330–335. https://doi.org/10.3945/ajcn.2010.29413, PMID 20519557

Stoll AL, Severus WE, Freeman MP, Rueter S, Zboyan HA, Diamond E et al (1999) Omega 3 fatty acids in bipolar disorder: a preliminary double-blind, placebo-controlled trial. Arch Gen Psychiatry 56(5):407–412. https://doi.org/10.1001/archpsyc.56.5.407, PMID 10232294

Sublette ME, Ellis SP, Geant AL, Mann JJ (2011) Meta-analysis of the effects of eicosapentaenoic acid (EPA) in clinical trials in depression. J Clin Psychiatry 72(12):1577–1584. https://doi.org/10.4088/JCP.10m06634, PMID 21939614

Tannock GW (2017) Understanding the gut microbiota. John Wiley & Sons, Hoboken, NJ

Tennant C (2002) Life events, stress and depression: a review of recent findings. Aust N Z J Psychiatry 36(2):173–182. https://doi.org/10.1046/j.1440-1614.2002.01007.x, PMID 11982537

Uauy R, Dangour AD (2006) Nutrition in brain development and aging: role of essential fatty acids. Nutr Rev 64((5 Pt 2)(suppl_2)):S24–S33. https://doi.org/10.1301/nr.2006.may.s24-s33, PMID 16770950; discussion S72

Wall R, Cryan JF, Ross RP, Fitzgerald GF, Dinan TG, Stanton C (2014) Bacterial neuroactive compounds produced by psychobiotics. In: Microbial endocrinology: the microbiota-gut-brain axis in health and disease, pp 221–239. https://doi.org/10.1007/978-1-4939-0897-4_10, PMID 24997036

Williams BB, Van Benschoten AH, Cimermancic P, Donia MS, Zimmermann M, Taketani M et al (2014) Discovery and characterization of gut microbiota decarboxylases that can produce the neurotransmitter tryptamine. Cell Host Microbe 16(4):495–503. https://doi.org/10.1016/j.chom.2014.09.001, PMID 25263219

Xu Y, Wang Z, You W, Zhang X, Li S, Barish PA et al (2010) Antidepressant-like effect of trans-resveratrol: involvement of serotonin and noradrenaline system. Eur Neuropsychopharmacol 20(6):405–413. https://doi.org/10.1016/j.euroneuro.2010.02.013, PMID 20353885

Yary T, Aazami S (2012) Dietary intake of zinc was inversely associated with depression. Biol Trace Elem Res 145(3):286–290. https://doi.org/10.1007/s12011-011-9202-y, PMID 21932045

Yunes RA, Poluektova EU, Dyachkova MS, Klimina KM, Kovtun AS, Averina OV et al (2016) GABA production and structure of gadB/gadC genes in lactobacillus and bifidobacterium strains from human microbiota. Anaerobe 42:197–204. https://doi.org/10.1016/j.anaerobe.2016.10.011, PMID 27794467

Zhang G, Ding H, Chen H, Ye X, Li H, Lin X et al (2013) Thiamine nutritional status and depressive symptoms are inversely associated among older Chinese adults. J Nutr 143(1):53–58. https://doi.org/10.3945/jn.112.167007, PMID 23173173

Zheng P, Zeng B, Zhou C, Liu M, Fang Z, Xu X et al (2016) Gut microbiome remodeling induces depressive-like behaviors through a pathway mediated by the host's metabolism. Mol Psychiatry 21(6):786–796. https://doi.org/10.1038/mp.2016.44, PMID 27067014

Dietary Modifications to Manage Stress 25

Tanuj Joshi, Mukesh Nandave, and Archana N. Sah

Abstract

Stress leads to several health complications. Stress results in poor dietary habits and poor dietary habits lead to stress. Thus, stress and diet go hand in hand. Stress, eating behaviors, diet, addiction to food, and dietary modifications are important concepts associated with the modern lifestyle. The relation of stress with diet is very important and good dietary modifications can help to beat stress. Food has a very significant effect on health and mood, and by eating a good diet, various physical and psychological problems can be countered. Food products like dietary fibers, vitamins, minerals, vegetables, fruits, etc. are very important to maintain good health. The key to long life and stress lies to a great extent in the diet and food we consume along with good eating behavior.

Keywords

Stress · Diet · Nutrients · Food addiction · Blueberries · Tea

T. Joshi · A. N. Sah (✉)
Department of Pharmaceutical Sciences, Bhimtal Campus, Kumaun University, Nainital, Uttarakhand, India
e-mail: drarchanansah@gmail.com

M. Nandave
Department of Pharmacology, Delhi Pharmaceutical Sciences and Research University (DPSRU), New Delhi, India

R. Tulsawani, D. Vohora (eds.), *Adaptation under Stressful Environments through Biological Adjustments and Interventions*,
https://doi.org/10.1007/978-981-99-7652-2_25

25.1 Introduction

There is an association of stress with the consumption of food. The pattern of food eating is changed due to stress. The quality of diet eaten by a stressed person is poor in nature (Abril-Ulloa et al. 2022). In today's world, eating habits have worsened with times. The food we eat nowadays is packed with high amount of animal fats, artificial colors, refined carbohydrates, sweeteners, and preservatives. Unhealthy food products and unhealthy eating practices result in an increase in the level of stress. The combination of unhealthy eating habits and stress will increase the incidence of diseases in future. The importance of food for healthy living has been rightly described many centuries ago by Hippocrates through the following lines: "Let food be your medicine and let medicine be your food." With proper eating habits, right supplementations, and stress reduction a person can live a long, healthy, and happy life (Gonzalez and Miranda-Massari 2014; Abril-Ulloa et al. 2022).

Hyperpalatable foods are being consumed by people on a regular basis these days. Hyperpalatable foods are those foods that are loaded with high sugar and fats and are highly addictive when consumed like potato chips. Stress that has become out of control leads to changes in eating patterns and in an increased consumption of hyperpalatable foods. Stress has important and interesting relation with addiction and addiction relapse (Yau and Potenza 2013). Mental health is affected by the food we consume and stress has a relationship with both mental health and our diet. There are associations between diet, stress, obesity, metabolic disorders, and mental health. Thus, it can be said that if we improve our understanding of the relationship between stress, diet, metabolic disorders and mental health many new miraculous treatments can be generated for a variety of diseases and disorders (Bremner et al. 2020).

Diet affects both physical and mental health (Sofi et al. 2008). Bad dietary modifications can induce mental disorders through direct effects on mood; interestingly, on the other hand, mental issues can lead to changes in dietary habits (Freeman and Rapaport 2011). The choice of what food we eat is driven by both physiological as well as psychological factors. Once bad quality foods are consumed, they affect our mood (Osdoba et al. 2015). Recently, researchers have studied the emotional outcomes due to consumption of various foods using different experimental settings (Osdoba et al. 2015). Hunger is associated with irritability and alertness whereas satiety is associated with calmness and sleepiness (Gibson 2006). Emotional responses are provoked by the taste of a food like its sweetness (Steiner 1974). Also, the feeling of guilt can be provoked after eating a calorie rich food (Macht et al. 2003). The past involvements with foods and cognitive roles associated with foods play an important role in food addiction and poor eating habits. In these cases, past experiences and memories with foods will affect our emotional responses towards a particular diet (Cardello et al. 2012; Walsh and Kiviniemi 2014; Wansink et al. 2007; Mojet and Köster 2002). There is also a concept of emotional eating. In emotional eating a person overeats as a coping mechanism to deal with negative emotions like stress, anxiety, and depression (Shehata and Abdeldaim 2023; Ganley 1989). During emotional eating, a person is unable to differentiate between physiological hunger from a desire to eat due to negative emotions (Bruch 1964). There exists a relation between the emotional state of a person and his desire to eat. Anger, sadness, and fear can all lead to changes in the eating patterns of a person. In these cases, there are both biological and psychological influences to eat. The food consumption is increased in people who are in a state of stress, pressure and anxiety and also in those experiencing other types of extreme mood and behavioral states (GÜNeŞEr and Him 2021). Thus, eating is adopted by a person as a method of escape and distraction from negative influences and emotions in life (Heatherton and Baumeister 1991). There are various factors which can influence the quality and pattern of the diet of people around the world. Some of these factors include age, adiposity, smoking, education, income, ethnicity, psychological factors, and ethnicity (Boynton et al. 2008; Giskes et al. 2011).

Stress and diet have a very important bond and the dietary influences on stress are individualized in nature (Khaled et al. 2020). Some studies have evaluated the influence of stress on energy consumption and these studies have given an idea that when people are under stress 20% of people eat the same amount of food whereas 40% eat more and 40% eat less amount of food when compared to the amount of food they consume in absence of stress (Torres and Nowson 2007; Block et al. 2009). The differences in the amount of food consumed by individuals in different studies can be due to factors like the type of stressor, variations in the levels of satiety and hunger and the stress exposure duration in various studies (Yau and Potenza 2013). It has been seen that acute/severe stress that is felt during upcoming exam or work deadline restrict the feeding habits whereas unhappy marriage, unemployment, long-term poverty, and other such related factors increase binge intake and food intake (Yau and Potenza 2013).

According to experimental studies people suffering from stress tend to eat hyperpalatable foods that are nutritionally low, high in calorie and full of fats even when they are not hungry or their body is not demanding food. Some examples of these foods are like cream, cheese, butter, full-fat products etc. (Oliver et al. 2000; Zellner et al. 2006; Rutters et al. 2009). There is an enhanced effect of stress in people who are obese as compared to normal weight people since obese people have a higher degree of insulin resistance as compared to normal weight people and there is a higher activation of reward regions of the brain under stress in obese people (Jastreboff et al. 2013; Yau and Potenza 2013). Perceived stress in young adults and students of universities leads to low diet quality as shown by various studies (Bayram and Bilgel 2008; Wichianson et al. 2009). It has been shown in experiments that young females who were stressed consumed a fattier diet than nonstressed females (Habhab et al. 2009; Wichianson et al. 2009). A study conducted by researchers showed an association between diet and perceived stress in women of reproductive age. It was found the females in high stress group consumed more sweets and fats than low stress group (Habhab et al. 2009). In another study, high fast-food consumption was linked with perceived stress in young women (Barrington et al. 2012; Khaled et al. 2020).

There are several ways in which we can defeat stress and one of them is by maintaining a proper diet. A diet that is rich in nutrients and is balanced can help in decreasing stress and enhancing the function of the brain. There are nutrients that can decrease the levels of adrenaline and cortisol that are elevated in stress. The substances present in food that have a stress reducing effects are vitamin C, B vitamins, complex carbohydrates, proteins, selenium, and magnesium. These nutrients have a great role in reducing stress. Besides, these nutrients there are other nutrients too that can counter stress. Thus dietary modifications play a very important role in the management of stress (Singh 2016).

25.2 Concepts Related to Diet and Eating Behaviors in Relation to Stress

25.2.1 Influences of Diet and Eating Habits on Physical and Psychological Health

Eating patterns and diet have a very important effect on our health and mood. In today's hectic life schedules, we are experiencing a variety of lifestyle associated diseases. There are numerous clinical evidences by which we can say that dietary modifications can prevent and manage several diseases that affect mankind like diabetes and cardiovascular diseases. Dietary modifications are also helpful in the production of psychiatric diseases and stress-related diseases (Yau and Potenza 2013; Bremner et al. 2020). Chronic stress responses are activated by early life stresses such as maternal separation in rats (Yau and Potenza 2013; Bremner et al. 2020). When energy rich food are consumed on a regular basis, it results in changes in carbohydrate and fat metabolism, appetite hormones, insulin sensitivity, neuronal networks, etc. Due to the modifications in all these factors cravings, salience, choice of food and selection, intake of food, and motivation are altered (Alsio et al. 2010; Habhab et al. 2009).

Uncontrolled and constant stress with time deregulates the HPA axis. This in turn affects the eating behavior and energy homeostasis (Adam and Epel 2007). Corticotropin releasing factor (CRF) and noradrenaline may inhibit appetite during stress whereas cortisol has an appetite promoting effect during recovery from stress (Takeda et al. 2004). Abdominal fat deposition is promoted by prolonged stress induced glucocorticoid secretion, this in synergy with insulin secretion can decrease activity of the HPA axis (Dallman et al. 2005). Those people who are under chronic stress eat more in acute stress conditions (Gibson 2006). These people show a liking for hyperpalatable energy rich sweetened and fatty foods (Torres and Nowson 2007; Warne 2009). There are various factors which mediate between stress and eating behavior. These factors affect people at individual levels and cannot be generalized. These factors lead to stress induced hyperphagia (Yau and Potenza 2013). One of these factors is weight and diet-related metabolism. Allostatic load is modified by weight-related metabolic changes (Ikeda et al. 1986).

Studies on animals have shown that most of the times the characteristic features of obesity are a decrease in the level of adipose signal or production of resistance at the receptor level (Ikeda et al. 1986). When there is a state of inadequate adipose signaling, it leads to negative feedback by reducing pleasant effect of food. This leads to a prolongation of food consumption and interferences in stopping the eating process of food. Also, the peripheral tissue sensitivity of skeletal muscle and fat towards glucocorticoids may be modified (Reynolds et al. 2002). Increase in weight, resistance towards insulin and consumption of diets, high in fats lead to weakened glucocorticoid responses towards stressful circumstances. Also, the peripheral catecholamine and autonomic responses are altered (Tyrka et al. 2012). This blunting of the natural responses towards stress may lead to eating behaviors that are non-homeostatic state in nature (Figlewicz et al. 2004). There is a strong association between gaining of weight and chronic stress in people with a high BMI rate in comparison to people with a low BMI rate who experience the same amount of stress (Block et al. 2009). In line with this fact, stress-related eating has a relationship with obesity in women (Laitinen et al. 2002). Overweight and obese people are more sensitive towards food cues particularly after dealing with stress.

In a study related to food eating behavior, it was found that the participants of the study who were lean and healthy showed a reduction in mean food craving and energy intake in the absence of hunger in both stressful and normal conditions (Lemmens et al. 2011). However, increased mean craving of food and energy intake of hyperpalatable foods was shown by visceral overweight participants in the absence of hunger in stress versus conditions of rest (Lemmens et al. 2011). In comparison to lean people, obese people showed a significant activation in the brain regions associated with reward like insula, striatum, and thalamus while being exposed to stress and favorite food cues (Jastreboff et al. 2013). Also, when obese people were compared to lean people, a positive correlation was found between level of insulin resistance and the activation of striatum and insula in response to both stressful conditions and favorite food cues (Jastreboff et al. 2013).

High levels of insulin resistance as seen in obese and overweight individuals result in a decrease in the ability of insulin to suppress motivational pathways, and this results in an increase in stress and food cue-related responses (Yau and Potenza 2013; Adam and Epel 2007). Emotional eating is a concept that is gaining much attention nowadays. It is known that food has rewarding properties, and due to this fact, hyperpalatable food serves as a sort of comfort food for a person under distress. When people undergo a negative affective state, his cravings for hedonically rewarding food like high sugar and high fat foods goes high whereas in a state of happiness, people prefer less palatable dry fruits (Garg et al. 2007).

25.2.2 Selection of Food Items Under Stress and the Role of Reward System in Relation to Diet and Eating Behaviors

The mesolimbic dopaminergic system is related to the feeling of reward in our body. There is a link between the activation of the mesolimbic dopaminergic system and the HPA axis (Groesz et al. 2012; Sinha 2008). Opioidergic and the dopaminergic system are the pathways that are commonly utilized by both the drugs and foods (Volkow et al. 2011; Koob and Volkow 2010). It has been observed that high fat diets and consumption of drugs leads to alteration in glucocorticoid, CRF, and noradrenergic activity to increase the sensitization of reward pathways. This leads to an increase in craving and intake of substances of abuse and hyperpalatable foods (Sinha 2008). Conditioning, control, and reward sensitivity have been associated with dopamine with respect to foods and substances of abuse. Cues regarding food and actual food have been associated with an increase in dopamine levels (Di Chiara and Imperato 1988). Progressive compulsive behaviors may occur in people due to neuroadaptation which is caused by constant stimulation of the dopaminergic reward pathways (Everitt et al. 2008). It was observed in rats that antagonists of dopamine or lesions associated with dopaminergic system decreased the response towards food and decreased the reward value of both sugar rich foods and drugs of abuse (Wise et al. 2011).

In a study, participants were subjected to acute stress during positron emission tomography. It was found that both cortisol release and stress in this experiment increased the release of dopamine from nucleus accumbens (Wand et al. 2007). It was also found that more dopamine was released in ventral striatum by people with more cortisol reactivity. Thus, by studying the above-mentioned facts, we can say that a strong interconnection exists between cortisol and dopamine (Pruessner et al. 2004). Leptin, insulin, orexin, and ghrelin are homeostatic regulators of energy balance associated with HPA axis, modulate the rewarding abilities of food, and regulate behaviors that are non-homeostatic (Pruessner et al. 2004; Yau and Potenza 2013). These neuropeptides interact with the dopaminergic systems by the help of cognate receptors on ventral tegmental area dopaminergic neurons and by this, interaction may regulate the intake of food (Yau and Potenza 2013).

There is an involvement of the limbic system with memories for highly emotional events, coding of rewards, and reward-cue-based learning and feeding. On the other hand, prefrontal cortex is involved with executive control functions and higher cognitive functions. It is also involved with regulation of impulses, emotions, desires, and cravings (Berthoud 2012). During stress, limbic system circuitry is highly stimulated and the activity of the prefrontal cortex is reduced, this promotes the so-called automatic behaviors that favor survival and lead to behaviors like being alert for food cues. Synaptic branching is increased in the anterior cingulate cortex and the amygdala by both acute and chronic stressors while simultaneously there is a decrease in synaptic contacts with the prefrontal regions and the hippocampus (Vyas et al. 2002). This process further helps in progression of the chronic stress network towards limbic-biased stress responses. However, food is an inexpensive resource that provides reward with hyperpalatable foods and this may motivate stress-related eating to control stress responses. Thus, stress and the rewarding effects of hyperpalatable foods have a very strong relationship (Sinha 2008; Yau and Potenza 2013).

25.3 Stress Management Through Dietary Modifications

The most important hurdle that is being faced by mankind these days which is leading to stress, mental complications, and poor health are poor dietary and nutritional choices. With the challenges of the fast-paced world we are living in, good dietary habits have taken a back seat. Nowadays, people are consuming less and less of the good and fresh Indian diet which is prepared in our homes since time immemorial and shifting more towards junk foods, packaged foods, and processed foods which are detrimental to health (Naidoo 2021).

Stress and poor diet have a very strong interconnection. As discussed in the previous sections, increased levels of stress are associated with consumption of hyperpalatable foods (Fuhrman 2018; Kaur and Kochr 2019; Naidoo 2021; Ashakiran 2012). Also, poor food habits can lead to mental complications like stress and anxiety. Thus, the proper management of stress can be done by selection of a good diet and eating behaviors. Various food consumed nowadays can lead to stress and anxiety (Fuhrman 2018; Kaur and Kochr 2019; Naidoo 2021; Ashakiran 2012).

Drinking excess amount coffee or other products containing caffeine can be very harmful to mental health. Excess caffeine can lead to precipitation or worsening of anxiety (Naidoo 2021). Caffeine leads to overstimulation of those brain areas that are involved in processing of threat. In a study, healthy male participants were given either a placebo capsule or 250 mg of caffeine. The participants were shown threatening or neutral faces and the blood flow in the various regions of their brains was measured. They found that due to the action of caffeine there was activation of the midbrain periaqueductal gray matter. This is a region of the brain that is activated if a predator is coming close to us (Naidoo 2021). Also, caffeine shuts down a region of the brain that helps in the regulation of anxiety. Majority of experimental studies have found that caffeine above 400 mg can increase anxiety. So, it is recommended that the daily intake of caffeine should be reduced by people to avoid anxiety problems (Naidoo 2021).

Consumption of alcohol has also been associated with mental problems (Naidoo 2021). Consumption of alcohol on a regular basis can impair the quality of sleep, and this effect becomes more pronounced if the person is suffering from anxiety also (Becker 2012; Chueh et al. 2019). The problem of alcohol consumption is also seen in people suffering from social anxiety disorders. It is seen that these people try to overcome the fears of social anxiety by self-medicating them with alcohol which they think can provide them with courage (Terlecki et al. 2014; Naidoo 2021). Many people use alcohol as means of coping with difficulties of life and means of escape from their life situations. Some people try hard to leave alcohol and these people should be very carefully counseled by health practitioners in leaving alcohol, since during the withdrawal process these people can suffer from anxiety. People consume a lot of sugars in the form cocktails. Anxiety is worsened by sugar and both anxiety and depression are worsened by consumption of alcohol (Naidoo 2021).

Artificial sweeteners can also damage human health. There is no nutritional value of artificial sweeteners. These sweeteners can increase the level of harmful bacteria in the intestine. They thus lead to anxiety and influence mood in a bad way.

Aspartame which is used as a sweetener has important association with anxiety. This sweetener should be avoided and used in less amounts (Choudhary and Lee 2018; Naidoo 2021). As discussed in the previous sections, junk food, processed food, and packaged foods all can lead to mental complications and increase the level of stress in humans. Thus, by seeing all the negative effects of poor diet and its association with stress, we can improve our health and reduce the amount of stress by positive dietary modifications and eating behaviors in our day-to-day life (Naidoo 2021).

There are several types of food that we can consume to reduce stress and move towards a better state of health. Cultured, aged, and fermented foods; dietary fibers; Omega 3 fatty acids; and vegetables and fruits all can contribute to a reduction in stress in our lives (Naidoo 2021; Singh 2016). Researchers have found that stress may be linked to a variety of diseases and disorders. Stress has been associated with cardiovascular diseases, diabetes, and lifestyle diseases. Since obesity is also a factor in the above conditions. Researchers are now concentrating on the role of stress on the eating behavior of a person. They are particularly focusing on the amount of food eaten by a person and type of food consumed by him with respect to stress (Singh 2016; Naidoo 2021).

Physiological requirements of a person undergoing stress are highly increased. Oxygen/energy demands and circulatory requirements are increased in stressful situations and thus more metabolic cofactors are required like vitamins and minerals. The person under stress requires a more nutritionally sound diet. However, when a person is under stress, the exact opposite thing happens and the person begins to go for a comfort diet consisting of high sugar and fatty foods. This results in a nutritionally deficient diet which further puts a load over our metabolism. Thus, a good diet comprising of balanced nutrients is a must to counter stress. The various types of dietary choices that can be useful in the management of stress and other mental complications and make a difference in our lives are as follows (Naidoo 2021; Singh 2016; Fuhrman 2018; Kaur and Kochr 2019; Ashakiran 2012).

25.3.1 Dietary Fibers

The consumption of dietary fibers leads to a decrease in inflammation throughout our body including the brain. It has been found that the level of inflammation in the brain is enhanced in anxiety (Salim et al. 2012). Dietary fiber rich diets have been found to be useful to reduce the risk of anxiety, stress, and depression. Lentils, beans, seeds, nuts, and healthy whole grains are sources of rich dietary fibers (Felger 2018; Naidoo 2021; Singh 2016).

25.3.2 Omega-3 Fatty Acids

Many clinical trials have evaluated the beneficial effects of omega-3 fatty acids on anxiety and inflammation. A study carried out by a team of researchers evaluated the effects of n-3 polyunsaturated fatty acids on medical students. In the study, anxiety levels were measured in these students during different stress periods. It was found in the study that those students who received polyunsaturated fatty acids showed a reduction in the level of anxiety than the control group (Kiecolt-Glaser et al. 2011; Naidoo 2021). The omega-3 fatty acids produce a reduction in anxiety through neurochemical and anti-inflammatory mechanisms that affect the brain (Selhub et al. 2014). Having sea food and fish rich in omega-3 fatty acids can decrease the level of anxiety. Algae, chia seeds, sea vegetables, and flax seeds are alternatives sources of omega-3 fatty acids for vegetarians. The formation of healthy nerve cells in the brain is an important function performed by omega-3 fatty acids. Omega-3 fatty acids are associated with a lower risk of depression in people (Naidoo 2021; Singh 2016).

25.3.3 Vitamins

Majority of the vitamins belonging to the B-vitamin family play an important role in the maintenance and development of the nervous system. The ill effects of deficiency of B-vitamins on the nervous system lead to the production of symptoms related to stress like lethargy, irritability, and depression. These vitamins also contribute in the maintenance of the stability of mood and energy by keeping the blood sugar levels in a balanced state. Pantothenic acid or vitamin B_5 is the most important B-vitamin that is useful in stress, and it is referred to as antistress vitamin (Singh 2016; Gheita et al. 2020). An easing effect is provided on mood by vitamin B_{12}. Foods like almond are a rich source of vitamin E and help in stimulating the immune system. Mental illnesses are produced due to deficiencies in folic acid levels. Thus, folic acid might be very useful in providing relief in anxiety, stress, panic, and even depression. Sources of folic acid are Asparagus, kidney beans, whole meal bread, broccoli, Brussels sprouts, green leafy vegetables, etc. (Singh 2016; Lee and Han 2018; Young et al. 2019; Sangle et al. 2020; Shulpekova et al. 2021).

Vitamin C also plays an important role in various important functions of the body. The levels of vitamin C are affected by both physical and mental stress. When the levels of vitamin C go down, the human body becomes more susceptible to infections and diseases. Also, the stress levels are increased due to vitamin C deficiency. Increase in vitamin C intake enhances the ability of the body to cope with stress and decreases the harmful effects of stress hormones. Vitamin C intake helps in a quick recovery from both physical and mental stress. Vitamin C supports the adrenal glands and aids in recovery of a person (Brody et al. 2002; Singh 2016; Plevin and Galletly 2020).

25.3.4 Proteins and Amino Acids

Theanine is an amino acid that is very helpful in decreasing physical and mental stress (Kimura et al. 2007). In synergy with caffeine, theanine helps in improving cognitive functions and in elevating the mood (Haskell et al. 2008). It has a tranquilizing effect, relaxes the brain, and decreases anxiety and stress. A very good source of theanine is tea (Kimura et al. 2007; Haskell et al. 2008; Singh 2016). Thus, amino acids and proteins can play important roles as antistress agents.

25.3.5 Cultured and Fermented Food Products

Yogurt, kimchi, and other such products are very useful for intestinal health and in turn very useful to reduce anxiety. Yogurt consumed in its natural form is best without the addition of sugar. Sources of fermented foods include apple-cider vinegar, sauerkraut, kimchi, kombucha, miso, kefir, tempeh, and apple-cider vinegar (Naidoo 2021; Singh 2016).

25.3.6 Minerals

There is an interrelation between stress and magnesium. Several studies have shown that deficiency of magnesium is involved in both psychological and physical stress, thus taking magnesium containing foods and supplements can help in fighting stress (Pickering et al. 2020). Selenium is also a mineral that might be important with respect to stress and fatigue and might help to reduce both (Chen et al. 2022; Singh 2016).

25.3.7 Vegetables and Fruits

Spinach is a wonderful green leafy vegetable that has miraculous effects on health. Spinach is rich in magnesium. It has antistress properties according to researchers (Beaulah et al. 2020; Son et al. 2018; Singh 2016). Oranges are a rich source of vitamin C and, vitamin C is a proven antioxidant and helps in scavenging free radicals and protects the human body against the harmful effects of stress. It also contains folic acid which is effective against stress as discussed before (Goes et al. 2012; Ahmadu et al. 2016; Singh 2016). Blueberries are very important source of antioxidants and nutrients. It is rich in vitamin C. The nutrients found in blueberries are great to reduce stress. The free radicals are scavenged by antioxidants present in blueberry (Shivembe and Ojinnaka 2017; Suleman et al. 2018). It also has a neuroprotective effect (Samani et al. 2023). There are many other vegetables and fruits that can be very useful to counter stress.

25.3.8 Miscellaneous Food Products

Spices like turmeric which can be stored for a long time are also useful in decreasing anxiety. The effect of turmeric is enhanced by adding a pinch of black pepper, since black pepper enhances the bioavailability of turmeric. As a result of this more turmeric will be available for utilization by the brain (Noorafshan et al. 2017; Naidoo 2021).

Chocolates are also stress busting food products. Dark chocolate is especially effective in decreasing stress. Chocolate relieves stress both by chemical mechanisms as well as by its emotional effects. Chocolates not only fights free radicals but also produces such a feeling of goodness which no other food product can mimic (Al Sunni and Latif 2014; da Silva et al. 2015; Salama and El-Sherif 2018). Cocoa beans are a rich source of theobromine. Also, cocoa beans contain such nutrients which are important for the physical and mental health (Botella-Martínez et al. 2021).

Chocolate helps in uplifting the mood in a positive way and has good cognitive effects on the brain. The above effects are due to the flavonoid content of the chocolate which are mainly present as epicatechin. Cocoa contains many types of bioactive compounds some of which are like tryptophan, tryptamine, phenylethylamine, serotonin, tyramine, etc. (Nehlig 2012). Endorphins are released due to the action of serotonin. Also, chocolate interacts with many neurotransmitters in the brain like serotonin, dopamine, and endorphins that are involved in appetite, mood regulation, and reward (Benton and Donohoe 1999). Chocolate is regarded as a comfort food that provides emotional support to the people however research studies have also established the role of chocolate as a psycho-active food (Nehlig 2012).

Although in the previous sections of the present chapter, it has been discussed that excess amount of caffeine can have harmful effects on the body but consumption of coffee in normal amounts can be beneficial to human health. Researchers have found that caffeine has a stimulatory role on the central nervous system. Caffeine is instrumental in enhancing alertness and decreasing fatigue. Besides these effects, caffeine enhances mood. When consumed in proper and appropriate doses, it is said to be safe for human health (Singh 2016; Quinlan et al. 2000; Nehlig et al. 1992). Walnuts have also been found to be effective in stress (Herselman et al. 2022). Eggs whites are associated with antioxidant effects and are useful in reducing mental fatigue (Oe et al. 2020). Tea contains an important amino acid theanine, and as discussed above, it is responsible for antistress activity. So, the consumption of tea is associated with antistress effects on humans (Unno et al. 2016; White et al. 2016).

25.4 Conclusion

In the present chapter, various aspects regarding diet, stress, and food products that help in reducing stress have been discussed. Stress has become a very important problem associated with the modern world. Many diseases and disorders can be linked with stress. Bad food habits, junk foods, packaged and processed food, all can lead to stress. Stress and diet have a strong connection with each other. If a person is under stress or feels emotionally weak, he starts consuming alcohol or starts consuming comfort food/hyperpalatable foods which are bad for health and they lead to further stress. Thus, stress leads to bad eating habits and in turn, these bad eating habits lead to stress. Thus, to counter stress a good diet plan and eating habits are very essential. The present chapter describes both the harmful consequences of stress/bad diet and also the beneficial effects of a good diet that can lead to a long, healthy, and stress-free life. By adopting the good old diet of the previous generations and the golden concepts of Ayurveda, we can eliminate stress from our lives significantly.

References

Abril-Ulloa V, Santos SPL, Morejon-Teran YA, Carpio-Arias TV, Espinoza-Fajardo AC, Vinueza-Veloz MF (2022) Stress and diet quality among Ecuadorian adults during the COVID-19 pandemic. A cross-sectional study. Front Nutr 9:924076

Adam TC, Epel ES (2007) Stress, eating and the reward system. Physiol Behav 91:449–458

Ahmadu S, Mohammed A, Buhari H, Auwal A (2016) An overview of vitamin C as an antistress in poultry. Malays J Vet Res 7:9–22

Al Sunni A, Latif R (2014) Effects of chocolate intake on perceived stress; a controlled clinical study. Int J Health Sci (Qassim) 8:393–401

Alsio J, Olszewski PK, Norback AH, Gunnarsson ZE, Levine AS, Pickering C et al (2010) Dopamine D1 receptor gene expression decreases in the nucleus accumbens upon long-term exposure to palatable food and differs depending on diet-induced obesity phenotype in rats. Neuroscience 171:779–787

Ashakiran DR (2012) Fast foods and their impact on health. J Krishna Inst Med Sci Univ 1:7–15

Barrington WE, Ceballos RM, Bishop SK, McGregor BA, Beresford SA (2012) Perceived stress, behavior, and body mass index among adults participating in a worksite obesity prevention program, Seattle, 2005-2007. Prev Chronic Dis 9:E152

Bayram N, Bilgel N (2008) The prevalence and socio-demographic correlations of depression, anxiety and stress among a group of university students. Soc Psychiatry Psychiatr Epidemiol 43: 667–672

Beaulah A, Rajamanickam C, Swaminathan V (2020) Nutritive values and importance of tropical green leafy vegetables in human diet—a review. Int J Curr Microbiol Appl Sci 9:656–669

Becker HC (2012) Effects of alcohol dependence and withdrawal on stress responsiveness and alcohol consumption. Alcohol Res 34:448–458

Benton D, Donohoe RT (1999) The effects of nutrients on mood. Public Health Nutr 2:403–409

Berthoud HR (2012) The neurobiology of food intake in an obesogenic environment. Proc Nutr Soc 71:478–487

Block JP, He Y, Zaslavsky AM, Ding L, Ayanian JZ (2009) Psychosocial stress and change in weight among US adults. Am J Epidemiol 170:181–192

Botella-Martínez C, Lucas-Gonzalez R, Ballester-Costa C, Pérez-Álvarez JÁ, Fernández-López J, Delgado-Ospina J et al (2021) Ghanaian cocoa (Theobroma cacao L.) bean shells coproducts: effect of particle size on chemical composition, bioactive compound content and antioxidant activity. Agronomy 11:401

Boynton A, Neuhouser ML, Sorensen B, McTiernan A, Ulrich CM (2008) Predictors of diet quality among overweight and obese postmenopausal women. J Am Diet Assoc 108:125–130

Bremner JD, Moazzami K, Wittbrodt MT, Nye JA, Lima BB, Gillespie CF et al (2020) Diet, stress and mental health. Nutrients 12:2428

Brody S, Preut R, Schommer K, Schurmeyer TH (2002) A randomized controlled trial of high dose ascorbic acid for reduction of blood pressure, cortisol, and subjective responses to psychological stress. Psychopharmacology 159:319–324

Bruch H (1964) Psychological aspects of overeating and obesity. Psychosomatics 5:269–274

Cardello AV, Meiselman HL, Schutz HG, Craig C, Given Z, Lesher LL et al (2012) Measuring emotional responses to foods and food names using questionnaires. Food Qual Prefer 24:243–250

Chen X, Zhang J, Li H, Liu W, Xi Y, Liu X (2022) A comprehensive comparison of different selenium supplements: mitigation of heat stress and exercise fatigue-induced liver injury. Front Nutr 9:917349

Choudhary AK, Lee YY (2018) Neurophysiological symptoms and aspartame: what is the connection? Nutr Neurosci 21:306–316

Chueh KH, Guilleminault C, Lin CM (2019) Alcohol consumption as a moderator of anxiety and sleep quality. J Nurs Res 27:e23

da Silva MN, Koslowsky Marder R, Farias Wohlenberg M, Funchal C, Dani C (2015) Total phenolic content and antioxidant activity of different types of chocolate, milk, semisweet, dark, and soy, in cerebral cortex, hippocampus, and cerebellum of Wistar rats. Biochem Res Int 2015:294659, 1

Dallman MF, Pecoraro NC, la Fleur SE (2005) Chronic stress and comfort foods: self-medication and abdominal obesity. Brain Behav Immun 19:275–280

Di Chiara G, Imperato A (1988) Drugs abused by humans preferentially increase synaptic dopamine concentrations in the mesolimbic system of freely moving rats. Proc Natl Acad Sci U S A 85: 5274–5278

Everitt BJ, Belin D, Economidou D, Pelloux Y, Dalley JW, Robbins TW (2008) Review. Neural mechanisms underlying the vulnerability to develop compulsive drug-seeking habits and addiction. Philos Trans R Soc Lond Ser B Biol Sci 363:3125–3135

Felger JC (2018) Imaging the role of inflammation in mood and anxiety-related disorders. Curr Neuropharmacol 16:533–558

Figlewicz DP, Bennett J, Evans SB, Kaiyala K, Sipols AJ, Benoit SC (2004) Intraventricular insulin and leptin reverse place preference conditioned with high-fat diet in rats. Behav Neurosci 118: 479–487

Freeman MP, Rapaport MH (2011) Omega-3 fatty acids and depression: from cellular mechanisms to clinical care. J Clin Psychiatry 72:258–259

Fuhrman J (2018) The hidden dangers of fast and processed food. Am J Lifestyle Med 12:375–381

Ganley RM (1989) Emotion and eating in obesity: a review of the literature. Int J Eat Disord 8:343–361

Garg N, Wansink B, Inman JJ (2007) The influence of incidental affect on consumers' food intake. J Mark 71:194–206

Gheita AA, Gheita TA, Kenawy SA (2020) The potential role of B5: a stitch in time and switch in cytokine. Phytother Res 34:306–314

Gibson EL (2006) Emotional influences on food choice: sensory, physiological and psychological pathways. Physiol Behav 89:53–61

Giskes K, van Lenthe F, Avendano-Pabon M, Brug J (2011) A systematic review of environmental factors and obesogenic dietary intakes among adults: are we getting closer to understanding obesogenic environments? Obes Rev 12:e95–e106

Goes TC, Antunes FD, Alves PB, Teixeira-Silva F (2012) Effect of sweet orange aroma on experimental anxiety in humans. J Altern Complement Med 18:798–804

Gonzalez MJ, Miranda-Massari JR (2014) Diet and stress. Psychiatr Clin North Am 37:579–589

Groesz LM, McCoy S, Carl J, Saslow L, Stewart J, Adler N et al (2012) What is eating you? Stress and the drive to eat. Appetite 58:717–721

Güneşer R, Him A (2021) Evaluation of perceived stress level and emotional eating behaviors of sports college students during COVID-19 pandemic. Sakarya Med J 11:434–441

Habhab S, Sheldon JP, Loeb RC (2009) The relationship between stress, dietary restraint, and food preferences in women. Appetite 52:437–444

Haskell CF, Kennedy DO, Milne AL, Wesnes KA, Scholey AB (2008) The effects of L-theanine, caffeine and their combination on cognition and mood. Biol Psychol 77:113–122

Heatherton TF, Baumeister RF (1991) Binge eating as escape from self-awareness. Psychol Bull 110:86–108

Herselman MF, Bailey S, Deo P, Zhou XF, Gunn KM, Bobrovskaya L (2022) The effects of Walnuts and Academic Stress on Mental Health, general well-being and the gut microbiota in a sample of university students: a randomised clinical trial. Nutrients 14:4776

Ikeda H, West DB, Pustek JJ, Figlewicz DP, Greenwood MR, Porte D Jr et al (1986) Intraventricular insulin reduces food intake and body weight of lean but not obese Zucker rats. Appetite 7: 381–386

Jastreboff AM, Sinha R, Lacadie C, Small DM, Sherwin RS, Potenza MNJDc. (2013) Neural correlates of stress-and food cue–induced food craving in obesity: association with insulin levels. Diabetes Care 36:394–402

Kaur H, Kochr R (2019) Nutritional challenges and health consequences of junk foods. Curr Res Diabetes Obes J 10:555796

Khaled K, Tsofliou F, Hundley V, Helmreich R, Almilaji O (2020) Perceived stress and diet quality in women of reproductive age: a systematic review and meta-analysis. Nutr J 19:92

Kiecolt-Glaser JK, Belury MA, Andridge R, Malarkey WB, Glaser R (2011) Omega-3 supplementation lowers inflammation and anxiety in medical students: a randomized controlled trial. Brain Behav Immun 25:1725–1734

Kimura K, Ozeki M, Juneja LR, Ohira H (2007) L-Theanine reduces psychological and physiological stress responses. Biol Psychol 74:39–45

Koob GF, Volkow ND (2010) Neurocircuitry of addiction. Neuropsychopharmacology 35:217–238

Laitinen J, Ek E, Sovio U (2002) Stress-related eating and drinking behavior and body mass index and predictors of this behavior. Prev Med 34:29–39

Lee GY, Han SN (2018) The role of vitamin E in immunity. Nutrients 10:1614

Lemmens SG, Rutters F, Born JM, Westerterp-Plantenga MS (2011) Stress augments food 'wanting' and energy intake in visceral overweight subjects in the absence of hunger. Physiol Behav 103:157–163

Macht M, Gerer J, Ellgring H (2003) Emotions in overweight and normal-weight women immediately after eating foods differing in energy. Physiol Behav 80:367–374

Mojet J, Köster E (2002) Texture and flavour memory in foods: an incidental learning experiment. Appetite 38:110–117

Naidoo U (2021) Eat to beat stress. Am J Lifestyle Med 15:39–42

Nehlig A (2012) The neuroprotective effects of cocoa flavanol and its influence on cognitive performance. Br J Clin Pharmacol 75:716–727

Nehlig A, Daval JL, Debry G (1992) Caffeine and the central nervous system: mechanisms of action, biochemical, metabolic and psychostimulant effects. Brain Res Brain Res Rev 17:139–170

Noorafshan A, Vafabin M, Karbalay-Doust S, Asadi-Golshan R (2017) Efficacy of curcumin in the modulation of anxiety provoked by sulfite, a food preservative, in rats. Prev Nutr Food Sci 22: 144–148

Oe M, Sakamoto H, Nishiyama H, Sasahara R, Masuda Y, Adachi M et al (2020) Egg white hydrolyzate reduces mental fatigue: randomized, double-blind, controlled study. BMC Res Notes 13:443

Oliver G, Wardle J, Gibson EL (2000) Stress and food choice: a laboratory study. Psychosom Med 62:853–865

Osdoba KE, Mann T, Redden JP, Vickers Z (2015) Using food to reduce stress: effects of choosing meal components and preparing a meal. Food Qual Prefer 39:241–250

Pickering G, Mazur A, Trousselard M, Bienkowski P, Yaltsewa N, Amessou M et al (2020) Magnesium status and stress: the vicious circle concept revisited. Nutrients 12:3672

Plevin D, Galletly C (2020) The neuropsychiatric effects of vitamin C deficiency: a systematic review. BMC Psychiatry 20:315

Pruessner JC, Champagne F, Meaney MJ, Dagher A (2004) Dopamine release in response to a psychological stress in humans and its relationship to early life maternal care: a positron emission tomography study using [11C]raclopride. J Neurosci 24:2825–2831

Quinlan PT, Lane J, Moore KL, Aspen J, Rycroft JA, O'Brien DC (2000) The acute physiological and mood effects of tea and coffee: the role of caffeine level. Pharmacol Biochem Behav 66:19–28
Reynolds RM, Chapman KE, Seckl JR, Walker BR, McKeigue PM, Lithell HO (2002) Skeletal muscle glucocorticoid receptor density and insulin resistance. JAMA 287:2505–2506
Rutters F, Nieuwenhuizen AG, Lemmens SG, Born JM, Westerterp-Plantenga MS (2009) Acute stress-related changes in eating in the absence of hunger. Obesity 17:72–77
Salama RM, El-Sherif NM (2018) Ameliorative effect of dark chocolate on brain and heart of chronic restraint stressed male albino rats. Austin J Anat 5:1076
Salim S, Chugh G, Asghar M (2012) Inflammation in anxiety. Adv Protein Chem Struct Biol 88:1–25
Samani P, Costa S, Cai S (2023) Neuroprotective effects of blueberries through inhibition on cholinesterase, tyrosinase, cyclooxygenase-2, and amyloidogenesis. Forum Nutr 3:39–57
Sangle P, Sandhu O, Aftab Z, Anthony AT, Khan S (2020) Vitamin B12 supplementation: preventing onset and improving prognosis of depression. Cureus 12:e11169
Selhub EM, Logan AC, Bested AC (2014) Fermented foods, microbiota, and mental health: ancient practice meets nutritional psychiatry. J Physiol Anthropol 33:2
Shehata WM, Abdeldaim DE (2023) Emotional eating in relation to psychological stress during COVID-19 pandemic: a cross-sectional study in faculty of medicine, Tanta University, Egypt. BMC Public Health 23:277
Shivembe A, Ojinnaka D (2017) Determination of vitamin C and total phenolic in fresh and freeze dried blueberries and the antioxidant capacity of their extracts. Integr Food Nutr Metab 4:1–5
Shulpekova Y, Nechaev V, Kardasheva S, Sedova A, Kurbatova A, Bueverova E et al (2021) The concept of folic acid in health and disease. Molecules 26:3731
Singh K (2016) Nutrient and stress management. J Nutr Food Sci 6:528
Sinha R (2008) Chronic stress, drug use, and vulnerability to addiction. Ann N Y Acad Sci 1141:105–130
Sofi F, Cesari F, Abbate R, Gensini GF, Casini A (2008) Adherence to mediterranean diet and health status: meta-analysis. BMJ 337:a1344
Son H, Jung S, Shin JH, Kang MJ, Kim HJ (2018) Anti-stress and anti-depressive effects of spinach extracts on a chronic stress-induced depression mouse model through lowering blood corticosterone and increasing brain glutamate and glutamine levels. J Clin Med 7:406
Steiner JE (1974) Discussion paper: innate, discriminative human facial expressions to taste and smell stimulation. Ann N Y Acad Sci 237:229–233
Suleman M, Khan A, Baqi A, Kakar MS, Samiullah AM (2018) Antioxidants, its role in preventing free radicals and infectious diseases in human body. Pure Appl Biol 8:380–388
Takeda E, Terao J, Nakaya Y, Miyamoto K, Baba Y, Chuman H, Kaji R, Ohmori T, Rokutan K (2004) Stress control and human nutrition. J Med Investig 51:139–145
Terlecki MA, Ecker AH, Buckner JD (2014) College drinking problems and social anxiety: the importance of drinking context. Psychol Addict Behav 28:545–552
Torres SJ, Nowson CA (2007) Relationship between stress, eating behavior, and obesity. Nutrition 23:887–894
Tyrka AR, Walters OC, Price LH, Anderson GM, Carpenter LL (2012) Altered response to neuroendocrine challenge linked to indices of the metabolic syndrome in healthy adults. Horm Metab Res 44:543–549
Unno K, Hara A, Nakagawa A, Iguchi K, Ohshio M, Morita A et al (2016) Anti-stress effects of drinking green tea with lowered caffeine and enriched theanine, epigallocatechin and arginine on psychosocial stress induced adrenal hypertrophy in mice. Phytomedicine 23:1365–1374
Volkow ND, Wang GJ, Baler RD (2011) Reward, dopamine and the control of food intake: implications for obesity. Trends Cogn Sci 15:37–46
Vyas A, Mitra R, Shankaranarayana Rao BS, Chattarji S (2002) Chronic stress induces contrasting patterns of dendritic remodeling in hippocampal and amygdaloid neurons. J Neurosci 22:6810–6818

Walsh EM, Kiviniemi MT (2014) Changing how I feel about the food: experimentally manipulated affective associations with fruits change fruit choice behaviors. J Behav Med 37:322–331

Wand GS, Oswald LM, McCaul ME, Wong DF, Johnson E, Zhou Y et al (2007) Association of amphetamine-induced striatal dopamine release and cortisol responses to psychological stress. Neuropsychopharmacology 32:2310–2320

Wansink B, Payne CR, North J (2007) Fine as North Dakota wine: sensory expectations and the intake of companion foods. Physiol Behav 90:712–716

Warne JP (2009) Shaping the stress response: interplay of palatable food choices, glucocorticoids, insulin and abdominal obesity. Mol Cell Endocrinol 300:137–146

White DJ, de Klerk S, Woods W, Gondalia S, Noonan C, Scholey AB (2016) Anti-stress, behavioural and magnetoencephalography effects of an L-Theanine-based nutrient drink: a randomised, double-blind, placebo-controlled, crossover trial. Nutrients 8:53

Wichianson JR, Bughi SA, Unger JB, Spruijt-Metz D, Nguyen-Rodriguez ST (2009) Perceived stress, coping and night-eating in college students. Stress Health 25:235–240

Wise A, O'Brien K, Woodruff T (2011) Are oral contraceptives a significant contributor to the estrogenicity of drinking water? Environ Sci Technol 45:51–60

Yau YH, Potenza MN (2013) Stress and eating behaviors. Minerva endocrinologica 38:255–267

Young LM, Pipingas A, White DJ, Gauci S, Scholey A (2019) A systematic review and meta-analysis of B vitamin supplementation on depressive symptoms, anxiety, and stress: effects on healthy and 'At-Risk' individuals. Nutrients 11:2232

Zellner DA, Loaiza S, Gonzalez Z, Pita J, Morales J, Pecora D et al (2006) Food selection changes under stress. Physiol Behav 87:789–793

The manufacturer's authorised representative in the EU is Springer Nature Customer Service Centre GmbH, Europaplatz 3, 69115 Heidelberg, Germany. If you have any concerns regarding our products, please contact ProductSafety@springernature.com

Printed and bound by CPI Group (UK) Ltd, Croydon, CR0 4YY
01/10/2026
02225957-0014